1st International Seminar on Non-Ideal Compressible-Fluid Dynamics for Propulsion & Power (NICFD 2016)

Journal of Physics: Conference Series Volume 821

Varenna, Italy
19 – 20 October 2016

ISBN: 978-1-5108-3906-9
ISSN: 1742-6588

Printed from e-media with permission by:

Curran Associates, Inc.
57 Morehouse Lane
Red Hook, NY 12571

Printed by Curran Associates, Inc. (2017)

For permission requests, please contact the Institute of Physics
at the address below.

Institute of Physics
Dirac House, Temple Back
Bristol BS1 6BE UK

Phone: 44 1 17 929 7481
Fax: 44 1 17 920 0979

techtracking@iop.org

Additional copies of this publication are available from:

Curran Associates, Inc.
57 Morehouse Lane
Red Hook, NY 12571 USA
Phone: 845-758-0400
Fax: 845-758-2633
Email: curran@proceedings.com
Web: www.proceedings.com

Table of contents

Volume 821

1st International Seminar on Non-Ideal Compressible-Fluid Dynamics for Propulsion & Power

19–20 October 2016, Varenna, Italy

Accepted papers received: 8 March 2017
Published online: 7 April 2017

Preface

Papers

IOP Conf. Series: Journal of Physics: Conf. Series **821** (2017) 011001 doi:10.1088/1742-6596/821/1/011001

Proceedings of the NICFD 2016 Seminar

October 19th and 20th, Villa Monastero, Varenna, Italy

Preface

This volume contains the papers presented at *NICFD 2016: 1st International Seminar on Non-Ideal Compressible-Fluid Dynamics for Propulsion & Power* held on the 19th and 20th of October, 2016, at Villa Monastero, Varenna, Italy. The conference website is `nicfd2016.polimi.it`.

NICFD 2016 promoted the exchange of scientific knowledge and fostered the interaction among researchers and professionals in the field of non-ideal compressible-fluid dynamics, with special emphasis on topics related to propulsion and power applications. The conference themes range from theoretical foundations to advanced numerical and experimental practices and applied technologies. Key themes are: experiments; fundamentals; numerical methods; optimization and UQ; critical and supercritical flows; turbulence and mixing; multi-component fluid flows; applications in ORC power systems; applications in supercritical CO_2 power systems; steam turbines; cryogenic flows; condensing flows in nozzles; cavitating flows; super/trans-critical fluids in space propulsion.

Four keynote lectures reviewed the state-of-the-art and illustrated future studies and applications: *Non-Ideal Compressible Fluid Dynamics: A Challenge for Theory* by Prof. Alfred Kluwick, TU Wien, Austria; *Cavitation Instabilities in Turbopump Inducers* by Prof. Seung Jin Song, Seoul National University, Korea; *Industrial applications of turbomachinery operating on non-ideal fluids* by Dr. Douglas Hofer, GE Global Research, USA; *Wall bounded turbulence in flows with strong property gradients* by Prof. Rene Pecnik, Delft University of Technology, The Netherlands.

The proceedings gather 27 papers reporting on cutting-edge research activities in NICFD for propulsion and power. The review paper by Prof. Alfred Kluwick opens the collection and provides the perspective on past discoveries, the context of present activities and the roadmap of future challenges in this new, exciting scientific field.

We would like to thank all the participants to NICFD 2016 and the members of the Scientific Committee, for providing guidance during the organisation of the conference and for managing the revision of the papers. We thank all reviewers for their invaluable help to attain the high scientific quality of the contributions collected here.

The conference was supported in part by the European Research Council under ERC Consolidator Grant N.617603, NSHOCK Project (`crealab.polimi.it`)

Varenna, October 2016

Prof. Alberto Guardone, Politecnico di Milano, Chairman
Prof. Piero Colonna, Delft University of Technology, Co-chair
Dr. Andrew Wheeler, Cambridge University, Co-chair
Giorgia Cammi, Politecnico di Milano, Program Manager
Davide Vimercati, Politecnico di Milano, Submission Manager

Invited keynote lectures

Non-Ideal Compressible Fluid Dynamics: A Challenge for Theory
by Prof. Alfred Kluwick, Wien Technical University, Austria

Abstract: The possibility that rarefaction as well as compression shocks may form in single phase vapours was envisaged first by H. A. Bethe (1942), *The theory of shock waves for an arbitrary equation of state*, Technical Report No. 545, Office of Scientific Research and Development. However, calculations based on the Van der Waals equation of state indicated that the former type of shock is possible only if the ratio of specific heat at constant volume c_v to universal gas constant R is larger than about 17.5 which he considered too large to be satisfied by real fluids. This conclusion was contested by P. A. Thompson, (1971) *A Fundamental Derivative in Gas Dynamics*, Physics of Fluids 14 pp. 1843–1849, who showed that the type of shock capable of forming in arbitrary fluids is determined by the sign of the thermodynamic quantity $\Gamma = v^3/(2a^2)(\partial^2 P/\partial v^2)_s$, which he referred to as the fundamental derivative of gasdynamics. Here v, p, s and a denote specific volume, pressure, entropy and the speed of sound. Thompson and co-workers also showed that the required condition for the existence of rarefaction shocks, that Γ takes on negative values, is indeed satisfied for several fluorocarbon vapours. This finding spawned a burst of theoretical studies elaborating on the unusual and often counterintuitive behaviour of flows with rarefaction shocks present. This produced results not only of a theoretical nature but results also suggesting the practical importance of Non-Ideal Compressible Fluid Dynamics (NICFD) in general. The present talk addresses some of the challenges encountered in connection with the theoretical treatment of the associated flow behaviour. Weakly nonlinear acoustic waves of small but finite amplitude serve as a starting point. Here mixed rather than strictly positive nonlinearity generates a wealth of phenomena not possible in perfect gases. Examples of steady flows where these nonclassical effects play a decisive role (and which may be useful also for future experimental work) are quasi 1D nozzle flows and transonic 2D flows past corners. The investigation of dissipative effects reveals completely different mechanisms for the regularisation of shock discontinuities in unconfined and confined flows. As a final example of nonclassical phenomena in NICFD the problem of laminar boundary layers with constant pressure gradient will be considered. In contrast to the classical Howarth problem solutions do not necessarily terminate in the form of a Goldstein singularity but may be continued beyond the point of zero wall shear. The local flow behaviour then is governed by an equation of Fisher type which describes the transition from strictly steady to unsteady periodic flow exhibiting early signs of the onset of turbulence.

Cavitation Instabilities in Turbopump Inducers
by Prof. Seung Jin Song, Seoul National University, Korea

Abstract: Turbopumps deliver liquid fuel and oxidizer into the main combustion chamber in liquid rocket engines and are usually composed of radial pumps driven by axial turbines. Turbopumps operate under extreme environments characterized by high speeds, high pressures, cryogenic temperatures, and steep temperature gradients. Therefore, turbopumps are the critical devices which determine the success of liquid rocket systems. To prevent cavitation in the pumps, turbopumps deploy inducers, and, yet, the inducers themselves undergo cavitation. This cavitation can have thermodynamic, hydrodynamic, and even rotordynamic consequences. This talk will present such issues which arise in turbopump development efforts.

Industrial applications of turbomachinery operating on non-ideal fluids
by Dr. Douglas Hofer, General Electric Global Research, USA

Abstract: Many industrial turbomachinery applications involve fluids that are not amenable to ideal gas assumptions. The classical application is steam turbines where both pressures above ~ 20 bar and condensing liquid in the low pressure turbine leads to significant departures from the ideal gas relationships. In aviation engines and industrial gas turbines the high temperatures result in dissociation effects that render the ideal gas assumptions invalid. The processing of hydrocarbons frequently involves the intersection of non-ideal fluids and turbomachinery with applications ranging from gas compression to multiphase pumping. The emergence of closed loop power cycles using supercritical CO_2 as a working fluid brings another set of challenges for the industry to understand including operation of turbomachinery very near the critical point and potential for liquid and solid condensation. This talk will touch on each of these areas and provide an overview of industrial practice and open questions where further research is needed.

Wall bounded turbulence in flows with strong property gradients
by Prof. Rene Pecnik, Technical University of Delft, The Netherlands

Abstract: Turbulent fluid flows with large gradients of thermo-physical properties can occur in a wide range of engineering applications, such as in flows with strong heat transfer at the wall, high-speed flows and in application with dense gases. The mean velocity, near-wall turbulent structures and mixing of scalar fields are greatly affected by near wall gradients in fluid properties. In order to study these effects, we will use direct numerical simulations with different constitutive relations for density and viscosity as a function of temperature, to mimic a wide range of fluid behaviours and to develop a generalised framework for studying scaling laws and turbulence modulations in variable property flows. The first part of the presentation will be on mean velocity scaling. Based on observations we will discuss a velocity scaling that is solely based on the semi-local Reynolds number. This scaling is able to collapse velocity profiles irrespective of near-wall density and viscosity gradients. In the second part, we will outline how near-wall density and viscosity gradients alter the Reynolds stress generation mechanism and inter-component energy transfer among turbulent stresses. In the last part of the presentation we will discuss how these effects alter heat transfer in turbulent pipe flows with fluids close to the vapour liquid critical point. In addition to that, we will also refer to recent advances in turbulence research for dense gases made by other researchers.

Scientific Committee of NICFD 2016

Review Statements

Each paper accepted for publication underwent two rounds of review by two anonymous reviewers. The review process of each paper was managed by two members of the Scientific Committee. The first round of reviews took place before the conference. During the conference, each delegate had access to a web portal to input her/his review for each presented paper, in either an anonymous or named way. The second round of reviews took place after the conference and considered the opinion of the two anonymous reviewers as well as the delegates'. Final acceptance was decided by the two managing members of the Scientific Committee together with the Conference Chairman.

Peer review statement

All papers published in this volume of *Journal of Physics: Conference Series* have been peer reviewed through processes administered by the proceedings Editors. Reviews were conducted by expert referees to the professional and scientific standards expected of a proceedings journal published by IOP Publishing.

Non-Ideal Compressible Fluid Dynamics: A Challenge for Theory.

A. Kluwick[1]

[1]Institute of Fluid Mechanics and Heat Transfer, Technical University Vienna.

E-mail: alfred.kluwick@tuwien.ac.at

Abstract.

The possibility that compression as well as rarefaction shocks may form in single phase vapours was envisaged first by Bethe (1942). However calculations based on the Van der Waals equation of state indicated that the latter type of shock is possible only if the specific heat at constant volume c_v divided by the universal gas constant R is larger than about 17.5 which he considered too large to be satisfied by real fluids. This conclusion was contested by Thompson (1971) who showed that the type of shock capable of forming in arbitrary fluids is determined by the sign of the thermodynamic quantity

$$\Gamma = \frac{v^3}{2c^2}\frac{\partial^2 p}{\partial v^2}\bigg|_s$$

to which he referred to as fundamental derivative of gas dynamics. Here v, p, s and c denote the specific volume, the pressure, the entropy and the speed of sound. Thompson and co-workers also showed that the required condition for the existence of rarefaction shocks, that Γ may take on negative values, is indeed satisfied for a number of hydrocarbon and fluorocarbon vapours. This finding spawned a burst of theoretical studies elaborating on the unusual and often counterintuitive behaviour of shocks with rarefaction shocks present. These produced both results of theoretical character but also results suggesting the practical importance of Non-Ideal Compressible Fluid Dynamics in general. The present paper addresses some of the challenges encountered in connection with the theoretical treatment of the associated flow behaviour. Weakly nonlinear acoustic waves of finite amplitude serve as a starting point. Here mixed rather than strictly positive nonlinearity generates a wealth of phenomena not possible in perfect gases. Examples of steady flows where these non-classical effects play a decisive role (and which may be useful also for future experimental work) are quasi one-dimensional nozzle flows and transonic two-dimensional flows past corners. The study of viscous effects concentrates on laminar flows of boundary layer type. Here non-classical phenomena are caused by the uncommon smallness of the Eckert number but also by the unconventional Mach number dependence on p in the external inviscid flow region.

1. Introduction and Motivation

In 1983 I spent half a year at the Virginia Polytechnic Institute and State University in Blacksburg Va. One day Mark Cramer stepped into my office, handed me a paper by Ph.Thompson entitled "A fundamental derivative in gas dynamics" and asked me to let him know my opinion. In this paper Thompson discusses the crucial effect of the curvature of isentropes in the pressure specific volume diagram for the dynamical behaviour of compressible fluids. Most conveniently this curvature is expressed in non-dimensional form

$$\Gamma = \frac{v^3}{2c^2}\frac{\partial^2 p}{\partial v^2}\bigg|_s$$

where p, v, s and c denote the pressure, specific volume, specific entropy and sound speed. The symbol Γ is the one used by Hayes (1954) [13] but the notion fundamental derivative of gas dynamics had been coined by Thompson to emphasis its importance and in particular the importance of its sign in most areas of gas dynamics. The role the second derivative of p with respect to v at constant s plays in the theory of shock waves was however recognised much earlier. Concentrating on weak shocks in arbitrary gases Duhem (1909) [8] demonstrated that the entropy jump across such shocks is proportional to a quantity $H(p_1, v_1)$ and the third power of the jump of the specific volume:

$$[s] = -\frac{1}{12T_1 v_1^3} H(p_1, v_1)[v]^3. \tag{1}$$

Here p_1, v_1, T_1 denote the values of p, v and the absolute temperature T immediately upstream of the shock. Referring to the second law of thermodynamics Duhem thus concludes that shocks are of compressive/expansive type if $H > 0/ < 0$. It is easily shown that $H(p, v)$ can be written also in the form

$$H(p, v) = v^3 \frac{\partial^2 p}{\partial v^2}\bigg|_s \tag{2}$$

and Duhem's finding thus is equivalent to the statement that shocks are of compressive/expansive type if $\partial^2 p/\partial v^2|_s$ is $> 0/ < 0$ first formulated by Bethe (1942) [2] who apparently was unaware of Duhem's work. An important further step towards the understanding of shocks in arbitrary fluids was taken by Becker (1922) [1] who, in his widely unknown Habilitationsschrift, showed that compression/expansion waves steepen if the curvature of isentropes is positive/negative and equally important that the resulting shocks automatically satisfy the second law of thermodynamics. Bethe's lasting contribution to the theory of shocks is due to the fact that he did not stop when he reached his conclusion but - as also did the Russian scientist Zel'dovich (1946) [30] - went on to search for fluids having regions of negative isentropic curvature in the p, v diagram. Specifically he showed that van der Waals gases exhibit such regions in the general neighbourhood of the thermodynamic critical point if the ratio of the specific heat at constant volume c_v and the universal gas constant R is larger than about 17.5

$$\beta = \frac{c_v}{R} \gtrsim 17.5 \tag{3}$$

which he immediately dismissed as being physically impossible. Despite this negative statement Thompson (1971) [28] continued to investigate the dynamical behaviour of such fluids and concludes that "for fluids with $\Gamma < 0$ the classical results of ordinary gas dynamics will be inverted". As an example Fig.1 taken from the original publication shows the time evolution of a pressure pulse of triangular shape.

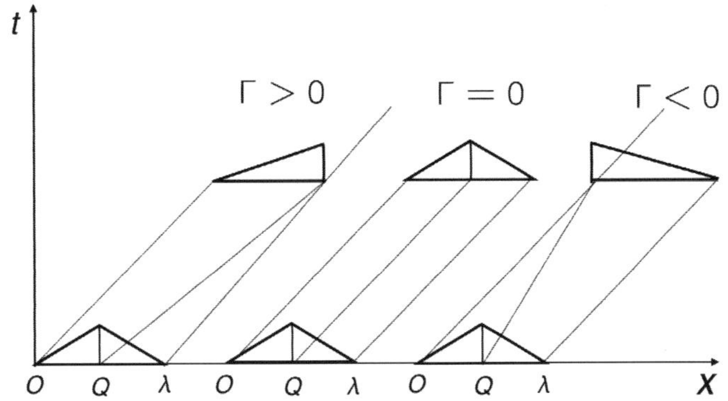

Figure 1. Time evolution of pressure pulse in the space (x) - time (t) diagram.

As pointed out before, compression waves steepen while rarefaction waves spread out if $\Gamma > 0$. In contrast, rarefaction waves steepen while compression wave spread out if $\Gamma < 0$. Therefore only compression shocks will form in a positive Γ fluid and only rarefaction shocks will form in a negative Γ fluid. Thompson then concluded: "the existence of rarefaction shocks is consistent with the tendency of rarefaction waves to steepen. Thus, if certain results are inverted, others necessarily follow. The interest in such phenomena is quite academic, however, if no real fluids with negative Γ exist. We believe that such fluids exist and will report specifically in the near future." Indeed theoretical evidence for the existence of such fluids was provided already in 1972 by Lambrakis and Thompson [23] and in more detail in 1973 by Thompson and Lambrakis [28]. In 1983 the list of negative Γ fluids included several dense gases such as hydrocarbons and fluorocarbons of practical interest. So, when Mark Cramer entered my office several days later I told him that I was interested in joint research in this area provided we concentrated on aspects of fluid mechanics rather than the search for further candidates of negative Γ fluids as in my opinion the existing literature formed a sufficiently strong and reliable basis for such efforts. This is supported also by more recent investigations by Colonna, Guardone and Nannan (2007) [3] which point to the existence of a new class of negative Γ fluids so-called Siloxanes. This then sets the stage for this paper which will be based on the premise that negative Γ fluids exist and will concentrate on the associated aspects of non-ideal compressible fluid dynamics.

2. Theoretical building blocks

2.1. *Shock formation*

As a starting point we go back to the problem of acoustic wave propagation and treat the formation of shocks in simple right running waves in more detail.

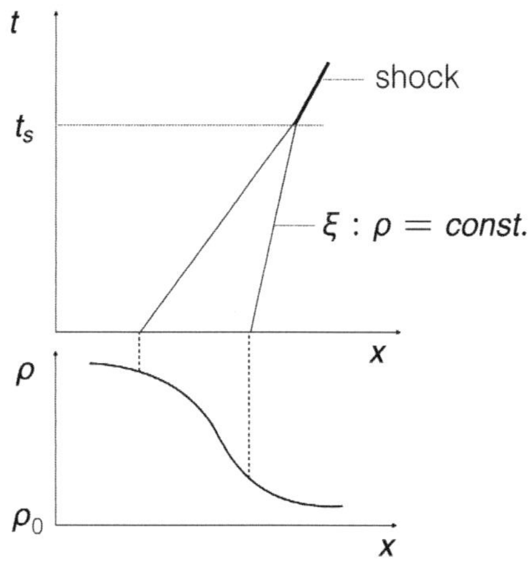

Figure 2. Shock formation by intersecting characteristics $\xi(x,t) = const.$

As sketched in Fig.2 we assume that the density distribution imposed at the initial time is a smooth function of the propagation distance x and as in classical gas-dynamics the density ρ is constant on wave fronts or characteristics $\xi(x,t) = const.$ propagating with speed v_w which depends on ρ only, Fig. 2. Making use of the relationship

$$\frac{dv_w}{d\rho} = \frac{c}{\rho}\Gamma \tag{4}$$

first derived by Becker (1922) [1] it is easily shown that neighbouring characteristics intersect at times $t = \frac{\rho}{c\Gamma} / \frac{d\rho}{d\xi}$ and the minimum value defines the shock formation time:

$$t_s = \min\left(\frac{\rho}{c\Gamma} \Big/ \frac{d\rho}{d\xi}\right). \tag{5}$$

If the fundamental derivative Γ is of $O(1)$ as in the case of dilute, i.e. perfect gases and if the wave carries small density disturbances

$$\frac{\rho - \rho_0}{\rho_0} = O(\epsilon), \quad \epsilon \ll 1 \tag{6}$$

t_s is of $O(1/\epsilon)$ as known from classical acoustics. If, however, Γ is small t_s increases drastically and tends to infinity as Γ tends to zero. We thus conclude that non-linear effects in dense gases having small values of Γ - positive or negative - are significantly weaker than in classical gases leading in turn to a delay of shock formation.

2.2. Wave evolution

Following the discussion of shock formation we next consider the evolution of the wave profile as a whole. To this end we express the requirement that ρ is constant on characteristics propagating with speed v_w as

$$\frac{\partial \rho}{\partial t} + v_w \frac{\partial \rho}{\partial x} = 0. \tag{7}$$

Further simplification is achieved by introducing the quantity $j(\rho)$:

$$\frac{dj}{d\rho} = v_w(\rho). \tag{8}$$

The wave evolution then is governed by the equation

$$\frac{\partial \rho}{\partial t} + \frac{\partial j(\rho)}{\partial x} = 0 \tag{9}$$

which has the form of a continuity equation. The physical meaning of $j(\rho)$ thus is obvious: it represents the perturbation mass flux carried by the wave. The equation for j has to be solved numerically in general. However analytical progress is possible if the density disturbances caused by the wave are small, Cramer and Kluwick (1984) [4], Kluwick (1991) [17]. Equally important $j(\rho)$ then has universal meaning, i.e., is of relevance not only for acoustic waves but also for most areas of gas dynamics.

2.3. Shock propagation

To complete the theoretical basis for the treatment of dense gas flows it is finally necessary to consider the process of shock propagation. Assuming one-dimensional steady flow as in Fig. 3 where w, p, v and h denote the flow velocity in the x direction, the pressure, the specific volume and the specific enthalpy. Subscripts 1 and 2 specify values immediately upstream and downstream of the shock while square brackets denote the jump of any physical quantity, say q: $[q] = q_2 - q_1$.

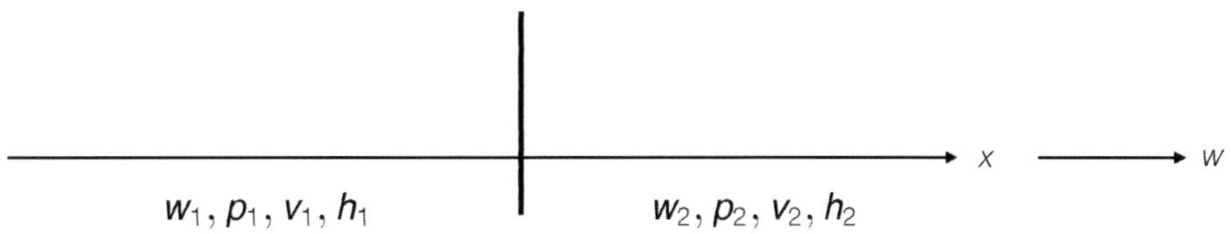

Figure 3. Shock propagation.

The balances of mass, momentum and energy in integral form then yield the celebrated Rankine-Hugoniot relationship

$$[h] = [p](v_1 + v_2)/2 \tag{10}$$

which contains thermodynamic quantities only and therefore holds for arbitrary fluids. Specific fluid properties enter via the caloric equation of state

$$h = h(p, v) \tag{11}$$

and when combined with the Rankine - Hugoniot relationship yields the shock adiabat

$$p_2 = f(p_1, v_1; v_2). \tag{12}$$

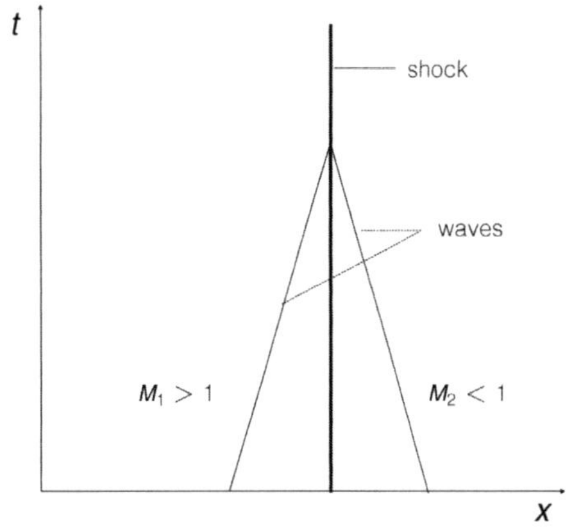

Figure 4. Wave speed ordering relationship

If one specifies values p_1, v_1 upstream of the shock and v_2 downstream the formal solution of the shock adiabat predicts the downstream value of the pressure p_2. But the question then arises: are all formal solutions also physically realisable? This is the so called admissibility problem. Of course, a necessary requirement is provided by the second law of thermodynamics which states that the specific entropy must not decrease during adiabatic changes of state. A second natural requirement is that shocks must be able to form mechanically through intersecting characteristics, Fig. 4. Therefore, the Mach number $M = w/c$ has to be larger than one upstream and smaller than one downstream of the shock which sometimes is summarised as the wave speed ordering relationship: $M_1 > 1 > M_2$.

As pointed out before, Thompson and his doctoral student Lambrakis, making use of more sophisticated equations of state than the van der Waals equation of state considered by Bethe, were the first to demonstrate the existence of real negative Γ fluids and thus also the possible existence of single phase rarefaction shocks. Honouring the pioneering work of Bethe, Zel'dovich and Thompson such fluids are now commonly termed Bethe-Zel'dovich-Thompson fluids or in short BZT fluids. Interestingly, the existence of such fluids is suggested already by pure theoretical reasoning independent of any assumptions concerning specific forms of equations of state, namely by taking into account that the specific internal energy $u(v,T)$ and the specific entropy $s(v,T)$ can be decomposed in this form:

$$u(v,T) = \phi(T) + \beta^{-1} f(v,T), \quad s(v,T) = \psi(T) + \beta^{-1} g(v,T). \tag{13}$$

Here $\phi(T)$ and $\psi(T)$ depend on the ideal gas heat capacity $c_{v\infty}$, i.e. the limiting value of c_v as v tends to infinity only and β is the parameter introduced by Bethe (1942) [2] which measures the relative magnitude of $c_{v\infty}$ evaluated at the critical point temperature and the universal gas constant R. For perfect gases and other gases of low molecular complexity β is of $O(1)$ and the dependence of u and s on v and T is equally important. However, for gases of moderate and high molecular complexity β is large and both quantities depend mainly on T with small corrections of $O(1/\beta)$ due to vatiations of v accounted for by the functions $f(v,T)$ and $g(v,T)$. Substitution of expressions (13) into the Rankine-Hugoniot relationship assuming β to be large then yields the result that temperature changes caused by shocks are small and of $O(1/\beta)$:

$$\frac{[T]}{T_1} = O(\beta^{-1}). \tag{14}$$

As a consequence, shock adiabats, isotherms and isentropes differ little in the p, v diagram. Therefore, in the definition of the fundamental derivative Γ derivatives at constant s can be approximated by derivatives at constant T

$$\Gamma = \frac{v^3}{2c^2}\left(\frac{\partial^2 p}{\partial v^2}\right)_s = \frac{v^3}{2c^2}\left(\frac{\partial^2 p}{\partial v^2}\right)_T + O(\beta^{-1}) \tag{15}$$

and since isotherms are known to have negative curvature in the general neighbourhood of the thermodynamic critical point it is expected that Γ will be negative there also.

As pointed out before β is large for gases with large specific heats and, therefore, such gases are candidates of BZT fluids. As an example Fig.5 displays results for van der Waals gases having values $1/\beta$ equal to 0.02 and 0.001. p and v are non-dimensional with their critical point values p_c and v_c. In both cases there exists a region of negative Γ bounded by the so called transition line $\Gamma = 0$. While for $1/\beta = 0.02$ the difference between shock adiabats, isotherms and isentropes is still clearly visible they almost collapse for the smaller value $1/\beta = 0.001$. For the treatment of shocks in BZT fluids it is essential to acknowledge the qualitative changes of the shock adiabat associated with increasing values of β, Fig.6. If β is of $O(1)$ shock adiabats are curved upwards as in the case of perfect gases. For large values of β, however, shock adiabats exhibit two inflexion points separated by a bulge with negative curvature.

To simplify the analysis we make use of the relationship

$$\frac{[p]}{[v]} = -\rho_1^2 w_1^2 = -\rho_2^2 w_2^2 \tag{16}$$

which relates the slope of the straight line, the so called Rayleigh line, which connects the upstream and downstream states on the shock adiabat with the square of the associated mass flux, Fig.7.

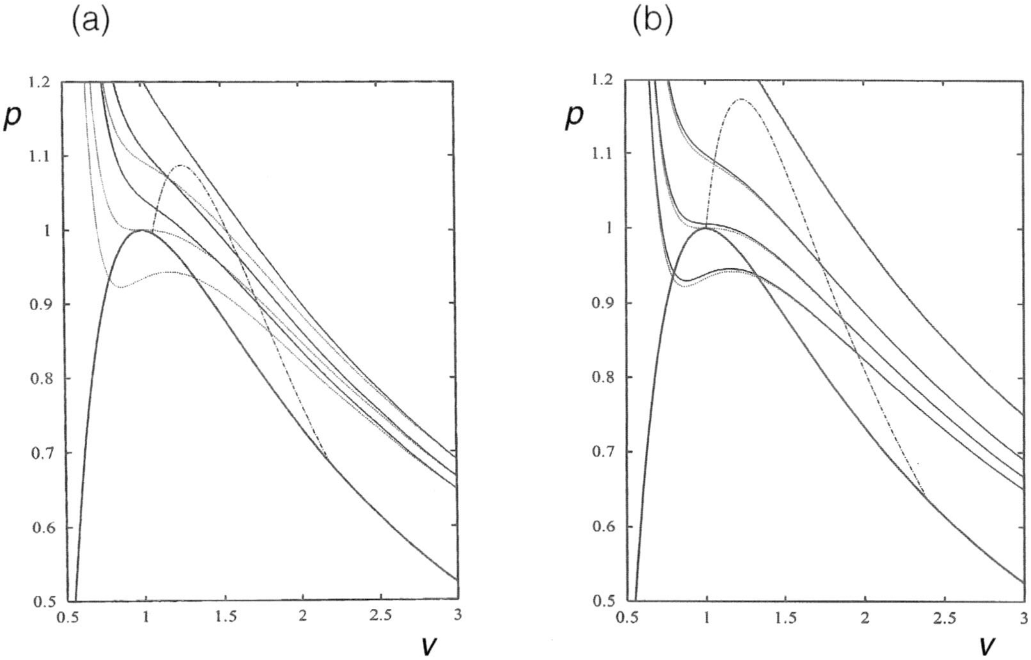

Figure 5. —— Isentropes, ---- shock adiabats and ········ isotherms for van der Waals gases with (a) $1/\beta = 0.02$ and (b) $1/\beta = 0.001$. **——** saturation line, -·-·- $\Gamma = 0$ locus.

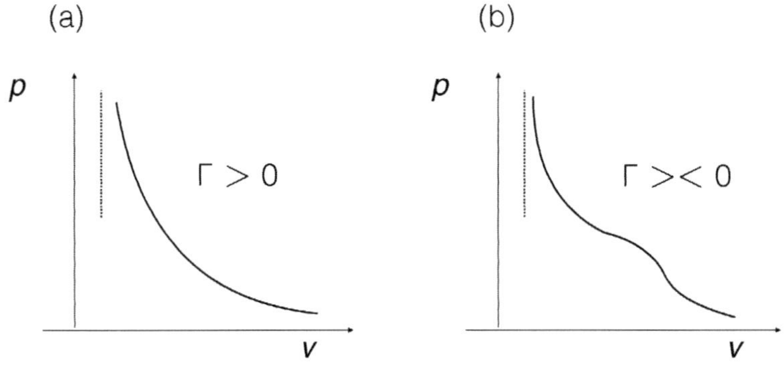

Figure 6. Qualitative shape of shock adiabats for (a) $\beta = O(1)$ and (b) $\beta \gg 1$.

For the following, two properties of the Rayleigh line are of importance, see e.g. Kluwick (2001) [21]. First, it can be shown that the wave speed ordering relationship is satisfied if the Rayleigh line does not cut the shock adiabat in interior points. Therefore, a shock with upstream and downstream pressures p_1, p_2 is admissible while a shock with

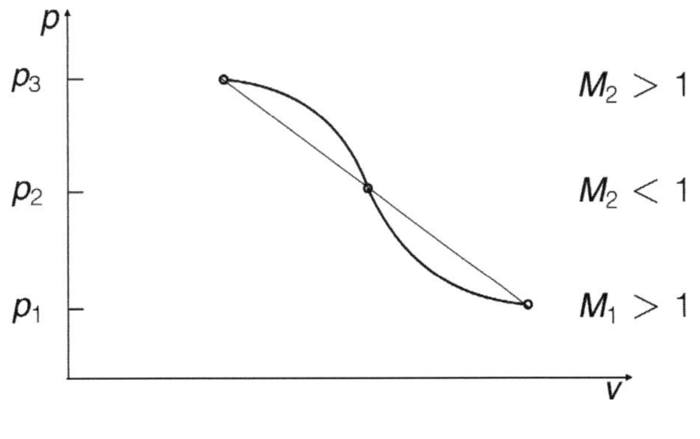

Figure 7. Admissible and inadmissible shocks. ▬ shock adiabat, —— Rayleigh line.

downstream pressure p_3 is inadmissible. Second, it can be shown that if the Rayleigh line is tangent to the shock adiabat in some point, Fig.8, then the Mach number M associated with this state is equal to one and the distribution of the specific entropy along the shock adiabat exhibits an extremum.

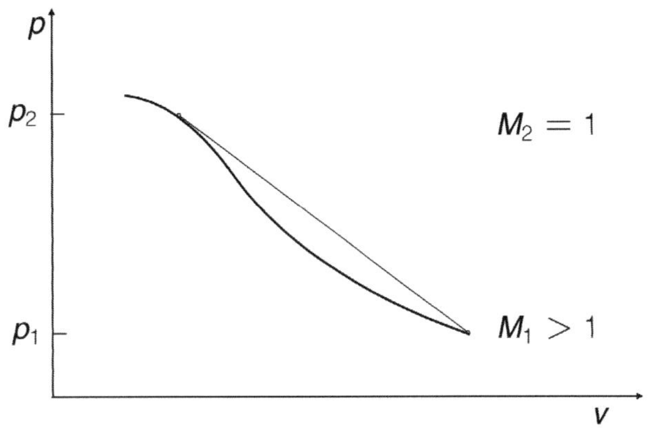

Figure 8. Shock with sonic downstream conditions. ▬ shock adiabat, —— Rayleigh line.

These results motivated Lax (1957) [24] and Oleinik (1959) [25] to propose the extended wave speed ordering relationship $M_1 \geq 1 \geq M_2$ which allows the existence of non-classical shocks with Mach number 1 upstream or downstream (sonic shocks) or even with Mach number 1 upstream and downstream (double sonic shocks). Representative examples are shown in Fig.9.

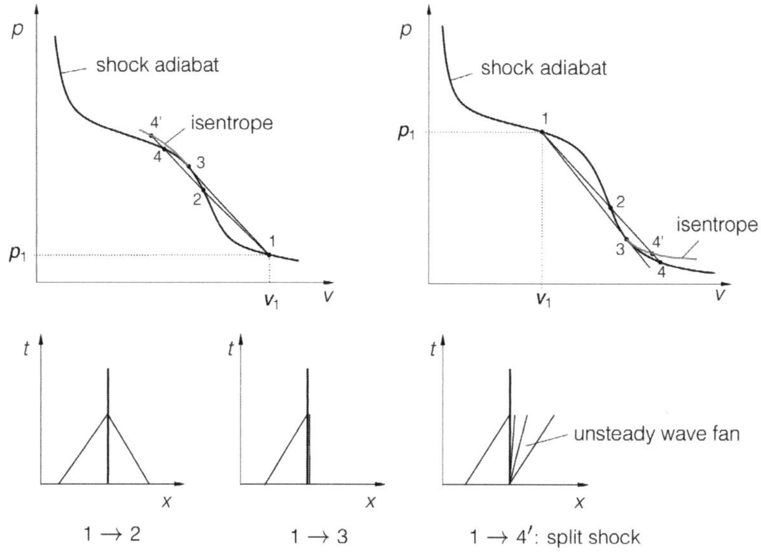

Figure 9. Admissible and inadmissible compression/rarefaction shocks

First let us consider compression shocks with an upstream state 1. Shocks leading to a downstream state 2 are clearly admissible as the associated Rayleigh line is located entirely above the shock adiabat so that the wave speed ordering relationship is satisfied as shown in the x, t diagram. By increasing the shock strength one approaches state 3 where the Rayleigh line is tangent to the shock adiabat indicating that a sonic downstream state has been reached: in the x, t diagram the shock front coincides with the wave front downstream of the shock. Also it is seen that single shocks of even higher strength, leading e.g. from state 1 to state 4 are inadmissible as the Rayleigh line then cuts the shock adiabat. By imposing a larger pressure increase we obtain the following structure: the shock from state 1 to state 3 with sonic downstream conditions is followed by a continuous compression of the fluid along the isentrope starting in point 3 which generates a compression wave fan in the space-time diagram and represents another non classical feature of BZT fluids. The resulting compound structure consisting of a sonic compression shock and an attached compression wave fan is termed a compression split shock.

The diagram on the right hand side of Fig.9 shows that the scenario resulting from the consideration of rarefaction shocks of variable strength is completely analogous to that obtained for compression shocks: by increasing the shock strength, i.e. by reducing the downstream pressure, a sonic downstream state 3 is reached eventually. Further decrease of the downstream pressure is possible only through a rarefaction split shock, i.e. through the combined action of a sonic rarefaction shock and an attached rarefaction wave fan.

3. Weakly nonlinear progressive waves

The results summarised in Section 2 provide the building blocks necessary to treat specific problems e.g. acoustic waves which served as the starting point of our considerations. To simplify these considerations we concentrate on waves which carry small density disturbances

$$\frac{\rho - \rho_0}{\rho_0} = \epsilon\rho_1 + O(\epsilon^2), \quad \epsilon \ll 1. \tag{17}$$

Such an approach yields a complete qualitative picture of all possible phenomena and allows for significant analytical progress. Taylor expansion of the perturbation mass flux $j(\rho)$ involves the values Γ_0, Λ_0, N_0 of the fundamental derivative and its first and second derivative with respect to ρ evaluated in the unperturbed state. By introducing suitably scaled quantities

$$\bar{\rho}_1 = \frac{\Lambda_0}{\Gamma_0}\rho_1, \quad \bar{j}_1 = \frac{\Lambda_0^2}{\Gamma_0^3}j_1, \tag{18}$$

one then obtains the remarkable result that the representation of $\bar{j}_1$ in terms of $\bar{\rho}_1$ depends on a single similarity parameter $\hat{N}_0$ only

$$\bar{j}_1 = \frac{1}{2}\bar{\rho}_1^2 + \frac{1}{6}\bar{\rho}_1^3 + \frac{\hat{N}_0}{12}\bar{\rho}_1^4, \quad \hat{N}_0 = \frac{N_0\Gamma_0}{\Lambda_0}. \tag{19}$$

Graphs $\bar{j}_1$ versus $\bar{\rho}_1$ are depicted in Fig.10 for two values of $\hat{N}_0$ including $\hat{N}_0 = 0$.

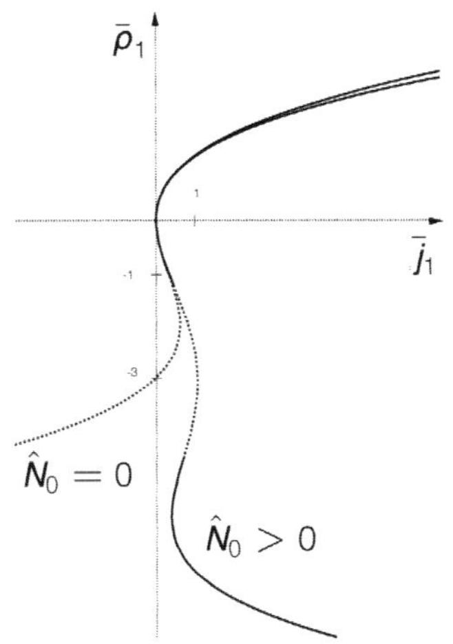

Figure 10. Representative examples of $\bar{j}_1$, $\bar{\rho}_1$ diagrams; —— $\Gamma > 0$, ······· $\Gamma < 0$.

By making use of the expression

$$\bar{\Gamma} = \frac{\Gamma}{\Gamma_0} = \epsilon^2\frac{d^2\bar{j}_1}{d\bar{\rho}_1^2} \tag{20}$$

we infer that inflexion points in Fig.10 separate regions of positive Γ (solid lines) and regions of negative Γ (dotted lines). As a representative example of progressive waves generated by localised disturbances we investigate the evolution of square pulses specified by initial conditions at $\tau = \epsilon^3 c_0 t/L = 0$, Fig. 11.

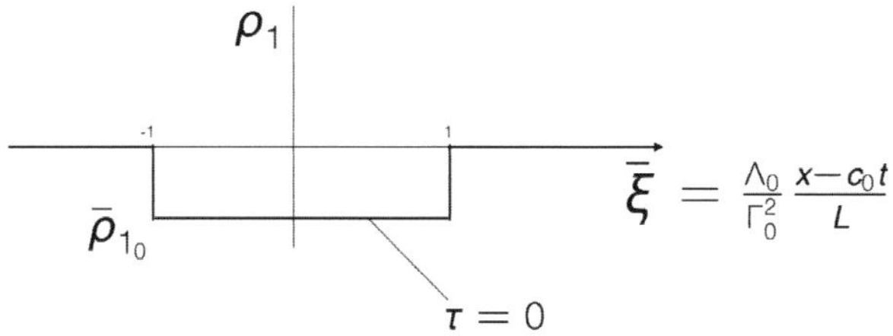

Figure 11. square pulse

Here $\bar{\rho}_1$ vanishes for values of the transformed propagation distance $\bar{\xi} > 1$, $\bar{\xi} < -1$ and assumes the constant value $\bar{\rho}_{1_0}$ in between. For $\bar{\rho}_{1_0} > -1$ the fluid remains in the positive Γ region and, consequently, the wave evolution is qualitatively similar to the predictions of classical nonlinear acoustics: a rarefaction wave fan forms at $\bar{\xi} = 1$, the pulse is terminated by a compression shock and the wave profile eventually approaches a linear density distribution as τ tends to infinity, Fig.12.
For $\bar{\rho}_{1_0} = -2.2$ the density disturbances inside the pulse are large enough for Γ to change sign which is seen to lead to significant changes of the wave pattern, Fig.13. Specifically it is found that a rarefaction wave fan at $\bar{\xi} = 1$ taking the fluid from the unperturbed to the perturbed state is no longer possible but requires the combined action of a sonic rarefaction shock and an attached rarefaction wave fan. Also the discontinuous increase of $\bar{\rho}_1$ at the wave tail no longer generates a single compression shock but results in the formation of a compression wave fan and a sonic compression shock. Both shock-fan combinations are weakly nonlinear versions of finite amplitude split shocks described in section 2.3. The sonic rarefaction shock is weakened by its interaction with the compression wave fan. It emanates characteristics in the tangential direction thereby generating a so-called precursor region. This has the distinguishing feature that the characteristics are no longer connected with the initial data but that information propagates along characteristics that cross the sonic rarefaction shock. Another interesting phenomenon arises from the collision of the sonic rarefaction and compression shocks at $\tau = \tau_c$. While according to the classical theory of weakly nonlinear acoustics two merging shocks result in a single shock of larger amplitude, the collision of a rarefaction and a compression shock is seen to be associated with the formation of a sharp spike in the density distribution as τ approaches τ_c and a sudden reduction of the shock strength at $\tau = \tau_c$.

We conclude the discussion of weakly nonlinear progressive waves by considering a square pulse of even larger negative amplitude, large enough to take the fluid across the whole negative Γ region. The resulting wave pattern then exhibits an astounding

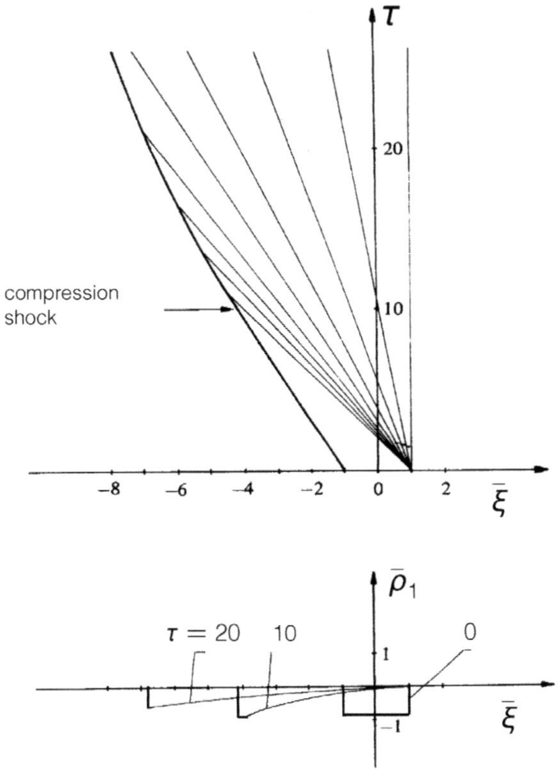

Figure 12. evolution of square pulse: $\hat{N}_0 = 0, \bar{\rho}_{1_0} = -0.9$

richness of wave fans, sonic shocks and double sonic shocks, Fig.14. The three examples of weakly nonlinear acoustic waves may be of limited value as far as practical applications are concerned. However, they demonstrate how the complexity of dense gas flows increases dramatically if non classical phenomena such as compression wave fans, sonic shocks and double sonic shocks come into play. Clearly, two and three dimensional unsteady flows are expected to exhibit even more complex behaviour which in part explains the difficulty to perform clean shock tube experiments where three dimensional disturbances can hardly be avoided. In the following we shall, therefore, concentrate on steady flows through Laval nozzles and two dimensional external flows which are of interest on their own and where unwanted disturbances may be controlled more easily in experimental work.

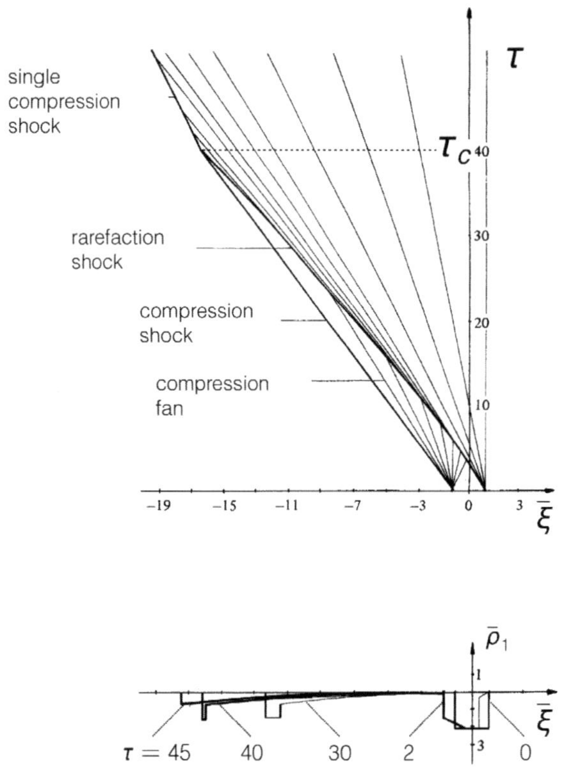

Figure 13. evolution of square pulse: $\hat{N}_0 = 0, \bar{\rho}_{1_0} = -2.2$

4. Weakly nonlinear steady flow

4.1. Flow through slender nozzles

Transonic quasi - one dimensional flow through slender nozzles, Fig.15, is governed by the equation

$$\frac{\partial \rho_1}{\partial \tau} + \frac{1}{2}\frac{\partial j_1}{\partial x} = -\frac{1}{2}\frac{dA_1}{dx} \tag{21}$$

where unsteady effects are included for the moment, Kluwick (1993) [18], Kluwick and St.Scheichl (1996) [19]. A_1 characterises the variation of the area of cross section relative to its value A_0 at the throat

$$\frac{A}{A_0} = 1 + \epsilon^4 A_1. \tag{22}$$

As before ϵ is a small perturbation parameter and that it enters the analysis at $O(\epsilon^4)$ reflects the extreme sensitivity of transonic flows with respect to changes of stream tube area. In the case of steady flow (21) can immediately be integrated to yield

$$j_1 + A_1 = Q \tag{23}$$

where Q is an integration constant. It is convenient to require $\rho_1 = 0$ if the local Mach number M equals 1 which leads to the following representation for the perturbation

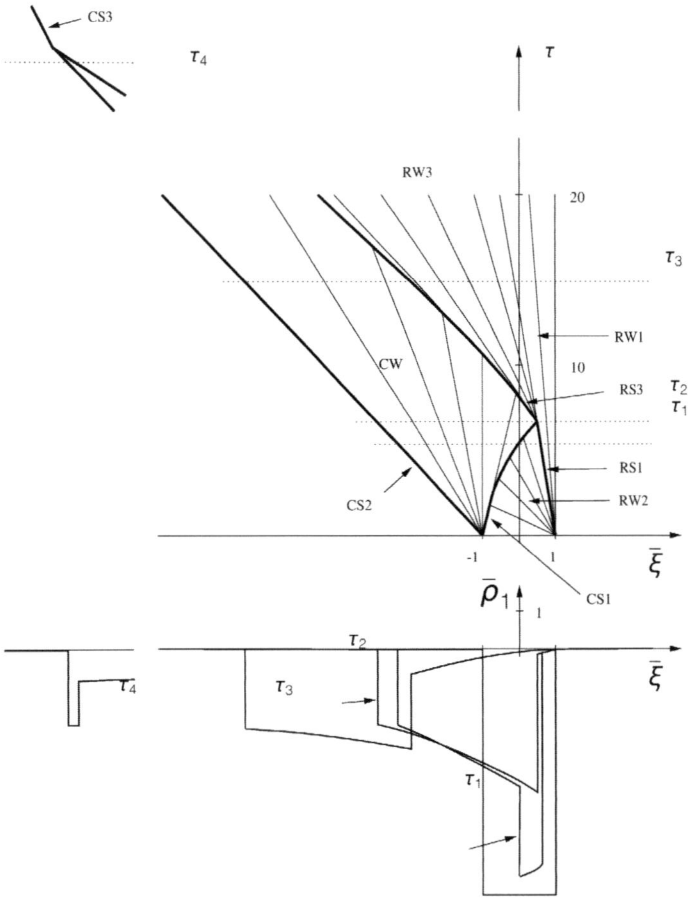

Figure 14. evolution of square pulse: $\hat{N}_0 = 0.342, \bar{\rho}_{1_0} = -6.5$

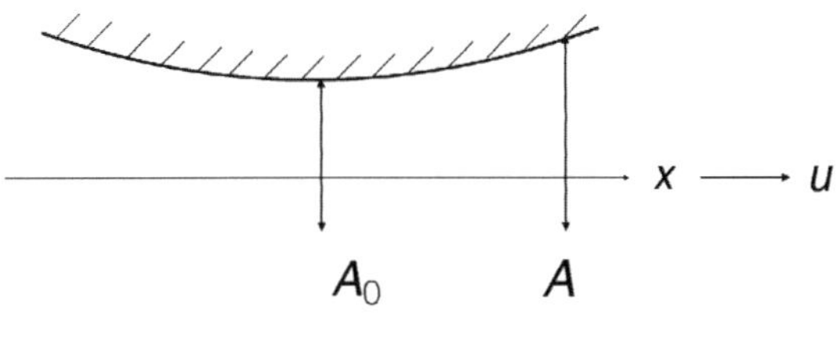

Figure 15. transonic slender nozzle flow

mass flux j_1 obtained by a Taylor series expansion:

$$j_1 = -\hat{\Gamma}_0\rho_1^2 - \frac{1}{3}\hat{\Lambda}_0\rho_1^3 - \frac{1}{12}\hat{N}_0\rho_1^4. \tag{24}$$

j_1 includes constants which are suitably scaled values of the fundamental derivative Γ and its first and second derivative with respect to the fluid density in the unperturbed state. We note that the above perturbation approach has recently been generalised in Vimercati (2015) [29], Guardone and Vimercati (2016) [12] where, as in the numerical study by Cramer and Fry (1993) [6], the exact expression for the perturbation mass flux is taken into account rather than its approximation by equation (24). As a first specific example we consider convergent-divergent Laval nozzles such that A_1 is a quadratic function of the distance x measured from the throat: $A_1 = Cx^2$. If $\hat{\Gamma}_0$ satisfies the inequality

$$\hat{\Gamma}_0 > \frac{3\hat{\Lambda}_0^2}{8\hat{N}_0} \tag{25}$$

the flow is of classical type where Γ is positive throughout and a sonic state is reached only once. Solutions form two branches of continuously accelerating or decelerating flows leading, respectively, from subsonic to supersonic or from supersonic to subsonic conditions. In addition, there exist flows with compression shocks present in the diverging part of the nozzle, Fig.16.

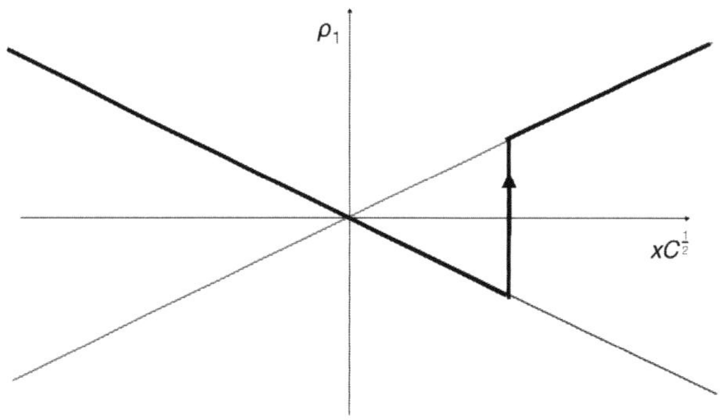

Figure 16. Laval nozzle: posible density distributions. $\hat{\Gamma}_0 = 24.06, \hat{\Lambda}_0 = 30.96, \hat{N}_0 = 19.22$.

Fig.17 shows how these conventional flow patterns change if $\hat{\Gamma}_0$ is reduced so that the flow then takes the fluid across the whole negative Γ region and there exist three sonic states rather than a single one, Kluwick(1993) [18]. It is seen that density distributions are no longer single-valued but exhibit regions of multi-valuedness which have to be eliminated by the insertion of shock discontinuities. Consequently, shock free flow is not possible but instead we observe flow configurations with one, two or even three shocks present. This naturally raises the question: can dense gases of BZT type be accelerated shock free from subsonic to supersonic speeds? To answer this question we formulate

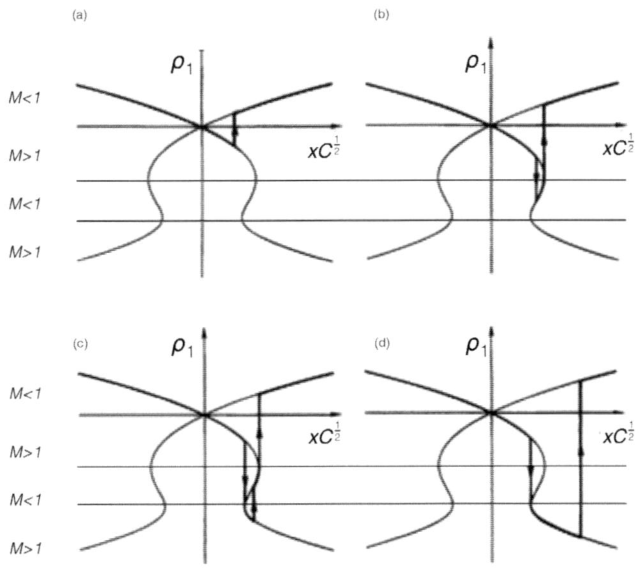

Figure 17. Laval nozzle: posible density distributions. $\hat{\Gamma}_0 = 4.91, \hat{\Lambda}_0 = 15.80$, $\hat{N}_0 = 17.67$.

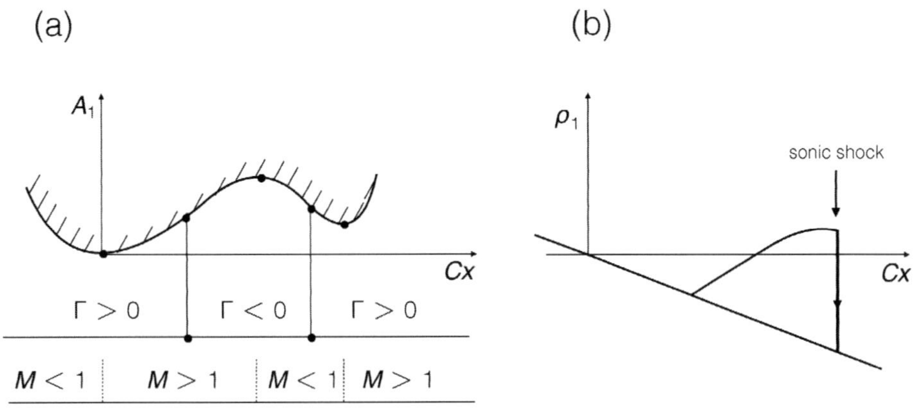

Figure 18. (a) Nozzle shape designed to generate a shock free transition from subsonic to supersonic flow conditions. (b) Associated density distributions. $\hat{\Gamma}_0 = 4.91$, $\hat{\Lambda}_0 = 15.80, \hat{N}_0 = 17.67$.

an inverse problem: we prescribe the desired density distribution, e.g. a linear one

$$\rho_1 = -Cx, \quad C > 0 \tag{26}$$

and use the continuity equation to determine the necessary nozzle shape. One then finds that a nozzle with two throats and an intervening anti- throat is required to achieve a subsonic-supersonic transition from a state with positive Γ to another positive Γ state which passes through the whole negative Γ region, Fig.18 a. However, one also observes that in addition to the prescribed density distribution there exists a second solution branch which exhibits a sonic shock, Fig.18b. So which of these solutions will be realised in practice? To my knowledge a definite answer is not known at present but calculations carried out with the aim to understand how the steady state solution is reached by means of an unsteady starting process governed by the full equation (21) appear to favour the solution with the sonic shock present, Kluwick and Scheichl (1996) [19]. To obtain a definitive answer however will require an extension of the present theory beyond the first order approximation in order to account for the differences between the density and pressure disturbances due to entropy changes caused by shocks. Alternatively the theory developed in Guardone and Vimercati (2016) [12], Vimercati (2015) [29] might be used to clarify this point.

4.2. Two dimensional external transonic flow

As a last example of inviscid gasdynamics we consider transonic external flows, Cramer and Tarkenton (1992) [5]. Specifically we concentrate on two dimensional ramp configurations as sketched in Fig. 19.

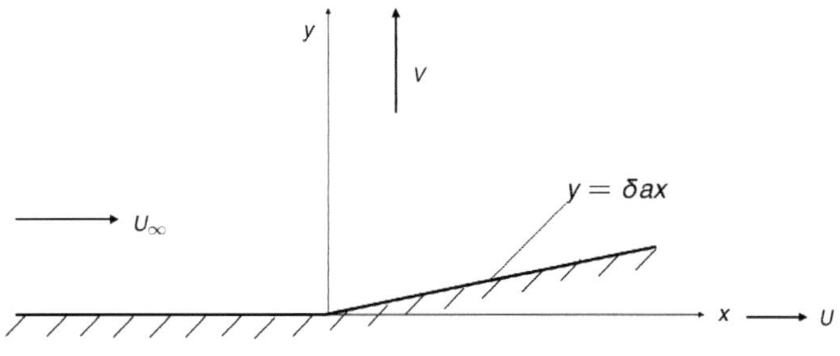

Figure 19. Transonic flow past compression/expansion ramps

Herein x, y and u, v denote Cartesian coordinates parallel and normal to the incoming flow and the corresponding velocity components. Parameters δ and a define the ramp angle. We assume that the ramp angle is small, i.e. we require that $a = O(1)$ while $\delta \ll 1$. The governing equations

$$\frac{\partial j}{\partial x} + \frac{\partial v}{\partial y} = 0, \quad \frac{\partial u}{\partial y} - \frac{\partial v}{\partial x} = 0, \quad j(u) = (1 - M_\infty^2)u - \Gamma_\infty u^2 + \frac{\Lambda_\infty}{3}u^3 \tag{27}$$

include the continuity equation and the requirement of irrotational flow. Five different flow regimes characterised by two non-dimensional similarity parameters

$$K_1 = \frac{M_\infty^2 - 1}{(\delta\Gamma_\infty)^{\frac{2}{3}}}, \quad K_2 = \frac{\delta^{\frac{2}{3}}\Lambda_\infty}{\Gamma_\infty^{\frac{4}{3}}} \tag{28}$$

can be identified, Kluwick and Cox (2016) [22]. For upstream supersonic flow considered here only K_1 is strictly positive while K_2 can be positive or negative.

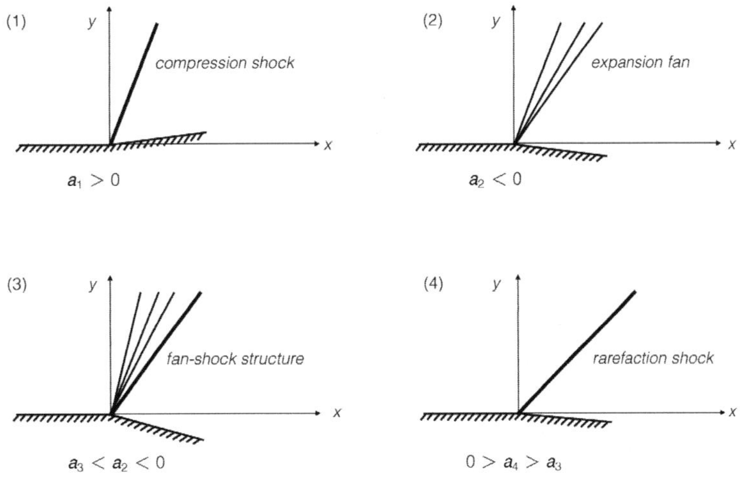

Figure 20. Possible wave patterns for $K_1 > 0, K_2 > 0$ fixed, $\Gamma_\infty > 0$, $a >< 0$

As a representative example Fig.20 displays results for $K_2 > 0$ and $\Gamma_\infty > 0$. For positive ramp angles $a > 0$ we observe a single compression shock while expansion ramps causing small negative turning angles $a < 0$ generate pure expansion fans in agreement with intuition based on classical gas dynamics. Non classical effects, however, arise if the negative ramp angle is increased which causes the emergence of a rarefaction shock with sonic upstream conditions. With ramp angle a decreasing even further the wave fan shrinks in size which finally results in the formation of a single rarefaction shock.

Fig.21 shows solutions of a different type of problem first posed by Guardone, Colonna, Casati and Rinaldi (2014) [11] where the ramp angle a is kept constant while the fluid properties upstream of the ramp vary. Specifically we consider an expansive ramp with $a = -1.6$ and assume $K_1 = 1$. As K_2 increases from -5.5 to 1.5 the flow pattern changes from a centred wave fan to a single shock in full qualitative agreement with the numerical results presented by Guardone et al. (2014) [11].

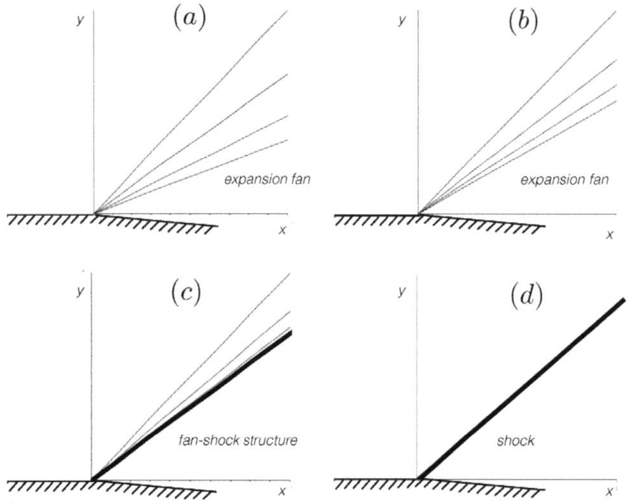

Figure 21. Possible wave patterns for $K_1 = 1, a = -1.6$ and K_2 varying between -5.5 and 1.5.

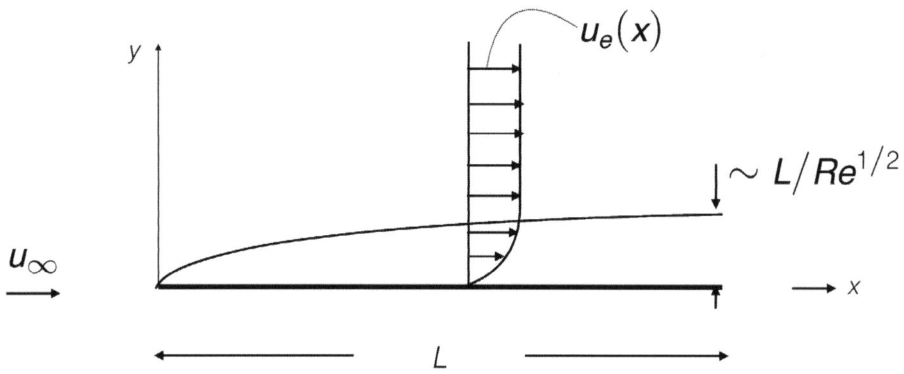

Figure 22. Laminar boundary layer on a flat plate

5. Laminar boundary layers

Following the outline of inviscid flows given in the preceding sections we now turn to a brief discussion of dissipative effects. To simplify this discussion I shall concentrate on laminar flows of boundary layer type. A representative example, see Fig.22, is given by the boundary layer which forms on a flat plate of length L in the limit of large Reynolds number.

$$Re = \frac{u_\infty L \rho_\infty}{\mu_\infty} \gg 1. \qquad (29)$$

Here u, ρ and μ denote the velocity component parallel to the plate, the density and the dynamic viscosity. Subscripts ∞ and e refer to quantities evaluated under free stream conditions and at the boundary layer edge. If the Reynolds number is scaled out as usual the nondimensional groups which enter the boundary layer equations are the Prandtl number, the coefficient of thermal expansion α_∞ non dimensional with T_∞ and the Eckert number

$$Pr = \frac{\mu_\infty c_{p\infty}}{\lambda_\infty}, \quad \alpha_\infty T_\infty, \quad Ec = \frac{u_\infty^2}{c_{p_\infty} T_\infty} \tag{30}$$

λ and c_p are, of course, the thermal conductivity and the specific heat at constant pressure. Similar to perfect gases the Prandtl number and $\alpha_\infty T_\infty$ are of $O(1)$ also for dense gases if the immediate neighbourhood of the thermodynamic critical point is excluded from the considerations. In contrast, however, the Eckert number no longer is of $O(M_\infty^2)$ but of $O(M_\infty^2/\beta)$. As pointed out before the quantity β introduced by Bethe increases with increasing molecular complexity and typical values for vapours of BZT fluids are in the range of 100 to 150. As a consequence dissipation caused by internal friction and heat conduction can be neglected even at moderately large supersonic Mach numbers. For flows past adiabatic walls this means that the temperature and thus also the fluid density is nearly constant across the boundary layer.

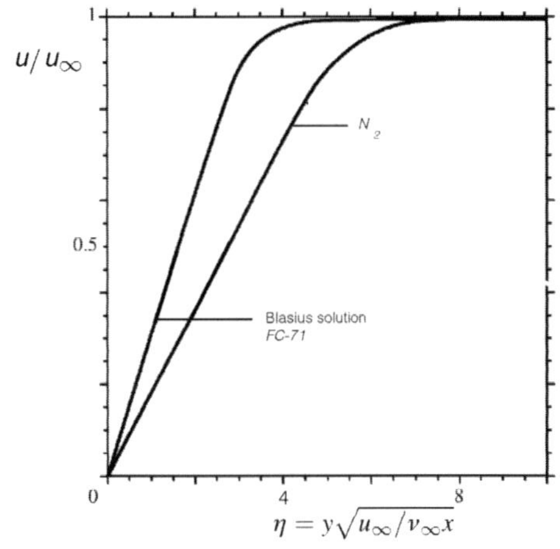

Figure 23. Flat plate boundary layers

As an example Fig.23 shows the velocity distribution across a flat plate boundary layer with zero pressure gradient at $M_\infty = 2$ for N_2 and the BZT fluid pf-trihexylamine ($C_{18}F_{39}N$) which I will continue to refer to using the manufacturers designation FC-71. In the case of N_2 dissipation causes the velocity profile to deviate substantially from the Blasius result for incompressible flows while the solution for FC-71 is indistinguishable from it within graphical accuracy, Kluwick (2000) [20]. Deviations from the classical boundary layer behaviour are caused not only by the smallness of the Eckert number but also by the unconventional gas dynamic properties of BZT fluids in the external inviscid flow region. An example is provided by the interaction of a dense gas boundary layer with an oblique shock. Both diagrams in Fig.24 display distributions of the skin friction

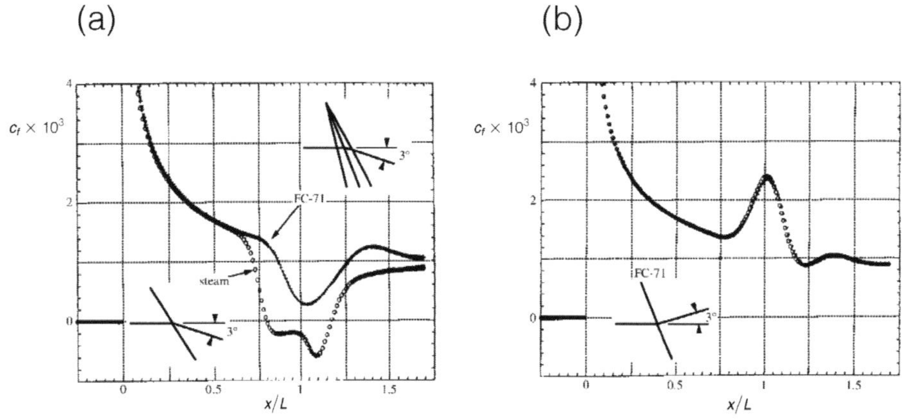

Figure 24. Flat plate boundary layer which interacts (a) with a weak compression wave and (b) with a weak rarefaction shock. $M_\infty = 2, T_\infty = 646.2K, P_\infty = 8.55atm$.

coefficient calculated by Cramer and Park (1999) [7]. In the case of steam the imposed deflection angle of 3 degrees leads to the formation of a compression shock which is strong enough to separate the boundary layer. In the case of FC-71 the same deflection angle leads to a smooth compression fan and the boundary layer remains attached. If the deflection angle is negative no shock discontinuity can form in a regular fluid such as steam but a rarefaction shock forms in the BZT fluid FC-71. Across this shock the fluid accelerates which in turn causes the wall shear to increase rather than to decrease.

Following this brief account of dense gas local interactions triggered by rapid pressure changes we turn to the last topic treated here: the effect of an adverse pressure gradient acting over a distance of $O(1)$ on the typical boundary layer length scale. A representative example is provided by linearly retarded flows. The streamwise velocity component in the external inviscid flow region then is of the form

$$u_e(x) = u_\infty(1 - \frac{x}{L}). \tag{31}$$

The incompressible version of this problem was investigated first by Howarth (1938) [15] and Hartree (1939) [14] who observed that the pressure increase associated with the imposed velocity decrease always leads to separation and that the numerical solution of the boundary layer equations cannot be continued beyond the point of zero wall shear. A related result holds for compressible flow past adiabatic walls if the fluid is a perfect gas or even a dense gas with low molecular complexity, Stewartson (1962) [26].

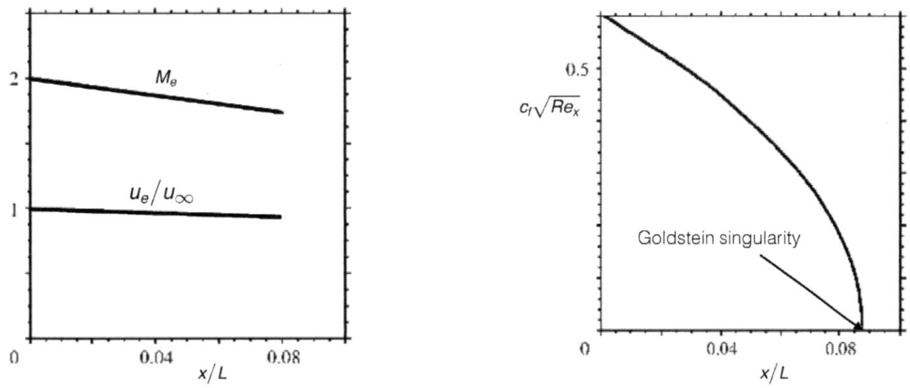

Figure 25. Linearly retarded flow for N_2: $v_\infty = 5.009v_c$, $T_\infty = 1.001T_c, M_\infty = 2$, $R/c_{v_\infty} = 0.39, Ec = 0.72$.

As an example we consider Nitrogen with free stream conditions listed in Fig.25 where ρ_c and T_c denote the critical point density and temperature. Decreasing values of u_e lead to decreasing values of the local Mach number M_e at the boundary layer edge. The associated pressure increase causes the local friction coefficient c_f to decrease also and the skin friction distribution eventually tends to zero in an irregular manner known as Goldstein singularity, Goldstein (1948) [10].

Fig.26 shows what happens if the regular fluid N_2 is replaced by the BZT fluid FC-71. As before c_f drops initially but this tendency comes to a halt and c_f starts to rise despite the fact the pressure gradient is still unfavourable. A more detailed investigation of the flow suggests that this unconventional feature is closely related to the Mach number distribution at the boundary layer edge. In contrast to N_2 the isentropic compression of FC-71 causes M_e to increase rather than to decrease. Evaluation of the continuity equation outside the boundary layer then suggests that the displacement body felt by the external inviscid flow may shrink rather that expand in the streamwise direction. This is confirmed by numerical calculations. The associated momentum influx into the boundary layer is then able to overcome the onset of separation and causes c_f to rise sharply. Eventually, however, M_e drops which in turn quenches this effect so that the formation of a separation singularity is inevitable. Nevertheless the results plotted in Fig.26 clearly demonstrate the potential of dense gas flows to delay flow separation. If the free stream value v_∞ of the specific volume is slightly increased and approaches the critical value $v_{\infty_c} = 5.01v_c$ one obtains the limiting case shown in Fig.27: the wall shear vanishes in a single point but immediately recovers; at the point of zero wall shear the distribution of the displacement thickness delta exhibits a sharp corner. To resolve these physically unrealistic slope discontinuities we perform a

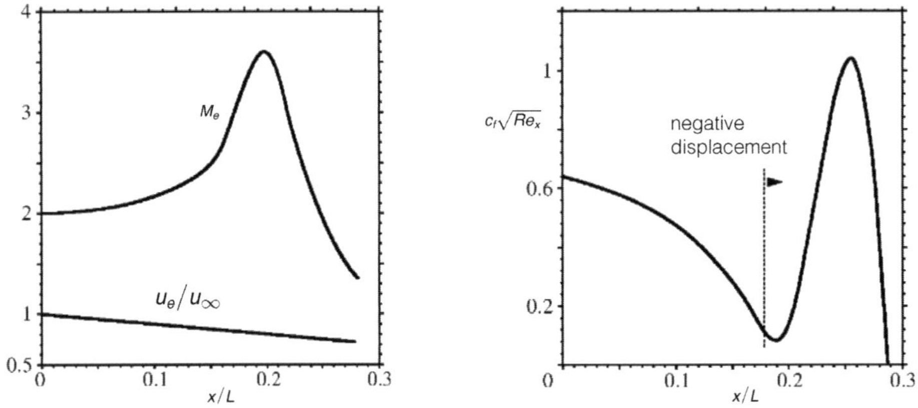

Figure 26. Linearly retarded flow for FC-71: $v_\infty = 5.009v_c$, $T_\infty = 1.001T_c$, $M_\infty = 2$, $R/c_{v_\infty} = 0.007, Ec = 0.015$.

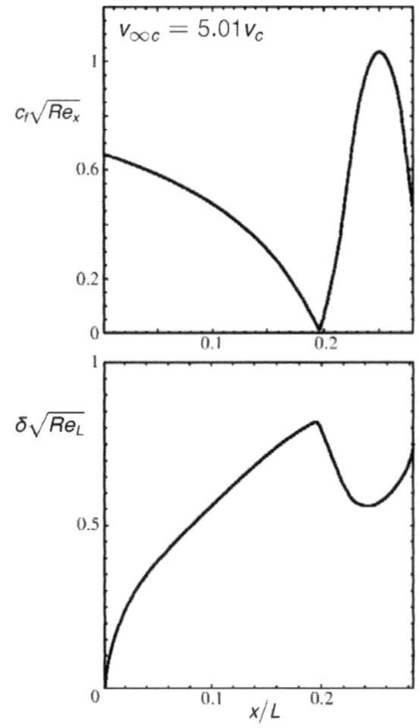

Figure 27. Linearly retarded flows for FC-71: $v_\infty = 5.01v_c$, $T_\infty = 1.001T_c$, $M_\infty = 2$, $R/c_{v_\infty} = 0.007, Ec = 0.015$.

local analysis based on the assumption that a further increase of v_∞ will result in the formation of a short separated flow region close to $x = x_s$ where classical boundary layer theory fails. Introducing stretched quantities X, A, and the parameter a which accounts for the difference between v_∞ and $v_{\infty c}$

$$a \propto Re^{\frac{2}{5}}(v_\infty - v_{\infty c}) \tag{32}$$

one finds that the scaled friction coefficient A(X) satisfies the integro- differential equation

$$A^2(X) - X^2 + 2a = \lambda \int_{-\infty}^{x} \frac{A''(\xi)}{\sqrt{X-\xi}} d\xi \tag{33}$$

where λ is a positive constant, Kinell and Kluwick (2003) [16]. Solutions are shown in Fig.28. As expected A decreases with increasing values a and eventually the formation

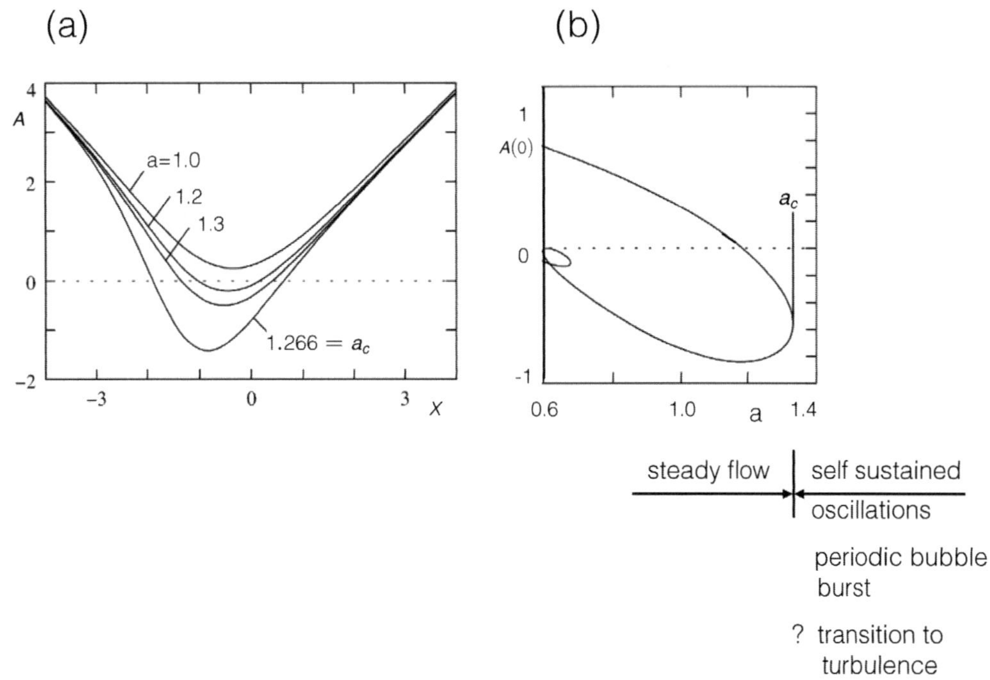

Figure 28. Marginally separated dense gas boundary layer: (a) wall-shear stress distributions for various values of a, (b) bifurcation diagram; transition from steady flow to self-sustained oscillations.

of a separated flow region where A is negative is observed, Fig.28a. Fig.28b displays the variation of $A(0)$ with a and it is seen that solutions of the interaction equation (33) form two branches: an upper branch where skin friction distributions are of the form depicted in Fig.28a and a lower branch which exhibits significantly larger recirculation zones. Even more important, however, it is seen that steady solutions exist up to a critical value $a_c \doteq 1.266$ only. An extended analysis which includes both steady and unsteady effects indicates that the passage through a_c leads the onset of self-sustained oscillations associated with periodic bubble burst. These oscillations may possibly be interpreted as first signs of transition to turbulence but this represents another challenge for future research in the field of NICFD.

6. Conclusions

The present paper starts from the premise that negative Γ Fluids, so called Bethe-Zel'dovich-Thompson (BZT) fluids, exist and concentrates on non-classical aspects which arise in this context. To simplify the discussion inviscid weakly non-linear flows were considered first. Such an approach allows for significant analytical progress but still yields a complete qualitative picture of all possible new phenomena including rarefaction shocks, sonic shocks, double sonic shocks, split shocks, compression wave fans, the delay of shock formation by decreasing values of Γ, etc.. This is achieved by introducing the concept of the perturbation mass flux which applies to both steady and unsteady flows and simplifies the derivation of the three building blocks needed for the general understanding of these unconventional flow features: shock formation, wave evolution, shock propagation. Specific examples considered which might be of interest also in connection with experimental work to demonstrate existence of in particular rarefaction shocks are unidirectional acoustic waves, quasi one-dimensional nozzle flow and transonic two-dimensional flow past compression/expansion ramps. Besides providing results which are of importance for practical applications the treatment of these flows highlights another advantage of the present perturbation analysis, the emergence of non-dimensional similarity parameters which identify regimes of qualitatively different flow behaviour.

Following the discussion of inviscid flows in sections 2, 3 and 4 viscous effects are considered in section 5 which concentrates on laminar flows of boundary layer type. Specifically it is found that the Eckert number which characterises the importance of dissipation in the energy equation may be small not only for small values of the free stream Mach number but also for moderately large supersonic values. Other non-classical features arise from the unconventional fluid properties in the inviscid external inviscid flow region which may delay flow separation in cases of viscous inviscid interaction but also if an adverse pressure gradient acts over a distance of order one on the typical boundary layer length scale.

Acknowledgement

I would like to thank Dr. E. A. Cox for his invaluable assistance during the preparation of this manuscript. Without his help the manuscript could not have been published in its present form.

References

[1] Becker R. (1922).Stoßwelle und Detonation. Zeitschrift für Physik 8: 321 - 362.

[2] Bethe H.A. (1942). The theory of shock waves for an arbitrary equation of state. Report No. 545. Office of Scientific Research and Development.

[3] Colonna P., Guardone A., Nannan N.R. (2007). Siloxanes: A new class of Bethe-Zel'dovich Thompson fluids. Phys. Fluids, 19(8): 086102-086113.

[4] Cramer M.S., Kluwick, A. (1984). On the propagation of waves exhibiting both positive and negative nonlinearity. J. Fluid Mech. 142: 9 - 37.

[5] Cramer M.S., Tarkenton G.M. (1992). Transonic flows of Bethe-Zel'dovich-Thompson fluids. J. Fluid Mech. 240: 197 - 228.

[6] Cramer M.S., Fry R.N. (1993). Nozzle flows of dense gases. Phys. Fluids A 5: 1246 - 1259.

[7] Cramer M.S., Park S. (1999). On the suppression of shock-induced separation in Bethe-Zel'dovich-Thompson fluids. J. Fluid Mech. 393: 1 - 21.

[8] Duhem P. (1909). On the propagation of shock waves in fluids. Z. Phys. Chem. 69: 169 - 186.

[9] Colonna P., Guardona A., Nannan N.R. (2007). Siloxanes: A new class of Bethe-Zel'dovich Thompson fluids. Phys. Fluids, 19(8): 086102-086114.

[10] Goldstein S. (1948). On laminar boundary-layer flow near a position of separation. Q. J. Mech. Appl. Math. 1: 43 - 69.

[11] Guardone A., Colonna P., Casati E., Rinaldi E. (2014). Non-classical gas dynamics of vapour mixtures. J. Fluid Mech. 741: 681 - 701.

[12] Guardone A., Vimercati D., (2016) Exact solutions to non-classical steady nozzle flows of Bethe - Zel'dovich - Thompson fluids, J. Fluid Mech. 800: 278 - 306.

[13] Hayes W.D. (1958). The Basic Theory of Gasdynamic Discontinuities, in H.W.Emmons, ed., Fundamentals of Gasdynamics, vol. 3, Princeton Series in High Speed Aerodynamics and Jet Propulsion, Princeton University Press, Princeton, N.J.

[14] Hartree D. R. (1939). A solution of the laminar boundary-layer equations for retarded flow. Mem. Aero. Res. Coun., London, no 2426 (spec. vol. I.).

[15] Howarth L. (1938). On the solution of the laminar boundary layer equations. Proc. R. Soc. Lond. A 164: 547 - 579.

[16] Kinell M., Kluwick A. (2003). On a new form of marginal separation. PAMM 3: 358 - 359.

[17] Kluwick A. (1991). Small-amplitude finite-rate waves in fluids having both positive and negative nonlinearity. In A. Kluwick, ed., Nonlinear Waves in Real Fluids, volume 315 of the International Centre for Mechanical Sciences, pp. 1 - 43. Springer Vienna.

[18] Kluwick A. (1993).Transonic nozzle flow of dense gases. J. Fluid Mech. 247: 661 - 688.

[19] Kluwick A., Scheichl St. (1996) Unsteady transonic nozzle flow of dense gases. J Fluid Mech. 310: 113 - 137.

[20] Kluwick A. (2000). Marginally separated flows in dilute and dense gases. Phil. Trans. R. Soc. Lond. A 358: 3169 - 3192.

[21] Kluwick A. (2001). Rarefaction Shocks. in: Handbook of Shock Waves, Vol 1, G. Ben-Dor, T. Elprin, O. Ingra (ed.); Academic Press, San Diego: 339-411.

[22] Kluwick A., Cox E. A. (2016). Steady transonic dense gas flow past a two-dimensional compression/expansion ramp; Talk: 24th International Congress on Theoretical and Applied Mechanics (ICTAM 2016), Montréal, Canada; in: "Book of Papers", IUTAM, (2016), 1 - 2.

[23] Lambrakis K.C., Thompson P. A. (1972). Existence of real fluids with a negative fundamental derivative Γ. The Physics of Fluids 15: 933 - 935.

[24] Lax P.D. (1957). Hyperbolic systems of conservation laws II. Comm. Pure Appl. Math. 19: 537 - 566.

[25] Oleinik O. (1959). Uniqueness and stability of the generalized solution of the Cauchy problem for a quasi-linear equation. Usp. Math. Nauk. 14: 165 - 170.

[26] Stewartson K. (1962). The behaviour of a laminar compressible boundary layer near a point of

zero skin friction. J. Fluid Mech. 12: 117 - 128.

[27] Thompson, P.A. (1971). A Fundamental Derivative in Gas Dynamics. Physics of Fluids 14: 1843 - 1849.

[28] Thompson P.A., Lambrakis K.C. (1973). Negative shock waves. J. Fluid Mech. 60 : 187 - 208.

[29] Vimercati D.(2015) Exact solutions to non-classical, quasi-1D nozzle flows of Bethe-Zel'dovich-Thompson fluids. Tesi di laurea, Politecnico di Milano

[30] Zel'dovich Ya.B. (1946). On the possibility of rarefaction shock waves. Zh. Eksp. Teor. Fiz. 4: 363 - 364.

[31] Zieher F. (1993). Kompressible Grenzschichtströmungen bei Zuständen in der Nähe des kritischen Punktes. Masters Thesis, Vienna University of Technology, 1 - 44.

Supersonic flow of non-ideal fluids in nozzles: An application of similitude theory and lessons for ORC turbine design and flexible use considering system performance

M White[1], A I Sayma[2] and C N Markides[1]

[1] Clean Energy Processes (CEP) Laboratory, Department of Chemical Engineering, Imperial College London, South Kensington Campus, London, SW7 2AZ, UK
[2] Department of Mechanical Engineering and Aeronautics, City University London, Northampton Square, London, EC1V 0HB, UK

Email: martin.white@imperial.ac.uk

Abstract. A significant improvement in the economy-of-scale of small-scale organic Rankine cycle (ORC) systems can arise from the appropriate design of components that can be manufactured in large volumes and implemented flexibly into a wide range of systems and potential applications. This, in turn, requires accurate predictions of component performance that can capture variations in the cycle conditions, parameters or changes to the working fluid. In this paper previous work investigating a modified similitude theory used to predict the performance of subsonic ORC turbines is extended to analyse the supersonic flow of organic fluids within 2D converging-diverging nozzles. Two nozzles are developed using a minimum length method of characteristics design model coupled to REFPROP. These are designed for R245fa and Toluene as working fluids with nozzle exit Mach numbers of 1.4 and 1.7 respectively. First, the nozzle performance is confirmed using CFD simulations, and then further CFD simulations are performed to evaluate the performance of the same nozzles over a range of different inlet conditions and with different working fluids. The CFD simulations are compared to predictions made using the original and modified similitude theories, and also to predictions made by conserving the Prandtl-Meyer function for the different operating conditions. The results indicate that whilst the modified similitude model does not accurately predict nozzle performance, conserving the Prandtl-Meyer function allows to predict the nozzle outlet Mach number to within 2% providing there is not a significant change in the polytropic index. Finally, the effect of working fluid replacement on the ORC system is discussed, and preliminary results demonstrate the possibility of matching a particular turbine to a heat source through optimal working fluid selection.

1. Introduction

The organic Rankine cycle (ORC) is a promising technology for the conversion of low temperature heat sources into mechanical work across a range of power outputs ranging from a few kW to the MW scale. However, despite successful commercialization for large-scale systems [1], implementation of smaller systems has not followed the same progress, primarily due to their high capital costs. Among other improvements such as in performance and reliability, improvements in their economy-of-scale can act as an important enabling factor in their widespread implementation. One approach to achieving this is to develop ORC systems that operate efficiently over a range of operating conditions, whilst using different working fluids. Although robust techniques do exist for the design of components that best match targeted operating conditions, the development of ORC components that can operate flexibly over a range of conditions requires methods to predict the performance of these components at design and off-design conditions.

Nomenclature

a	Speed of sound, m/s	β	Relative flow angle, °
b	Blade height, mm	θ	Flow turning angle, °
c	Fluid velocity, m/s	ρ	Density, kg/m^3
D_h	Hydraulic diameter, mm	μ	Viscosity, Pa s
h	Enthalpy, J/kg	ν	Prandtl-Meyer function, °
L	Nozzle length, mm	ν_ts	Isentropic velocity ratio
k	Polytropic index	η_n	Stator isentropic efficiency
$\dot{m}$	Mass flow rate, kg/s	Δh_s	Isentropic enthalpy drop, J/kg
Ma	Mach number		
N	Rotational speed, rpm	**Subscripts**	
o_th	Throat width, mm	*	Choked conditions
P	Pressure, kPa	0	Original operating point
R	Nozzle construction radius, mm	01	Total inlet conditions
Re	Reynolds number	2	Rotor inlet/nozzle outlet
T	Temperature, K	3	Rotor outlet
u	Rotor blade velocity, m/s	c	Condensation
Z_n	Stator blade number	d	Design
α	Absolute flow angle, °	s	Isentropic conditions

Within an ORC system, the turbine is arguably a critical component since it is responsible for generating the power within the system. ORC turbine design is complicated since the behaviour of organic fluids can significantly deviate from that of an ideal gas, thus requiring complex equations of state. Furthermore, at operation close to their critical point, organic fluids could exhibit non-classical fluid dynamic behaviour leading to expansion shocks and compression fans [2,3]. In addition, low values for the speed of sound can also result in supersonic flows at the rotor inlet and possibly at the rotor outlet. For rotor inlet Mach numbers only slightly exceeding unity, a conventional converging stator blade can be implemented with the required supersonic conditions being achieved in the rotor-stator interspace [4]. However, at higher Mach numbers, supersonic stators are required. These are constructed with a subsonic converging section that accelerates the flow to the choked conditions followed by a diverging section that expands the flow isentropically to the desired Mach number. A number of papers have addressed the design of these stators, for example Wheeler and Ong [5] who developed a stator design model using the Method of Characteristics, and Pasquale et al. [6] who constructed a stator using Bezier splines and optimized the geometry using a genetic algorithm.

Alongside the development of design methods, studies have focused on supersonic flows in ORC stators using computational fluid dynamics (CFD) [7-10]. These simulations couple either research flow solvers, or commercial CFD codes, with suitable equations of state. In the absence of experimental data for validation, these solvers were evaluated qualitatively against design models, or experimental data for ideal gas turbines. Recently, Galiana et al. [3] presented experimental results investigating trailing edge losses in ORC turbines in a Ludwieg tube. This, along with further developments in other organic vapour test facilities [11,12], is critical for the validation of CFD tools used to simulate supersonic flows in ORC turbine components.

Clearly, the added complexities of using non-ideal working fluids mean that it is not suitable to apply conventional ideal gas turbine performance prediction models, such as similitude theory, directly to ORC turbines. This has been attempted, for example in Ref. [13], but the errors introduced by using simplified methods based on the ideal gas law has been demonstrated recently [14]. The similitude concept has been applied to ORC turbines [15], but this is generally for turbine design, rather than off-design performance predictions. Previously, the present authors investigated the application of similitude theory for the prediction of subsonic ORC turbine performance and developed a modified similitude model for this purpose [16]. Equation 1 is a simplified form of this model assuming a given turbine design with a fixed geometry, where Δh_s is the isentropic enthalpy drop across the turbine, η is the turbine isentropic efficiency, $\dot{m}$ is the mass flow rate, N is the rotational speed, and a^*, ρ^* and μ are the speed of sound, density and dynamic viscosity at the stator throat. Recently, this model has been used to show how the economy-of-scale of a small subsonic ORC turbine can be improved through appropriate operation and working fluid selection [17].

$$\left[\frac{\Delta h_s}{a^{*2}}, \eta\right] = f\left(\frac{\dot{m}}{\rho^* a^*}, \frac{\rho^* a^*}{\mu}, \frac{N}{a^*}\right) \tag{1}$$

The aim of this paper is to extend the similitude concept by considering working fluid replacement in supersonic ORC turbines, through the design and analysis of two converging-diverging nozzles for R245fa and Toluene using a minimum length nozzle design model. Nozzle performance is investigated using CFD simulations. Subsequently, methods to predict nozzle performance following a change in the working fluid are hypothesized; these involve using the modified similitude theory to predict the nozzle mass flow rate and isentropic enthalpy drop, and predicting the nozzle performance by conserving the Prandtl-Meyer function. These predictions are compared to further CFD simulations. Finally, the implications of fluid replacement on ORC system performance are discussed with the ultimate goal of improving the current economy-of-scale.

2. Method of characteristics and inviscid 2D nozzle design

To design a supersonic nozzle an inviscid 2D minimum length nozzle design model, based on the method of characteristics (MoC), is coupled to real gas equations of state to account for non-ideal fluid properties [18]. Aldo and Argrow [19] were the first to investigate minimum length nozzles for organic fluids, and the model developed here is similar. Although more sophisticated nozzle design methods have been developed more recently, for example [6], the focus of this paper is to investigate the effects of working fluid replacement on nozzle performance, rather than generating optimized geometries. Therefore, the simpler, minimum length method was deemed to be suitable. For a differential fluid element undergoing a supersonic expansion, the amount of flow turning, $d\theta$, is given by Equation 2, where c is the fluid velocity. For a perfect gas with a constant ratio of specific heats, Equation 2 can be integrated analytically to determine the Prandtl-Meyer function. However, for a non-ideal gas Equation 2 can be integrated numerically.

$$d\theta = (\sqrt{\mathrm{Ma}^2 - 1}/c)\, dc \tag{2}$$

Before considering an ORC radial inflow turbine stator, this paper studies the flow within 2D converging-diverging nozzles. This is done not only to validate the design model, but also as a preliminary investigation into scaling effects in supersonic flows of organic fluids. Using the total inlet conditions and Mach numbers defined in Table 1 for R245fa and Toluene, the MoC model was used to design the diverging sections of two converging-diverging nozzles. These working fluids were selected since they are suitable working fluids for low temperature and high temperature ORC applications respectively [1]. The total inlet conditions were defined as 95% and 101% of the critical pressure and temperature respectively to ensure operation near the critical point where non-ideal gas effects are expected. The turbine outlet pressure was defined based on an assumed ORC condensation temperature T_c. Defining the rotor inlet absolute flow angle (α_2 = 74.5°), rotor inlet relative flow angle (β_2 = -21.8°), and isentropic velocity ratio ($\nu_{ts} = u_2/\sqrt{2\Delta h_s}$ = 0.68; where u_2 is the rotor inlet blade velocity and Δh_s is the turbine isentropic total to static enthalpy drop) enables the rotor inlet velocity triangle to be determined. Finally, assuming a stator isentropic efficiency (η_n = 0.9) defines the rotor inlet conditions. The mass flow rates correspond to a radial inflow turbine with a rotor diameter of 80 mm.

Table 1. Thermodynamic cycles and turbine designs used in this study.

	T_{01} [K]	P_{01} [kPa]	$\dot{m}$ [kg/s]	Ma_2	P_2 [kPa]	b_2 [mm]	T_c [K]
R245fa	431	3469	1.76	1.4	1138	2.38	323
Toluene	598	3920	0.45	1.7	667	1.20	383

The accuracy of the MoC model depends on the number of lines used to construct the expansion fan, and the number of elements used in the numerical integration of Equation 2. A sensitivity study using n = 50, 100 and 250 was conducted for both of these parameters and it was found that 100 elements were sufficient, with maximum errors for the wall coordinates of 0.05% and 0.25% for R245fa and Toluene respectively, relative to n = 250. Figure 1 shows the final nozzle designs. The parameters used to construct the rest of the nozzle are shown in Figure 2 and Table 2. The throat width, o_{th}, is calculated by Equation 3 with b_2 the rotor inlet blade height (Table 1), Z_{n} the number of turbine stator vanes (Z_{n} = 16), and ρ^* and a^* the static choked throat conditions. Equation 4 gives the design Reynolds number where D_{h} is the throat hydraulic diameter.

$$o_{\text{th}} = \dot{m}/(\rho^* a^* b_2 Z_{\text{n}}) \tag{3}$$

$$\text{Re}_{\text{d}} = (\rho^* a^* D_{\text{h}})/\mu \tag{4}$$

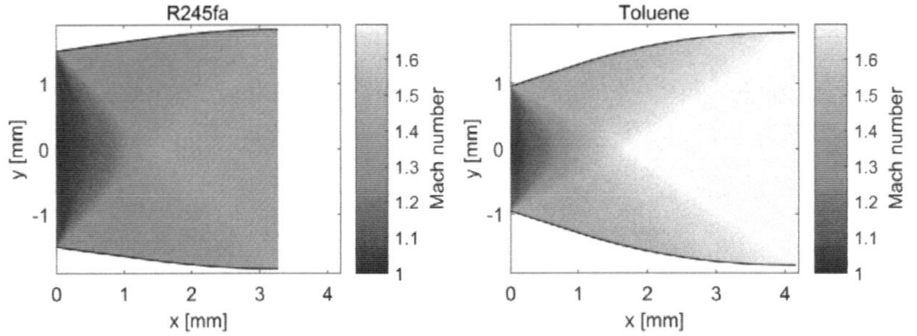

Figure 1. Minimum length supersonic nozzles for the expansion of R245fa to Ma = 1.4 and Toluene to Ma = 1.7, obtained using the MoC model.

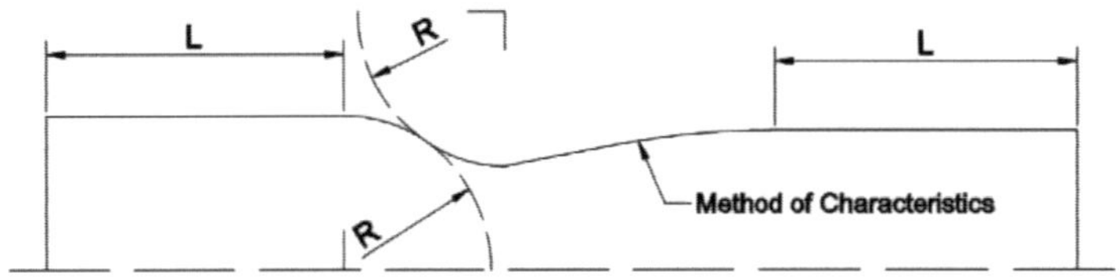

Figure 2. Geometry of the converging-diverging nozzles.

Table 2. Design parameters for the converging-diverging nozzles.

	o_{th} [mm]	R [mm]	L [mm]	Re_{d}
R245fa	2.97	2.23	8.91	2.52×10^6
Toluene	1.89	1.41	5.66	1.09×10^6

3. CFD simulation setup and MoC validation

A steady-state quasi-2D CFD simulation was setup in ANSYS CFX to validate the two nozzle designs. The inlet conditions were set to the total inlet conditions, the outlet was set to a supersonic outlet and the k-ω SST turbulence model with automatic wall function was selected. The meshes were constructed from 160×10^3 elements. For each nozzle design two additional meshes, consisting of 40×10^3 and 360×10^3 elements, were also constructed. It was found that the percentage error in the nozzle efficiency predicted by the original mesh compared to the finest mesh was less than 0.1% for both nozzles, thus confirming the suitability of the original meshes. Working fluid properties were accounted for by generating fluid property tables using REFPROP prior to the simulation, and a sensitivity study considering the resolution of these fluid property tables was conducted with table sizes of 50, 100 and 250. The results show that the percentage errors in the nozzle efficiency predicted for a table size of 100 compared to a table size of 250 were 0.07% and 0.04% for the R245fa and Toluene nozzles respectively. The CFD results for the R245fa and Toluene nozzles, both with a mesh size of $160\text{x}10^3$ elements and fluid property table size of 100, are shown in Figure 3.

Overall, the MoC model has generated supersonic nozzles that expand the two fluids to the desired Mach numbers. Weak oblique shocks are found downstream of the diverging supersonic sections and this leads to small bumps in the midline Mach number distribution. This effect is attributed to the nozzle wall boundary layers, which effectively reduce the nozzle flow area. Boundary layer effects are not considered within the MoC model, therefore explaining the small discrepancies observed. The model can be improved by extending the nozzle wall by the boundary layer displacement thickness. However, given that the outlet Mach numbers agree to within 1%, this modification was deemed unnecessary.

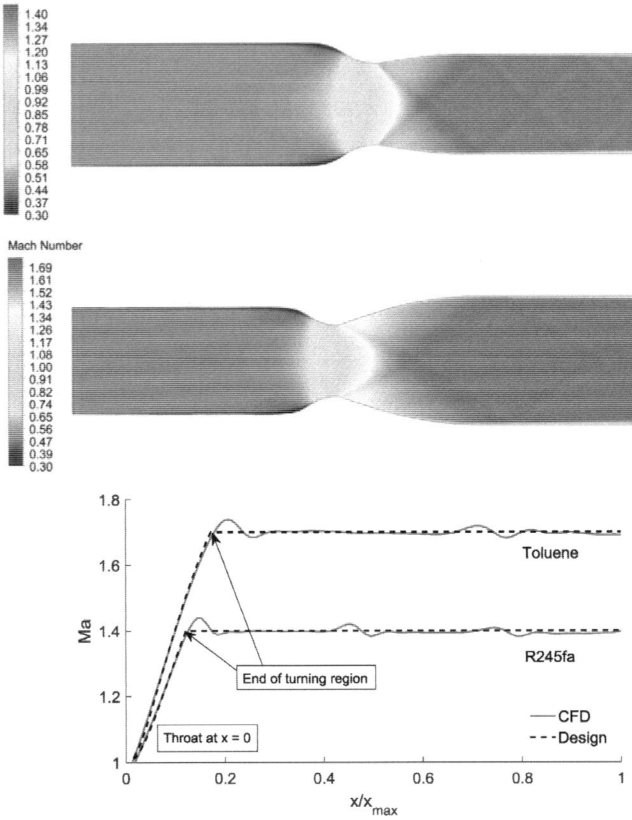

Figure 3. Validation of the MoC nozzle design model: R245fa nozzle CFD simulation (top); Toluene nozzle CFD simulation (middle); comparison between the midline Mach number distributions obtained using the MoC model and CFD simulation. Note that MoC results are only valid for $x > 0$.

4. Scaling of the supersonic nozzle

Having setup CFD simulations for both nozzles, the analysis can be extended to investigate alternative operating conditions. The first investigation considers a change in the total inlet conditions. Moving down the isentropes that correspond to the total inlet conditions defined in Table 1, alternative inlet conditions were defined at 75%, 50% and 25% of $\mathrm{Re_d}$ (Cases 1 to 3 in Table 3). Cases 4 to 9 correspond to changing the working fluid. For these cases it was desirable to keep Re constant, such that the effect of changing the working fluid and Re effects can be isolated from one another. Therefore, for each working fluid an isentrope was constructed that coincides with an operating point at 95% and 101% of the fluid critical pressure and temperature respectively. The throat Reynolds number was then calculated for various points along this isentrope and the nozzle inlet conditions were selected to ensure that Re remained constant. From Table 3 it can be seen that in some cases it was not possible to maintain Re. In these cases, the inlet conditions were set to the conditions that resulted in the Reynolds number that was closest to the original design point.

Using these alternative inlet conditions, a number of CFD simulations can be performed to establish nozzle performance. However, the purpose of this work is to also develop simple methods to predict nozzle performance following these changes. The first method applies the modified similitude model (Equation 1), and hypothesizes that for the same reduced head and flow coefficients, the nozzle performance will remain the same. Here the reduced head coefficient is defined as the isentropic total-to-static enthalpy drop across the nozzle normalized by the speed of sound squared $(h_{01} - h_{2s})/a^2$. In the conventional similitude model the total inlet speed of sound, a_{01}, is used whilst in the modified similitude model the choked throat speed of sound, a^* is used. Using a^* is expected to be more accurate, since the choked mass flow rate will be conserved. However, the use of a_{01} is included here as further validation of the modified similitude model. Similarly, the mass flow rate is related to the flow coefficient, $\dot{m}/\rho a$.

A second method considers the Prandtl-Meyer function, ν. Since ν is central to the supersonic nozzle design process, it is hypothesized that nozzle performance can be determined by maintaining the same amount of flow turning within the nozzle following a change in the operating conditions. For each case, the value of ν for a range of Mach numbers is shown in Figure 4.

Table 3. Alternative inlet conditions for the R245fa and Toluene nozzles.

Case	Working fluid	R245fa nozzle			Toluene nozzle		
		T_{01} [K]	P_{01} [kPa]	Re/Re_d	T_{01} [K]	P_{01} [kPa]	Re/Re_d
1	R245fa/Toluene	411	2366	0.75	574	2684	0.75
2	R245fa/Toluene	390	1470	0.50	553	1666	0.50
3	R245fa/Toluene	364	677	0.25	526	763	0.25
4	R245fa	**431**	**3469**	**1.00**	412	2438	1.00
5	Toluene	598	3920	0.77	**598**	**3920**	**1.00**
6	MDM	570	1344	0.57	570	1344	0.74
7	R1234yf	368	3046	1.00	349	2109	1.00
8	Isopentane	465	3209	0.85	457	2811	1.00
9	R123	461	3479	0.99	439	2470	1.00

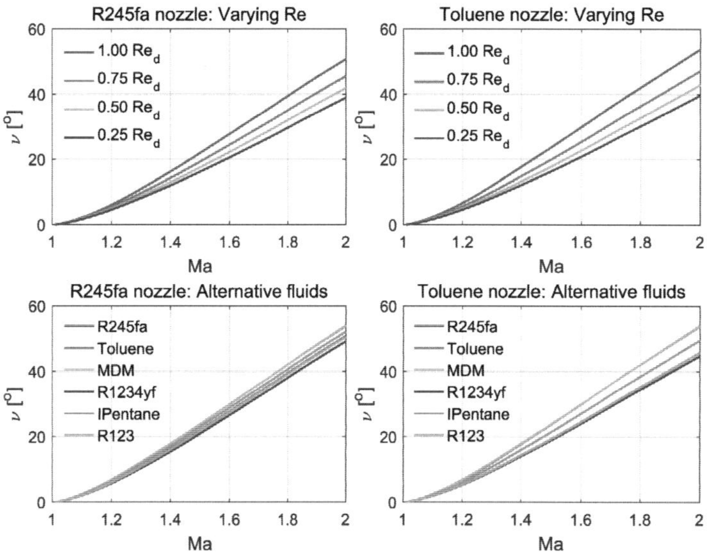

Figure 4. Prandtl-Meyer functions for: R245fa nozzle alternative inlet conditions (top-left); Toluene nozzle alternative inlet conditions (top-right); R245fa nozzle alternative fluids (bottom-left); Toluene nozzle alternative fluids (bottom-right).

For both nozzles, operating with the design fluid but at lower Reynolds numbers results in a reduction in ν for expansion to the same Mach number. Similarly, a change in the working fluid also causes a change in ν for expansion to the same Mach number. It is found that the lines for Toluene and MDM coincide, indicating that both fluids exhibit similar flow turning under similar expansion conditions. The same is found for the R245fa and R123 lines. Ultimately, Figure 4 could be used to predict the Mach number distribution within the nozzle for alternative conditions. By example, consider the centreline Mach number distribution obtained for the R245fa nozzle (Figure 3). For each x-coordinate the corresponding Mach number can be converted into a value for ν. Then for an alternative inlet condition each value of ν can be converted into the equivalent Mach number that would be required to maintain that same amount of flow turning within the nozzle.

At this point, it is suitable to discuss the work of Wheeler and Ong [5] who proposed an alternative method of obtaining ν to avoid the numerical integration of Equation 2. They assumed that during an isentropic expansion across a known pressure ratio the working fluid obeys the polytropic law P/ρ^k = constant, where k is the polytropic index and can be obtained numerically using a linear regression. It was shown that as k reduces,

ν increases, resulting in a greater flow turning for expansion to the same Mach number. Using the ν conservation method described in the previous paragraph the static outlet pressures for the cases described in Table 3 can be predicted, and the values for the polytropic index can be obtained (Table 4). These values confirm the results shown in Figure 4, with cases with a higher k value corresponding to lower ν values. Furthermore, the close proximity of the MDM and Toluene lines, and the R245fa and R123 lines, are confirmed by their similar k values. Finally, it should be noted that cases 1 to 3 also correspond to significant changes in k. Therefore, when the CFD results for these cases are evaluated it will be important to identify whether any deviations are attributed to the change in Reynolds number, or the change in k.

To conclude this section, a final remark is made regarding Wheeler and Ong's work [5]. Through calculating ν based on k it was demonstrated how the shape of a supersonic nozzle changes with changes in k. Considering this, it is expected that a significant change in k will require a significantly different nozzle design. In the context of developing small-scale ORC turbines to be used within multiple applications the important question then becomes how much of a deviation in k can be tolerated before the performance of the nozzle deviates significantly from the design point.

Table 4. Polytropic index values for the similitude cases corresponding to conservation of ν.

Case	R245fa	Toluene	Case	R245fa	Toluene	Case	R245fa	Toluene
1	0.876	0.901	4	0.790	0.923	7	0.823	0.944
2	0.947	0.962	5	0.733	0.813	8	0.757	0.869
3	1.010	1.011	6	0.714	0.802	9	0.797	0.925

5. CFD results for alternative inlet conditions

CFD simulations were also performed for the cases defined in Table 3, resulting in 9 simulations per nozzle. For each simulation the centreline Mach number distribution was determined, along with the mass flow rate, mass averaged values for T_{01} and P_{01}, and the area averaged values for P_2 and Ma_2. This facilitated the calculation of the nozzle efficiency η_n (Equation 5), which ranged between 0.983 and 0.987, and 0.984 and 0.988 for the R245fa and Toluene nozzles respectively. For both nozzles the lowest efficiency was found for the $0.25Re_\mathrm{d}$ cases, whilst the highest efficiencies were found for the highest Re_d. This was expected since lower Reynolds numbers imply thicker nozzle wall boundary layers, and a greater viscous loss.

$$\eta_\mathrm{n} = (h_{01} - h_2)/(h_{01} - h_{2\mathrm{s}}) \tag{5}$$

Using the CFD results, the reduced head and flow coefficients were calculated using both total ($\dot{m}/\rho_{01}a_{01}$) and choked ($\dot{m}/\rho^*a^*$) conditions. The flow coefficient based on total inlet conditions ranged between 68.9×10^{-6} and 79.7×10^{-6} for the R245fa nozzle, and 2.27×10^{-6} and 2.62×10^{-6} for the Toluene nozzle. By comparison the flow coefficient based on the choked conditions was found to be 112×10^{-6} for all cases for the R245fa nozzle, and 3.68×10^{-6} for all cases for the Toluene nozzle. This shows that the similitude model based on total conditions cannot predict the choked mass flow rate following a change in the working fluid or inlet conditions. By comparison, the choked conditions can accurately predict the mass flow rate. However, the head coefficients based on the choked conditions ranged between 1.08 and 1.32 for the R245fa nozzle, and 1.73 and 2.34 for the Toluene nozzle. Therefore, since the head coefficient is not the same, these results indicate that the modified similitude model cannot fully predict the supersonic nozzle performance following a change in the inlet conditions. It is therefore necessary to investigate scaling using ν.

Considering the R245fa nozzle operating at the design point (Case 4), the midline Mach number distribution obtained from CFD for that particular case can be scaled to the other 8 cases defined in Table 3 by conserving ν. This provides the predicted midline Mach number distribution for that nozzle operating under each alternative condition. These predictions can then be compared to the midline Mach number distributions obtained directly from CFD for that particular nozzle and alternative inlet condition, thereby confirming whether the conservation of ν can be used to predict nozzle performance. To conduct a thorough assessment on the suitability of the scaling method, the process was repeated using each CFD simulation as the initial starting point which is scaled to the other 8 cases. Overall, this resulted in 144 comparisons (9×8 per nozzle) between the predictions made by conserving ν and the CFD results obtained directly from CFD. Quantitative comparisons were done using three parameters. The first parameter k/k_0 is the change in the polytropic index where k is the polytropic index corresponding to the alternative inlet condition, and k_0 is the polytropic index associated with the original simulation used during the ν conservation. Similarly, the second parameter $\mathrm{Re}/\mathrm{Re}_0$ is the change in Reynolds numbers where Re is the Reynolds number associated with the alternative inlet conditions, and Re_0 is the Reynolds number associated with the original simulation. The final parameter was defined as the maximum percentage error between the prediction and the CFD result.

Of the 144 comparisons, Figure 5 displays 4 of these comparisons in more detail. The solid lines correspond to the predictions made by conserving ν and the dashed lines correspond to the CFD results. These comparisons have been selected to isolate the effects of varying k, and varying Re, from each other. The left plot displays comparisons made for the R245fa nozzle when $\mathrm{Re}/\mathrm{Re}_0 = 1$, whilst the right plot displays comparisons made for the R245fa nozzle when $k/k_0 = 1$.

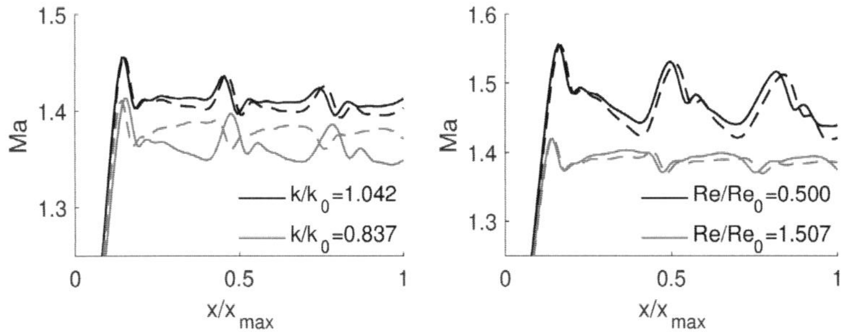

Figure 5. Comparison between centreline Mach number predicted by conserving ν (full) and CFD (dashed): R245fa nozzle, $\mathrm{Re}/\mathrm{Re}_0 = 1$ (left); R245fa nozzle, $k/k_0 = 1$ (right).

Firstly, it is noted that in some cases more uneven Mach number distributions are found compared to the flatter Mach number distributions shown in Figure 3. This is because a significant change in the inlet conditions can result in either under or over expansion within the nozzle, resulting in more pronounced variations in the Mach number in the straight section downstream of the diverging nozzle. Furthermore, since each case was used as a reference case to be scaled to the other 8 cases, there are situations where an off-design simulation is used during the conservation of ν (i.e. the $\mathrm{Re}/\mathrm{Re}_0 = 0.5$ case). Secondly, it is observed from the right plot that the ν conservation method accurately predicts the midline Mach number distribution following a change in Re, provided that k does not change. However, the left plot shows the predictions made by conserving ν are much more sensitive to a change in the polytropic index. Whilst the results for the R245fa nozzle with $k/k_0 = 1.042$ show good agreement, the $k/k_0 = 0.837$ case shows a much greater deviation. The CFD results indicate that for $k/k_0 < 1$ the flow accelerates quicker within the diverging section of the nozzle than predicted by the ν conservation method. Although not demonstrated in Figure 5, it was also observed that for $k/k_0 > 1$ the CFD results indicate the opposite, with the flow taking longer to reach the maximum Mach number. Interestingly though, both results predict the same maximum Mach number within the nozzle.

The results from all 144 comparisons are summarized in Figure 6, which is a contour plot showing the maximum percentage error in the Mach number predicted by conserving ν compared to the CFD simulation results. This plot indicates that the ν conservation method accurately predicts the midline Mach number distribution when $k/k_0 = 1$ and $\mathrm{Re}/\mathrm{Re}_0 = 1$. Furthermore, within the ranges of $0.9 < k/k_0 < 1.1$ and $0.5 < \mathrm{Re}/\mathrm{Re}_0 < 2.0$ the ν conservation method can predict the midline Mach number distribution to within 2%. Therefore, this investigation into the scaling effects in supersonic nozzles suggests that providing the change in k and Re following a change in the inlet conditions or working fluid is within this limit, the ν conservation method can accurately predict nozzle performance.

6. Implications of working fluid replacement on ORC system performance

Finally, it is important to discuss the implications of working fluid replacement on the performance of the whole ORC system. Considering the CFD results shown for the $k/k_0 = 1.042$ and $\mathrm{Re}/\mathrm{Re}_0 = 1.507$ cases in Figure 5, these results show a relatively flat Mach number distribution. This suggests that the flow is not significantly under or over turned at the end of curved supersonic section such that significant variations in the Mach number downstream of this point are not observed, therefore suggesting the nozzle is operating close to optimal conditions. These $k/k_0 = 1.042$ and $\mathrm{Re}/\mathrm{Re}_0 = 1.507$ cases correspond to the R245fa nozzle operating with R1234yf and Isopentane respectively. A similar result is also obtained when this nozzle operates with R123. Therefore, based on the known nozzle outlet static pressures and absolute flow velocities, and assuming that the nozzle design can be incorporated into a suitable stator design that will deliver the flow at the desired flow angle (i.e. $\alpha_2 = 74.5°$), the turbine blade velocity for each case can be selected to ensure the same relative flow angle at the rotor inlet (i.e. $\beta_2 = -21.8°$) using Equation 6. Based on this, and assuming that the rotor operates at the same isentropic velocity ratio (i.e. $\nu_{ts} = 0.68$), the enthalpy at the turbine outlet following an isentropic expansion h_{3s} (Equation 7), and therefore, the turbine pressure ratio and ORC condensation temperature can be obtained. Finally, conserving the flow coefficient ($\dot{m}/\rho^* a^*$) supplies the mass flow rate, and in turn the expected turbine power.

$$u_2 = \mathrm{Ma}_2 a^* (\sin\alpha_2 - \cos\alpha_2 \tan\beta_2) \tag{6}$$

$$h_{3s} = h_{01} - (1/2)(u_2/\nu_{ts})^2 \tag{7}$$

This results from this analysis are plotted against the turbine inlet temperature in Figure 7. Ultimately, the results demonstrate how the same turbine could be used in applications with different heat sources by matching the working fluid to the heat source, thus providing an indication of the possibility of working fluid replacement in supersonic ORC turbines. This is a promising result that extends previous work completed for subsonic ORC turbines [17]. It should of course be noted that a slight penalty in terms of an increasing condensation temperature is observed, and future efforts should investigate this more thoroughly.

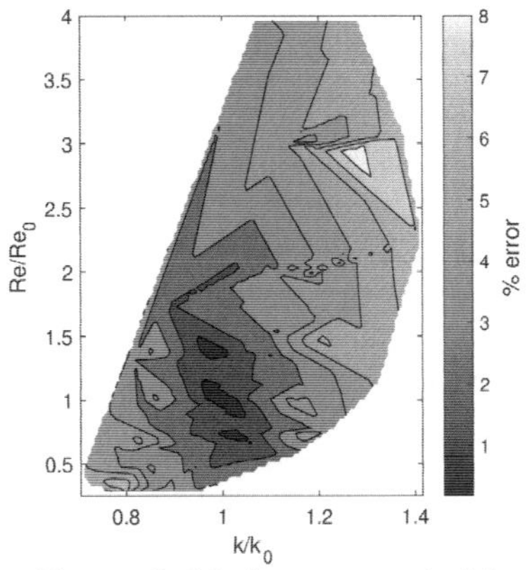

Figure 6. Maximum error in Ma predicted by conserving ν, compared to the CFD results, as a function of k/k_0 and $\mathrm{Re}/\mathrm{Re}_0$.

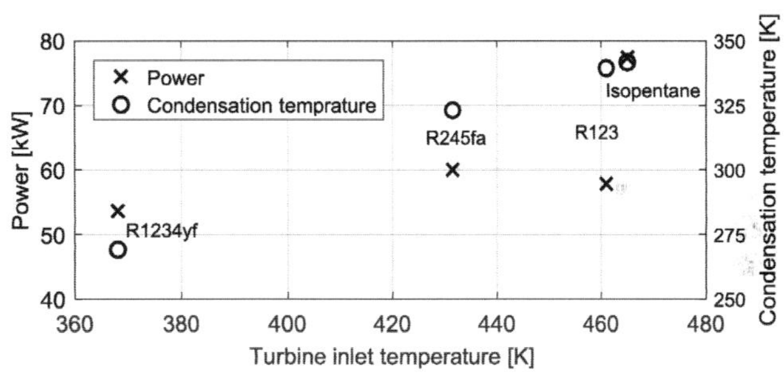

Figure 7. The impact of working fluid replacement on the overall ORC system performance assuming the same flow conditions within the turbine supersonic stator.

7. Conclusions

To realise the widespread implementation of small-scale ORC technology it is necessary to reduce system costs. One way to do this is to develop ORC components that can operate flexibly and with high efficiency with different working fluids, over a wide range operating conditions and in a variety of applications. This paper has investigated the effect of the inlet conditions and of the working fluid on the performance of supersonic converging-diverging nozzles, as simplified models of turbine flow paths. Two supersonic nozzles have been developed using a minimum length method of characteristics design model and the nozzle performance has been verified using CFD. To investigate nozzle performance following a change in the operating conditions, 9 operating cases were defined for each nozzle covering a range of nozzle Reynolds numbers and values for the polytropic index k (P/ρ^k = constant). Predictions for the nozzle performance were made using the similitude model, and also by conserving the amount of flow turning (the Prandtl-Meyer function). Whilst the modified similitude model accurately predicted the nozzle mass flow rate, it is unsuitable for predicting the nozzle outlet conditions. However, it was found that conserving the Prandtl-Meyer function can lead to reasonably accurate predictions (to within 2%) of the nozzle outlet conditions provided the percentage change in k is less than 10%. Finally, by assuming the nozzle design can be implemented into a suitable turbine design and then ORC system, the possibility of matching the same turbine with different heat sources through working fluid selection has been demonstrated.

References

[1] Colonna, P., Casati, E., Trapp, C., Mathijssen, T., Larjola, J., Turunen-Saaresti., T., and Uusitalo, A., 2015, "Organic Rankine cycle power systems: from the concept to current technology, applications and an outlook for the future", J Eng Gas Turb Power, **137**(10).

[2] Thompson, P., 1971, "Fundamental derivative in gas-dynamics", Physics of Fluids, **14**(90), pp. 1843.

[3] Galiana, F. J. D., Wheeler, A. P. S., and Ong, J., 2015, "A study on the trailing-edge losses in organic Rankine cycle turbines", ASME Turbo Expo 2015, 15-19th June, Montreal Canada.

[4] Moustapha, H., Zelesky, M. F., Baines, N. C., and Japiske, D., 2003, *Axial and Radial Turbines*, Concepts NREC, Inc.

[5] Wheeler, A. P. S. and Ong, J., 2013, "The role of dense gas dynamics on organic Rankine cycle turbine performance", J Eng Gas Turb Power, **135**(10).

[6] Pasquale, D., Ghidoni, A., and Rebay, S., 2013, "Shape Optimization of an Organic Rankine Cycle Radial Turbine Nozzle", J Eng Gas Turb Power, **135**(4).

[7] Hoffren, J., Talonpoika, T., Larjola, J., and Siikonen, T., 2002, "Numerical Simulation of Real-Gas Flow in a Supersonic Turbine Nozzle Ring", J Eng Gas Turb Power, **124**(4).

[8] Harinck, J., Turunen-Saaresti, T., Colonna, P., Rebay, S., and van Buijtenen, J., 2010, "Computational study of a high-expansion ratio radial organic Rankine cycle turbine stator", J Eng Gas Turb Power, **132**(5).

[9] Harinck, J., Pasquale, D., Pecnik, R., Buijtenen, J. V., and Colonna, P., 2013, "Performance improvement of a radial organic Rankine cycle turbine by means of automated computational fluid dynamic design", P I Mech Eng A-J Pow, **227**(6), pp. 637-645.

[10] Wheeler, A. P. S., and Ong, J., 2014, "A study on the three-dimensional unsteady real-gas flows within a transonic ORC turbine", ASME Turbo Expo 2014, 16-20th June, Dusseldorf, Germany.

[11] Spinelli, A., Guardone, A., Cozzi, F., Carmine, M., Cheli, R., Zocca, M., Gaetani, P. and Dossena, V., 2015, "Experimental observation of non-idea nozzle flow of siloxane vapor MDM", 3rd International Seminar on ORC Power Systems, Brussels, Belgium.

[12] Mathijssen, T., Gallo, M., Casati, E., Guardone, A., and Colonna, P., 2015, "Wave speed measurements in non-ideal compressible flows using flexible asymmetric shock tube (FAST)", 3rd International Seminar on ORC Power Systems, Brussels, Belgium.

[13] Cameretti, M. C., Ferrara, F., Gimelli, A., and Tuccillo, R., 2015, "Employing micro-turbine components in integrated solar-MGT-ORC power plants", ASME Turbo Expo 2015, 15-19th June, Montreal, Canada.

[14] Wong, C. S., and Krumdieck, S., 2015, "Scaling of gas turbine from air to refrigerants for organic Rankine cycles using similarity concept", J Eng Gas Turb Power, **138**(6).

[15] Astolfi, M., and Macchi, E., 2015, "Efficiency correlations for axial-flow turbines working with non-conventional fluids", 3rd International Seminar on ORC Power Systems", 12-14th October, Brussels, Belgium.

[16] White, M., and Sayma, A. I., 2015, "The application of similitude theory for the performance prediction of radial turbines within small-scale low-temperature organic Rankine cycles", J Eng Gas Turb Power, **137**(12).

[17] White, M., and Sayma, A.I., 2015, "The impact of component performance on the overall cycle performance of small-scale low temperature organic Rankine cycles", 9th International Conference on Compressors and their Systems, London, UK.

[18] Lemmon, E. W., Huber, M. L., and McLinden, M. O., 2013, *NIST Standard Reference Database 23: Reference Fluid Thermodynamic and Transport Properties-REFPROP*, Version 9.1, National Institute of Standards and Technology, Standard Reference Data Program, Gaithersburg.

[19] Aldo, A. C., and Argrow, B. C., 1995, "Dense gas-flow in minimum length nozzles", Journal of Fluids Engineering, **117**(2), pp. 270-276.

IOP Conf. Series: Journal of Physics: Conf. Series **821** (2017) 012003 doi:10.1088/1742-6596/821/1/012003

Non-ideal compressible-fluid effects in oblique shock waves

G. Gori, D. Vimercati and A. Guardone

[1] Department of Aerospace Science and Technology, Politecnico di Milano
Via La Masa 34, 20156, Milano, Italy

E-mail: alberto.guardone@polimi.it

Abstract. The non-monotone dependence of the speed of sound along adiabatic transformations is demonstrated to result in the admissibility of non-ideal increase of the flow Mach number across oblique shock waves, for pre-shock states in close proximity of the liquid-vapour saturation curve. This non-ideal behaviour is primarily associated with a less-than-unity value of the fundamental derivative of gasdynamics and, therefore, non-ideal shock waves are expected to be observed in flows of fluids with moderate molecular complexity. The simple yet qualitatively sound van der Waals model is used to confirm the present findings and to provide exemplary non-ideal shock waves.

1. Introduction

The gasdynamics of single-phase flows of molecularly complex fluids in close proximity of the liquid-vapour equilibrium, for pressures on the order of those corresponding to the critical point, can significantly differ from its dilute-gas counterpart. The dynamic behaviour of compressible-fluid flows is determined by the so-called fundamental derivative of gasdynamics Γ, see reference [1], namely

$$\Gamma \equiv 1 + \rho c \left(\frac{\partial c}{\partial P}\right)_s, \tag{1}$$

where ρ is the fluid density, P the pressure, c the speed of sound and s the specific entropy. The fundamental derivative plays a central role in delineating the qualitative behaviour of fluid flows in many contexts, including, for instance, shock formation in unsteady flows, steady isentropic flow, steady duct flow with friction and steady oblique shock waves, which is the focus of the present work [1, 2, 3, 4, 5].

Conventional treatments of gasdynamics assume, either explicitly or implicitly, that $\Gamma > 1$. This is the case, for instance, for the ideal gas model, for which the fundamental derivative is a constant larger than unity. From the qualitative point of view, the behaviour of any fluid with constant $\Gamma > 1$ resembles that of ideal gases, see for instance reference [6]. It is well-known, however, that fluids with moderately or highly complex molecular structure exhibit non-ideal gasdynamic behaviour in a finite vapour-phase thermodynamic region [2, 3, 1, 7, 8]. In the non-ideal fluid-dynamic context, it is standard practice to distinguish between the cases $\Gamma \leq 0$ and $0 < \Gamma \leq 1$. In the former case, one rather speaks of non-classical gasdynamics, as opposed to classical gasdynamics if $\Gamma > 0$, because unconventional phenomena such as expansion shocks, isentropic compression fans, split shocks and composite waves are allowed, see for instance references [1, 8, 9, 10, 11, 12, 13]. In the present work, however, the discussion will be limited to the non-ideal classical case, namely $0 < \Gamma \leq 1$, where shock waves are of the ordinary compressive type and isentropic fans carry an expansion.

Among the other effects, the non-ideal increase of the speed of sound upon isentropic expansion has important consequences on the variation of thermodynamic quantities, such as the speed of sound, along the Hugoniot locus of a given initial state, i.e. the locus of states that can possibly be connected to the initial state by means of a shock wave. The aim of this work is to provide evidence that the non-monotone dependence of the speed of sound along the Hugoniot locus can result in the non-ideal increase of the flow Mach number across oblique shock waves. The simple van der Waals model, which gives a qualitatively accurate description of the fluid behaviour in the thermodynamic region of interest in this investigation, is used for the purpose.

The characterisation of non-ideal compressible-fluid flows is not only of theoretical interest, but also is relevant to applications involving molecularly complex fluids in their dense-gas regime, such as machinery operating in Organic Rankine Cycle (ORC) and supercritical CO_2 power systems, turbogenerators, refrigerators and many others, see reference [14, 15, 16, 17]

The present work is organised as follows. In section 2, the Rankine-Hugoniot jump relations across surfaces of discontinuity are recalled along with the admissibility requirements. Exemplary oblique shock configurations for the van der Waals model of siloxane fluid MDM, are presented and discussed in section 3. Section 4 presents a numerical assessment of non-ideal streaming flows over a wedge, using the open-source CFD solver SU2, which was recently extended to deal with non-ideal thermodynamics. Section 5 presents the concluding remarks.

2. Jump relations and shock admissibility

The Euler equations of gasdynamics admit solutions including finite jumps across vanishing-thickness layers, namely surfaces of discontinuity. In a coordinate system moving with the local velocity of the discontinuity, the relations expressing the conservation of mass, momentum and energy locally assume the form

$$[\rho u_n] = 0, \tag{2a}$$
$$[\rho u_n \boldsymbol{u}_t] = 0, \tag{2b}$$
$$[P + \rho u_n^2] = 0, \tag{2c}$$
$$[\rho u_n h^t] = 0, \tag{2d}$$

where $[X] = X_B - X_A$ denotes the difference between a quantity evaluated at the post-shock state (subscript B) and at the pre-shock state (subscript A). In the above expressions, u_n is the normal component of the the velocity vector $\boldsymbol{u}$ and $\boldsymbol{u}_t$ is the tangential velocity vector, both with respect to the surface of discontinuity, and $h^t = h + u^2/2$ is the total enthalpy, where h is the fluid enthalpy and u is the magnitude of the velocity. System (2) represents the well-known Rankine-Hugoniot jump relations linking the states across a surface of discontinuity. Discontinuities having nonzero mass flux $j = \rho u_n$ are referred to as shock waves (in contrast to contact discontinuities, for which $j = 0$) and are the case of interest in this work. The Rankine-Hugoniot relations require that the tangential velocity and the total enthalpy are constant across shock waves (in the coordinate system in which the jump relations assume the form (2)). System (2) can be rearranged to obtain a single scalar equation involving the pre-shock pressure P_A and specific volume v_A and the post-shock specific volume v_B, namely

$$h(P_B, v_B) - h(P_A, v_A) - \tfrac{1}{2}(P_B - P_A)(v_B + v_A) = 0. \tag{3}$$

The above equation determines the locus of thermodynamic states that can be connected by a shock wave. Setting a generic post-shock state $P = P_B$ and $v = v_B$, the curve $P^{\mathrm{RH}}(v; P_A, v_A)$

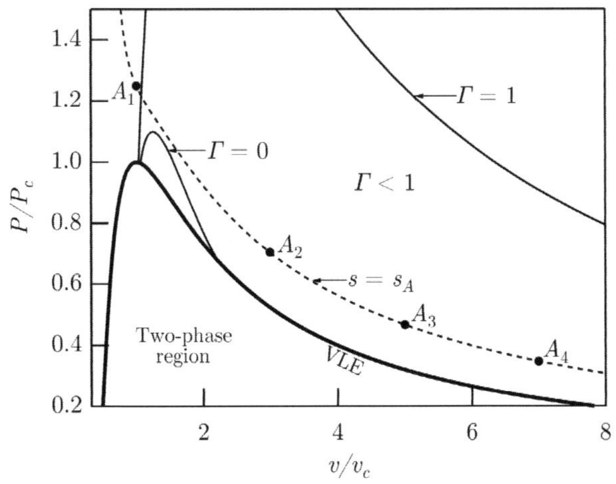

Figure 1: Pressure versus specific volume diagram computed from the polytropic van der Waals model of siloxane fluid MDM, illustrating the thermodynamic regions $\Gamma < 1$ and $\Gamma < 0$. Also shown are the four pre-shock states of the exemplary oblique shock configurations, chosen along the isentrope $s = s_A$.

defined by equation (3) is known as the *shock adiabat* centered on state A. Moreover, equations (2a) and (2c) may be combined to give

$$-\frac{P_B - P_A}{v_B - v_A} = j^2. \tag{4}$$

If the pre-shock state is fixed, and so is j^2, then the above expression defines a straight line $P^{\mathrm{R}}(v; P_A, v_A, u_{An})$, in the P-v plane, referred to as *Rayleigh line* centered on state A. The intersections between the shock adiabat and the Rayleigh line correspond to values of the post-shock specific volume that satisfy (2).

System (2) must be supplemented with admissibility criteria that exclude the unphysical solutions. By the second law of thermodynamics, the jump of the specific entropy across a shock wave must not be negative:

$$[s] \geq 0. \tag{5}$$

In addition, admissible shocks must satisfy the condition

$$M_{A,n} \geq 1 \geq M_{B,n}, \tag{6}$$

where additional subscritp n indicates the normal component, which is known as speed-ordering relation and ensures that the shock front is stable, see [18, 19]. On a geometrical basis, fulfilment of condition (6) requires

$$\left.\frac{dP^{RH}}{dv}\right|_A \geq \left.\frac{dP^R}{dv}\right|_A \quad \text{and} \quad \left.\frac{dP^{RH}}{dv}\right|_B \leq \left.\frac{dP^R}{dv}\right|_B. \tag{7}$$

As pointed out in reference [20], the entropy inequality and the speed ordering condition are sufficient to rule out inadmissible shock waves if the shock adiabat is convex, insofar as both criteria can be simultaneously satisfied only by compression ($[P] > 0$) shocks.[1]

[1] The shock adiabat is convex in the P-v plane as a direct consequence of $\Gamma > 0$. The proof, see [20], relies on the impossibility of the existence of acoustic points (at which the Rayleigh line is tangent to the shock adiabat) along the shock adiabat for $[v] \neq 0$.

Case	M_A	P_A/P_c	ρ_A/ρ_c	T_A/T_c	Γ_A
A_1	2	1.250	1.000	1.063	1.674
A_2	2	0.704	0.333	1.037	0.667
A_3	2	0.467	0.200	1.027	0.829
A_4	2	0.347	0.143	1.021	0.887

Table 1: Pre-shock states of the exemplary oblique shock configurations for the polytropic van der Waals model of siloxane fluid MDM. All the cases considered have the same pre-shock Mach number $M_A = 2$ and pre-shock entropy $s_A = s(1.250P_c, v_c)$, where subscript c denote the properties at critical point.

3. Non-ideal Oblique Shock-waves

In this section, we discuss non-ideal gasdynamic effects in steady oblique shock waves, such as those generated in a streaming steady flow over a wedge. The wedge prescribes a flow deflection angle $\theta \leq \theta_{\max}$, which is realized by a planar shock wave of angle β which originates from the corner and separates the two uniform regions A and B, upstream and downstream of the shock, respectively. The condition $\theta = \theta_{max}$ corresponds to the maximum deflection that the flow can sustain across a planar attached shock wave. If $\theta > \theta_{\max}$, the shock wave detaches and it is no longer a planar front.

The following treatment is intended to provide evidence of non-ideal gasdynamic effects across planar shock waves in flows of fluids experiencing $0 < \Gamma \leq 1$. Exemplary cases will be investigated in which the magnitude of the Mach number increases as the flow passes through the shock wave, namely $M_B > M_A$. From the theoretical point of view, this is possible in any fluid flow in the regime $\Gamma < 1$, because the speed of sound no longer increases monotonically with increasing density along the shock adiabat. As a result, the magnitude of the tangential component of the Mach number may possibly increase, and such an increase may not be compensated by the necessary decrease in the normal Mach number. The latter claim is confirmed in the following by means of the polytropic (i.e., with constant isochoric specific heat c_v) van der Waals model [21], which is one of the simplest, yet qualitatively correct, thermodynamic models allowing for a non-constant Γ. As pointed out by Kluwick [22], the condition $c_v/R \gtrsim 3.78$ is to be satisfied in order that a thermodynamic region featuring $\Gamma < 1$ exists in the vapor phase. Figure 1 shows a portion of the non-ideal thermodynamic region $\Gamma < 1$ for the polytropic van der Waals model of siloxane fluid MDM (corresponding to $c_v/R = 57.69$), which will be used throughout the present work, together with the nonclassical region $\Gamma < 0$. The present investigation will be limited to oblique shocks that exhibit non-ideal, yet classical, features. Therefore, we will naturally focus on shock waves whose adiabat lies entirely in the region $\Gamma > 0$. The analysis of steady oblique shock waves in the regime of mixed nonlinearity, where nonclassical wave fields are expected to occur [1, 8, 12], is left for future investigations.

A set of exemplary cases for the polytropic the van der Waals model of siloxane fluid MDM is examined and used to investigate the dependence of the properties of oblique shock waves on the upstream state (P_A, v_A, M_A). Four different upstream states are considered, as described in table 1 and shown in figure 1. In order to simplify the following treatment, the selected configurations exhibit the same upstream values of the Mach number and of the entropy, namely $M_A = 2$ and $s_A = s(1.25P_c, v_c)$. Firstly, we consider oblique shocks originating form pre-shock state A_1, which is located in the region $\Gamma > 1$ at density values higher than those corresponding to $\Gamma < 1$. Hence, $\Gamma_{A_1} > 1$ and, with reference to figure 2a, the compressive branch of shock adiabat H_1 centred on state A_1 is entirely located in the region $\Gamma > 1$. The endpoint denoted as N_1 corresponds the post-shock state of the normal shock from state A_1. Figure 3a shows the variation of the ratios c_A/c_B, M_B/M_A and $M_{B,n}/M_{A,n}$ with the shock angle β. Owing to relation (2b), the tangential component of the velocity (with respect to the oblique shock) is conserved across the shock and, therefore, $c_A/c_B = M_{B,t}/M_{A,t}$, where the additional subscript

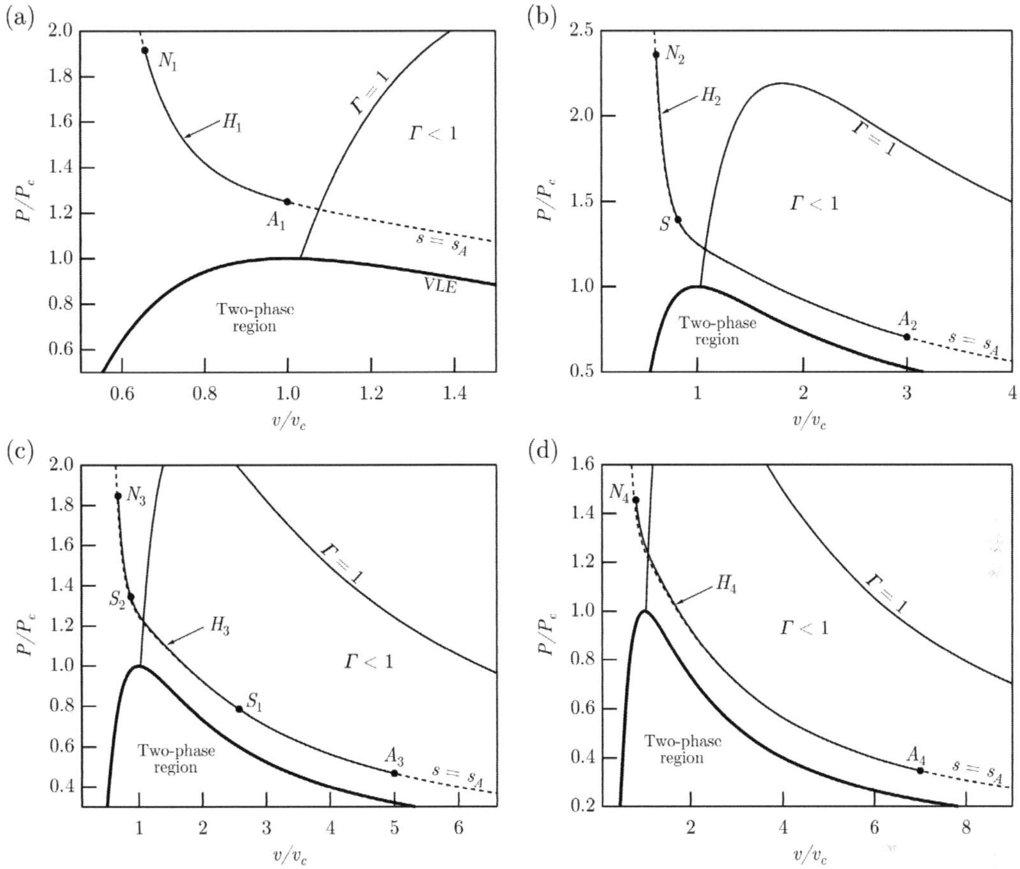

Figure 2: Pressure versus specific volume diagrams illustrating shock adiabats (curves labelled H) originating from pre-shock states corresponding to (a) case A_1, (b) case A_2, (c) case A_3 and (d) case A_4, as computed from the polytropic van der Waals model of siloxane fluid MDM. The endpoints labelled as N correspond to the normal shock configuration. Also shown are the post-shock states S at which $M_B = M_A$.

t denotes the tangential component. The range of shock angles spans from the minimum value $\beta_{\min} = \text{asin}(1/M_A)$, corresponding to an acoustic wave, up to $\beta = \pi/2$, corresponding to a normal shock wave. Note that β may also be used to parametrize the shock adiabat in place of v (except for the trivial shock $[v] = 0$ and $M_{B,n} = 1$), because

$$dv = 2\text{cotg}\beta \frac{M_{B,n}^2}{1 - M_{B,n}^2} \left(1 + \frac{[v]}{2v_B} G_B\right) [v] d\beta,$$

where $G = v(\partial P/\partial e)_v$ is the Gruneisen parameter and condition $1 + [v]G_B/(2v_B) = 0$ is the singularity that limits the maximum density increase across compressive shock waves, see for instance Kluwick [23]. Hence, along the compressive branch of the shock adiabat, the post-shock specific volume decreases with increasing shock angle. As shown in figure 3a, c_B increases and, therefore, $M_{B,t}/M_{A,t}$ decreases with increasing β or, equivalently, decreasing v. Because $M_{B,n} < M_{A,n}$ as required by the speed ordering relation, $M_B < M_A$ for each possible oblique

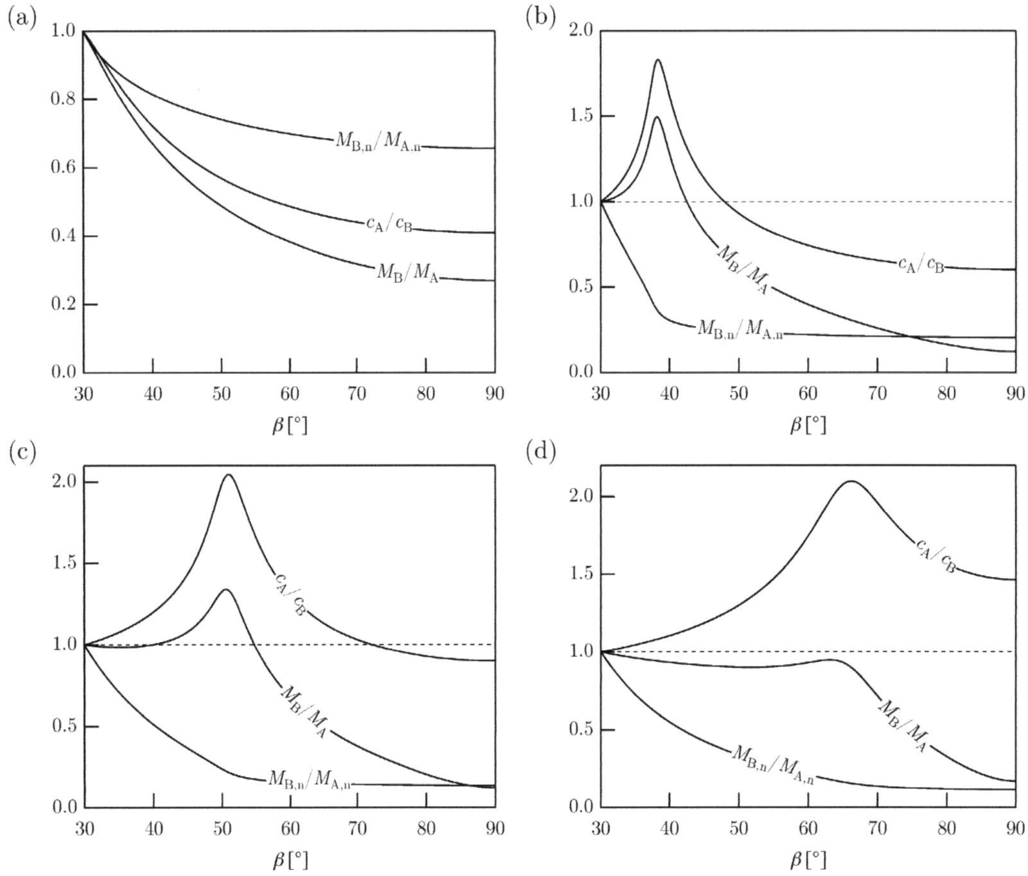

Figure 3: Variation of the speed of sound ratio c_A/c_B, normal Mach number ratio $M_{B,n}/M_{A,n}$ and Mach number ratio M_B/M_A with shock angle β for (a) case A_1, (b) case A_2, (c) case A_3 and (d) case A_4, as computed from the polytropic van der Waals model siloxane fluid MDM. The shock angle spans from the minimum value $\beta = \text{asin}(1/M_A)$, corresponding to an acoustic wave, up to the maximum value $\beta = \pi/2$, corresponding to a normal shock wave.

shock configuration with upstream state A_1. These observations could be anticipated and reasonably justified by neglecting the entropy rise across the shock wave, so that the shock adiabat would coincide with the isentrope passing through the pre-shock state. In this limit, the speed of sound would necessarily increase with increasing density because $\Gamma > 1$. However, this is true even if we take into account the entropy rise whenever condition $(\partial c/\partial s)_v > 0$ is satisfied, as it is the case within the polytropic van der Waals model. Moreover, as discussed by Menikoff and Plohr [12], $(\partial c/\partial s)_v > 0$ is expected to hold to occur for most real materials in their pure phases.

Next, state A_2, with $\Gamma_{A_2} < 1$, is considered. With reference to figure 2b, a wide portion of the shock adiabat H_2 lies inside the region $\Gamma < 1$. As is expected due to the almost negligible entropy rise together with $\Gamma_B < 0$, c_B initially decreases with increasing shock angle, as shown in figure 3b, where $c_B < c_A$ up to 47.9° and a pronounced local minimum if found at 38.3°. As a result, there exists a range of shock angles in which the increase in tangential component of the Mach number is sufficient to compensate the necessary decrease in the normal Mach number.

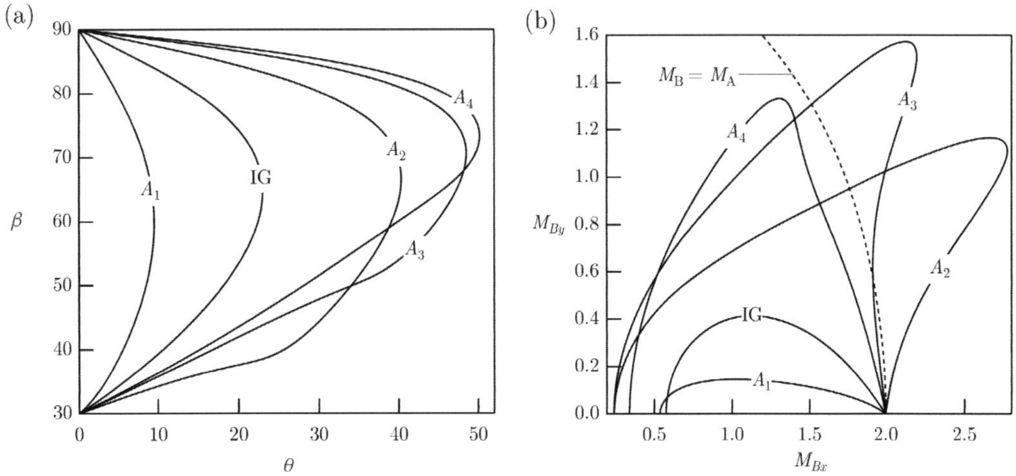

Figure 4: Diagrams illustrating (a) the variation of the flow deflection angle θ with shock angle β and (b) the polar of the Mach number, where $M_{Bx} = M_B \cos\theta$ and $M_{By} = M_B \sin\theta$, as computed from the polytropic van der Waals model of siloxane fluid MDM. Also shown are the curves, labeled IG, that correspond to the polytropic ideal gas case with the same upstream Mach number M_A used in the computation of the non-ideal cases.

The post-shock state characterized by $M_B = M_A$ occurs at 42.5° and it is denoted as point S in figure 2b. In the range between $\beta = \beta_{\min}$ and $\beta = 42.5°$ the post-shock Mach number exceeds the pre-shock Mach number. The local peak in the post-shock Mach number is found at 38.2° and corresponds to the ratio $M_B/M_A = 1.5$.

If the pre-shock density is further lowered, a larger portion of the shock adiabat in embedded in the region $\Gamma < 1$, as shown in figure 2c for the case corresponding to pre-shock state A_3. As a result, the range of shock angles where $c_B < c_A$ also increases, see 3c. For the present configuration, $c_B < c_A$ up to 72.1° with a local maximum at 51.0°. However, the evolution of the Mach number is somewhat more complex if compared to the previous case. The Mach number initially decreases with increasing shock angle, as the increase of $M_{B,t}$ is overcompensated by the decrease of $M_{B,n}$. At $\beta = 35.2°$, the post-shock Mach number shows a local minimum and subsequently increases up to the pronounced peak ($\beta = 50.6°$) associated with the local minimum in the post-shock speed of sound. The two post-shock states that satisfy $M_B = M_A$ are shown in figure 2c, where they are denoted as points S_1 and S_2. State S_1 corresponds to a shock angle $\beta = 39.8°$ and it is located inside the region $\Gamma < 1$; post-shock state S_2 occurs at $\beta = 54.7°$ and, similarly to state S of case A_2, it is located in the thermodynamic region $\Gamma > 1$.

In our last exemplary case, we consider pre-shock state A_4. With reference to figure 2d, the shock adiabat H_4 is almost entirely embedded in the region $\Gamma < 1$ and, as a result, $c_B < c_A$ over the complete range of shock angles, as shown in 3d. However, the local minimum in the post-shock speed of sound now occurs at such large values of β that the necessary decrease in the normal component of the Mach number across the shock wave is sufficient to compensate the non-ideal increase in the tangential component of the Mach number. Therefore, $M_B < M_A$ for each possible oblique shock configuration originating from state A_4.

The variation of the flow deflection angle θ with the shock angle β for each of the configurations considered is shown in figure 4a, where it is compared with the curve labelled IG, that corresponds to the polytropic ideal gas case and, as is well-known, is independent of the

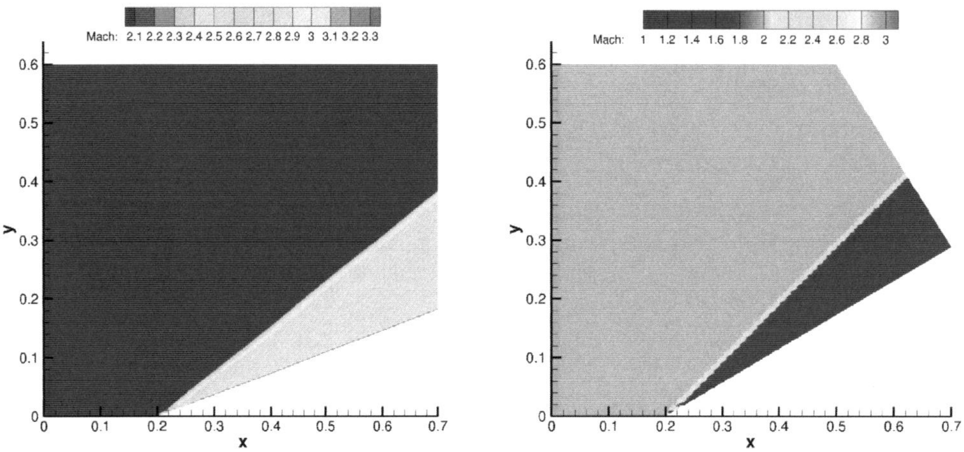

Figure 5: Left picture shows the computed Mach number field for a turning angle of $\theta = 20°$, while right picture presents the same result for a turning angle of $\theta = 30°$.

pre-shock thermodynamic state P_A, v_A [6]. The curves computed by means of the polytropic van der Waals model exhibit substantial differences depending on the pre-shock thermodynamic state. In particular, the maximum turning angle $\theta_{\max}$ that the flow can sustain across a planar attached shock waves varies from 9.5° of case A_1 up to 50.2° of case A_4. In contrast, $\theta_{\max} = 23.0°$ for the polytropic ideal gas.

Finally, the same results can be conveniently presented also in term of the shock polar for the Mach number, namely a plot of $M_{By} = M_B \sin\theta$ versus $M_{Bx} = M_B \cos\theta$, as shown in figure 4b for each non-ideal configuration considered and the polytropic ideal gas counterpart.

4. Numerical Results

In this section numerical simulations are reported to illustrate exemplary flow conditions, chosen among those discussed in the previous, theoretical section. Numerical simulations were run using SU2, an open-source Computational Fluid Dynamic (CFD) solver, which was recently extended to non-ideal and non-classical flows [24]. SU2 contains an embedded thermodynamic library which counts several different equation of state but can also exploit external library such as FluidProp. In the numerical simulations presented hereinafter only the polytropic van der Waals equation of state was considered, so that they can be compared to the results presented in the previous sections. In figure 5, exemplary computed flow fields for a turning angle of $\theta = 20°$, on the left-hand side, and $\theta = 30°$, on the right side, are reported. The flow enters the domain at left vertical boundary, where the state A_2 (with reference to table 1) is prescribed, and exits through the right boundary at supersonic speed. On the upper and lower boundaries an inviscid wall condition is imposed. The domain for the higher value of turning angle has a non-vertical outlet boundary to help the convergence of the numerical simulation. Both picture clearly show the presence of a shock-wave induced by the slope discontinuity at the lower all. Note that in the numerical simulations, an artificial internal shock structure is formed in which overshoots of the flow variables are observed. These are numerical artefacts common to all shock-capturing methods.

Figure 6 shows reduced density and pressure values for a fluid entering the domain at conditions A_2, taken along a horizontal line crossing the domain at $y = 0.2$. From figure 6 it is possible to verify that density and pressure increase for each different value of turning

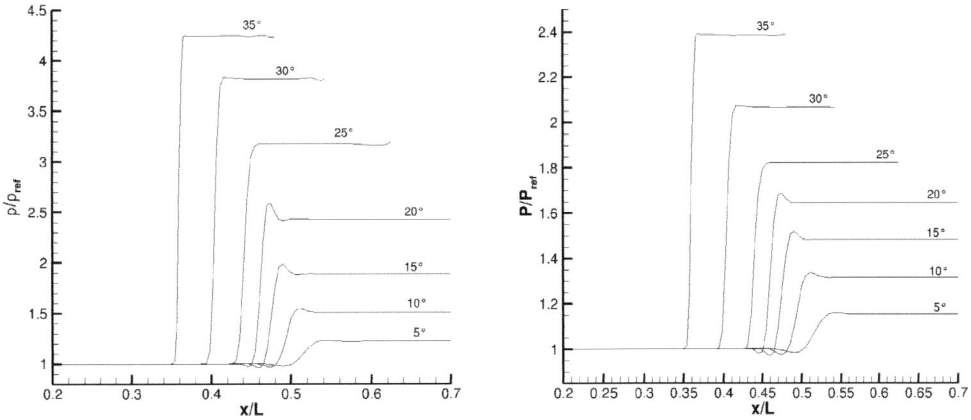

Figure 6: Left picture shows reduced density values ($\rho_{\mathrm{ref}} = 85.6067$ kg/m^3) while right picture shows the pressure ($P_{\mathrm{ref}} = 996220$ Pa) trends for several, different, turning angles. The density and pressure profiles were extracted along a line crossing the domain at $y = 0.2$.

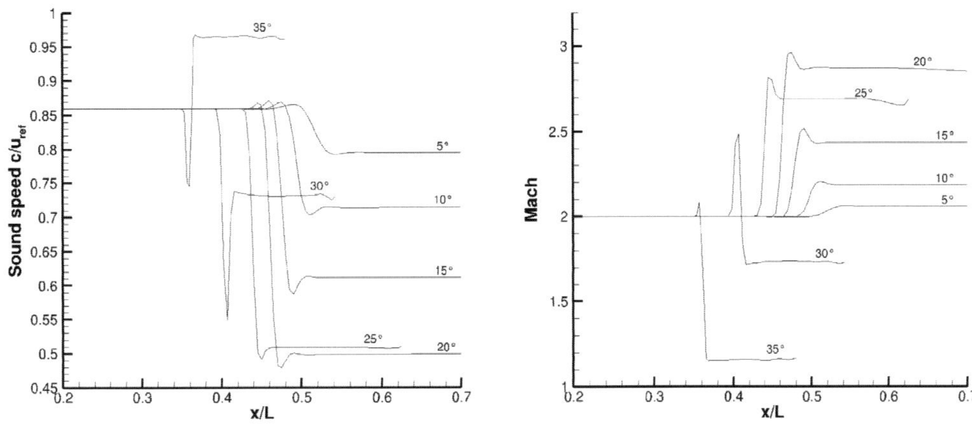

Figure 7: Left picture shows reduced speed of sound values ($u_{\mathrm{ref}} = 107.876$ m/s) while right picture shows the Mach trends for several, different, turning angles. The speed-of-sound and Mach profiles were extracted along a line crossing the domain at $y = 0.2$.

angle. This means that the shock maintains its compressive nature and the flow can be still categorised as non-ideal. In figure 7, speed of sound and Mach number trends are reported. From this graph it is possible to observe the non-ideal effect causing the Mach number to rise through an oblique shock-wave, in agreement with theoretical predictions. For this particular pre-shock thermodynamic state A_2, non-ideal phenomena are revealed in numerical simulations for a set turning angle which spans from 5 to 25 degrees. The highest M_B/M_A ratio is reached at $\theta = 20°$ while for $\theta = 30°$ and $\theta = 35°$ the flow does not present unusual features, as predicted by theory.

5. Conclusions

Steady oblique shock waves were investigated in the non-ideal thermodynamic framework. It was shown that, if the pre-shock state is located in close proximity of the non-ideal thermodynamic region $\Gamma < 1$, the speed of sound need not increase with pressure along the shock adiabat. The exemplary computations were performed using the polytropic van der Waals model of siloxane fluid MDM, for which the post-shock Mach number was found to increase up to 1.5 times the pre-shock Mach number. The present analysis has a sound theoretical basis, owing to the fact that simple van der Waals model is known to predict the correct qualitative behaviour in the single-phase thermodynamic region close to liquid-vapour equilibrium (sufficiently far from the critical point for critical phenomena to be negligible). Therefore, non-ideal shock waves are expected to be predicted also by more complex thermodynamic models, such as the Peng-Robinson [25], the Martin-Hou [26, 27] or the Span-Wagner [28, 29]. A more complete study is left for future investigations. The experimental verification of the occurrence of non-ideal shock waves can be devised in test rigs working with fluids of high or even moderate molecular complexity, though certain practical difficulties, mostly related to thermal stability of the working fluid, may be raised. Future work includes the definition of the thermodynamic admissibility region for non-ideal shock waves and the extension to flows of fluids exhibiting negative nonlinearity.

6. References

[1] Thompson P A 1971 *Phys. Fluids* **14** 1843–1849
[2] Bethe H A 1942 The theory of shock waves for an arbitrary equation of state Technical paper 545 Office Sci. Res. & Dev.
[3] Zel'dovich Y B 1946 *Zh. Eksp. Teor. Fiz.* **4** 363–364
[4] Weyl H 1949 *Comm. Pure Appl. Math.* **2** 103–122
[5] Hayes W D 1958 *Fundamentals of gasdynamics* (*High speed aerodynamics and jet propulsion* vol 3) ed Emmons H W (Princeton, N.J.: Princeton University Press) pp 416–481
[6] Thompson P A 1988 *Compressilbe Fluid Dynamics* (McGraw-Hill)
[7] Lambrakis K C and Thompson P A 1972 *Phys. Fluids* **15** 933–935
[8] Thompson P A and Lambrakis K C 1973 *J. Fluid Mech.* **60** 187–208
[9] Cramer M S and Kluwick A 1984 *J. Fluid Mech.* **142** 9–37
[10] Cramer M S and Sen R 1986 *Phys. Fluids* **29** 2181–2191
[11] Cramer M S and Sen R 1987 *Phys. Fluids* **30** 377–385
[12] Menikoff R and Plohr B J 1989 *Rev. Mod. Phys.* **61(1)** 75–130
[13] Bates J W and Montgomery D C 1999 *Phys. Fluids* **11** 462–475
[14] Colonna P, Harinck J, Rebay S and Guardone A 2008 *J. Propul. Power* **24** 282–294
[15] Congedo P, Corre C and Cinnella P 2011 *Comput. & Fluids* **49** 290–301
[16] Wheeler A and Ong J 2013 *J. Eng. Gas Turbines Power* **135** 102603
[17] Colonna P, Casati E, Trapp C, Mathijssen T, Larjola J, Turunen-Saaresti T and Uusitalo A 2015 *J. Eng. Gas Turb. Power* **137** 100801–1–19
[18] Lax P D 1957 *Commun. Pur. Appl. Math.* **10** 537–566
[19] Oleinik O 1959 *Usp. Mat. Nauk* **14** 165–170
[20] Landau L D and Lifshitz E M 1987 *Fluid mechanics, 2nd edn* (Pergamon Press)
[21] van der Waals J D 1873 *Over de Continuïteit van den Gas - en Vloeistoftoestand (on the continuity of the gas and liquid state)* Ph.D. thesis Leiden University
[22] Kluwick A 2004 *Acta Mech.* **169** 123–143
[23] Kluwick A 2001 *Handbook of Shock Waves* (Academic Press) chap 3.4. Rarefaction shocks, pp 339–411
[24] Vitale S, Gori G, Pini M, Guardone A, Economon T D, Palacios F, Alonso J J and Colonna P 2015 *AIAA Paper* **2760**
[25] Peng D Y and Robinson D B 1976 *Ind. Eng. Chem. Fundam.* **15** 59–64
[26] Martin J J and Hou Y C 1955 *AIChE J.* **1** 142–151
[27] Martin J J, Kapoor R M and De Nevers N 1959 *AIChE J.* **5** 159–160
[28] Span R and Wagner W 2003 *Int. J. Thermophys.* **24** 1–39
[29] Span R and Wagner W 2003 *Int. J. Thermophys.* **24** 41–109

IOP Conf. Series: Journal of Physics: Conf. Series **821** (2017) 012004 doi:10.1088/1742-6596/821/1/012004

A one-dimensional analytical calculation method for obtaining normal shock losses in supersonic real gas flows

Maximilian Passmann[1], Stefan aus der Wiesche[1], and Franz Joos[2]
[1] Department of Mechanical Engineering, Muenster University of Applied Sciences, 48565 Steinfurt, Germany
[2] Laboratory of Turbomachinery, Department of Power Engineering, Helmut-Schmidt University, 22043 Hamburg, Germany

E-mail: max.passmann@fh-muenster.de

Abstract.
The calculation of isentropic flow and normal shock waves of real gases are important, especially in the preliminary design of turbo-machinery and test rigs. In an ideal gas, the relations for one-dimensional isentropic flow and normal shock waves are well known and can be found in standard textbooks. However, for fluids exhibiting strong deviations from the ideal gas assumption universal relations do not exist due to complex equations of state. This paper presents a analytical method for the prediction of isentropic real gas flows and normal shock waves, based on the Redlich-Kwong (RK) equation of state. Explicit expressions based on a series expansion for describing isentropic flow of Novec^{TM} 649 are compared to Refprop data and ideal gas equations. For moderate pressures the RK method is in very good agreement with the Refprop data, while the ideal gas equations fail to predict the real gas behaviour. The same observations are made for normal shock calculations, where both real gas methods yield very close results. Especially the predicted stagnation pressure losses across a shock wave are in excellent agreement.

Nomenclature

Latin Symbols

a	Speed of sound	[m/s]
A	Cross sectional area	[m^2]
A,B,C	Constants in Eq(8)	[–]
c	Velocity	[m/s]
c_v	Const. volume heat capacity	[J/(kgK)]
h	Specific enthalpy	[kJ/kg]
M	Mach number	[–]
M_m	Molar mass	[kg/kmol]
p	Pressure	[Pa]
P	Shock strength	[–]
R	Specific gas constant	[J/(kgK)]
R_{univ}	Universal gas constant	[J/(kgK)]
s	Specific entropy	[J/(kgK)]
T	Temperature	[K]
u	Internal energy	[m/s]
v	Specific volume	[m^3/kg]
Z	Compressibility factor	[–]

Greek Symbols

α	Constant in Eq(1)	[–]
β	Constant in Eq(1)	[–]
ϵ	Error	[%]
ζ	Constant in Eq(15)	[–]
∂	Partial derivative	[–]
Γ	Fundamental derivative	[–]
γ	Adiabatic exponent	[–]

Subscripts

0	Stagnation condition
1	Value upstream from shock wave
2	Value upstream from shock wave
c	Value at critical point
ig	Ideal gas state
rg	Real gas state

1. Introduction

Despite the rapid advancements in computational fluid dynamics (CFD), a design and optimization approach solely based on CFD simulations is still time consuming. Therefore one-dimensional methods, especially in the preliminary design phase, are important. In an ideal gas, the relations for one-dimensional isentropic flow and normal shock waves are well known and can be found in standard textbooks. However, for fluids exhibiting strong deviations from the ideal gas assumption, universal relations do not exist. It is the purpose of this paper to develop a method for the description of one-dimensional isentropic flow and normal shock losses in supersonic real gas flows, using the analytical Redlich-Kwong (RK) equation of state (EoS).

Fluids exhibiting real gas behaviour are of special interest for organic Rankine Cycle (ORC) applications. An ORC process is a thermodynamic cycle, based on the classical steam Rankine cycle that utilizes organic fluids instead of water as working fluid. The key parameter in terms of organic fluid properties is a low boiling point, making the usage of low-temperature heat sources possible. In recent years, research has focused on so called dry fluids, with an overhanging saturation curve, which is favourable to avoid blade erosion [1].

The reliable design of ORC components requires both numerical and experimental work. For this reason, the Laboratory for Thermal and Power Engineering at Muenster University of Applied Sciences, Germany is building up a closed loop wind tunnel for organic vapors, aiming at the investigation of real-gas effects in nozzles and axial turbine blade cascades [2]. However, to determine the working point of the test rig, the overall losses at operating conditions have to be estimated. The dominating contribution to the total losses occurs in the test section. Details on its design can be found in [3, 4]. To get a better quantitative estimate about the pressure losses and hence the required pressure rise to drive the wind tunnel, a normal shock wave in the test section was considered. The computational method to calculate these losses for an organic vapour are developed and tested in this contribution. Another important technical application for normal shock waves is the correction of total pressure measurements from supersonic Pitot-tubes.

The paper is structured as follows. After a brief literature review on existing real gas flow models, the present approach is described in detail. First, the thermodynamic real gas model is presented, followed by methods to describe isentropic flow and normal shock calculations. The results are then presented and discussed.

2. Literature Review

A number of models for describing the steady one-dimensional isentropic flow of a real gas can be found in the literature. An excellent historical review of isentropic flow models has been published by Sullivan [5]. Reimer [6] and Johnson [7] presented isentropic relations based on modified ideal gas relations that accurately describe fluid behaviour. However, in both cases only slight deviations from ideal gas behaviour occurred, which can be attributed to the fluids and the range of operating conditions selected for the study. Kouremenos and Kakatsios [8] chose a similar approach and defined explicit relations approximating real gas behaviour, with a similar mathematical form as the ideal gas equations. Later, they extended this work to the description of thermodynamic properties using the RK EoS [9]. One of the few experimental works has been carried out by Bier et al. [10]. They investigated the flow of the refrigerant R22 in an axially symmetric de Laval nozzle for Mach numbers up to $M = 2.2$. The experimental results were compared to one-dimensional computations using an advanced EoS as well as an approximation procedure, that could reproduce all measured values with deviations within a few percent.

A different approach on isentropic flow makes use of the fundamental derivative Γ as defined by Thompson [11]. The fundamental derivative of gas dynamics is a quantitative measure of the variation of the speed of sound with respect to density in an isentropic change of state [12]. Thompson and Sullivan [13] reported a method for predictions of the sonic conditions in a real gas utilizing the fundamental derivative. The results are relatively simple relations based on a Taylor series expansion. However, the authors remark, that while the application of the fundamental derivative leads to a simple and effective description of isentropic real gas flows, the calculation of Γ itself is not necessarily simple. Colonna et al. [12] investigated the possibility of computing the fundamental derivative using different EoS. They found non-analytical or reference EoS to be superior to analytical formulations. Especially in the vicinity of the critical point, analytical EoS fail to predict the correct physical behaviour.

One of the few analytical approaches to calculate normal shock waves of real gases was published by Kouremenos [14]. He derived a non-linear equation using the RK EoS, but this first study lacks information on the results and accuracy of the method. Later on, Kouremenos et al. [15] presented a modified calculation procedure and investigated the influence of different EoS and stagnation conditions on normal shock waves in steam. They conclude, that the RK EoS provides accurate solutions for moderate pressures, but becomes increasingly inaccurate for higher pressures and temperatures.

3. Thermodynamic Real Gas Model

The description of the thermodynamic behaviour of a real gas requires a suitable EoS. The most accurate type are non-analytic or fundamental EoS [16], explicit in Helmholtz energy. Equations in this form are capable of describing the thermodynamic behaviour of pure substances with an accuracy that probably exceeds measurement uncertainties [16]. Unfortunately, this type of EoS has to be solved numerically and cannot be used to derive simple working equations.

For the present work the widely used Redlich-Kwong EoS [17] has been selected. The equation in the form $p(T, v)$ is represented by

$$p = \frac{R \cdot T}{v - \beta} - \frac{\alpha}{T^{\frac{1}{2}}} \cdot \frac{1}{v \cdot (v + \beta)}. \tag{1}$$

Herein, p, v, T are pressure, specific volume, and temperature, and R is the specific gas constant $R = R_{univ}/M_m$. The constants α and β are functions of pressure and temperature at the critical point (p_c, T_c) and are given by

$$\alpha = 0.42748 \cdot \frac{R^2 \cdot T_c^{\frac{5}{2}}}{p_c}, \qquad \beta = 0.08664 \cdot \frac{R \cdot T_c}{p_c}. \tag{2}$$

If the compressibility factor Z is close to unity, the ideal gas EoS is reproduced by Eq 1 [18].

A full description of the thermodynamic behaviour using a cubic EoS requires additional analytical expressions for enthalpy and entropy. However, it can be shown, that a fluid model consisting of the three equations $p(T, v)$, $h(T, v)$, and $s(T, v)$ is thermodynamically equivalent to a fundamental equation of state [18]. For a thermal EoS with the independent variables temperature and volume, the enthalpy of a real gas is defined as [18]

$$h(T, v) = u(T, v) + p(T, v) \cdot v, \tag{3}$$

where $u(T, v)$ describes the specific internal energy, given by

$$u(T, v) = u_{ig} + \int_{T_{ig}}^{T} c_v^{ig}(T)\,\mathrm{d}T + \int_{\infty}^{v} \left[T \left(\frac{\partial p}{\partial T} \right)_v - p \right] \mathrm{d}v. \tag{4}$$

Herein, u_{ig} is the specific internal energy of the corresponding ideal gas state ($p_{ig} \to 0$). The first integral in Eq (4) describes the temperature dependency of u for an ideal gas state, whereas the second integral accounts for deviations from ideal behaviour. The specific entropy of a real gas can be dealt with in a similar manner, resulting in the expression

$$s(T, v) = s_{ig} + \int_{T_{ig}}^{T} c_v^{ig}(T) \frac{\mathrm{d}T}{T} + R \ln \frac{v}{v_{ig}} + \int_{\infty}^{v} \left[\left(\frac{\partial p}{\partial T} \right)_v - \frac{R}{v} \right] \mathrm{d}v, \tag{5}$$

where s_{ig} is the specific entropy of the corresponding ideal gas state ($p_{ig} \to 0$). The selection of the reference point is arbitrary. In the present work, internal energy u_{ig} and entropy s_{ig} were set to zero at $T_{ig} = 273.15\,\mathrm{K}$ and $p_{ig} = 1\,\mathrm{Pa}$.

For the speed of sound a holds the exact relation [18]

$$a^2 = -v^2 \left(\frac{\partial p}{\partial v} \right)_s = v^2 \left[\frac{T}{c_v} \left(\frac{\partial p}{\partial T} \right)_v^2 - \left(\frac{\partial p}{\partial v} \right)_T \right]. \tag{6}$$

Equation 6 shows that the speed of sound can be derived from the thermal EoS $p(T, v)$ and the real gas isochoric specific heat capacity $c_v(T, v)$:

$$c_v(T, v) = c_v^{ig}(T) + T \int_{\infty}^{v} \left(\frac{\partial^2 p}{\partial T^2} \right)_v \mathrm{d}v. \tag{7}$$

Herein, the ideal gas specific heat capacity $c_v^{ig}(T)$ is approximated by a second-order polynomial

$$c_v^{ig}(T) = A + B \cdot T + C \cdot T^2. \tag{8}$$

The constants A, B, and C can be taken from the literature or from a fit procedure. For this paper, Refprop data was used. Equation 8 was fitted by means of a least-squares method and reproduces the exact values within 0.15 %. After integration Eq (3) and Eq (5) become:

$$s(T,v) = s_{ig} + A\cdot\ln\frac{T}{T_{ig}} + B\cdot(T - T_{ig}) + \frac{C}{2}\cdot(T^2 - T_{ig}^2) + R\cdot\ln\frac{v}{v_{ig}} + \frac{\alpha}{2bT^{\frac{3}{2}}}\cdot\ln\frac{v}{v+\beta} + R\cdot\ln\frac{v-b}{v}, \quad (9)$$

$$h(T,v) = u_{ig} + A\cdot(T - T_{ig}) + \frac{B}{2}\cdot(T^2 - T_{ig}^2) + \frac{C}{3}\cdot(T^3 - T_{ig}^3) + \frac{3\alpha}{2\beta T^{\frac{1}{2}}}\cdot\ln\frac{v}{v+\beta} + \frac{RTv}{v-\beta} - \frac{\alpha}{T^{\frac{1}{2}}v(v+\beta)}. \quad (10)$$

Thus, Eq (1,9,10) represent a closed thermodynamic real gas model, which has been used for the following investigations.

4. Isentropic Flow Relations

From a practical point of view, simple working formula for the description of isentropic changes in a real gas are desirable. A method for obtaining relations in the form p/p_0, T/T_0, v/v_0, and A^*/A in terms of Mach number is outlined in the following.

The governing equations of steady one-dimensional isentropic flow are the continuity equation

$$c_1 \cdot v_2 \cdot A = c_2 \cdot v_1 \cdot A = const., \quad (11)$$

and the energy equation

$$h_1 + \frac{c_1^2}{2} = h_2 + \frac{c_2^2}{2} = h_0 = const., \quad (12)$$

where c is the velocity, v the specific volume, and h and s the specific enthalpy and entropy.

In the following, an isentropic expansion process proceeding from a known set of stagnation conditions p_0, T_0, v_0 along an isentrope $s_0 = const.$ to an arbitrary point p_1, T_1, v_1 is considered. Furthermore, the stagnation or reservoir conditions are assumed to be known. Consequently, the isentrope s_0 passes through this point. To describe this isentrope we calculate the fluid properties along this path for an arbitrary number of points n_i. The local Mach number M_i at each point is defined by

$$\mathrm{M}_i = \frac{c_i}{a_i}. \quad (13)$$

Herein, the local speed of sound a_i can be obtained from Eq 6 and the local velocity c_i is related to the enthalpy by

$$c_i = \sqrt{2\cdot(h_0 - h_i)}. \quad (14)$$

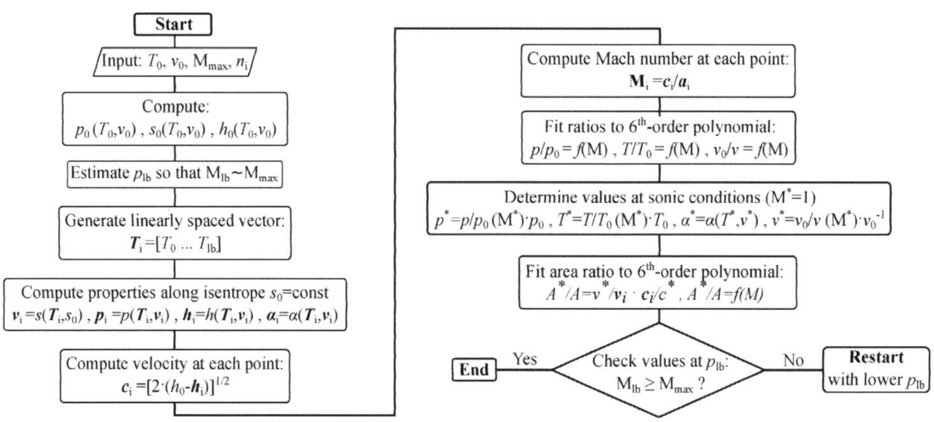

Figure 1. Flow-chart showing the computational procedure for deriving one-dimensional isentropic relations based on a sixth-order polynomial. Bold variables are vector quantities.

Pressure, temperature, specific volume and area ratios are then fitted to a polynomial in terms of Mach number by means of a least-squares fit. The present work has shown, that simple ideal-gas-like formulations, as used for instance by Kouremenos and Kakatsios [8], fail to predict real gas behaviour of fluids like Novec^{TM}649. Therefore the equations were extended by means of a series expansion. A sixth-order polynomial was found to be sufficient to describe real gas behaviour in the present case. The polynomials $F(\mathrm{M})$ are of the generic form:

$$F(\mathrm{M}) = \sum_{i=0}^{6} \zeta_i \cdot \mathrm{M}^i. \tag{15}$$

Herein, ζ_i describe the fitted constants.

The computational procedure is summarised in Fig 1. The correct value of the lower pressure limit for the fit p_{lb} has to be determined iteratively, based on a prescribed maximum Mach number M_{max}. It is worth noting, that this algorithm is independent of the EoS and can be used with cubic as well as fundamental EoS.

5. Normal Shock Calculations

A shock wave is a discontinuity in a very small region that occurs between two boundaries in supersonic flow. Normal shock waves are perpendicular to the flow direction and transform a supersonic flow to a subsonic flow regime. A schematic sketch of a normal shock wave is shown in Fig 2 (left). Figure 2 (right) shows the idealised process in the h-s-diagram. Shock waves are dominated by viscous and heat transfer effects. Therefore the assumption of isentropic flow is no longer valid. As a result from the second law of thermodynamics the entropy across a shock wave has to increase, i.e. $s_{02} > s_{01}$.

The description of a normal shock wave involves five variables, i.e. static pressure, temperature, specific volume, enthalpy, and velocity. Therefore in addition to the thermal EoS $p = p(T, v)$, enthalpy equation $h = h(T, v)$, continuity of mass Eq (11), and conservation of energy Eq (12), a fifth equation is necessary to compute the changes across a normal shock. This is achieved by introducing the conservation of momentum:

$$p_1 + \frac{c_1^2}{v_1} = p_2 + \frac{c_2^2}{v_2}. \tag{16}$$

Herein, p describes the static pressure, c is the velocity, and v the specific volume.

For an ideal gas, this set of equations can be solved in closed form, resulting in the normal shock or Rankine-Hugoniot equations [19]. For non-ideal fluids this is usually impossible due to the complexity of the EoS. A way of finding an analytical solution is by means of cubic EoS.

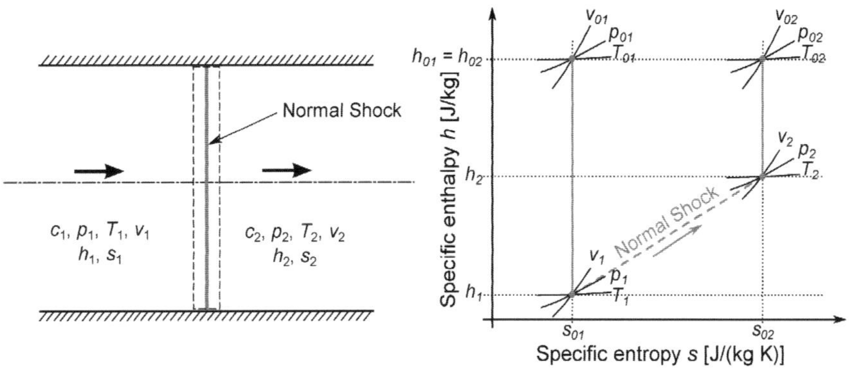

Figure 2. Schematic sketch of an idealised normal shock wave (left) and changes across a normal shock wave in the specific enthalpy versus entropy plane (right).

Using Eq (1) and Eq (10) for thermal and enthalpy EoS, in principle allows for a solution of one analytical expression containing only one unknown, i.e. $v_2 = f(T_2)$. However, the present study has shown that the resulting expression becomes quite cumbersome and not very reliable to solve. Far better results were obtained from a system of two non-linear equations with two unknowns T_2 and v_2 that can be solved numerically using a root finding algorithm, such as the Newton-Raphson method [20]. Supplying the corresponding ideal gas values as initial data, results in a quick convergent behaviour. Using the conservation of momentum in combination with the RK EoS (Eq (1)) to eliminate the unknown pressure p_2 results in:

$$0 = p_1(T_1, v_1) - \frac{R \cdot T_2}{v_2 - \beta} + \frac{\alpha}{T_2^{\frac{1}{2}} \cdot v_2 \cdot (v_2 + \beta)} + \frac{c_1^2}{v_1} - \frac{v_2 \cdot c_1^2}{v_1^2}. \quad (17)$$

Similarly the energy equation, Eq (12), in combination with the enthalpy equation, Eq (10), can be used to obtain the second expression:

$$0 = h_1(T_1, v_1) - u_{ig} + A \cdot (T_2 - T_{ig}) + \frac{B}{2} \cdot (T_2^2 - T_{ig}^2) + \frac{C}{3} \cdot (T_2^3 - T_{ig}^3) + \frac{3\alpha}{2\beta T^{\frac{1}{2}}} \cdot \ln \frac{v_2}{v_2 + \beta} + \frac{R T v_2}{v_2 - \beta} - \frac{\alpha}{T_2^{\frac{1}{2}} v_2 (v_2 + \beta)} + \frac{c_1^2}{2} - \frac{v_2^2 \cdot c_1^2}{2 \cdot v_1^2}. \quad (18)$$

The numerical solution of Eq (17) and Eq (18) provides T_2 and v_2. All other properties downstream from the shock wave can be obtained from the RK EoS (Eq (1)) and enthalpy and entropy equations (Eq (10,9)).

6. Results and Discussion

The calculation procedures presented in this paper are tested with the fluorinated ketone Novec^{TM}649, which is the working fluid for the closed loop wind tunnel [2]. Figure 3 shows an h-s-diagram for Novec^{TM}649 [21]. Of special interest for the present study are stagnation conditions in the superheated region with pressure and temperature in the range of $p_0 = 4...6 \cdot 10^5$ Pa and $T_0 = 370...420$ K. This range was chosen because it represents the possible operating conditions of the planned closed loop wind tunnel. The fluid possesses the desired overhanging saturation curve, as mentioned in Sec 1. Deviations from ideal gas behaviour can already be observed for moderate pressures, i.e. at $5 \cdot 10^5$ Pa and 390 K, the compressibility factor is already as low as $Z = 0.829$.

In the first step, accuracy and consistency of the RK real gas model were compared with the reference EoS (McLinden et al. [22]), which is implemented in Refprop [21]. For stagnation conditions of $5 \cdot 10^5$ Pa and 390 K an exemplary comparison between both EoS is presented in Fig 4. The RK predictions show deviations mostly lower than 10 %. The biggest errors occur in the prediction of the specific volume, with a maximum deviation of 20.9 %. Errors increase for higher pressures, near the saturation line, and in the vicinity of the critical point. However, for the range of interest in this work, the RK EoS is a reasonable approximation of the real fluid behaviour. It should also be noted, that some initial tests with other cubic EoS, such as the formulations by Aungier [23], Soave [24], and Peng and Robinson [25] showed no significant improvement over the RK EoS.

In the next step, the isentropic relations were fitted to the fluid properties according to the procedure described in Sec 4 and Fig 1. Isentropic expansions using the RK and Refprop EoS were compared to each other over a wide range of stagnation conditions, up to the critical point region. Table 1 shows the maximum error for the RK EoS with respect to the corresponding Refprop data. The results show, that the prediction of the specific volume is afflicted with the highest deviations up to 40 % near the critical point, followed by the static pressure with deviations up to 18 %. The lowest errors were observed for the temperature with maximum deviations around 2 %. Furthermore, the results indicate, that the RK EoS becomes less reliable for higher pressures and temperatures, especially for expansions from stagnation conditions above $p_0 = 12 \cdot 10^5$ Pa, errors higher than 10 % are being observed.

Deviations were also found to increase in the vicinity of the saturation curve. As shown in Tab 1, for a stagnation pressure of $5 \cdot 10^5$ Pa three different stagnation temperatures were considered. For $T_0 = 378$ K, the point closest to the saturation curve, the deviations are larger than for the two points with higher temperatures, that are farther away from the saturation curve. A similar behaviour, albeit with higher errors, has been observed for higher pressures. For the range of interest in this work ($p_0 = 4...6 \cdot 10^5$ Pa, $T_0 = 370...420$ K), the RK EoS has been found to deliver re-

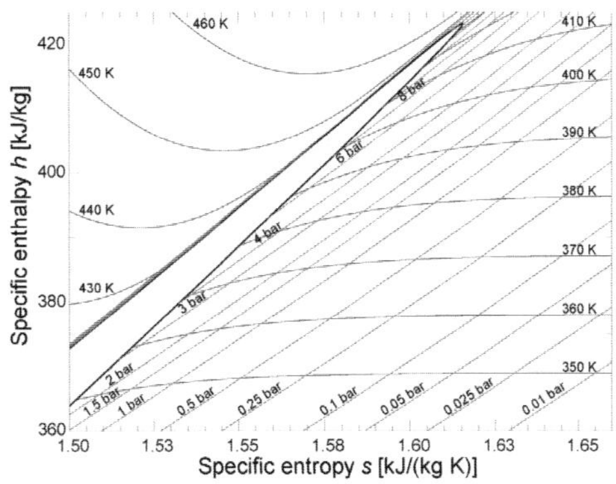

Figure 3. Specific enthalpy h vs. specific entropy s plot with lines of constant temperature (blue) and pressure (red) for Novec^{TM}649 calculated from [21].

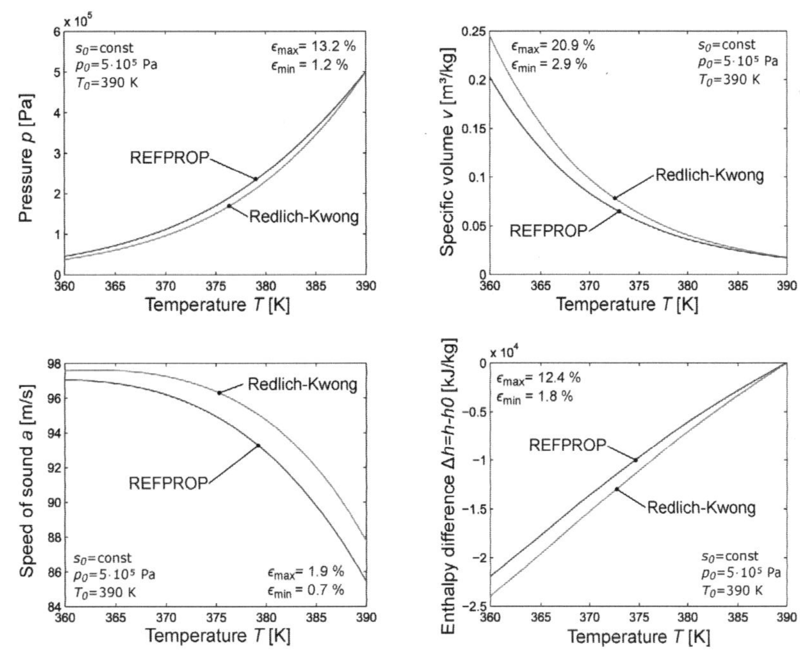

Figure 4. Comparison of isentropic lines (s_0=const) for Novec^{TM}649 calculated from Redlich-Kwong and Refprop [21] for stagnation conditions of $p_0 = 5 \cdot 10^5$ Pa, $T_0 = 390$ K and maximum and minimum error for Redlich-Kwong predictions.

Table 1. Maximum error of isentropic fits for the RK EoS with respect to Refprop data for expansions from different stagnation conditions up to a Mach number of $M_{max} = 2.2$. Critical pressure and temperature for Novec^{TM}649 are $p_c = 18.69 \cdot 10^5$ Pa and $T_c = 441.81$ K.

p_0/p_c	T_0/T_c	p_0	T_0	Error p/p_0	Error T/T_0	Error v_0/v
–	–	MPa	K	%	%	%
0.9952	1.0027	1.86	443	17.59	1.80	39.48
0.9096	0.9891	1.7	437	11.73	2.07	21.51
0.8561	0.9846	1.6	435	2.41	1.75	17.53
0.6421	0.9506	1.2	419	1.90	1.33	9.49
0.5350	0.9280	1	410	1.71	1.18	8.19
0.3210	0.8737	0.6	386	1.13	0.83	5.21
	0.8556		378	1.73	0.74	4.89
0.2675	0.8714	0.5	385	1.04	0.63	3.85
	0.8827		390	0.98	0.57	2.88
0.2140	0.8375	0.4	370	1.02	0.63	4.62
0.1605	0.8148	0.3	360	0.92	0.52	3.48

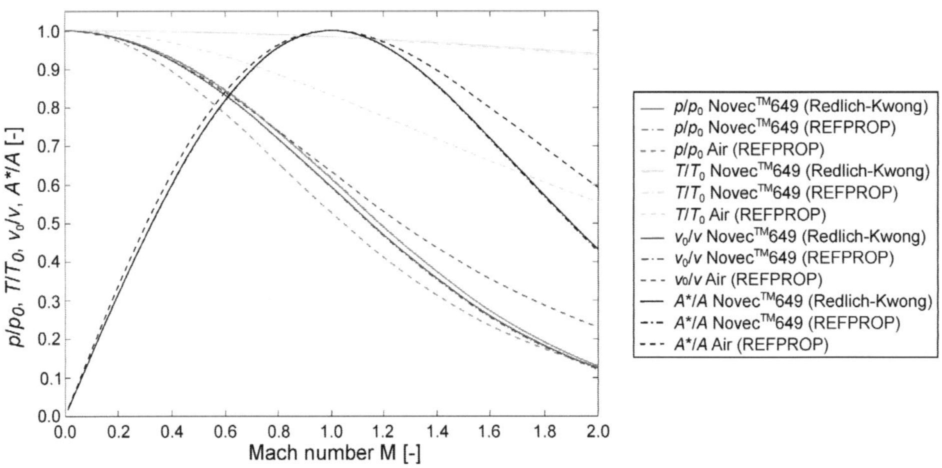

Figure 5. Comparison of isentropic relations (s_0=const) for Novec^{TM}649 and air for stagnation conditions of $p_0 = 5 \cdot 10^5$ Pa and $T_0 = 390$ K.

liable predictions with maximum deviations for pressure and temperature mostly around 1 % and slightly higher values up to 5 % for the specific volume. As an example, Fig 5 presents the isentropic flow relations for Novec^{TM}649 for stagnation conditions of $5 \cdot 10^5$ Pa and 390 K. The results obtained from RK and Refprop are in good agreement, with maximum deviations given in Tab 1.

Finally, the method for calculating the normal shock was tested, using the following procedure: For a given set of stagnation conditions p_0, T_0 the fluid properties for a specified Mach number M_1 are computed, using the isentropic relations described in Sec 4. The next step is the normal shock calculation. In case of the RK EoS the procedure described in Sec 5 is followed, whereas for the Refprop case three governing equations (Eq (11,12,16)) in combination with the EoS by McLinden et al. [22] were solved numerically, using a Newton-Raphson method. Subsequently, the stagnation conditions behind the shock are determined from the EoS and the results are checked for consistency, i.e. the stagnation enthalpy has to remain constant $h_{02} = h_{01}$, whereas the entropy has to increase $s_{02} > s_{01}$.

Figure 6 shows the variation of M_2, p_1/p_2, p_{02}/p_{01}, T_1/T_2 and v_2/v_1 in terms of upstream

Mach number M_1. The results for the RK EoS were obtained from the procedure described above, while the solutions with Refprop were computed by directly solving the three governing equations (Eq (11,12,16)) in combination with the EoS by McLinden et al. [22]. A stable convergent behaviour was achieved by setting the corresponding ideal gas solution as initial values for the iterative solution. Again, excellent agreement was found between RK and Refprop predictions, while the ideal gas formulations fail to predict the non-ideal fluid behaviour.

An additional parameter in the description of normal shock waves is the strength of a shock wave P, which is defined as the ratio of the increase in static pressure to the initial static pressure upstream from the shock wave [19]:

$$P = \frac{p_2 - p_1}{p_1} = \frac{p_2}{p_1} - 1. \tag{19}$$

The results shown in Fig 7 (right) indicate, that normal shock waves calculated from real gas formulations are weaker than the solutions obtained from the ideal gas equations. However, the absolute loss in stagnation pressure for the real gas model is up to 40 % higher than the corresponding losses for an ideal gas. Of major interest is also the consistency of the calculations.

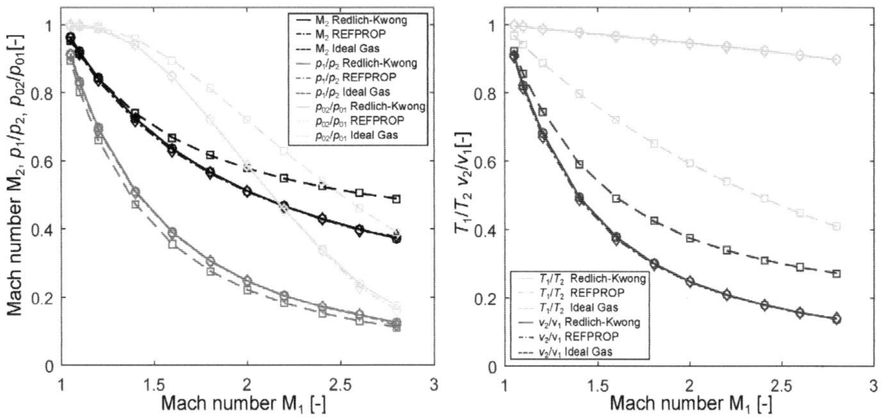

Figure 6. Variation of M_2, p_1/p_2 (left) and T_1/T_2, v_2/v_1 (right) in terms of M_1 for normal shock waves in Novec^{TM}649. Stagnation conditions were set at $p_0 = 5\,\text{bar}$ and $T_0 = 390\,\text{K}$.

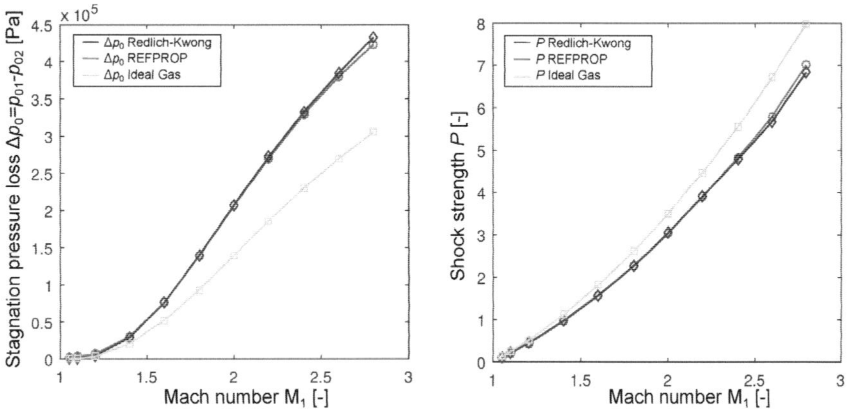

Figure 7. Variation of shock losses (left) and shock strength (right) in terms of M_1 for normal shock waves in Novec^{TM}649. Stagnation conditions were set at $p_0 = 5\,\text{bar}$ and $T_0 = 390\,\text{K}$.

Tests with different EoS, i.e. using Refprop for the description of isentropic flow and the RK EoS for normal shock calculations, resulted in inconsistencies and non-physical solutions.

7. Conclusion and Outlook

A method for calculating one-dimensional isentropic flow and the normal shock of a real gas is presented. A thermodynamic real gas model based on the cubic RK EoS accurately predicts the fluid properties of Novec™649 for moderate pressures, but becomes less accurate for higher pressures and near the critical point. Simple working formulae based on a sixth-order polynomials are sufficient to describe the isentropic flow of real gases. The agreement with Refprop data is excellent, while the ideal gas formulation fails to predict real gas behaviour. A system of two non-linear equations, using the RK EoS, is solved to calculate normal shock waves. The results are compared to Refprop and ideal gas data. While both real gas models are in very good agreement, the ideal gas shows strong deviations. Compared to the ideal gas model, both real gas models predict higher absolute pressure losses, but lower values for the shock strength.

This paper shows, that a analytical real gas model based on the RK EoS accurately predicts isentropic flow and normal shocks for moderate pressures. Therefore it can be used as an alternative to a purely empirical approach using Refprop. Although, the prediction of the specific volume by the RK model is afflicted with relatively high errors, the normal shock losses are in very good agreement with Refprop data. The model has been found to be reliable for moderate pressures and temperatures, higher errors occur near the critical point region. Also the behaviour at supercritical conditions has to be investigated.

The results of the present paper suggest the following areas for future research: So far only Novec™649 has been considered. An extension to other fluids is necessary to further investigate the reliability of the method. While initial tests with other cubic EoS have not shown significant improvement over the RK EoS, the investigation of other cubic EoS (e.g. Soave [24]) could further improve the method. Subsequently, an extension to oblique shock waves will be of interest.

8. References

[1] Chen H, Goswami D Y and Stefanakos E K 2010 *Renewable and Sustainable Energy Reviews* **14** 3059–3067
[2] Reinker F, Hasselmann K, aus der Wiesche S and Kenig E Y 2016 *ASME Journal of Engineering for Gas Turbines and Power* **138** 052601
[3] Hasselmann K, Reinker F, aus der Wiesche S and Kenig E Y 2015 *ASME/JSME/KSME 2015 Joint Fluids Engineering Conference* (American Society of Mechanical Engineers) pp V001T15A002–V001T15A002
[4] Passmann M, Reinker F, Hasselmann K, aus der Wiesche S and Joos F 2016 *ASME Turbo Expo 2016: Turbine Technical Conference and Exposition* (American Society of Mechanical Engineers) pp 1–8
[5] Sullivan D 1981 *ASME Journal of Fluids Engineering* **103** 258–267
[6] Reimer R M 1964 *ASME Journal of Basic Engineering* **86** 169–173
[7] Johnson R C 1964 *ASME Journal of Basic Engineering* **86** 519–525
[8] Kouremenos D A and Kakatsios X K 1985 *Forschung im Ingenieurwesen* **51** 169–174
[9] Kouremenos D and Kakatsios I X 1991 *Forschung im Ingenieurwesen* **57** 25–27
[10] Bier K, Ehrler F and Kissau I G 1977 *Forschung im Ingenieurwesen* **43** 175–184
[11] Thompson P A 1971 *Physics of Fluids* **14** 1843–1849
[12] Colonna P, Nannan N, Guardone A and Van der Stelt T 2009 *Fluid Phase Equilibria* **286** 43–54
[13] Thompson P and Sullivan D 1977 *ASME Journal of Fluids Engineering* **99** 217–225
[14] Kouremenos I D A 1986 *Forschung im Ingenieurwesen A* **52** 23–31
[15] Kouremenos I D, Kakatsios I X and Krikkis R 1990 *Forschung im Ingenieurwesen* **56** 54–57
[16] Poling B E, Prausnitz J M, John Paul O and Reid R C 2001 *The properties of gases and liquids* vol 5 (McGraw-Hill New York)
[17] Redlich O and Kwong J N 1949 *Chemical Reviews* **44** 233–244
[18] Baehr H D and Kabelac S 2012 *Thermodynamik: Grundlagen und technische Anwendungen* (Springer-Verlag)
[19] Shapiro A H 1953 *New York: Ronald Press, 1953-54* **1**
[20] Press W H 2007 *Numerical recipes 3rd edition: The art of scientific computing* (Cambridge University Press)
[21] Lemmon E W, Bell I H, Huber M L and McLinden M O 2015 *NIST Reference Fluid Thermodynamic and Transport PropertiesREFPROP, Beta Version* **9.1.2.15**
[22] McLinden M O, Perkins R A, Lemmon E W and Fortin T J 2015 *Journal of Chemical & Engineering Data* **60** 3646–3659
[23] Aungier R 1995 *ASME Journal of Fluids Engineering* **117** 277–281
[24] Soave G 1972 *Chemical Engineering Science* **27** 1197–1203
[25] Peng D Y and Robinson D B 1976 *Industrial & Engineering Chemistry Fundamentals* **15** 59–64

IOP Conf. Series: Journal of Physics: Conf. Series **821** (2017) 012005 doi:10.1088/1742-6596/821/1/012005

Non-Ideal Compressible-Fluid Dynamics of Fast-Response Pressure Probes for Unsteady Flow Measurements in Turbomachinery

G Gori[1], P Molesini[1], G. Persico[2] and A. Guardone[1]

[1] Department of Aerospace Science and Technology, Politecnico di Milano
Via La Masa 34, 20156, Milano, Italy
[2] Department of Energy, Politecnico di Milano
Via Lambruschini 4, 20156, Milano, Italy

E-mail: alberto.guardone@polimi.it

Abstract. The dynamic response of pressure probes for unsteady flow measurements in turbomachinery is investigated numerically for fluids operating in non-ideal thermodynamic conditions, which are relevant for e.g. Organic Rankine Cycles (ORC) and super-critical CO_2 applications. The step response of a fast-response pressure probe is investigated numerically in order to assess the expected time response when operating in the non-ideal fluid regime. Numerical simulations are carried out exploiting the Non-Ideal Compressible Fluid-Dynamics (NICFD) solver embedded in the open-source fluid dynamics code SU2. The computational framework is assessed against available experimental data for air in dilute conditions. Then, polytropic ideal gas (PIG), i.e. constant specific heats, and Peng-Robinson Stryjek-Vera (PRSV) models are applied to simulate the flow field within the probe operating with siloxane fluid octamethyltrisiloxane (MDM). The step responses are found to depend mainly on the speed of sound of the working fluid, indicating that molecular complexity plays a major role in determining the promptness of the measurement devices. According to the PRSV model, non-ideal effects can increase the step response time with respect to the acoustic theory predictions. The fundamental derivative of gas-dynamic is confirmed to be the driving parameter for evaluating non-ideal thermodynamic effects related to the dynamic calibration of fast-response aerodynamic pressure probes.

1. Introduction

Time resolved measurements of fluid properties in highly unsteady flows are nowadays fundamental for turbomachinery design and development. Indeed, several phenomena related to the turbine stator-rotor interactions, such as wake transport and rectification, wake-induced boundary layer transition, vortex migrations and boundary layer shock-wave interaction significantly affect the overall stage efficiency and have other consequences, including noise and thermal stresses [1]. Among the most relevant instrumentations currently employed in experiments, fast-response aerodynamic pressure probes stand out for their capability to measure both flow velocity and pressure field with a dynamic response of several tens of kHz, thus allowing to resolve the unsteadiness downstream of turbomachinery rotors (see for example [2, 3, 4, 5, 6, 7]). These devices went through an intensive development during the last decades

[8, 9, 10] and their designs considerably differ from conventional pneumatic probes used for measuring steady cascade performances.

The development of fast-response aerodynamic pressure probes has mostly profited from the installation of micro piezoresistive sensors and the finalization of robust flow reconstruction techniques to resolve the flow field by means of multiple measurements taken at different times, allowing higher frequency responses and lower blockage effects due to installation of single hole pressure probes that reduce the probe heads size [11, 10]. Piezoresistive transducer can be flash-mounted [9] or incapsulated within the sensor to enhance the probe strength [12, 8]. The latter configuration implies the overall dynamics to be influenced by the resulting line-cavity system, connecting the external domain to the piezoresistive transducer. Its geometrical configuration is of major importance to determine the dynamic properties of fast-response aerodynamic pressure probes, which can vary considerably with different internal designs [10]. The probe response also depends on the thermo-physical properties of the fluid and accurate models were proposed by relying on the so-called ideal, dilute gas assumption, which delivers acceptable predictions in most working conditions [10].

Molecularly complex fluids operating in non-ideal conditions are employed in turbomachinery applications in the frame of Organic Rankine Cycle (ORC) [13, 14]. In these conditions, the fluid behaviours can significantly depart from ideal gas model predictions [15, 16], typically in the close proximity of the saturation curve [17, 18]. To extend the applicability of measurement techniques devised for dilute gas flow to non-ideal compressible-fluid flows within ORC turbines, a shortcoming in the available analytical and experimental data calls for a careful investigation of the physics of measurement devices in the limit of non-ideal flows. Such an extension procedure is not straightforward in the non-ideal, compressible-fluid regime, due to e.g. the non-ideal dependence of the speed of sound on the density along isentropic transformation, high compressibility and possibly phase transition. In these conditions, the fluid thermodynamics is to be computed via complex, namely non-ideal, Equations of State (EoS).

The aim of this work is to provide a dynamic characterization of a selected fast-response pressure probe geometry operating in the non-ideal fluid regime. To this end, the open-source software suite SU2 [19], which has been recently extended to deal with Non-Ideal Compressible-Fluid dynamics (NICFD) [20] is used to carry out step response simulations, where a finite pressure perturbation, namely a shock wave, is applied at the inlet of the line-cavity system and the resulting pressure inside the cavity is evaluated to produce a time signal. This configuration reproduces the experimental setup implemented at Politecnico di Milano for the characterization of fast-response pressure probes in dilute air [10, 21]. The selected non-ideal fluid is siloxane MDM (octamethyltrisiloxane), which is currently used in relatively high enthalpy ORC systems [22]. The NICFD solver embedded in SU2 is based on the Peng-Robinson Stryjek-Vera (PRSV) thermodynamic model, which provides a good trade-off between accuracy, computational cost and uncertainties, because it relies on a limited number of parameters [23]. Dynamic characterizations is carried out by using both the dilute ideal gas (PIG) and the PRSV models.

The simulated operating conditions are based on two applications that involve siloxane fluid MDM. The first operating condition corresponds to test conditions to be observed within the Test-Rig for Organic Vapors (TROVA) at Politecnico di Milano [24, 25, 26]. The experiment is aimed at observing for the first time non-ideal nozzle flows relevant to ORC applications. The second test condition is taken from an existing axial organic Rankine stator nozzle, first presented in [16] and further studied recently in [27]. Operating conditions of the two cases are gathered in table 1.

This work is organized as follows: section 2 presents a description of the selected fast-response aerodynamic pressure probe, the dynamic characterization methods and the assessment of the numerical method against available experimental data for air in dilute; section 3 presents the

Table 1. Operating conditions for MDM. The corresponding speed of sound values are computed both using the PIG and the PRSV models. The subscript 0 indicates the operating conditions.

Operating Condition	P [bar]	T [C]	$c_0^{\mathrm{PIG}}[m/s]$	$c_0^{\mathrm{PRSV}}[m/s]$
A (TROVA)	25	310.3	144.5	117.8
B (Stator nozzle [28])	8	270.5	139.5	96.7

thermodynamic model and discusses possibly observable non-ideal effects; section 4 reports the results of the dynamic characterization carried out with MDM along with a comparison between PIG and PRSV models previsions; finally section 5 contains the conclusion.

2. Numerical modelling of fast-response pressure probes

Single-hole pressure probes are applied for 2D or 3D dynamic measurements as virtual three-hole or virtual five-hole probes. Indeed, the yaw angle cannot be resolved instantaneously, though the flow unsteadiness can be resolved through ensemble-averaging methods, which allow the field reconstruction. Reference [10] presents the design and the analysis of fast-response aerodynamic pressure probes at Politecnico di Milano. The sensor denoted as CYL1 has been chosen for this work, given its geometrical simplicity (see figure 1 for details) and the availability of experimental data. The probe is set to measure the total pressure. CYL1 features a relatively simple geometry, which was shown to be properly represented by a 2D model, by virtue of comparisons with experimental data [10]. For this reason, CYL1 was chosen as case study in order to provide a fast tool to study the acoustic behaviour of line-cavity systems when employing non-ideal fluids. Figure 1 provides a 2D representation of the computational domain for CYL 1 simulations. Experimental data regarding the CYL1 step response are available from reference [10]. The probe was equipped with a square shaped sensor membrane of 0.9 mm and covered by a RTV coating. The cavity diameter is 1.4 mm, the length of the line is 0.2 mm, whereas the line diameter is 0.3 mm. A corrected line width of 0.236 mm is used for 2D numerical simulation so that the actual cross sectional area is preserved.

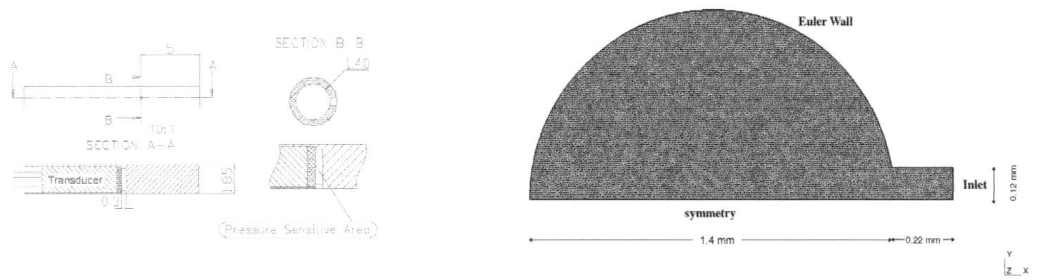

Figure 1. CYL1 geometry from reference [10] and its 2D representation for the CFD simulations.

2.1. *Available methods for the prediction of the step response*

Since in the configuration of interest here the piezoresistive transducer is encapsulated within the probe head, the fluid dynamics of the resulting line-cavity system is of major importance to determine the overall dynamic characteristics. Apart from experimental assessments, there exist several techniques to estimate the fundamental parameters which characterize the step response of an aerodynamic pressure probe, such as analytical methods and numerical simulations of different fidelity levels. The latter is chosen to simulate and analyze the step response when working with non-ideal fluids, although analytical models may also be addressed to interpret CFD results.

Line-cavity systems are usually modelled as Helmholtz resonators, which behave as second-order linear systems. However, available analytical models fail to predict the dynamic characteristics of CYL1, whose response was found to be associated to an over-damped second order dynamic system or a first order dynamic system [10]. The geometrical assumptions upon which analytical models rely are indeed not fulfilled for the geometry considered here, as the line is much shorter than the axial dimension of the cavity. When the cavity filling process is predominant with respect to the oscillating behaviour of the line, a first-order step response can be predicted according to the resistor-capacitor circuit analogy [29], where the line impedance $R = \rho_0 c_0 / S_n$ is the resistance and the capacity is $C = V/(\rho_0 c_0^2)$. Here, V represents the cavity volume and S_n the line cross section. The thermodynamic reference quantities are the speed of sound c_0 and the density ρ_0, which are assumed to remain constant during the process and correspond to the operating condition values. As the flow is initially at rest, ρ_0 and c_0 represent a reference both for total and static quantities and they will be denoted by the subscript 0 throughout the paper. The speed of sound is defined as $c^2 = (\partial P/\partial \rho)_s$, where s is the specific entropy per unit mass. The resulting time constant is given by:

$$\tau = \frac{1}{RC} = \frac{V}{S_n c_0}, \tag{1}$$

The driving parameter for the line-cavity system dynamic characterization is the speed of sound, therefore molecular complexity of the fluid is expected to play a significant role to define the step response. The expression (1) is referred to as acoustic model (AM). The AM provides a fair approximation of CYL1's step response if the pressure perturbation is limited. This means that a high ratio between the external pressure to be measured (p_{ext}) and the operating pressure (p_0) $\beta = p_{\text{ext}}/p_0$ causes significant departures from AM predictions due viscous and thermal effects and possibly to pressure losses ascribable to non-linear wave propagation and turbulence. The working fluid is also expected to influence the validity of the AM, since second-order thermodynamic effects may become relevant compared to the diluite gases case.

Unsteady CFD simulations can be employed to simulate the entire characterization experiment, as they predict the flow field on the overall domain. CYL1's peculiar geometry is exploited to carry out 2D simulations, reducing considerably the computational cost. The open-source CFD software SU2 is considered for this work.

2.2. Numerical simulations for dilute air at room temperature

As a first step towards the simulation of non-ideal compressible flows inside the pressure probe CYL1, the capability of the open-source code SU2 [19, 20] to accurately address this unsteady problem is first assessed. To this end, a simulation is carried out for air, modelled as a PIG. The numerical results are assessed against the available experimental data. The experimental setup used during the dynamic calibration of CYL1 (originally performed in the low-pressure shock-tube of Politecnico di Milano) is reproduced: the operating condition (low pressure section) prescribes a pressure of 1 bar and a temperature of 290 K. The high pressure section of the shock tube is about 2 bar, resulting in a nominal shock wave of 0.5 bar, which corresponds to a total pressure jump of 0.59 bar. However, the measured total pressure perturbation is 0.31 bar, resulting in a total pressure ratio $\beta = 1.31$. This is mainly due to the partial opening of the diaphragm [21]. Given that CYL1 was tested for measuring stagnation conditions, a Riemann boundary condition consisting in the definition of the total pressure and the total temperature is applied at the probe inlet. The total temperature is found via Rankine-Hugoniot relations and thermodynamic model expressions. At the initial time, the boundary condition represents a finite thermodynamic perturbation, namely a shock wave. Three unstructured meshes have been tested (10 000, 25 000 and 50 000 elements) and three time-step intervals are used (0.1 μs, 0.05 μs and 0.025 μs) for both Euler and Navier-Stokes simulations. The latter have are carried out by employing the Spalart-Allmars turbulence model [20]. The dual-time-stepping technique is applied to address the unsteady simulation, where a second-order Roe scheme is applied for the spatial integration of the convective terms, the Averaged Gradient method is used for the computation of the viscous terms and the Backward Euler scheme is adopted for the integration in time.

A fair convergence is obtained both for Euler and Navier-Stokes simulations. The viscous contribution was found to be negligible as in [10], thus only Euler simulations are performed for MDM.

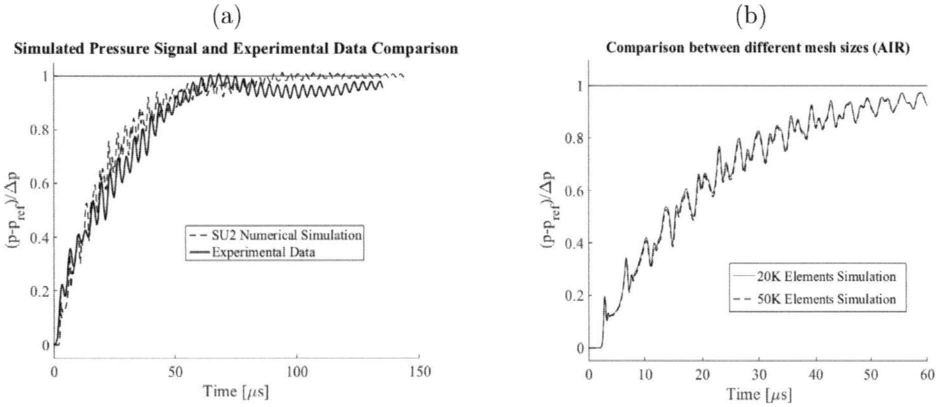

Figure 2. (a) Two dimensional SU2 Euler simulation (performed with the finer mesh of 50000 elements and time-step 0.05 μs) compared to experimental data available from reference [10, 30] (sampling time is 0.5 μs). Non dimensional values of the pressure at the center of the cavity are reported. (b) Comparison between numerical simulations performed with different mesh sizes.

Euler simulations results are compared with experimental data in figure 2a. A good agreement is observed as far as settling time and high frequency behaviour are concerned. Small differences can be observed once the pressure signal reaches the step value. This discrepancy is consistent with the fact that the experimental data is affected by the shock tube fundamental harmonic (about 5-10 kHz) [21, 30]. Different mesh sizes consistently reproduce the experimental data, as shown in figure 2b. A first-order linear identification is carried out for the pressure step response by adopting the least square method, resulting in a time constant value of 21.3 μs, whereas the AM gives 19.1 μs. This difference may be imputed to viscous, thermal and possibly non-linear effects that are not taken into account by the acoustic approximation.

3. Non-ideal effects in fast-response pressure probes

Non-ideal effects in fast-response pressure probes have three major interplaying sources, namely non-linear wave propagation, turbulence and thermodynamic non-linearity. The latter can be modelled by considering complex thermodynamic models accounting for non-ideal fluid effects.

3.1. Thermodynamic models

Thermodynamic models are needed in compressible-fluids CFD simulations to provide relations that connect pressure, temperature and transport parameters to the fluid-dynamic conservative variables (density, momentum density and total energy density). The majority of CFD software employs the perfect gas law (PIG) which is based on the assumptions of dilute gas. The PIG model has been proven to be reliable for a large range of engineering applications, though from textbook thermodynamics it is well known that all fluids exhibit non-ideal behaviour in the close proximity of the saturation curve, the compressibility factor $z = Pv/RT$ being significantly different from unity. More accurate thermodynamic models must be employed in these conditions to properly descrive the fluid properties. Cubic equations of state (EoS) such as the PRSV model [31], allow more accurate predictions of non-ideal phenomena. The behaviour of the fluid is related to the value of the fundamental derivative of gas-dynamic [32], defined as

$$\Gamma = 1 + \frac{\rho}{c}\left(\frac{\partial c}{\partial \rho}\right)_s . \tag{2}$$

The fundamental derivative is usually greater than unity ($\Gamma = (\gamma+1)/2$ for perfect gases, where γ is the ratio of the specific heats), though regions where $0 < \Gamma < 1$ or even $\Gamma < 0$ are near the saturation curve, giving origin to non-ideal and non-classical phenomena, such as rarefaction shock waves [17]. The value of Γ is expected to play a major role on the dynamic response of the pressure probe, since it is intimately related to the rate of variation of the speed of sound due to pressure perturbations. Hence, the PIG model can possibly deliver unaccurate results concerning the response of the sensor for certain operating conditions, while non-ideal model may provide more reliable final results, at the cost of increased computational effort.

The PRSV model has been recently introduced within the SU2 environment [20]. The model is defined by the following relations:

$$P(T,v) = \frac{RT}{v-b} - \frac{a\alpha^2(T)}{v^2+2bv-b^2} \quad \text{and} \quad e(T,v) = \phi(T) - \frac{a(k+1)}{b\sqrt{2}}\alpha(T)\tanh^{-1}\frac{b\sqrt{2}}{v+b} \tag{3}$$

where pressure P and internal energy per unit mass e are expressed as functions of the specific volume v and the temperature T. R is the gas constant and $\phi(T) = \lim_{v\to\infty} e(T,v)$ is the internal energy per unit mass in dilute conditions. The coefficient a represents the action of inter-molecular forces and b is the co-volume. A complete definition of all parameters can be found in [31]. They depend on several quantities, such as critical pressure ($p_c = 14.15$ bar), critical temperature ($T_c = 564.1$ K), acentric factor ($\omega = 0.529$), gas constant ($R = 35.2$ J/kg/K) and reference specific heat ratio ($\gamma_\infty = 1.018$).

3.2. Non-ideal effects on the dynamic characterization of the pressure probe

The AM (1), discussed in section 2.1, provides estimates of the time constant describing CYL1's step response, which will be later shown to agree well with the CFD simulation results. However, non-linear effects might cause a large departure from the estimation of the model, due to pressure losses associated to non-linear wave propagation within the feeding line and to thermodynamic properties variations within the cavity. Density and speed of sound variations are expected to play the most relevant role for distinguishing the pressure probe step response when working with air and MDM. Thermodynamic derivatives

$$\left(\frac{\partial c}{\partial P}\right)_s = \frac{\Gamma - 1}{\rho c} \quad \text{and} \quad \left(\frac{\partial \rho}{\partial P}\right)_s = \frac{1}{c^2}$$

suggest that peculiar values for the initial reference state, the molecular complexity and the fundamental derivative of gas-dynamic may possibly lead to departures from the linear AM (1). By simple differentiation it can be inferred that density and speed of sound variations act in opposite directions, since an increase of density is expected to delay the response, whereas an increase of the speed of sound is expected to accelerate it. In the following section, numerical simulation results will be addressed with the aid of these considerations and differences between thermodynamic models will be highlighted.

4. Dynamic characterization results

Simulations have been performed with unstructured meshes of 10 000, 20 000 and 50 000 elements to ensure convergence in space, while time-step intervals of 0.1 μs, 0.25 μs and 0.5 μs were tested to ensure convergence in time.

4.1. MDM and air response comparison under the PIG model

(*a*)

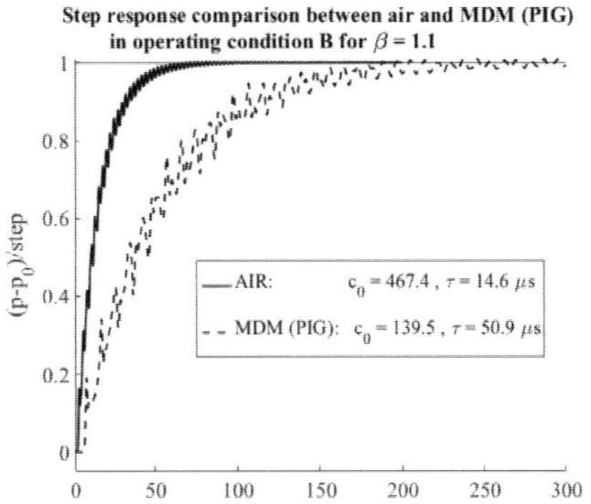

(*b*)

	Air	MDM (PIG)
c_0 [m/s]	467.4	139.5
$\frac{\Delta c}{c_0}\%$	1.35	0.08
$\frac{\Delta \rho}{\rho_0}\%$	7.10	9.82
τ_{SU2} [μs]	14.6	50.9
τ_{AM} [μs]	14.0	46.8

Figure 3. Comparison between the step response of air and MDM (PIG model) in operating condition B for a pressure perturbation $\beta = 1.1$. The percentage variations of speed of sound and density are computed by considering the values of the perturbed state and the reference values (e.g. $\Delta c = c_{\text{ext}} - c_0$).

Figure 3 compares the simulated step response of CYL1 when employing air and MDM, both modelled as ideal gases. The operating condition is state B (reported in table 1) for both fluids and a perturbation $\beta = 1.1$ is applied. Both responses are typical of a first-order system, which indicates that molecular complexity does not affect the qualitative characteristics of the flow field inside the pressure probe. However, the time needed to reach the step value is very different. Indeed, first-order system identification gives $\tau = 14.6$ μs for air and $\tau = 50.9$ μs for MDM, while the AM gives $\tau = 14.0$ μs and $\tau = 46.8$ μs. Molecular complexity implies lower speed of sounds for MDM, thus its step response is much slower, as expected. The relative difference between the actual time constant and the AM prediction is higher for MDM than for air. This might be explained by considering again the higher molecular complexity of MDM with respect to air: speed of sound variation of MDM is indeed extremely limited ($\Gamma \to 1$) and its influence in compensating the density increase is negligible with respect to air.

Op. Cond.	β	c_0 [m/s]	$\frac{\Delta \rho}{\rho_0}\%$	$\frac{\Delta c}{c_0}\%$	τ_{SU2} [μs]	τ_{AM} [μs]	Error %
A	1.23	144.5	22.94	0.19	53.5	45.2	15.6
B	1.23	139.5	22.92	0.17	55.5	46.8	15.7
B	1.1	139.5	9.82	0.08	50.9	46.8	8.02

Table 2. Comparison between the step response results of MDM modelled as a PIG.

Table 2 reports selected results concerning three different simulations performed with MDM. It can be inferred that the operating conditions poorly influence the step response when working with high molecular complexity fluid modeled as ideal gases. This is a direct consequence of the fact that $\Gamma \to 1$, thus the speed of sound is similar even for different operating conditions. The time constants provided by the simulations in operating conditions A and B with the same perturbation $\beta = 1.23$ are indeed very similar. Moreover, the error with respect to the AM is identical, again stressing the concept. As expected, by considering the same operating condition (B) and two different pressure perturbations ($\beta = 1.1$ and $\beta = 1.23$), the relative error with respect to the AM is higher for the latter, as non-linear effects are greater.

4.2. PIG model vs PRSV model simulations

(*a*)

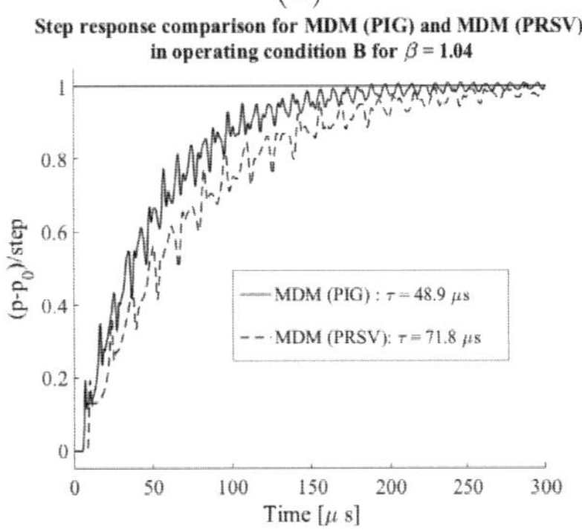

(*b*)

	Air	MDM$^{(PIG)}$	MDM$^{(PRSV)}$
c_0 [m/s]	467.4	139.5	96.7
$\frac{\Delta c}{c_0}\%$	0.54	0.04	-2.15
$\frac{\Delta \rho}{\rho_0}\%$	2.78	3.81	5.68
τ_{SU2} [μs]	14.5	48.9	71.8
τ_{AM} [μs]	14.0	46.8	67.5
$\frac{(\tau_{SU2}-\tau_{AM})}{\tau_{AM}}\%$	3.7	4.2	5.9

Figure 4. Comparison between MDM (described as an PIG) and MDM (described as a PRSV gas) in operating condition B for $\beta = 1.04$. Results for air in the same conditions are also shown.

Figure 4a reports the CYL1's step response of MDM modelled by means of PIG and PRSV models in operating condition B for $\beta = 1.04$. Table 4b reports the thermodynamic properties variations and the time constants. It is remarkable that the AM provides fair predictions, since the acoustic approximation becomes more and more accurate as the pressure perturbation decreases. A slightly larger error is observable for the case of MDM modelled as a PRSV gas, which might be due to the more pronounced density variation. It should be noted that PRSV model gives a completely different value of c_0 with respect to the PIG model, suggesting that in these operating conditions non-ideal effects are non negligible.

When β increases, the differences between the PIG and PRSV models simulations become greater, as figure 5a demonstrates. Step responses are reported for operating condition B, $\beta = 1.23$ and $\beta = 1.2$. The derived time constants are $\tau = 55.5$ μs and $\tau = 84.4$ μs, while the errors with respect to AM predictions are 15.5% and 20%, respectively. This suggests that if the PIG model is employed, the performances of the pressure probe in terms of frequency response are overestimated for two main reasons, namely the overestimation of the speed of sound value in the actual operating condition and the underestimation of the speed of sound variation due to the pressure perturbation. The fundamental derivative of gas-dynamic computed by means of PRSV model relations is indeed very different from unity in operating condition B ($\Gamma = 0.63$) and its value preludes to a reduction of the speed of sound, which causes the step response to be slower. This effect could become dramatic as β increases. For instance, a perturbation $\beta = 1.37$ in operating condition B cause a difference with respect to the AM prediction as high as 33%. The operating conditions are indeed very important for the dynamic characterization. The last example is figure 5b, which reports the step response of MDM described as a PRSV gas in operating condition A and B for $\beta = 1.2$. The differences between the two responses are quite large and the differences with respect to the AM predictions are at odds, since in

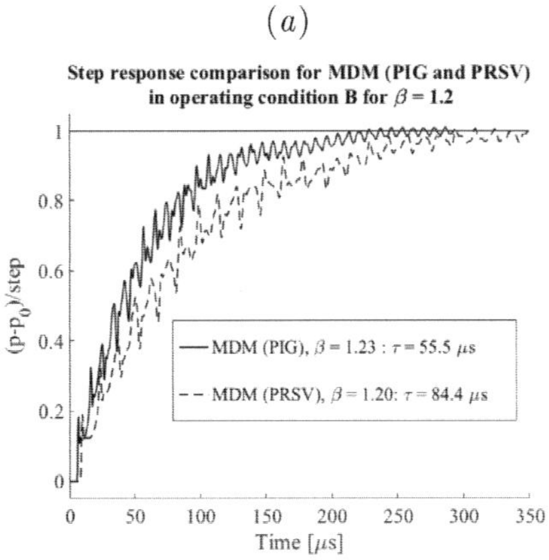

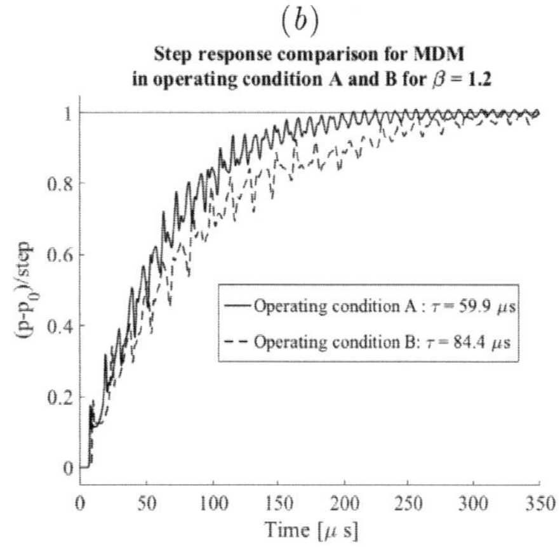

Model	Op. Cond.	c_0 [m/s]	$\frac{\Delta\rho}{\rho_0}\%$	$\frac{\Delta c}{c_0}\%$	τ_{SU2} [μs]	τ_{AM} [μs]	Error %
PIG	B	139.5	22.92	0.17	55.5	46.8	15.7
PRSV	B	96.7	33.53	-12.22	84.4	67.5	20.0
PRSV	A	117.8	8.96	17.96	59.9	55.45	7.4

Figure 5. (a) Comparison of the step responses between MDM (described as PIG) and MDM (described as PRSV gas) in operating condition B for $\beta = 1.23$ and $\beta = 1.2$, respectively. (b) Comparison between MDM (described as PRSV gas) in operating condition A and B for $\beta = 1.2$.

operating condition A the line impedance augmentation (caused by the density increase) is almost compensated by the speed of sound increase within the cavity.

5. Conclusion

The dynamic calibration of a selected fast-response aerodynamic pressure probe working with flows in non-ideal regime has been addressed. The open source CFD software suite SU2 has been used to simulate the flow flied inside the probe, which can be configured as a line-cavity system. Thanks to the availability of experimental results from reference [10], it has been possible to assess the SU2 simulations through comparison. The NICFD solver of SU2 has been exploited to simulate dynamic characterization experiments with siloxane MDM. Both the constant-specific-heat ideal gas (PIG) and the Peng-Robinson-Stryjek-Vera (PRSV) models have been considered to describe the fluid and results have been compared. The step response of the probe is typical of a first-order system, thus the system time constant has been considered as a representative variable to compare the results.

Both models deliver higher time constants for MDM with respect to air for the same pressure perturbation and operating condition. This is possibly due the relatively low value of the speed of sound, because of the molecular complexity of MDM is higher. Molecular complexity plays also a role in determining the departure of the actual time constant from the AM predictions: for a fixed β, non-linear effects play a stronger role as molecular complexity increases.

Two main observations can be made by comparing the two considered thermodynamics models. These are the different operating speed of sound that the models predict and the different rates of variation of the thermodynamic properties during the filling process. Even if the former was somehow corrected by means of an *ad hoc* choice of the specific heat ratio (something that is not always possible, because $\gamma > 1$ from thermodynamic stability), the latter cannot be addressed if the PIG model is retained. PIG model results are particularly questionable in regions where Γ goes below unity. Non-ideal regime ($\Gamma < 1$) and low-speed-of-sound regions (such as operating condition B) appear to be the most critical configuration for the application of fast-response pressure probes, as the combined effects of density and speed of sound variations cause a huge increase of the system time constant.

Acknowledgments
This research is supported by ERC Consolidator Grant N. 617603, Project NSHOCK, funded under the FP7-IDEAS-ERC scheme.

References

[1] Sharma O P, Pickett G F and Ni R H 1992 *ASME J. Turbomachinery* **114** 79–90
[2] Miller R J, Moss R W, Ainsworth R W and Horwood C K 2003 *ASME J. Turbomachinery* **125** 1–13
[3] Schlienger A, Kalfas A I and Abhari R S 2005 *ASME J. Turbomachinery* **127**
[4] Gaetani P, Persico G, Dossena V and Osnaghi C 2007 *ASME J. Turbomachinery* **129** 580–590
[5] Porreca L, Kalfas A I and Abhari R S 2009 *J. Propulsion and Power* **25**
[6] Persico G, Gaetani P and Osnaghi C 2009 *ASME J. Turbomachinery* **131**
[7] Schennach O, Woisetschlager J, Paradiso B, Persico G and Gaetani P 2010 *ASME J Turbomachinery* **132**
[8] Kupferschmied P, Köppel P, Gizzi W, Roduner C and Gyarmathy G 2000 *Meas. Sci. Technol.* **11** 1036
[9] Ainsworth R W, Miller R J, Moss R W and Thorpe S J 2000 *Measurement Science and Technology* **11** 1055–1076
[10] Persico G, Gaetani P and Guardone A 2005 *Meas. Sci. Technol.* **16** 1741–1750
[11] Brouckaert J F 2007 *J. of Power and Energy* **126**
[12] Brouckaert J F 2000 *XV Bi-Annual Symposium on Measuring Techniques in Transonic and Supersonic Flows in Cascades and Turbomachines* pp 21–22
[13] Colonna P, Casati E I M, Trapp C, Mathijssen T, Larjola J, Turun-Saaresti T and UUsitalo A 2015 *J. Eng. Gas Turbines Power* **137** 100801–1–19
[14] Macchi E and Astolfi M 2016 *Organic Rankine Cycle (ORC) Power Systems* (Woodhead Publishing)
[15] Congedo P, Corre C and Cinnella P 2011 *Comput. & Fluids* **49** 290–301 ISSN 00457930
[16] Colonna P, Harinck J, Rebay S and Guardone A 2008 *J. Propul. Power* **24** 282–294
[17] Nannan N R, Guardone A and Colonna P 2012 1–7
[18] Wheeler A P S and Ong J 2013 *J. Eng. Gas Turbines Power* **135** 102603–1–9
[19] Palacios F, Colonno M F, Aranake A C, Campos A, Copeland S R, Economon T D, Lonkar A K Lukaczyk T W, Taylor T W R and Alonso J J 2013 *51st AIAA Aerospace Sciences Meeting and Exhibit*
[20] Vitale S, Gori G, Pini M, Guardone A, Economon T D, Palacios F, Alonso J J and Colonna P 2015 *22nd AIAA Computational Fluid Dynamics Conference* AIAA Paper 2760
[21] Persico G, Gaetani P and Guardone A 2005 *Meas. Sci. Technol.* **16** 1751–1759
[22] Angelino G and Colonna P 1998 *Energy* **23** 449–463
[23] Guardone A, Vigevano L and Argrow B M 2004 *Phys. Fluids* **16** 3878–3887
[24] Spinelli A, Pini M, Dossena V, Gaetani P and Casella F 2013 *ASME J. Eng. Gas Turb. Power* **135** 042303
[25] Guardone A, Spinelli A and Dossena V 2013 *ASME J. Eng. Gas Turb. Power* **135** 042307
[26] Spinelli A, Cozzi F, Zocca M, Gaetani P, Dossena V and Guardone A 2016 *Proceedings of the ASME 2016 Turbo Expo, Soul* GT2016-57357
[27] Pini M, Persico G, Dossena V and Gaetani P 2015 *J. Eng. Gas Turbines and Power* **137**
[28] Colonna P, Rebay S, Harinck J and Guardone A 2006 *ECCOMAS CFD 2006* (Egmond aan Zee, NL)
[29] Trusler J P M 1991 *Physical Acoustics and Metrology of Fluids* The Adam Hilger Series on Measurement Science and Technology (Bristol, England: Adam Hilger (IOP Publishing Ltd.))
[30] Gaetani P, Persico G and Guardone A 2008 *J. Fluid Mech.* **602**
[31] Stryjek R and Vera J H 1986 *Can. J. Chem. Eng.* **64** 323–333
[32] Thompson P A 1971 *Phys. Fluids* **14** 1843–1849 ISSN 00319171

Computation of tail probabilities for non-classical gasdynamic phenomena

N. Razaaly, P.M. Congedo

INRIA Bordeaux Sud-Ouest, 200 Rue de la Vieille Tour, 33405 Talence, France

E-mail: nassim.razaaly@inria.fr,pietro.congedo@inria.fr

Abstract. This paper presents a novel method for computing the tail probability of a given quantity of interest by using only a low number of evaluations of the computer model, representing the problem of interest under uncertainties. This method is then applied to the study of rarefaction shock waves (RSW) in a dense-gas shock tube. It is well-known in literature that the prediction of a RSW is highly sensitive to uncertainties on the initial flow conditions. The objective of this work is to compute a very accurate estimation of the RSW probability of occurrence in a shock tube configuration.

1. Introduction

Several theoretical and numerical studies have shown that dense-vapor transonic flows of substances formed by complex organic molecules could feature phenomena such as rarefaction shock waves and compression fans. Fluids that might exhibit nonclassical gasdynamic phenomena are called BZT fluids from the name of the three scientists who first theorized their existence. These anomalies occur when the fundamental derivative of gasdynamics $\Gamma = 1 + \frac{\rho}{a}\left(\frac{\partial a}{\partial \rho}\right)_s$ with ρ the fluid density, a the sound speed and s the entropy, becomes negative between the upper saturation curve and the $\Gamma = 0$ contour. Such a region is often referred to as the *inversion zone* and and the $\Gamma = 0$ contour is called the *transition line.*

The experimental proof of nonclassical gasdynamic effects in flows of dense vapors has been the subject of several studies [2, 3, 1], most of all focused on a shock tube configuration for generating a rarefaction shock wave (RSW). These works show that nonclassical gasdynamic effects are generally very weak with respect to compression shock waves, and can occur only in a limited range of conditions. As shown for instance in [4] and [5], the accuracy of the thermodynamic model has a strong influence on the simulation of nonclassical phenomena, to the point that their presence can depend on the accuracy with which fluid model parameters are determined.

In [1], an algorithm has been presented to handle stochastic inverse problem with high efficiency and a reduced computational cost. This algorithm has then been applied to the design of a challenging scientific experiment involving the compressible flow of a dense gas in the FAST shock tube facility at TU Delft. It was shown that the unconventional rarefaction shock wave is very sensitive to uncertainties in the initial experimental conditions.

This paper is focused to cure some of the issues emerged in [1]. The difficulties to accurately compute the tail probability and the associated computational cost are tackled with a novel uncertainty quantification techniques, based on an interpolation technique and an importance sampling algorithm. The final algorithm is then applied to the simulation of the TU Delft experiments.

The paper is organized as follows. Section 2 is devoted to the description of the various numerical tools needed in the study: the CFD solver for dense gas flows simulations, the thermodynamic models, and the uncertainty quantification approach. In section 2.4, the stochastic method is presented. Section 3 illustrates some results obtained on the computation of the tail probability of the RSW. The closing section summarizes the conclusions that can be drawn from the present work.

2. Methodology and tools

2.1. CFD solvers for dense gas flows

The NZDG code [1] solves the quasi-1D Euler equations with second-order accuracy in time and space. The convective fluxes are discretized using the Roe numerical flux and a second-order limited MUSCL variable reconstruction. The Roe average for dense gas flows is computed with the simplified approach proposed in [6] and the slope limiter introduced in the linear reconstruction is of the Van Albada type. Second-order accuracy in time and robust time-integration are achieved using a three-level implicit formula to approximate the physical time-derivative, within a dual-time sub-iterative approach to solve the resulting non-linear system.

2.2. Thermodynamic models

Siloxane D6 is the fluid currently chosen for the TU Delft experiment. The $Peng - Robinson - Strijek - Vera$ (PRSV) cubic equation of state (EoS) is considered to describe its thermodynamic behavior. The robustness of this equation with respect to more complex and potentially more accurate multi-parameter equations of state of the Span-Wagner type ([5],[7],[4] has been discussed in [8] and [9]. Peng and Robinson (1976) proposed a cubic EoS of the form:

$$p = \frac{RT}{v-b} - \frac{a}{v^2 + 2bv - b^2}. \tag{1}$$

where p and v denote respectively the fluid pressure and its specific volume, a and b are substance-specific parameters related to the fluid critical-point properties p_c and T_c. To achieve high accuracy for saturation-pressure estimates of pure fluids, the temperature-dependent parameter a in Eq.(2.1) is expressed as :

$$a = \left(0.457235 R^2 T_c^2 / p_c^2\right) \cdot \alpha\left(T\right) \tag{2}$$

while

$$b = 0.077796 RT_c / p_c \tag{3}$$

These properties are not completely independent, since the EoS should satisfy the conditions of zero curvature and zero slope at the critical point. Such conditions allow computing the critical compressibility factor $Z_c = (p_c v_c)/(RT_c)$ as the solution of a cubic equation. The corrective factor α is given by:

$$\alpha\left(T_r\right) = \left[1 + K\left(1 - T_r^{0.5}\right)\right]^2 \tag{4}$$

with

$$K = K_0 + K_1\left(1 + T_r^{0.5}\right)(0.7 - T_r)\ for\, T_r < 1\ \ (K_1 = 0, for\, T_r \geq 1) \tag{5}$$

and

$$K_0 = 0.378893 + 1.4897153\omega - 0.1713848\omega^2 + 0.0196554\omega^3. \tag{6}$$

The parameter ω is the fluid acentric factor and K_1 is obtained from saturation-pressure fitting of experimental data. The caloric behavior of the fluid is approximated through a power law for the isochoric specific heat in the ideal gas limit:

$$c_{v,\infty}(T) = c_{v,\infty}(T_c)\left(\frac{T}{T_c}\right)^n \tag{7}$$

with n a material-dependent parameter.

2.3. *Sources of Uncertainty : initial conditions of the experiment*

The initial conditions (IC) for the experiment (and for the simulation) are prescribed according to the theory described in [10], to maximize the Mach number of the rarefaction shock wave and facilitate the wave detection. The maximum achievable precision in controlling the temperature and pressure values in the charge tube has been estimated from the measurement instruments and hardware specifications to be 0.4% for the pressure and 0.1% for the temperature.

In [1], an automatic procedure has been set up to detect an initial left state satisfying the aforementioned requirements : in the first step, a discretization of the p-v plane has been generated by moving along isobaric and isentropic curves; in the second step, the uncertainty region is computed using the given uncertainty levels on initial pressure and temperature and a Monte-Carlo approach; in the third step, it is verified the uncertainty region does not cross the saturation curve (to avoid liquid-vapor mixture) and that the chosen point is located as close as possible to the saturation curve. This analysis yields an initial left state, denoted P1, represented in figure 1. The uncertainty region for P1 has a single point in common, without crossing, with the maximal allowed saturation curve computed with the TD uncertainties taken into account.

2.4. *Stochastic method*

In the following, we describe a new stochastic method specifically designed to accurately compute very low probability subject to multiple failure regions. The idea is to construct a metamodel of the quantity of interest specially refined on the failure branches, and then resort to a sampling-based method on the metamodel in order estimate the probability of failure.

2.4.1. Introduction Given a probabilistic model, described by its physical n-dimensional random vector $\mathbf{Y}$ with its probability density function (PDF) $f_{\mathbf{Y}}$ and a performance function J representing the system response, failure is usually defined as the event $\mathcal{G} = \{J(\mathbf{y}) \leq 0\}$ so that the failure probability is defined as follows:

$$p_f = \mathbb{P}([J(\mathbf{Y}) \leq 0]) = \mathbb{E}_{\mathbf{Y}}[\mathbb{1}_{\mathcal{G}}(\mathbf{Y})] = \int_{\mathbb{R}^n} \mathbb{1}_{\mathcal{G}}(\mathbf{y}) f_{\mathbf{Y}}(\mathbf{y}) d\mathbf{y} \tag{8}$$

where $\mathbb{1}_{\mathcal{G}}$ is the failure indicator function, being equal to one if $J(\mathbf{y}) \leq 0$, and zero otherwise. J is called the performance function in this paper. An isoprobabilistic transformation T (e.g. Rosenblatt transform) is used to define the standard random variables X and the performance function G in the standard space as $G(X) = J(T^{-1}(X))$ with $X = T(Y)$.

We recall that $X \sim \mathcal{N}(0, I_n)$ is the standard normal random vector of $\mathbb{R}^n$. The failure domain $\mathcal{F}$ in the standard space is then defined as $\mathcal{F} = \{\mathbf{x} \in \mathbb{R}^n \;\; s.t. \;\; G(x) < 0\}$. The failure probability reads

$$p_f = \mathbb{P}([G(\mathbf{X}) \leq 0]) = \mathbb{E}[\mathbb{1}_{\mathcal{F}}(\mathbf{X})] = \int_{\mathbb{R}^n} \mathbb{1}_{\mathcal{F}}(\mathbf{x}) f_{\mathbf{X}}(\mathbf{x}) d\mathbf{x} \tag{9}$$

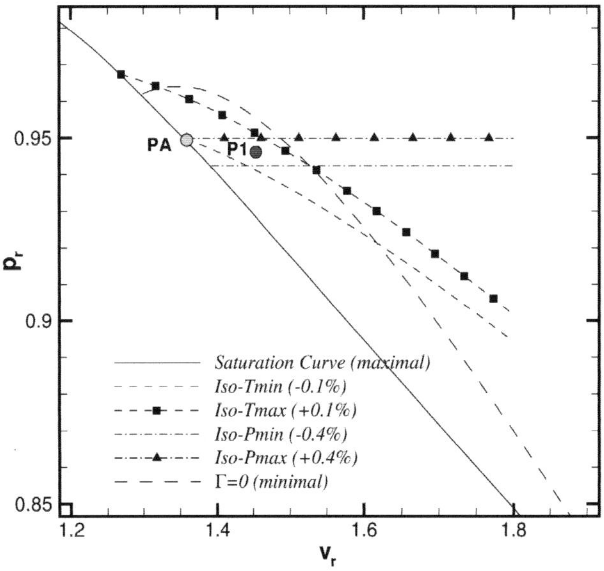

Figure 1: Robust point (P1) obtained by means of the automatic procedure, and uncertainty region.

Computing failure analysis in the standard space simplifies the analysis since, the random variables are decorrelated and normalized.

A typical approach to the estimation of the failure probability (Eq. 9) is that of resorting to a crude Monte Carlo (MC) scheme. It yields to an unbiased and convergent estimation of p_f, with a low convergence rate. It usuaally requires a prohibitive number of simulations.

Various methods have been proposed in the literature to adress this problem.

A first family of method aims to reduce the variance estimator in order to increase the convergence rate. Importance sampling techniques [11] [12] are part of this family, and have been developed over the past few decades to shift the underlying distribution towards the failure region so as to gain information from rare events more efficiently. The success of the method relies on a prudent choice of the importance sampling density, which undoubtedly requires knowledge of the system in the failure region.

The second class of methods relies on the substitution of the original performance function by a surrogate model within a sampling-based scheme; a metamodel is in general orders of magnitude faster to be evaluated. In such metamodel-based approaches, Monte-Carlo sampling (AK-MCS [17]) or Importance Sampling (IS) techniques (AK-IS [13], MetaAK-IS^2 [14], KAIS [15], AK-SS [16]) are then used directly on the surrogate. Here, we propose a new method for the metamodel construction. It makes the use of the learning function used in AK-MCS [17], k-Means clustering algorithm [18], and a Metropolis-Hastings MCMC sampling method [19]. It is designed to fit with multiple failure regions, and very low probability. It provides few advantages compared to other metamodels refinement:

- A parameter is used to avoid points clustering and also give the opportunity for the user to indirectly control the refinement cost.

- The different failure-branches are refined back and forth during the process. So if the number of DOE required during the metamodel building is too high (impossibility to build a metamodel for instance), the resulting metamodel will be "equally" refined in all branches.
- It focuses on the limit state performance function.

This section is organized as follows. The concept of probabilistic classification and of importance sampling are presented in Sections 2.4.2 and 2.4.3, respectively. The metamodel refinement strategy is detailed in subsection 2.4.4. An academic test-case for validation is discussed in subsection 2.4.5.

2.4.2. Probabilistic classification using Gaussian Processes A metamodel is built from a *design of experiments* (D.o.E.), a set of computed experiments denoted by $\mathcal{X} = \{\mathbf{x}_1, ..., \mathbf{x}_m\}$, belonging to the support $\mathcal{D}_{\mathbf{X}}$ of $\mathbf{X}$. The performance function G is assumed to be a sample path of an underlying Gaussian Process (GP) denoted by $\tilde{G}$. The *best linear unbiased estimation* (BLUE, see [20]) of $\tilde{G}$ at point $\mathbf{x}$ is shown to be a Gaussian random variate $\hat{G}(\mathbf{x}) \sim \mathcal{N}(\mu_{\hat{G}(\mathbf{x})}, \sigma_{\hat{G}(\mathbf{x})})$, where $\mu_{\hat{G}(\mathbf{x})}, \sigma_{\hat{G}(\mathbf{x})}$ are given by the Gaussian Process algorithm. Further details can be found in [21].
GP provides both a surrogate for the limit-state function $g(\mathbf{x})$ which is denoted by $\mu_{\hat{G}(\mathbf{x})}$ and an epistemic prediction uncertainty which is characterized by its variance $\sigma^2_{\hat{G}(\mathbf{x})}$.

We introduce the *probabilistic classification function*:

$$\pi(\mathbf{x}) = \mathcal{P}[\hat{G}(\mathbf{x}) \leq 0] = \phi\left(\frac{0 - \mu_{\hat{G}(\mathbf{x})}}{\sigma_{\hat{G}(\mathbf{x})}}\right), \text{ if } \mathbf{x} \notin \mathcal{X} \quad (10)$$

where the probability measure $\mathcal{P}[\cdot]$ refers to the Gaussian nature of the GP predictor $\hat{G}(\mathbf{x})$ and ϕ denotes the *cumulative density function* of the one-dimensional standard normal law. It may be interpreted as the probability that the predictor $\hat{G}(\mathbf{x})$ is negative with respect to the epistemic uncertainty. We will later use π as a surrogate for $\mathbb{1}_{\mathcal{F}}$.
We also introduce the meta-failure domain $\tilde{\mathcal{F}}$ in the standard space, defined as $\tilde{\mathcal{F}} = \left\{\mathbf{x} \in \mathbb{R}^n \;\; s.t. \;\; \mu_{\hat{G}(x)} < 0\right\}$.

2.4.3. Importance Sampling and Quasi-Optimal Density Let h be a PDF dominating $\mathbb{1}_{g\leq 0} f_{\mathbf{X}}$. The failure probability may be rewritten as follows:

$$p_f = \int_{\mathbb{R}^n} \mathbb{1}_{g\leq 0}(\mathbf{x}) \frac{f_{\mathbf{X}}(\mathbf{x})}{h(\mathbf{x})} h(\mathbf{x}) d\mathbf{x} = \mathbb{E}_h[\mathbb{1}_{g\leq 0}(\mathbf{X}) \frac{f_{\mathbf{X}}(\mathbf{X})}{h(\mathbf{X})}] \quad (11)$$

It easily leads to the *importance sampling estimator*:

$$\hat{p}_{f\ IS} = \frac{1}{N} \sum_{k=1}^{N} \mathbb{1}_{g\leq 0}(\mathbf{x}^{(k)}) \frac{f_{\mathbf{X}}(\mathbf{x}^{(k)})}{h(\mathbf{x}^{(k)})} \quad (12)$$

where $\{\mathbf{x}^{(1)}, ..., \mathbf{x}^{(N)}\}$ is a set of samples drawn from the instrumental density h. According to the central limit theorem, this estimation is unbiased and its quality may be measured by means of its variance of estimation which reads

$$Var[\hat{p}_{f\ IS}] = \frac{1}{N-1}\left(\frac{1}{N} \sum_{k=1}^{N} \mathbb{1}_{g\leq 0}(\mathbf{x}^{(k)}) \frac{f_{\mathbf{X}}(\mathbf{x}^{(k)})^2}{h(\mathbf{x}^{(k)})^2} - \hat{p}_{f\ IS}^2\right) \quad (13)$$

The corresponding coefficient of variation is defined as $\hat{\delta}_{p_f} = \frac{\sqrt{Var[\hat{p}_{f\ IS}]}}{\hat{p}_{f\ IS}}$.

This variance is zero (optimality of the IS estimator) when the instrumental PDF is chosen as $h^*(\mathbf{x}) = \frac{\mathbb{1}_{g\leq 0}(\mathbf{x}) f_{\mathbf{X}}(\mathbf{x})}{p_f}$.

However, this PDF involves p_f in its denominator, so it is not implementable in practice. Here, it is proposed to use the probabilistic classification function (Eq. 10) as a surrogate for the real indicator function $\mathbb{1}_{g\leq 0}(\mathbf{x})$. The proposed quasi-optimal PDF reads $\hat{h}^*(\mathbf{x}) = \frac{\pi(\mathbf{x}) f_{\mathbf{X}}(\mathbf{x})}{\mathbb{E}_{\mathbf{X}}[\pi(\mathbf{X})]}$, where $\mathbb{E}_{\mathbf{X}}[\pi(\mathbf{X})]$ can be evaluated with crude Monte-Carlo method.
Markov chain Monte Carlo (*Metropolis-Hastings sampler*) simulation [19] is used in the next subsubsection to sample points drawn from from the quasi-optimal instrumental PDF $\hat{h}^*$, that are more likely to be close to limit-state.

2.4.4. Metamodel Refinement Strategy The efficiency of the approach mostly relies on the optimality of the instrumental PDF $\hat{h}^*$. Thus it is proposed here to adaptively refine the probabilistic classification function so that the quasi-optimal instrumental PDF $\hat{h}^*$ converges towards its optimal counterpart h^*.

A learning function U [14] associated to the GP metamodel $\hat{G}$ is introduced. For a given $\mathbf{x}$, $U(\mathbf{x})$ is defined as

$$U(\mathbf{x}) = \frac{|\mu_{\hat{G}(\mathbf{x})}|}{\sigma_{\hat{G}(\mathbf{x})}} \tag{14}$$

$\phi(U(\mathbf{x}))$ is the probability that $\mathbf{x}$ is correctly classified by the predictor. So if N points $\mathbf{x}(i)$ are available,

$$\mathbf{x}_0 = \arg\min_i(U(\mathbf{x}(i))) \tag{15}$$

is the point in correspondence of which the classification is the most uncertain. In order to avoid DOE clustering, a constraint is added, stating that $\mathbf{x}_0$ should be distant of at least d_{min} from all existing DOE. The algorithm proceeds as follows:

(i) *Initial DOE and metamodel definition*: sample m_0 points generated by a Latin Hypercube sampling.
(ii) *Sampling*: Sample a Monte-Carlo population $\mathcal{N} = \{x^{(1)}, ..., x^{(N)}\}$, adapting N so a coefficient target of variation is reached.
(iii) *Global Point Selection*: Refine and update the metamodel selecting the best point among $\mathcal{N}$ according to the learning function U.
(iv) *Classification*: Classify $\mathcal{N}$ between failure and safe points.
(v) *Seed Selection*: Use k-Means [18] clustering algorithm on the set of failure points. Its K clusters centroids are considered: $(x_1^-, ..., x_K^-)$. Set $k = 1$.
(vi) *Unimodal MCMC Sampling*: Sample a set of N_{MCMC} points $\mathcal{S}_k^{MCMC}$ drawn from the PDF $h^*_{\hat{G}}$ using the modified MCMC Metropolis-Hastings sampler, with the seed initialized at x_k^-. Burning and thining procedure are used. Remove the first N_{MCMC}^{min} values obtained, and then, accept only one sample every s_{MCMC} samples.
(vii) *Enrich D.o.E. at step k*: Select the best point $\{\tilde{x}_k^-\}$ (if exists) among $\mathcal{S}_k^{MCMC}$ according to the learning function U. Update the metamodel.

(viii) *Stopping criterion* If $k \neq K$: set $k = k + 1$ and loop back to step vi.
If $k = K$ and $\bigcup_{k=1}^{K} \{\tilde{x}_k^-\} = \varnothing$, stop the metamodel refinement algorithm.
Otherwise, set $i = i + 1$ and loop back to step ii.

The metamodel refining procedure stops when no more points can be added. In order to evaluate the probability of failure once the metamodel is refined, it is possible to resort to Monte-Carlo method if p_f is low enough. Otherwise, it is proposed to use the Importance Sampling method, with a gaussian mixture density as the Importance Sampling Density. MCMC points are sampled from different seeds in parallel, and then classified using k-Means algorithm. The gaussian mixture parameters: weights, means, covariance are then empirically computed from those classified MCMC points.

2.4.5. Validation: 2D analytic example with four failure regions We test the method in case of four failure regions. The performance function [14] [22] in the standard space reads:

$$G(x_1, x_2) = \min \left\{ \begin{array}{c} 3 + \frac{(x_1-x_2)^2}{10} - \frac{x_1+x_2}{\sqrt{2}} \\ 3 + \frac{(x_1-x_2)^2}{10} + \frac{x_1+x_2}{\sqrt{2}} \\ x_1 - x_2 + \frac{7}{\sqrt{2}} \\ -(x_1 - x_2) + \frac{7}{\sqrt{2}} \end{array} \right\} \tag{16}$$

where x_1, x_2 are the realizations of two independent standard Gaussian random variables.

The results in Table 1, are compared to those reported in [14] [17]: crude MC, FORM, DS, Subset, SMART, MetaAK-IS^2 and AK-MCS+U. The method proposed shows a good accuracy, with less performance calls than the other methods.

Method	N_{calls}	$\hat{p}_f$	$\hat{\delta}_f$
Crude MC	781,016	2.24×10^{-3}	2.23%
FORM	7	1.35×10^{-3}	
DS	1800	2.22×10^{-3}	
Subset	600,000	2.22×10^{-3}	1.5%
SMART	1035	2.21×10^{-3}	
MetaAK-IS^2	138	2.22×10^{-3}	1.7%
AK-MCS+U	96	2.23×10^{-3}	
MetaAL-OIS	**15 + 61**	$\mathbf{2.23 \times 10^{-3}}$	**0.09%**

Table 1: Comparison of the performances of the MetaAL-OIS with several algorithms of literature[14].

Figure 2 compares the limit-state obtained by the metamodel, to the true limit-state.

3. Application to the shock tube

The meta-model refinement of the meta-IS algorithm is applied to the estimation of the failure probability of the experiment, *i.e.* the probability of the pre-shock Mach number, in the reference frame attached to the shock, to be less than one. Indeed, this Mach associated to the rarefaction shock wave is required to be larger than one for the shock to be well measured. In this configuration, the performance function can be then defined as follows

$$G(p_L, p_R) = M - 1, \tag{17}$$

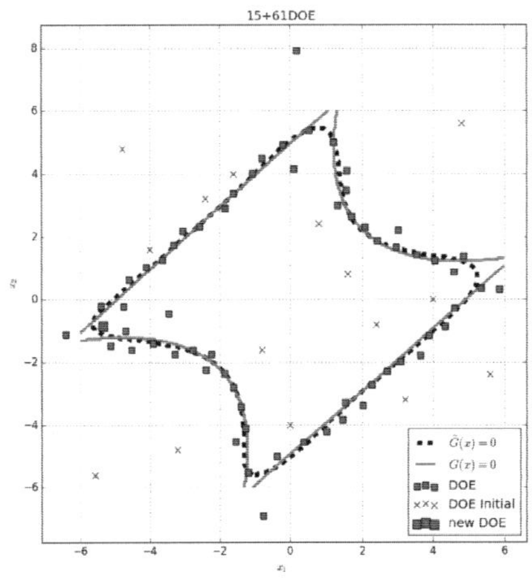

Figure 2: 2D analytic example with four failure regions: Refined Metamodel with 76 points.

p_L, p_R being respectively the left and right pressure. M denotes the Mach number associated to the RSW. The failure event is

$$p_f = \mathbb{P}(G(p_L, p_R) < 0) = \mathbb{P}(M(p_L, p_R) < 1) \tag{18}$$

where p_L, p_R are assumed to be two independent adimensioned random variables (Table 2), according to the uncertainty assessment.

Variable	Distribution	Minimum	Maximum
p_L	Uniform	0.908	0.984
p_R	Uniform	0.099	0.108

Table 2: Probabilistic model for the shock tube

The application of the algorithm presented in the previous section permits to compute the metamodel of the Mach number associated to the RSW, indicated as $\hat{M}$ in the following. By sampling the metamodel, several analysis are possible in terms of statistics evaluation. Results about the tail probability computations and some statistics associated to the Mach number are summarized in table 3.

Note that the failure probability, *i.e.* $M < 1$, is of 43.5%, where the variance of the estimator amounts to the 0.16%. This means that the tail probability is very well computed, and indicates a very large failure probability in terms of occurrence of a RSW during the experiment.

A plot of the probability density function (PDF) of the Mach number is shown 3, where some of the properties in terms of expected value, variance and 90% confidence interval, are reported in Table 3. Also, plots of the metamodel refined in the standard and its contour in the physical spaces are shown in Figures 4a and 4b. Note that the PDF features a not-gaussian behavior, it is very skewed towards the lower Mach, confirming the large probability that the experiment is not effective because of the low value of the Mach. This is a confirmation of the results obtained in [1], even if the shape of the PDF and the associated failure probability is slightly different. This could be associated to the limited number of samples used in [1] for computing the Polynomial-chaos based metamodel, which could deteriorate the computation of the low-probability events. Note that the presented tool provides a very fast estimation of the failure of the experiment, permitting potentially the investigation of a large set of conditions at a very low computational cost.

N_{calls}	$\hat{p}_f$	$\delta_{\hat{p}_f}$
20 + 7	0.435	0.16 %
$\mu_{\hat{M}}$	$\sigma^2_{\hat{M}}$	90% confidence interval
1.0072006	5.271×10^{-4}	0.9745-1.0466

Table 3: Tail probability estimation and Mach number statistics

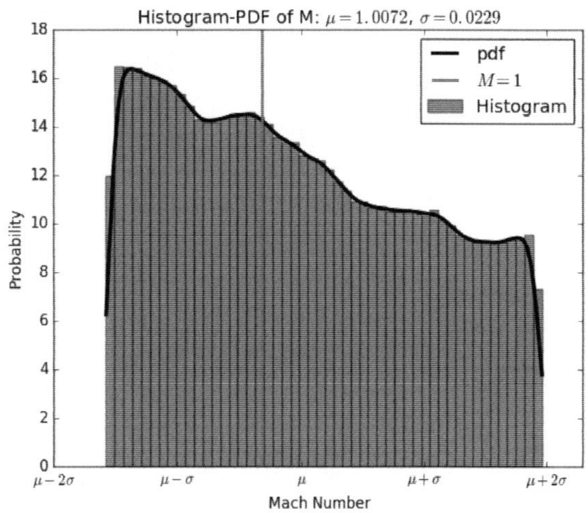

Figure 3: PDF of the Mach Number associated to the Rarefaction Shock Wave.

4. Conclusions

This paper illustrates a novel method for computing the tail probability of a given quantity of interest by using only a low number of evaluations of the computer model. This algorithm is applied to the prediction of the occurrence of a rarefaction shock wave (RSW) in a shock tube configuration, since the computation of the tail probability of the Mach associated to the RSW is very sensitive to the initial conditions uncertainty.

Results show that the occurrence of the RSW is around the 56%, which is quite low and makes questionable the repeatability of the experiment for the chosen initial conditions. This highlights the importance of assessing uncertainty when weak effects, as those associated to a non-classical gasdynamics, have to be captured. Otherwise, the risk is to overestimate the quality of the numerical experiments, which could be very sensitive to some conditions, and provide in practice low-probability results. The low number of evaluations, which are needed to compute the tail probability, illustrates the potentialities of the proposed stochastic method. Also, it should be mentioned that there is no limitation to apply this stochastic method, taking account for uncertainties with the parameters of the thermodynamic model adopted to describe the fluid.

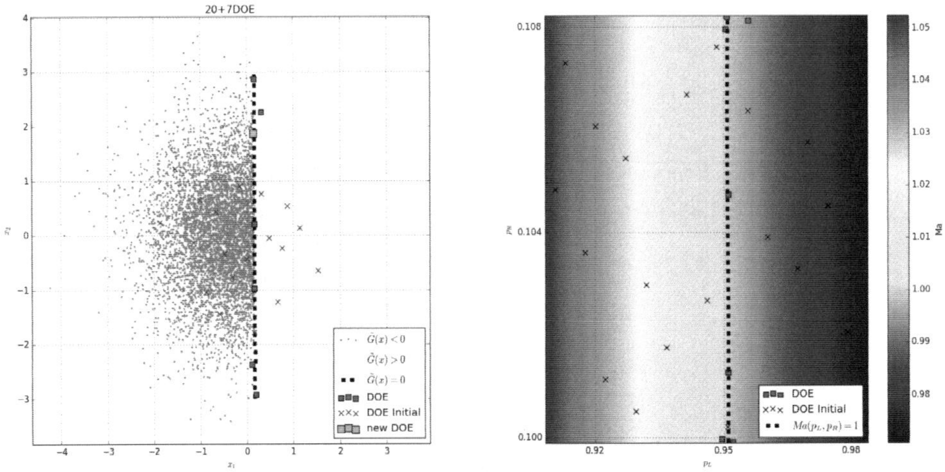

(a) Monte-Carlo population classified by the metamodel $\tilde{G}$ in the standard Space

(b) Metamodel Contour in the physical space: Mach number as function of p_L and p_R

Figure 4: Shock Tube: Refined Metamodel.

Future works will be oriented to the optimization of the geometry and the initial conditions in order to maximize the probability of occurrence of the RSW.

References

[1] Congedo PM, Colonna P, Corre C, Witteveen J and Iaccarino G 2012 *Computer Methods in Applied Mechanics and Engineering* **213** 314-326
[2] Colonna P, Guardone A, Nannan NR and Zamfirescu C 2008 *J. Fluid Eng.-T. ASME.* **130** 1-6
[3] Colonna P, Guardone A, Nannan NR and Van der Stelt TP 2009 *Fluid Phase Equilib.* **286** 43-54
[4] Colonna P, Nannan NR and Guardone A 2008 *Fluid Phase Equilib.* **263** 115-130
[5] Colonna P, Nannan NR, Guardone A and Lemmon EW 2006 *Fluid Phase Equilib.* **244** 193-211
[6] Cinnella P 2006 *Computers and Fluids* **35** 1264-1281
[7] Colonna P, Guardone A and Nannan NR 2007 *Phys. Fluids* **19086102** 1-12
[8] Cinnella P, Congedo PM, Pediroda V and Parussini L 2010 *International Journal of Engineering Systems Modelling and Simulation* **2** 12-24
[9] Congedo PM, Corre C and Martinez JM 2011 *Comput. Methods Appl. Mech. Engrg.* **200** 216-232
[10] Guardone A, Zamfirescu C and Colonna P 2010 *J. Fluid Mech.* **642** 127-146
[11] Rubinstein RY 1981 *Simulation and the Monte-Carlo method* (New York: Wiley)
[12] Fishman GS 1996 *Monte-Carlo: concepts, algorithms, and applications* (New York: Wiley)
[13] Echard B, Gayton N, Lemaire M and Relun N 2013 *Reliability Engineering and System Safety* **111** 232-240
[14] Cadini F, Santos F and Zio E 2014 *Reliability Engineering and System Safety* **131** 109-117
[15] Zhao H, Yue Z, Liu Y, Gao Z and Zhang Y 2015 *Applied Mathematical Modelling* **39** 1853-66
[16] Huang X, Chen J and Zhu H 2016 *Structural Safety,* **59** 86-95
[17] Echard B, Gayton N and Lemare M 2011 *Reliability Engineering and System Safety,* **33** 145-154
[18] MacQueen J 1967 *Proceedings of the Berkeley symposium on mathematical statistics and probability* **1** 281-97
[19] Au SK and Beck J 2001 *Prob. Eng. Mech.,* **16** 263-77
[20] Santner T, Williams B and Notz W 2003 *The design and analysis of computer experiments* (New York: Springer)
[21] Rasmussen CE and Williams C 2006 *Gaussian Processes for Machine Learning* (MIT Press)
[22] Dubourg V, Sudret B and Deheeger F 2013 *Probabilistic Engineering Mechanics,* **33** 47-57

Hybrid Explicit Residual Distribution Scheme for Compressible Multiphase Flows

Paola Bacigaluppi, Rémi Abgrall and Tulin Kaman

Institut für Mathematik, Universität Zürich, Winterthurerstrasse 190, CH-8057 Zürich, Switzerland

E-mail: paola.bacigaluppi@math.uzh.ch, remi.abgrall@math.uzh.ch, tulin.kaman@math.uzh.ch

Abstract. The aim of this work is the development of a fully explicit scheme in the framework of time dependent hyperbolic problems with strong interacting discontinuities to retain high order accuracy in the context of compressible multiphase flows. A new methodology is presented to compute compressible two-fluid problems applied to the five equation reduced model given in Kapila et al. (*Physics of Fluids 2001*). With respect to other contributions in that area, we investigate a method that provides mesh convergence to the exact solutions, where the studied non-conservative system is associated to consistent jump relations. The adopted scheme consists of a coupled predictor-corrector scheme, which follows the concept of residual distributions in Ricchiuto and Abgrall (*J. Comp. Physics 2010*), with a classical Glimm's scheme (*J. Sci. Stat. Comp. 1982*) applied to the area where a shock is occurring. This numerical methodology can be easily extended to unstructured meshes. Test cases on a perfect gas for a two phase compressible flow on a Riemann problem have verified that the approximation converges to its exact solution. The results have been compared with the pure Glimm's scheme and the expected exact solution, finding a good overlap.

1. Introduction

Due to the high interest in phenomena of shock propagation concerning areas such as engines, high explosives and energy, the domain of multiphase flows counts continuously increasing contributions. The main focus is on the ability of models to approximate the interfaces of mixed compressible flows. In particular, as outlined in [1], two main streams of approximations have been developed in the past years: one investigates flows using single pressure and velocity, while the other focuses on flows with multiple pressures and velocities. One drawback of treating multiple pressures and velocities is the fact that to each phase a velocity and pressure is associated, meaning that each phase has its own momentum and energy equations. This results in very large systems. Moreover, these systems are in the overall non-conservative and therefore it is not easy to determine explicit jump relations. One example of such non-equilibrium models is given by the seven equation system of Baer and Nunziato [2]. To simplify the model, it is possible to assume the shock propagation with a large interfacial area, meaning that single pressure and velocity can be assumed. This stiff mechanical relaxation has been widely investigated in [3, 4, 1] and many more. The following paper treats a five equation reduced model, as also many contributions as [1, 4, 5, 6, 7, 8, 9, 10] do. This system originates from Kapila et al. [11], where they take the formal limit [12] of the Baer and Nunziato seven equation model when the

relaxation parameters tend simultaneously to infinity, and thus permitting to assume a single pressure and velocity. The non-conservative character of the scheme due to the equation for the volume fraction keeps the question open how to provide compatible jump relations [13] to be able to obtain a sufficient accurate and robust approximation that converges to the exact solution. The possibility to simply consider conservative equations has been shown to induce large errors in the pressure and velocity computation as outlined with emphasis in [14, 15, 16]. Many methods have been studied in order to be able to treat the non-conservative system's jump relations (see e.g. [4]). Colella [17] presented the idea to apply Glimm's scheme [18] for continuous parts of the flow, and switch to Godunov scheme when large pressure jumps occur. Similarly, even though in contrast to the approach of Colella, Abgrall et al. [19] observed the ability of Glimm's scheme to capture a shock and the weakness of that scheme to propagate some numerical error in smooth areas, and coupled Glimm's scheme to the Roe scheme.
In the following paper, we have coupled Glimm's scheme with an explicit residual distribution (ERD) scheme, and the switch between one scheme and the other is performed by means of a shock detector, capturing large momentum jumps. We examine the set of equations of Kapila et al. [11] in one dimension. The ERD scheme is a predictor-corrector method, following the concept of residual distributions in Ricchiuto and Abgrall [20]. The advantage of this formalism is that the time step is dictated by a CFL-like condition, contrarily to the original Baer- and Nunziato-like system. In general, a further advantage of the residual distributions is given by the absence of a Riemann solver, which results particularly practical for conservative set of equations as there is no need in the exact understanding of the Riemann solution. Moreover, the residual distribution formulation allows to see the component of the discretization as an error between two different approximations. In particular, the residual distribution is computed considering a Lax Friedrich's scheme [20]. For higher order accuracy in space, a limiter, designed similarly to a positive stream-wise invariant [21] with a filtering term taken from Harten and Yee [22], is applied. This paper is divided into four sections. In section 2 we give some definitions and assumptions for the problem set and recall the Kapila's system of equations. In section 3 we describe the explicit residual distribution scheme and in the next section how it is coupled with Glimm's scheme. Finally, we compare our approximation to the original Glimm's method with respect to an exact solution and present our conclusive remarks.

2. Mathematical model

The considered modeling equations are designed for hyperbolic compressible two-phase flows in one dimension. Among the multiphase models, Bear and Nunziato [2] gave a generic formulation of a seven equation model with heat and mass transfer, which consists of three conservation equations for each phase k that read

$$\begin{cases} \frac{\partial(\alpha_k\rho_k)}{\partial t} + \frac{\partial(\alpha_k\rho_k\, u_k)}{\partial x} = \pm\rho_{tot}\dot{Y} \\ \frac{\partial(\alpha_k\rho_k\, u_k)}{\partial t} + \frac{\partial(\alpha_k\rho_k\, u_k^2+\alpha_k P_k)}{\partial x} = P_I\frac{\partial\alpha_k}{\partial x} \pm \lambda(u_2-u_1) \pm u_I\rho_{tot}\dot{Y} \\ \frac{\partial(\alpha_k E_k)}{\partial t} + \frac{\partial(\alpha_k(E_k+P_k)u_k)}{\partial x} = P_I u_I\frac{\partial\alpha_k}{\partial x} \pm \lambda u_I(u_2-u_1) \pm \mu P_I(P_2-P_1) \pm Q \pm E_I\frac{\rho_{tot}}{\rho_I}\dot{Y} \end{cases} \tag{1}$$

and a transport equation of the volume fraction $\frac{\partial\alpha_1}{\partial t} + u_I\frac{\partial\alpha_1}{\partial x} = \mu(P_1-P_2) + \eta Q + \frac{\rho_{tot}}{\rho_I}\dot{Y}$.

The terms α_k, ρ_k, u_k, P_k, E_k represent respectively the volume fraction, the density, the velocity, the pressure and the energy of each phase. Quantities with the subscript I correspond to interface terms, while those denoted by tot are given by mixture terms. In particular, the following mixture relations hold for the volume fraction $\sum_k \alpha_k = 1$ and the total density $\sum_k \alpha_k\rho_k = \rho_{tot}$. Moreover, in (1) $\dot{Y}$ represents the mass transfer term, while Q is associated to the heat transfer term. The coefficients μ and λ represent the dynamic compaction viscosity and the relaxation velocity parameter respectively [9].

2.1. The five equation reduced model

In this study we consider calorically perfect flows, s.t. terms associated to the heat transfer Q are neglected. Due to its high complexity we make the assumption to consider a very large interface and let μ and λ tend simultaneously to infinity. We retrieve a stiff mechanical relaxation of (1) as performed in [11] and [12]. The mass transfer $\dot{Y}$ is in the following neglected. The system shown in (1) is reduced to the well known five-equation model

$$\begin{cases} \frac{\partial \alpha_1}{\partial t} + u\frac{\partial \alpha_1}{\partial x} - K\frac{\partial u}{\partial x} = 0 \\ \frac{\partial(\alpha_1\rho_1)}{\partial t} + \frac{\partial(\alpha_1\rho_1\, u)}{\partial x} = 0 \\ \frac{\partial(\alpha_2\rho_2)}{\partial t} + \frac{\partial(\alpha_2\rho_2\, u)}{\partial x} = 0 \\ \frac{\partial(\rho_{tot}\, u)}{\partial t} + \frac{\partial(\rho_{tot}\, u^2+P)}{\partial x} = 0 \\ \frac{\partial E_{tot}}{\partial t} + \frac{\partial(E_{tot}+P)u}{\partial x} = 0 \end{cases} \tag{2}$$

with the total energy given by the internal and kinetic energy $E_{tot} = e + \frac{1}{2}\rho_{tot}u^2$, where the total internal energy is defined as $e = \sum_k \alpha_k e_k$. In particular, the internal energy of phase k is defined as $e_k = \rho_k \epsilon_k$, and ϵ_k corresponds to the specific internal energy of phase k. Further, we introduce a parameter K s.t. $K = \frac{\alpha_1\alpha_2(\rho_2 c_2^2 - \rho_1 c_1^2)}{\alpha_1\rho_2 c_2^2 + \alpha_2\rho_1 c_1^2}$, where $c_k^2 = \left(\frac{\partial P_k}{\partial \rho_k}\right)\big|_{Entropy\ in\ k}$ is the squared speed of sound associated to each phase.
The total speed of sound is given by the Wood velocity, which is a weighted harmonic mean of the speed of sound of each phase $\frac{1}{\rho_{tot}c_{tot}^2} = \frac{\alpha_1}{\rho_1 c_1^2} + \frac{\alpha_2}{\rho_2 c_2^2}$.
Note that due to the relaxation of the parameters μ and λ, we retrieve $u_1 = u_2 = u$ as well as $P_1 = P_2 = P$.

In order to close the system (2), we consider in the following two equations of state for a stiffened gas. The pressure is thus defined as $P(\rho_k, e_k) = P_1(\rho_1, e_1) = P_2(\rho_2, e_2) = (\gamma_k - 1)e_k - \gamma_k P_{\infty,k}$ for each phase k, with $P_{\infty,k}$ the reference pressure (see [9] for further details).

2.2. Jump relations for shock waves

The relation between the initial and final state across a shock (see also [4]) for (2) are given by the shock jump conditions of Rankine Hugoniot. In particular, given a left state V_L and a right state V_R we define the generic $\Delta V = V_L - V_R$ and have the relations

$$\begin{cases} \Delta\epsilon + \overline{P}\Delta\tau = 0, \\ \Delta P = -m^2\Delta\tau \end{cases} \tag{3}$$

We have $\epsilon = \frac{e}{\rho_{tot}}$, $\overline{P} = \frac{1}{2}(P_L + P_R)$, $\tau = \sum_k Y_k\tau_k$, with the mass fraction of each phase defined as $Y_k = \frac{\alpha_k\rho_k}{\rho_{tot}}$ and the specific volume as $\tau_k = \frac{1}{\rho_k}$. The relative velocity is given by $m = \rho_{tot}(u - \sigma)$, where σ is the shock speed. These relations guarantee the conservation of the mixture.

3. An Explicit Residual Distribution (ERD) scheme for second order accuracy

The adopted scheme is a predictor-corrector method, with the same spirit as the one of Ricchiuto et al. [20]. In the following we recall briefly the scheme.

Given an initial solution V_h^0 at $t_0 = 0$, we consider the treated system to be in the form

$$\partial_t V + \nabla\cdot\mathcal{F}(V) = A\cdot\nabla V \quad \text{on } \Omega\times[0,T] \tag{4}$$

with $V = [\alpha_1,\ \alpha_1\rho_1,\ \alpha_2\rho_2,\ u\rho_{tot},\ E_{tot}]^T$, $\mathcal{F}(V) = [0,\ \alpha_1\rho_1 u,\ \alpha_2\rho_2 u,\ \rho_{tot}u^2 + P,\ (E_{tot}+P)u]^T$. The first row of A reads $A_{1,s} = [-u,\ -\frac{K\,u}{\rho_{tot}},\ -\frac{K\,u}{\rho_{tot}},\ \frac{K}{\rho_{tot}},\ 0]^T$ and $A_{r,s} = 0$, where $s = 1,..,5$ and $r = 2,..,5$.

We consider the time interval $[0, T]$ divided into N intervals s.t $0 = t_0 < t_1 <...< t_n <...< t_N = T$ and denote the numerical approximation as $V_h^n = V_h(t_n)$. Further, we divide the interval $[t_n, t_{n+1}]$ in P sub-intervals, s.t. $t_n = t_{n,0} < t_{n,1} <...< t_{n,m} <...< t_{n,P} = t_{n+1}$. Let $\Delta t_{n,m} = t_{n,m+1} - t_{n,m}$ and let $V_h^{n,m}$ denote the approximation to $V_h(t_{n,m})$.
We discretise the spatial domain Ω with a grid denoted by Ω_h, and denote by Q a generic element of the mesh and h the generic characteristic mesh size.
Following the standard derivation of the Galerkin finite element method, we consider a spatial domain Ω divided in cells, with Δx the spatial step between two neighboring nodes, and on this grid we consider the space of all the piecewise linear polynomials spanned by the basis functions φ_i corresponding to node $i \in \Omega_h$. In particular these basis functions have to verify the condition of $\varphi_i(x_j, y_j) = \delta_{i,j}$, $\forall i, j \in \Omega_h$, and $\sum_{j \in Q} \varphi_j(x, y) = 1$, $\forall Q \in \Omega_h$. The approximation of V will be denoted by V_h and given by $V_h = \sum_{i \in \Omega_h} V_i \varphi_i = \sum_{Q \in \Omega_h} \sum_{i \in Q} V_i \varphi_i$.

We define the residual to be $r_h = \frac{V_h^\star - V_h^n}{\Delta t_{n,\star}} + \nabla \cdot \mathcal{F}_h(V_h^n, V_h^\star) - A(V_h^n, V_h^\star) \cdot \nabla(V_h^n, V_h^\star)$ and introduce a fluctuation term $\phi^Q = \int_Q r(u_h)$. Denoting by ϕ_i^Q a local nodal residual of node $i \in Q$, we have to fulfill $\phi^Q(V_h^n) = \sum_{i \in Q} \phi_i^Q$, $\forall Q \in \Omega_h$, $\forall \Omega_h$ meaning also that the fluctuations are distributed to ϕ_i^Q through a bounded distribution matrix coefficient β_i^Q, s.t.

$$\phi_i^Q = \beta_i^Q \phi^Q, \text{ with } \sum_{j \in Q} \beta_j^Q = 1. \tag{5}$$

In general, β_j^Q determines the distribution strategy and, therefore, the applied residual distribution scheme. The reader may refer to [23] for further details on the generic residual distribution scheme.

We choose now as specific case the sub-intervals to be $P = 2$, s.t. the scheme is designed as a predictor-corrector method, and update the time step via an initialization at $t_0 = 0$ of $\Delta t = \text{CFL} \cdot \Delta x$ which is then updated by computing $\Delta t = \text{CFL} \cdot \min_i \left(\frac{\Delta x_i}{|u_i| + c_i}\right)$.
For the predictor we have

$$|C_i| \frac{V_i^{n+\frac{1}{2}} - V_i^n}{\Delta t_{n,1}} + \sum_{Q \in Q_i} \phi_i^Q(V_h^n) = 0, \quad \forall i \in \Omega_h \tag{6}$$

As for the corrector, it contains a fluctuation computed on the $V_i^{n+\frac{1}{2}}$ and reads

$$|C_i| \frac{V_i^{n+1} - V_i^{n+\frac{1}{2}}}{\Delta t_{n,2}} + \sum_{Q \in Q_i} \phi_i^Q(V_h^n, V_h^{n+\frac{1}{2}}) = 0, \quad \forall i \in \Omega_h \tag{7}$$

where $|C_i|$ is the area of the median dual cell C_i obtained by joining the gravity centers of the cells with the midpoints of the edges meeting in i (see [20]).

The nodal fluctuation ϕ_i^Q is computed by a Lax-Friedrichs scheme

$$\phi_j = \phi_i^Q(V_h^n, V_h^\star) = \frac{\Delta x}{2}\Big(V_j^\star - V_j^n\Big) + \Delta t\Big(\frac{\psi}{2} + \theta(\overline{V})(\widetilde{V}_j - \overline{V})\Big), \tag{8}$$

with with $j = i, i+1$ and $\theta(\overline{V}) = 2\big(|\overline{u}| + \overline{c_{tot}}\big)$ where $V^\star$ assumes accordingly V^n in case of the predictor, while $V^{n+\frac{1}{2}}$ for the corrector. The quantities in (8) are defined as arithmetic averages with $\overline{V} = \frac{1}{2}\Big(\sum_j \frac{1}{2}(V_j^n + V_j^\star)\Big)$ and $\widetilde{V}_j = \frac{1}{2}\big(V_j^n + V_j^\star\big)$ for $j = i, i+1$.
Vector ψ is the difference of fluxes occurring between nodes on the same cell given by

$$\psi = \begin{pmatrix} \overline{u}\big[\frac{1}{2}(\alpha_1^n + \alpha_1^\star)_{i+1} - \frac{1}{2}(\alpha_1^n + \alpha_1^\star)_i\big] - \overline{K}\big[\frac{1}{2}(u^n + u^\star)_{i+1} - \frac{1}{2}(u^n + u^\star)_i\big] \\ \frac{1}{2}(F^n + F^\star)_{i+1} - \frac{1}{2}(F^n + F^\star)_i \end{pmatrix} \tag{9}$$

with the flux $F = [\alpha_1 \rho_1 u,\ \ \alpha_2 \rho_2 u,\ \ \rho_{tot} u^2 + P,\ \ (E_{tot} + P)u]^T$.
Note that quantities such as $\overline{u}, \overline{K}$ and $\overline{c_{tot}}$ are defined the same way as $\overline{V}$.

In general, (8) can also be expressed as

$$\phi_j = \sum_{s=1}^{5} l_s(\phi_j) R_s, \quad \text{with } j = i, i+1 \tag{10}$$

where $l_s(\phi_j)$ are the eigenforms corresponding to ϕ_j for each eigenvector R_s.
Since by definition in (5), we have to distribute the residual to the different nodes which are part of a cell, we rewrite ϕ_j as $\phi_j^\star = \beta_j\, \phi$.

Recalling (10), we introduce a sort of positive stream-wise invariant (PSI scheme) with some stabilization, which has to enforce invariance along the characteristics as explained in [20, 21].

For each eigenvector R_s, with $s = 1,..,5$ and reads for $j = i, i+1$ the eigenform of (10) is changed to $\phi_j^\star = \sum_s (l_s(\phi))_j^\star R_s$, where

$$(l_s(\phi_j))^\star = \begin{cases} l_s(\phi) \left[\dfrac{\max\left(\frac{l_s(\phi_j)}{l_s(\phi)}, 0\right)}{\max\left(\frac{l_s(\phi_i)}{l_s(\phi)}, 0\right) + \max\left(\frac{l_s(\phi_{i+1})}{l_s(\phi)}, 0\right)} \pm \dfrac{\eta}{2} \dfrac{\lambda_s}{\varphi^\star(\lambda_s)} \right], & \text{if } l_s(\phi) > 0 \\ l_s(\phi_j) & \text{else} \end{cases} \tag{11}$$

with $l_s(\phi) = l_s(\phi_i) + l_s(\phi_{i+1})$, while η is a parameter that switches on and off the stabilization and accepts values between $0.05 \leq \eta \leq 0.125$ according to [22] and, further, λ_s the eigenvalues associated to R_s.

The function $\varphi^\star(\lambda_s)$ provides an entropy fix and is designed according to the least dissipative function among the three point entropy satisfying TVD schemes of Harten-Yee [22], given by

$$\varphi^\star(x) = \begin{cases} |x| & \text{if } |x| \geq \epsilon, \quad \epsilon = 0.01 \\ \frac{x^2 + \epsilon^2}{2\epsilon} & \text{else} \end{cases} \tag{12}$$

In particular we refer the reader to [20] for the proof of second order accuracy of the scheme applied to a different problem.

4. A Hybrid Glimm-Explicit Residual Distribution (GERD) scheme

4.1. Glimm's scheme

Glimm's scheme [18, 24] is implemented according to the description in Colella [17] where the wave propagation is modeled by means of solving a Riemann problem, with consistent jump relations as shown in Eq. (3) (see also [4]). The approximation at each step is obtained by sampling the solution with the van der Corput random generator.

4.2. Coupled scheme

The ERD scheme, as also shown in Figure 4, approximates accurately the solution everywhere, besides the shock front, where the numerical dissipation affects the solution. On the other side, as shown in the limit test case in Figure 1-4 and as also outlined by many authors in this field (see [17], [19]), Glimm's scheme works perfectly for shock fronts, whereas gets some numerical oscillations in smooth regions. Therefore, we couple both schemes, so that the approximation is computed with the ERD scheme everywhere, with exception to the shock front. The criterion with which the switch is chosen is inspired partly by [17], partly by [19], but is intrinsically different and reads

$$\theta_i^n = \max\left((p_{i+1}^n - p_i^n) \cdot (u_{i+1}^n - u_i^n),\ (p_{i+1}^\star - p_i^\star) \cdot (u_{i+1}^\star - u_i^\star) \right) \tag{13}$$

IOP Conf. Series: Journal of Physics: Conf. Series **821** (2017) 012007 doi:10.1088/1742-6596/821/1/012007

The idea is to check when the momentum becomes strictly greater than zero, since in that case a shock front is occurring. The solution n is first approximated by the ERD scheme to get $\star$, and then (13) is evaluated. In case $\theta_i^n > 0$ Glimm's scheme is applied.

5. Numerical results

To validate the presented scheme we consider a typical very severe test case of a shock tube problem with two different mixtures and very high pressure jumps across the interface. This benchmark has been chosen due to its popularity among many authors as in [1, 7, 4, 3], etc. We are dealing with a Riemann problem, where a tube of 1 meter length is divided into two chambers by an interface at $x = 0.6m$ and initial conditions as outlined in Table 1.
For the solution V_i^{n+1}, Glimm's scheme is applied in case $\theta_i^n > 0$, else the approximation is given by the ERD scheme.
On the left and on the right of the interface a mixture of epoxy and spinel is set up, with on the left and right $u = u_1 = u_2 = 0$ $[\frac{m}{s}]$, while on the left a pressure $P_L = P_{1L} = P_{2L} = 2 \cdot 10^{11}$ $[Pa]$ and on the right $P_R = P_{1R} = P_{2R} = 1 \cdot 10^5$ $[Pa]$. Further, we choose to display the solution at $t = 29\mu s$.

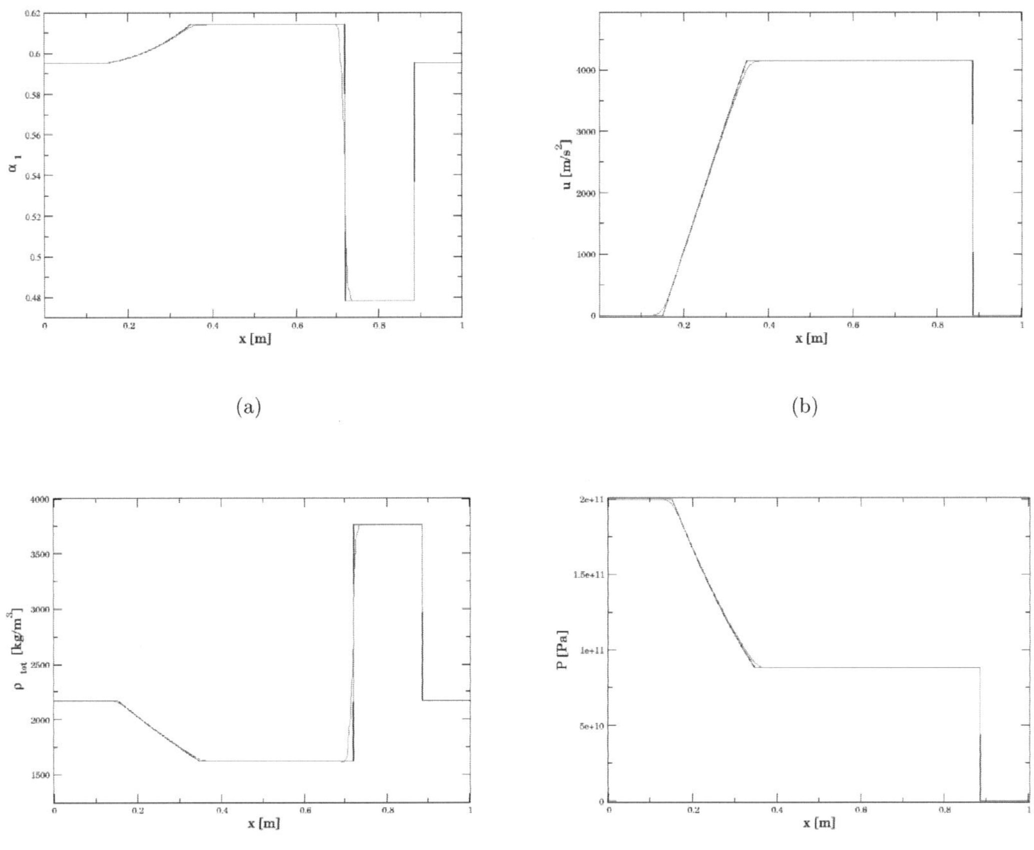

Figure 1. Comparison between the expected solution (black), the pure Glimm scheme (red) and the GERD scheme (green) with $\Delta x = 0.002m$. (a) Shows the volume fraction of phase 1, (b) the velocity $u = u_1 = u_2$, (c) the total density ρ_{tot} and (d) the pressure $P = P_1 = P_2$ along the shock tube.

Table 1. Initial fluid properties and parameters for the test case.

Phase	Fluid	α_1	$\rho\ [\frac{kg}{m^3}]$	γ	$P_\infty\ [Pa]$
1	Epoxy	0.5954	1185	2.43	$5.3 \cdot 10^9$
2	Spinel	0.4046	3622	1.62	$141 \cdot 10^9$

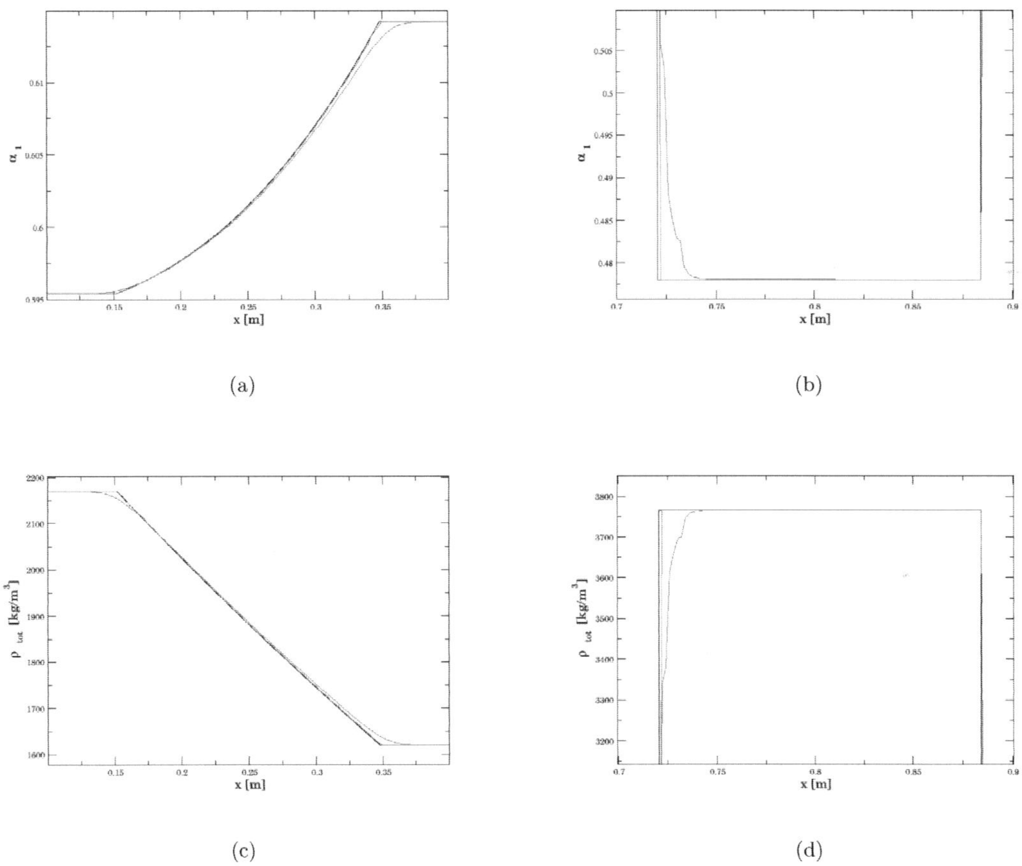

Figure 2. Details of Figure 1. Comparison between the expected solution (black), the pure Glimm scheme (red) and the GERD scheme (green) with $\Delta x = 0.002m$. (a) Shows the volume fraction of phase 1 between $x = [0.1, 0.4]m$, (b) the volume fraction of phase 1 between $x = [0.7, 0.9]m$, (c) the total density ρ_{tot} between $x = [0.1, 0.4]m$ and (d) between $x = [0.7, 0.9]m$.

We consider the expected solution to be given by Glimm's scheme for a very fine mesh with $\Delta x = 0.0001m$. This expected solution is compared on a grid with $\Delta x = 0.002m$ with the pure Glimm's scheme as described in [17] and as briefly recalled in this paper. The pure ERD scheme and the GERD scheme are then compared on the same coarse mesh as Glimm. Finally we display the comparison between the pure Glimm's scheme and the GERD one on a fine mesh with $\Delta x = 0.0001m$. In these simulations the ERD and the GERD scheme have the parameter $\eta = 0.05$. We consider the CFL=0.45.
In Figure 1 the total density, velocity and pressure as well as the volume fraction of phase 1 with

respect to the position in the shock tube for $\Delta x = 0.002m$ are shown. The pure Glimm and the GERD scheme show in the overall a good overlap also w.r.t. the expected solution. Clearly, also from the details of Figure 1 displayed in Figure 2 and Figure 3, it is possible to observe that indeed the pure Glimm results to have some oscillations in the smooth area after $x = 0.3m$, while captures sharply the shock front even with a coarse spatial discretisation. The GERD scheme is, on the contrary, able to approximate without oscillations the expected solution and the error is just due to the coarse mesh. Indeed, following this observation, in Figure 5 it is possible to see for the volume fraction of phase 1 that the GERD method converges to the expected solution. Finally we see that there are no delays or time displacements and the presented method reveals itself as very robust, being able to capture the correct approximation for extremely large jump conditions. As a last comparison, we show in Figure 4 the pure Glimm and ERD scheme with $\Delta x = 0.002m$ and compare them. While the approximation along the shock tube, away from the shock front, results to be well approximated by the ERD, and more subject to oscillations for the pure Glimm, we observe that Glimm is able to approximate very well the shock, while ERD scheme fails due to the fact that arithmetic averages are being considered. For the other quantities, as the pressure, displayed in Figure 4 (d)-(f), we see that there is not a relevant error between the Glimm and the ERD scheme. This confirms our initial statement that the non-conservative equation of the volume fraction represents a difficulty for shock capturing.

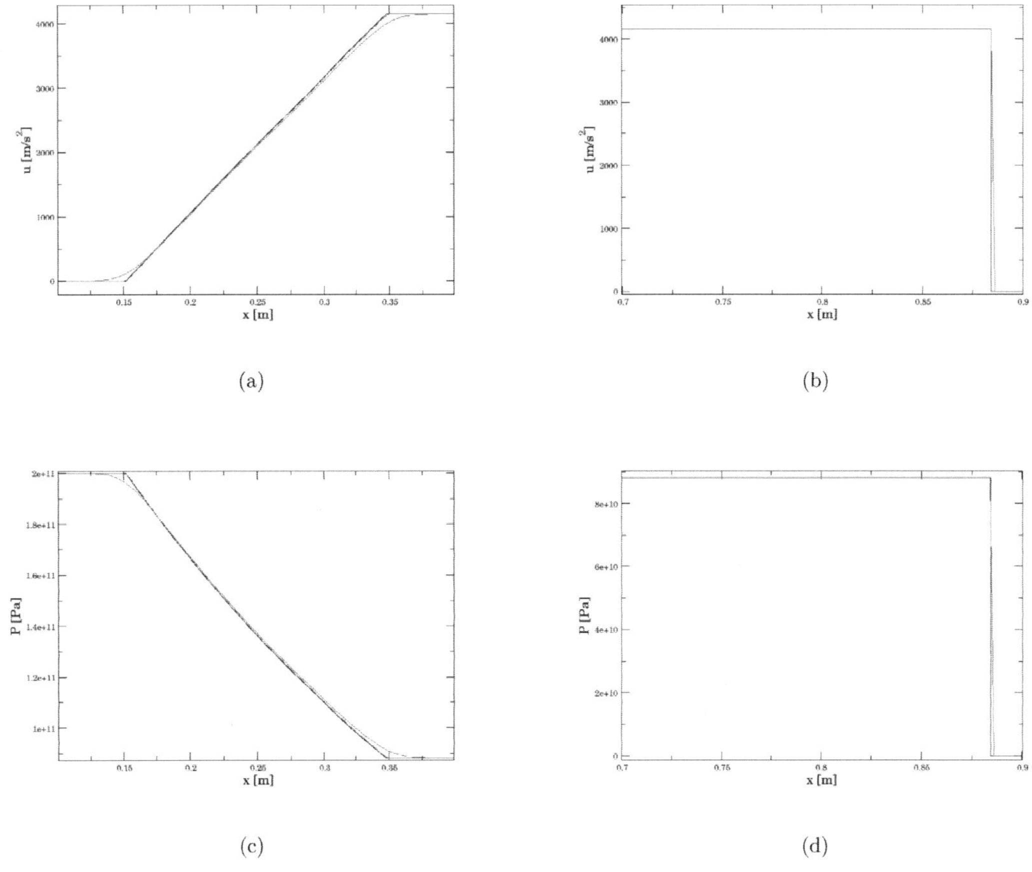

Figure 3. Details of Figure 1. Comparison between the expected solution (black), the pure Glimm scheme (red) and the GERD scheme (green) with $\Delta x = 0.002m$. (a) Shows the velocity $u = u_1 = u_2$ between $x = [0.1, 0.4]m$, (b) the velocity $u = u_1 = u_2$ between $x = [0.7, 0.9]m$, (c) the pressure $P = P_1 = P_2$ between $x = [0.1, 0.4]m$ and (d) the pressure $P = P_1 = P_2$ between $x = [0.7, 0.9]m$.

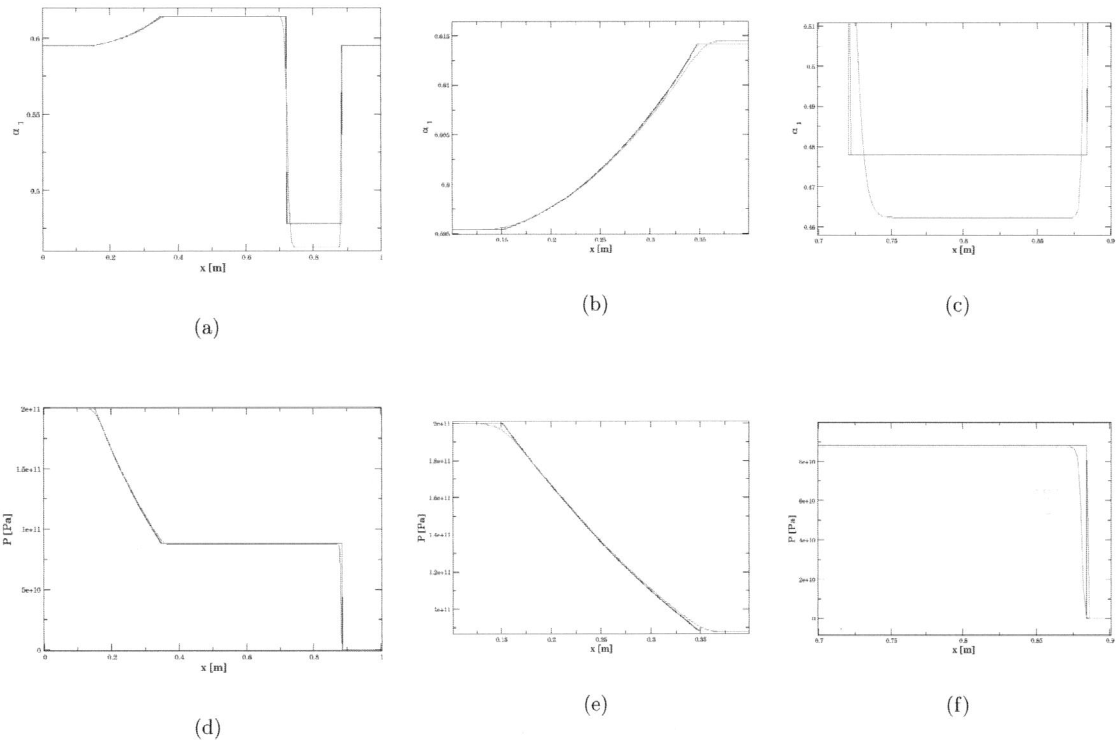

Figure 4. Comparison between expected (black), pure Glimm (red) and pure ERD (green) scheme with $\Delta x = 0.002m$. (a), (b), (c), Show respectively the volume fraction and details between $x = [0.1, 0.4]m$ and $x = [0.7, 0.9]m$, and (d),(e),(f) the pressure with details between $x = [0.1, 0.4]m$ and $x = [0.7, 0.9]m$.

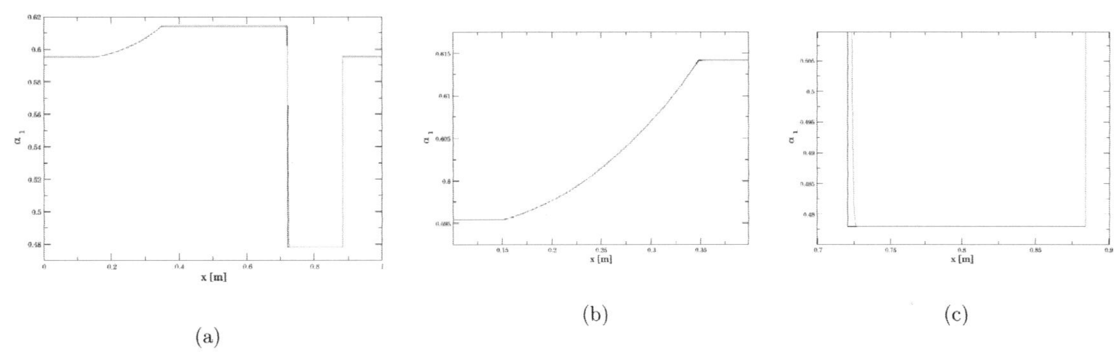

Figure 5. Comparison between expected (black) and GERD scheme (red) with $\Delta x = 0.0001m$. (a),(b),(c) Show the volume fraction of phase 1 with the details between $x = [0.1, 0.4]m$ and $x = [0.7, 0.9]m$.

6. Conclusions

We have presented an ERD method for the approximation of a compressible, one dimensional hyperbolic multiphase problem given by the Kapila et al. five equation reduced model.
Since the system is non-conservative, arithmetic averages of the volume fraction on the shock front do not make sense and result in a local numerical dissipation. On the other hand, Glimm's scheme permits to compute the exact solution of Riemann problems, but results to be less accurate then the ERD method where the solution is not displaying a shock. The idea of the paper has been thus to investigate the possible advantages of a coupled Glimm's scheme with the ERD scheme. To our knowledge the first attempt to check a hybrid scheme with Glimm and Roe on the five equation model has been performed only by Abgrall et al. in [19]. To check the robustness of the scheme a very severe test case has been considered, showing good overlaps with the expected solution. With respect to other contributions in that area, we are, moreover, currently extending the methodology to that particular non-conservative system, using a specific Roe-type averaging that respects all the invariants of the original system.

Acknowledgments

This work was funded partly by the SNF grant # 200021_153604/1 of the Swiss National Foundation.

References

[1] Petitpas F, Franquet E, Saurel R and Metayer O L 2007 *J. Comp. Phys.* **225** 2214
[2] Baer M and Nunziato J 1986 *Int. J. Multiphase Flow* **12** 861
[3] Marsh S P 1980 *LASL Shock Hugoniot Data* (University of California Press)
[4] Saurel R, Metayer O L, Massoni J and Gavrilyuk S 2007 *Shock Waves* **16** 209
[5] Saurel R, Petipas F and Abgrall R
[6] Saurel R, Petitpas F and Berry R A 2009 *J. Comp. Phys.* **228** 1678
[7] Petitpas F, Saurel R, Franquet E and Chinnayya A 2009 *Shock Waves* **19** 377
[8] LeMartelot S, Nkonga B and Saurel R 2013 *J. Comp. Phys.* **255** 53
[9] Rodio M G and Abgrall R 2015 *Int. J. of Heat and Mass Transfer* **89** 1386
[10] Rodio M G, Congedo P and Abgrall R 2015 *GAMM 86th Annual Scientific Conference* (Lecce, Italy)
[11] Kapila A, Menikoff R, Bdzil J, Son S and Stewart D 2001 *Phys. Fluids* **13** 3002
[12] Murrone A and Guillard H 2005 *J. Comp. Phys.* **202** 664
[13] Godlewski E and Raviart P A 2013 *Numerical approximation of hyperbolic systems of conservation laws* vol 118 (Springer Science & Business Media)
[14] Saurel R and Abgrall R 1999 *J. Comp. Phys.* **150** 425
[15] Abgrall R 1996 *J. Comp. Phys.* **125** 150
[16] Karni S 1994 *J. Comp. Phys.* **112** 31
[17] Colella P 1982 *J SIAM J. Sci. Stat. Comp.* **3** 76
[18] Glimm J, Marshall G and Plohr B 1984 *Advances in Applied Mathematics* **5** 1
[19] Abgrall R and Kumar H 2014 *Comm. Comp. Phys.* **15** 1237
[20] Ricchiuto M and Abgrall R 2010 *J. Comp. Phys.* **229** 5653
[21] Struijs R 1994 Ph.D. thesis TU Delft, Delft University of Technology
[22] Yee H, Warming R and Harten A 1985 *J. Comp. Phys.* **57** 327
[23] Deconinck H and Ricchiuto M 2007 *Residual Distribution Schemes: Foundations and Analysis* vol 3 (John Wiley & Sons, Ltd)
[24] Sod G A 1977 *J. Fluid Mech.* **83** 785–794

Numerical investigation of high pressure condensing flows in supersonic nozzles

L. Azzini, M. Pini

Faculty of Aerospace engineering,
Delft University of Technology,
Kluyverweg 1, 2629 HS, Delft, The Netherlands

E-mail: `l.azzini@tudelft.nl, m.pini@tudelft.nl`

Abstract.
High-pressure non-equilibrium condensing flows are investigated in this paper through a quasi-1D Euler model coupled to the method of moments for the physical characterization of the dispersed phase. Two different numerical approaches, namely the so-called (a) the mixture and (b) continuum phase model, are compared in terms of computational efficiency and accuracy. The results are verified against experimental data of high-speed condensing steam measured at high pressure (100.7 bar).

The analysis demonstrates that Model (b) markedly outperforms the mixture model in terms of computational cost, while retaining comparable accuracy. However, both models, in their original formulation, lead to considerable deviations in the nucleation onset prediction as well as an overestimation of the average droplet radius.

A further investigation is then conducted to figure out the main physical parameters affecting the condensation process, i.e. the surface tension, the growth rate and the nucleation rate. It is eventually inferred that applying proper correction to these three quantities allows to obtain best fit with the experimental data. A final calculation is carried out to show the dependence of these three correcting factors from the thermodynamic conditions of the mixture.

1. Introduction

High-speed flows usually condense at non-equilibrium conditions, namely when the vapor reaches thermodynamic states below saturation without any actual formation of liquid droplets. The physical characterization of such mixtures is extremely relevant for a vast number of industrial applications. For example, non-equilibrium condensation is responsible of significant aerodynamic efficiency decay in steam turbines [1]. Recently, several studies have been devoted to investigate supercritical CO2 condensation [2] for applications in compressor stages (carbon capture and sequestration, enhanced oil recovery, power cycles). Moreover, rapid expansions from supercritical states through the liquid-vapor dome can be exploited to enhance the efficiency of Organic Rankine Cycles [3].

Especially for application at high reduced pressures, two-phase flow models for condensation have to meet several requirements. In particular, a complex equation of state must be employed to account for the vapor non ideality. Note that the classical nucleation theory makes use of the perfect gas assumption for the derivation of the nucleation and the growth rate. Additionally, the reliability of the transport models, and in particular of the surface tension, is very questionable when approaching non-ideal thermodynamic regions and this strongly affects the physics of condensation. On another note, if the models are employed for design purposes, the simulations must be also computationally efficient. To the authors' knowledge, there is currently a lack of models complying with all these requirements.

Multiple models and numerical schemes potentially applicable to metastable condensation are described in detail in literature [4]. However, the great majority of them have been validated with steam-water experiments at low pressure [4, 5, 6]. It should be noted that the so called "high-pressure" validation is traditionally made with the experimental data by Bakhtar [7], and even in this case P_0 is lower than 35 bar, with a corresponding reduced pressure of only 0.15.

The aim of the present work is to investigate non-equilibrium condensing flows in converging-diverging nozzles at pressures much higher than those traditionally considered for steam-water. Since the full droplet spectra is not of interest in this study, the prediction of the second phase is performed through the well-established method of moments [8] in terms of averaged properties. Furthermore, only homogeneous condensation is of interest in this study, as the presence of foreign contaminants does not commonly trigger heterogeneous condensation in high-speed flows within supersonic nozzles.

The numerical model [9] consists of a quasi-1D Euler solver supplemented by two additional transport equations for the description of the second phase. Two different formulations are used for the main flow, i.e. the so called mixture and continuum phase conservation laws [9]. The results are compared with the experimental observations by [1].

The paper is organised as follows: the first section describes the governing equations of the model. The second section presents the numerical scheme employed for their resolution. The third section shows the comparison with the experimental observations and debates the results. Finally, the last section illustrates three different dimensionless corrections to the original model aimed at increasing its accuracy at high-pressures.

2. Governing equations

The characterization of a quasi-1D condensing flow requires a minimum of 8 variables: two thermodynamic properties for each phase, the velocities, the liquid mass fraction and the average droplet radius. As presented in [9], in order to reduce the number of equations down to 5 the following assumptions are made: i) the phases are in mechanical equilibrium, ii) the vapor velocity is equal to the liquid one and iii) the droplets temperature is determined through a simplified capillarity model. These assumptions are largely employed in literature [5, 6] and proved to be adequate for low pressure steam test cases.

The model adopted in this work consists of three Eulerian conservation laws and two additional transport equations taken from the moments theory [9]. All relations are reported in the next paragraphs.

2.1. Mass, momentum and energy balance

This work carries on the comparison between the two main approaches described in [9], i.e. the employment of the mixture conservation laws and the continuum phase balance equations. The paper refers to these models as (a) and (b) respectively hereafter. The Eulerian set of equations for Model (b) is

$$\begin{cases} \frac{\partial \rho_v}{\partial t} + \frac{\partial(\rho_v v_v)}{\partial x} = S_v - \rho_v v_v \frac{\partial A_c}{\partial x}\frac{1}{A_c}, \\ \frac{\partial \rho_v v_v}{\partial t} + \frac{\partial\left(\rho_v v_v^2 + P_v\right)}{\partial x} = S_v v_v - \rho_v v_v^2 \frac{\partial A_c}{\partial x}\frac{1}{A_c}, \\ \frac{\partial(\rho_v e_{0,v})}{\partial t} + \frac{\partial(\rho_v h_{0,v})}{\partial x} = S_v h_{0,l} - \rho_v h_{0,v} \frac{\partial A_c}{\partial x}\frac{1}{A_c}, \end{cases} \tag{1}$$

where ρ, v, P, e_0, h_0 are the density, velocity, pressure, total energy and total enthalpy respectively, A_c is the cross sectional area of the nozzle and S_v is the source term accounting for mass exchange across the interface vapor-liquid. The work refers to the properties of the continuum phase and the mixture with the pedex c and m respectively. The interested reader can refer to [5] for the derivation of such equations. On the other hand, the relations for Model (a) are obtained from the previous system using mixture properties and setting the source term S_v equal to 0.

The great restriction of Model (b) is that the equations are derived assuming a negligible volume fraction of liquid [5]. The last assumption can be adequate for low-pressure steam expansions, due to the very high difference in specific volume between the liquid and the vapor phase. On the other hand, the densities of the two phases get closer when moving towards the critical point, and the approximation on the droplet volume can lead to inaccuracies in the final solution. The majority of literature [6, 8] adopts Model (a) as the validity of the mixture equations is general and fluid-independent.

However, the study in [9] proved Model (b) to be up to six times more efficient in terms of computational time. The main reason is that Model (a) requires an additional iterative procedure to determine the mixture properties, significantly increasing the simulation cost [6, 9]. This work carries out a second analysis on a high pressure test case to choose the optimum solver in terms of both accuracy and code efficiency.

2.2. Moments equations for the disperse phase

The formulation adopted by the majority of the literature was presented by Hill [8], and involves 4 moments equations instead of the two theoretically required. Even though this choice leads to an increase in the computational cost, [8] claims that such model can reach a higher accuracy in the final results. However, the recent work in [9] presented a new 2-equation approach that proved to give comparable results with respect to Hill's model for low pressure steam. Therefore, the present work adopts this last solution and tests the accuracy of such numerical approach at high-pressure. The additional equations employed are

$$\frac{\partial}{\partial t}(\rho_m \mu_0) + \frac{\partial}{\partial x}(\rho_m \mu_0 v_m) + \rho_m \mu_0 v_m \frac{\partial A_c}{\partial x}\frac{1}{A_c} = \rho_m J(R_*), \tag{2}$$

$$\frac{\partial}{\partial t}(\rho_m \mu_3) + \frac{\partial}{\partial x}(\rho_m \mu_3 v_m) + \rho_m \mu_3 v_m \frac{\partial A_c}{\partial x}\frac{1}{A_c} = \rho_m J(R_*) R_*^3 + 3\rho_m R^2 G N, \tag{3}$$

where μ_0 and μ_3 are the moments of order -0 and -3, defined as

$$\mu_0 = \int_0^\infty f dr = N, \qquad \mu_3 = \int_0^\infty R^3 f dr = N R^3, \tag{4}$$

in which f is the radial derivative of the droplet number density [8], N is the number of droplets per unit of mixture mass, R and R_* the average and the critical radius respectively, J the nucleation rate and G the growth rate.

2.3. Closure models

All the missing terms in Eq.(2) and (3) must be modelled in order to close the equation system. In particular, the nucleation rate J and the growth rate G are modelled as in [9]. The empirical parameter β that enters in the model for G is here set to 1. Finally, the critical radius is evaluated as

$$R_* = \frac{2\sigma}{\rho_l \Delta\mathcal{G}}, \tag{5}$$

where $\Delta\mathcal{G}$ is free Gibbs energy variation due to the condensation process.

2.4. Thermodynamic model

The required thermodynamic properties for the vapor phase are retrieved by an in-house thermophysical library [10]. On the contrary, the liquid properties, i.e. density and enthalpy, are directly implemented following the models in [11]. For the sake of accuracy the Gibbs free energy variation also includes the liquid contribution. $\Delta\mathcal{G}$ is defined as the sum of two contributions

$$\Delta\mathcal{G} = \Delta\left(\mathcal{G}_v - \mathcal{G}_l\right) = \Delta\mathcal{G}_v - \Delta\mathcal{G}_l. \tag{6}$$

To maintain a fluid-independent approach, $\Delta\mathcal{G}_v$ is evaluated through the same commercial thermodynamic library adopted for the continuum phase, whereas $\Delta\mathcal{G}_l$ is found as in [11]. The speed of sound for the mixture is evaluated as in [9]. Furthermore, the final mixture properties are found by means of the quasi-Newton algorithm. Finally, the surface tension is evaluated as in [12]: this model was compared with experimental measurements in equilibrium conditions up to the critical temperature, showing a maximum deviation of $0.4 \cdot 10^{-3} N/m$.

3. Numerical resolution

The conservation laws for both Model (a) and (b) are discretized by means of finite volumes schemes, whereas the moments equations are rewritten as

$$\frac{\partial}{\partial t}U + \frac{\partial}{\partial x}F = Q, \tag{7}$$

in which

$$U = \begin{bmatrix} \rho_m\mu_0 \\ \rho_m\mu_3 \end{bmatrix}, \qquad Q = \begin{bmatrix} \rho_m J(R_*) \\ \rho_m J(R_*)R_*^3 \end{bmatrix} - U \cdot v_m \frac{\partial A_c}{\partial x}\frac{1}{A_c}, \tag{8}$$

$$F = \begin{bmatrix} \rho_m\mu_0 v_m \\ \rho_m\mu_3 v_m - \sum_{i=0}^{x_{out}} 3\left(\left(\rho_m\mu_0\right)^{\frac{1}{3}}\left(\rho_m\mu_3\right)^{\frac{2}{3}} G \cdot \frac{\Delta x}{x_{outlet}-x_{inlet}}\right)_i \cdot \left(x - x_{inlet}\right) \end{bmatrix}, \tag{9}$$

where x_{inlet} and x_{outlet} are the inlet and outlet abscissae of the domain and Δx is the cell dimension. The system (7) is discretized via the upwind numerical scheme in [9]. The code adopts a segregated approach, i.e. the droplets properties are kept as constant during the resolution of the Eulerian equations. Implicit time integration is adopted to reach convergence.

4. Validation

The model was already validated for low pressure by using the well-established Moore nozzle A test case. For the sake of brevity, the work refers to [9] for a more detailed analysis. Results show a well agreement with regard to static pressure trend along the nozzle as well as a deviation from the experimental diameter of less than 1%. In order to validate the model at high pressure, the present study considers two steam expansions from [1], namely the tests No.18C and 18B. The next paragraph reports the comparison with the measurements.

4.1. High pressure investigation

The first test-case considered is No.18C, with total inlet conditions $P_0 = 100.7bar$ and $T_0 = 615.2K$ ($P_r = 0.46$, $T_r = 0.95$). The simulations were made on a 1000-cell mesh with implicit time integration. The expansion of highly superheated steam at same total pressure [1] is taken as reference.

Fig.1a shows the expansion in the P-T chart obtained with Model (a). Due to the very high speed of the steam, the onset of condensation manifests for a temperature 15 K lower than the corresponding T_{sat} at same pressure. A rapid change in the steam temperature and pressure is observed close to the nozzle throat after inception of condensation. The reason behind this behaviour is here explained: i) the number of drops rapidly increases from 0 to $3.2 \cdot 10^{16} kg^{-1}$, ii)

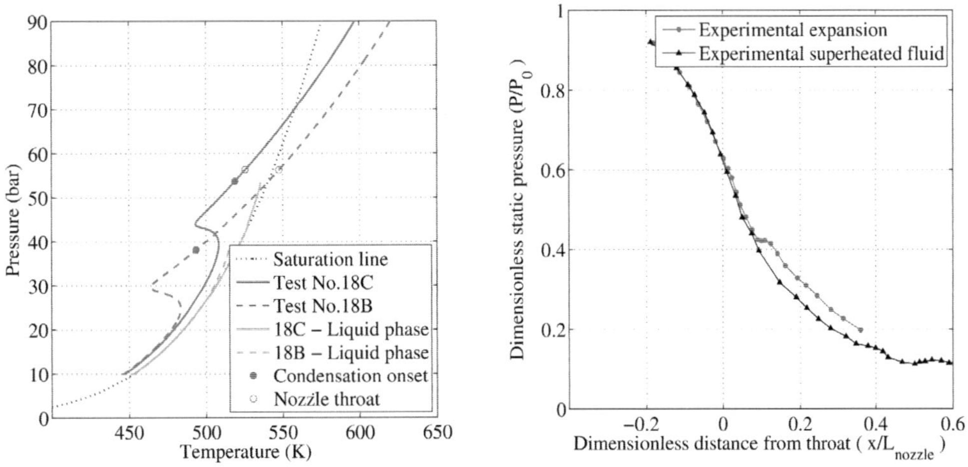

(a) Expansion on the P-T chart, test No.18B-C

(b) Static pressure field, experimental

Figure 1: Nozzle expansion in the P-T chart and pressure data (Test No.18C)

a large amount of energy in the form of latent heat is released and, as a consequence, iii) the total temperature of the fluid rises. The simplified equation for ideal gas [13] states that the static pressure increment depends on the total temperature and the cross sectional area variation. In this case, due to the latent heat released, the dominant term is the first one, thus, the expansion rate decreases with respect to the one at the nozzle inlet. Note that, depending on the test-case, the release of latent heat can even lead to a considerable increase of the static pressure, as clearly shown for the low pressure test case in [9]. After the metastable condensation started, the release rate of the latent heat decreases, and the total temperature variation becomes lower. The fluid is already in the divergent section of the nozzle, thus, the cross sectional area A_c is increasing. Therefore, the vapor continues the expansion and the static pressure rapidly reduces. Furthermore, the effect of metastabilities vanishes with time, and the vapor phase tends to reach equilibrium conditions: the steam approaches the saturation line at the nozzle outlet.

Given the assumptions of the model, the liquid phase is in mechanical equilibrium with the steam, thus, both phases have the same static pressure. Moreover, the temperature of the droplets is determined through a capillarity model as a function of the average radius R. The resulting thermodynamic state is close to the saturation line from the condensation onset till the nozzle outlet. Therefore, the droplets can be considered as saturated liquid.

Fig.1b depicts the experimental pressure field along the nozzle. Fig.2 reports the final solution obtained in terms of pressure and average radius. Both Model (a) and (b) provide similar results in terms of static pressure distribution (Fig.2a). The solution at the nozzle inlet and outlet is close to the measurements. However, a considerable deviation is observed in the condensation onset with respect to experimental observation. The characteristic pressure bump caused by the latent heat is observed in correspondence of a dimensionless static pressure of around 0.62 (62 bar) instead of 0.42 (42 bar) as for the experimental data (Fig.1b). Fig.2b depicts the average droplets radius along the nozzle. The trend for Model (a) is similar to the experimental curve, however the results obtained are almost twice the nominal values measured. The prediction for Model (b) is very close to the one of the first model (the difference is around 1%), even though the radius increases faster along the nozzle.

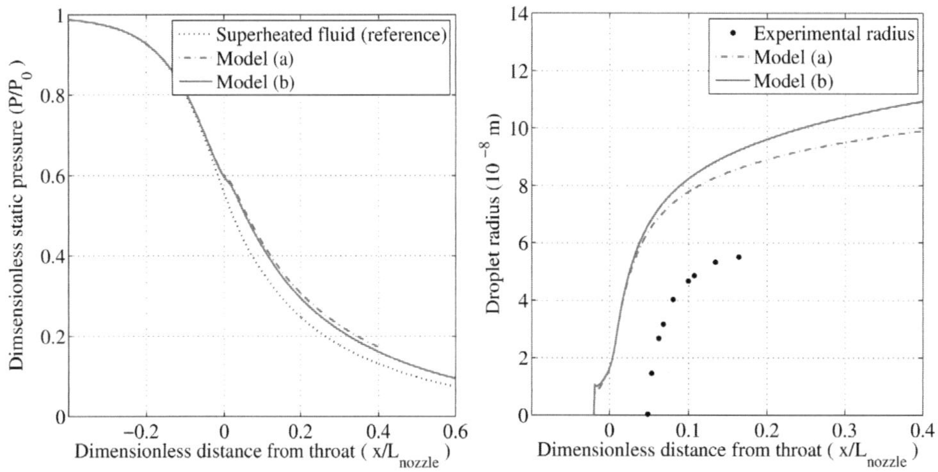

(a) Static pressure field along the nozzle (b) Averaged droplet radius

Figure 2: Comparison between Model (a), Model (b) and the results in [1], $x_{throat} = 0$

	Initial field	Single Phase	Model (a)	Model (b)
Physical time	Constant sol.	5 min, 47 s	83 min, 31 s	11 min, 29 s
	Superheated sol.	-	20 min, 20 s	8 min, 1 s

Table 1: Simulation time for single-phase, Model (a) and (b), explicit time integration, CFL=1

4.1.1. Comparison between model (a) and (b) In analogy with [9], a comparison is carried out between Model (a) and (b) in terms of computational cost and numerical stability. A first analysis is made with explicit time integration and CFL = 1. The simulations are stopped after a decrease of 4 orders of magnitude of the residual vector components. A single-phase simulation with highly superheated vapor at same total pressure is taken as benchmark.

Two different tests are made changing the initial motion field, i.e. (i) constant solution along the nozzle or (ii) superheated solution. Table 1 shows the physical time required by the benchmark, Model (a) and Model (b) respectively. For the first case, the comparison shows that the simulation time for Model (b) is two times higher than the single-phase one. On the other hand, it is worth pointing out that Model (a) requires a computational time that is more than 15 times greater with respect to the benchmark. As already observed by [9], the quasi-Newton algorithm adopted represents a large penalty for Model (a). Even for the second test, in which the initial motion field is close to the final solution, the presence of such iterative procedure leads to a time that is more than 2 times higher than Model (b). To test the numerical stability of the two models, a second analysis on the maximum CFL allowable is made. The tests are made on the same 1000-cell mesh with implicit time integration and constant CFL. The initial motion field is again the constant solution along the nozzle. Table 2 shows the final result.

Note that the maximum CFL allowable for all the models is relatively low, as it is kept as constant during the simulation. The introduction of a time-dependent CFL allows for a maximum value of more than 30 for both Model (a) and (b).

	Single Phase	Model (a)	Model (b)
Max CFL allowable	11	4	9

Table 2: Maximum CFL for single-phase, Model (a) and (b), implicit time integration

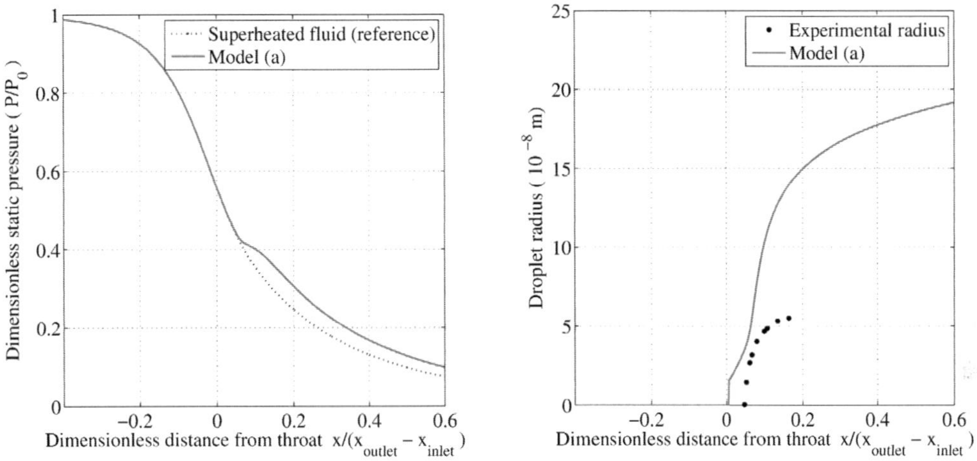

(a) Static pressure field along the nozzle

(b) Droplet radius along the nozzle

Figure 3: Static pressure and radius for Model (a) with $r_\sigma = 1.5$, $x_{throat} = 0$

5. Discussion

One of the key properties affecting the characteristics of metastable condensation is the surface tension σ. The underestimation of this parameter leads to a substantial anticipation of condensation onset resulting in fairly large difference in pressure and temperature trends. The work in [14] introduced a new parameter σ^* defined as

$$\sigma^* = r_\sigma \sigma, \tag{10}$$

in which r_σ is an empirical correction incorporating the effect of droplets curvature. It is reported that in the case $r_\sigma > 1$ the condensation onset is delayed when compared to $r_\sigma = 1$. However, the change in the nucleation starting point comes along with a considerable increase in the droplet radius. To gain insight of this behaviour, a simulation was performed with $r_\sigma = 1.5$. Fig.3 shows the pressure field and the radius obtained. Additionally, Fig.4 depicts the trend for the nucleation and the growth rate.

It can be observed that J decreases of one order of magnitude (Fig.4a): Eq.(5) shows that an increase in σ^* reflects on the critical radius value. Furthermore, the nucleation rate has an exponential dependence from R_*^2 and σ. As a consequence, J falls down from $1.3 \cdot 10^{24} kg^{-1}s^{-1}$ to $1.8 \cdot 10^{23} kg^{-1}s^{-1}$, and the number of drops N significantly decreases. At the same time, the higher critical radius required for the nucleation causes a delay in the condensation onset, thus, an increase in the steam degree of subcooling ΔT_{sub}. Due to the linear dependence of the growth rate from ΔT_{sub}, G reaches a value 1.5 times higher than that one obtained with $r_\sigma = 1$. Ultimately, the combined effect of the droplets number reduction and the increase of G leads to exceedingly high values of the average radius (Fig.3b).

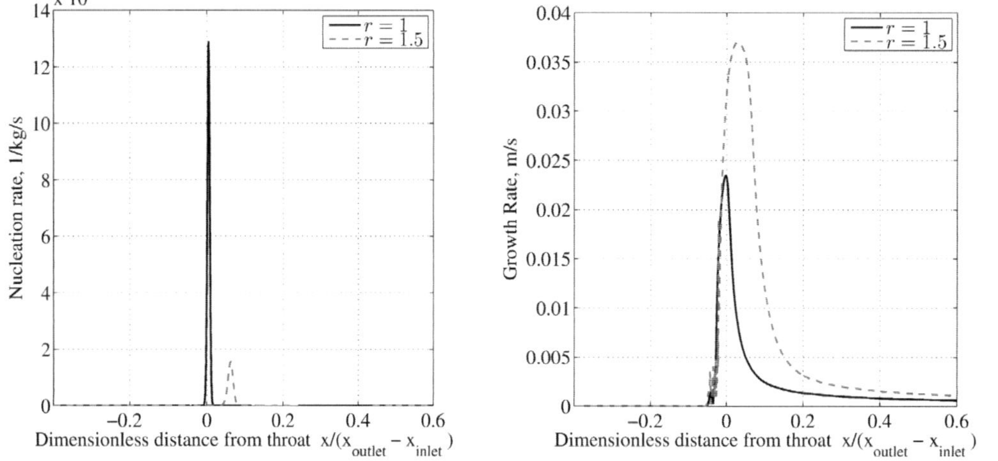

(a) Nucleation rate along the nozzle

(b) Growth rate along the nozzle

Figure 4: Comparison between nucleation and growth rate for $r_\sigma = 1$ and $r_\sigma = 1.5$, $x_{throat} = 0$

5.1. Correction factors for J and G

Based on the previous considerations, it was hypothesized that additional corrections were needed to the model in order to better fit the experimental values of the average radius. In analogy with what done for the surface tension two additional factors r_J, r_G are then introduced. Multiple tests were carried out on the test case No.18C to empirically determine the values of all the three coefficients for the best data fit (Table 3). The analysis showed that the variation of the nucleation rate slightly affects the location of the condensation onset (columns (4) and (5) in Table 3). In particular, the reduction of J causes a nucleation delay, in agreement with the behaviour observed in the previous section. However, the influence of this correction factor is rather limited: the change of r_J in the range 1 to 0.3 shifts the pressure bump from 0.44 to 0.42. Moreover, the corresponding radius variation is also minimal, from $5.7 \cdot 10^{-8}m$ to $5.5 \cdot 10^{-8}m$ (the value is taken in correspondence of the last experimental point in [1]). As observed in columns (3) and (4) in Table 3, the growth rate has a minor influence on the pressure field with respect to the surface tension. However, it considerably affects the average radius (from $13 \cdot 10^{-8}m$ to $5.7 \cdot 10^{-8}m$). The best fit with the experimental data is obtained for $r_\sigma = 1.4$, $r_J = 0.3$, $r_G = 0.23$. Fig.5 shows the achieved pressure field and average radius.

	$r_\sigma = 1$ $r_J = 1$ $r_G = 1$	$r_\sigma = 1.4$ $r_J = 1$ $r_G = 1$	$r_\sigma = 1.4$ $r_J = 1$ $r_G = 0.23$	$r_\sigma = 1.4$ $r_J = 0.3$ $r_G = 0.23$	Exp
Pressure bump location (Dimensionless static pressure)	0.62	0.46	0.44	0.42	0.42
Average radius $(\frac{x}{Lnozzle} = 0.16)$, $10^{-8}m$	8.6	13	5.7	5.5	5.5

Table 3: Determination of r_σ, r_J, r_G: condensation pressure and average radius, Test No.18C

It is worth pointing out that the corrections are approximatively equal to 1 for the low-pressure test case. In particular, the final solution presents no deviation in the condensation

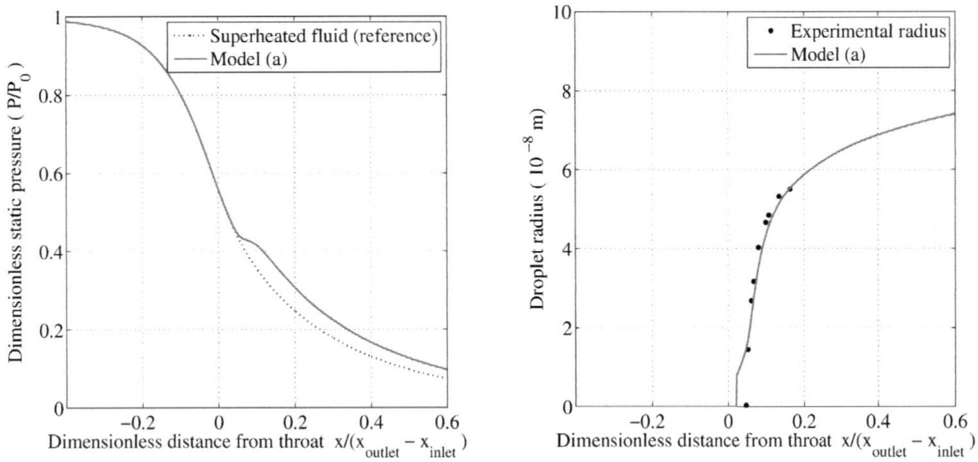

(a) Static pressure field along the nozzle

(b) Droplet radius along the nozzle

Figure 5: Static pressure and radius, Model (b), $r_\sigma = 1.4$, $r_J = 0.3$, $r_G = 0.23$, $x_{throat} = 0$

starting point, and a very high accuracy on the radius ([9]) with respect to the measurements. Additionally, the droplet dimension is comparable for both the low and high pressure test case ($5.0 \cdot 10^{-8}m$ and $5.5 \cdot 10^{-8}m$ as average radius respectively at the nozzle outlet). Therefore, it can be inferred that r_σ, r_J, r_G are not only a function of the droplet curvature as in [14], but they arguably show a dependence from the fluid thermodynamic conditions. To support this hypothesis, a second simulation was carried out on the test No.18B in [1], characterized by the same total pressure (100.7 bar) but a different total temperature (638.53 K) (Fig.1a). Table 4 reports the final results. As condensation starts in the same thermodynamic region at a slightly lower pressure and temperature, it is expected to find values of r_σ, r_J, r_G closer to 1 with respect to the previous case 18C. This hypothesis was confirmed by the results, which gave final values of $r_\sigma = 1.35$, $r_J = 0.35$, $r_G = 0.26$. A further investigation is required to gain insight of i) the influence of pressure and temperature on these parameters and ii) whether a correction of σ, J, G could be sufficient to increase the accuracy of the numerical model closer to the critical point of fluids.

	$r_\sigma = 1$ $r_J = 1$ $r_G = 1$	$r_\sigma = 1.4$ $r_J = 0.3$ $r_G = 0.25$	$r_\sigma = 1.35$ $r_J = 0.35$ $r_G = 0.25$	$r_\sigma = 1.35$ $r_J = 0.35$ $r_G = 0.26$	Exp
Pressure bump location (Dimensionless static pressure)	0.43	0.29	0.30	0.30	0.30
Average radius $\left(\frac{x}{Lnozzle} = 0.45\right)$, $10^{-8}m$	7.6	6.5	6	6.2	6.2

Table 4: Determination of r_σ, r_J, r_G: condensation pressure and average radius, Test No.18B

6. Conclusions

In this paper, high-pressure condensing steam flows were simulated using the method of moments. A comparison in terms of accuracy, computational cost and numerical stability was carried out between two alternative numerical approaches proposed in [9], employing the mixture or the continuum phase conservation laws, i.e. Model (a) and (b) respectively. The study showed the superiority of Model (b) compared to Model (a) in terms of robustness and computational efficiency even for high-pressure condensation. Nonetheless, both models were found quite inaccurate in terms of static pressure trend, due to a considerable deviation in the location of condensation onset. Furthermore, the average radius was found almost two times larger than the measurements. New correction factors were then proposed and implemented in the models to improve accuracy. In particular, the work introduced three correction factors for the surface tension, the nucleation rate and the growth rate model r_σ, r_J, r_G. It was observed that the rise of the surface tension significantly affects the final solution, causing a delay in the condensation onset as well as an increase of the average radius. Additionally, the droplet dimension are significantly affected by the growth rate G. On the other hand, J is the parameter with less influence, as shown in Table 3. The values of r_σ, r_J, r_G were determined empirically to obtain the best fit with the experimental data, reaching almost the same accuracy as in the low-pressure test case in [9]. A second analysis was carried out on the so-called No.18B (Table 4) to confirm the dependence of r_σ, r_J and r_G on the fluid thermodynamic conditions. Future efforts will be devoted to establish a physical correlation for these three parameters exploiting the data in [1] aiming at improving the reliability of the model for metastable condensation of steam (and possibly other fluids) close to the critical point, while maintaining high computational efficiency.

References

[1] Gyarmathy, G., 2005. "Nucleation of steam in high-pressure nozzle experiments". *Proceedings of the Institution of Mechanical Engineers, Part A: Journal of Power and Energy,* ***219***(6), jan, pp. 511–521. SAGE Publications.

[2] Lettieri, C., Yang, D., and Spakovszky, Z., 2015. "An investigation of condensation effects in supercritical carbon dioxide compressors". *J. Eng. Gas Turbines Power,* ***137***(8), feb, p. 082602. ASME International.

[3] Elliot, D., 1982. "Theory and tests of two-phase turbines". *Jet Propulsion Laboratory pubblication, California Institute of Technology, Pasadena, California*(81-105).

[4] White, A. J., 2003. "A comparison of modelling methods for polydispersed wet-steam flow". *International Journal for Numerical Methods in Engineering,* ***57***(6), pp. 819–834. Wiley-Blackwell.

[5] Gerber, A., and Kermani, M., 2004. "A pressure based eulerian–eulerian multi-phase model for non-equilibrium condensation in transonic steam flow". *International Journal of Heat and Mass Transfer,* ***47***(10-11), may, pp. 2217–2231. Elsevier BV.

[6] Dykas, S., and Wróblewski, W., 2012. "Numerical modelling of steam condensing flow in low and high-pressure nozzles". *International Journal of Heat and Mass Transfer,* ***55***(21-22), oct, pp. 6191–6199. Elsevier BV.

[7] Bakhtar, F., and Zidi, K., 1990. "Nucleation phenomena in flowing high-pressure steam part 2: theoretical analysis". *ARCHIVE: Proceedings of the Institution of Mechanical Engineers, Part A: Journal of Power and Energy 1990-1996 (vols 204-210),* ***204***(41), jun, pp. 233–242. SAGE Publications.

[8] Put, F., 2003. "Nucleation of steam in high-pressure nozzle experiments". PhD thesis, 19 Sepember. PhD Thesis, University of Twente, Enschede, The Netherlands. Url http://www2.ts.ctw.utwente.nl/rob/TS_publications/PDF/F.Put.pdf.

[9] Azzini, L., der Stelt, T. V., and Pini, M., 2016. "Numerical investigation of metastable condensing flows with an implicit upwind method". *VI European Conference on Computational Fluid Dynamics ECCOMAS, 5-10 June, Crete, Greece.*

[10] Colonna, P., and der Stelt, T. V., 2004. Fluidprop: a program for the estimation of thermo physical properties of fluids. Delft University of Technology, The Netherlands, http://www. FluidProp. com.

[11] Wagner, W., et al., 2008. "IAPWS industrial formulation 1997 for the thermodynamic properties of water and steam". pp. 7–150. Springer Science Business Media.

[12] Vargaftik, N. B., Volkov, B. N., and Voljak, L. D., 1983. "International tables of the surface tension of water". *Journal of Physical and Chemical Reference Data,* ***12***(3), p. 817. AIP Publishing.

[13] Greitzer, E. M., Tan, C. S., and Graf, M. B., 2004. *Internal Flow.* Cambridge University Press (CUP). http://dx.doi.org/10.1017/cbo9780511616709.

[14] Lai, D. S., and Kadambi, J. R., 1990. "Monodisperse droplet generation by spontaneous condensation of steam flow in nozzles". *Particulate Science and Technology,* ***8***(1-2), jan, pp. 55–75. Informa UK Limited.

Numerical simulation of cavitating flows under uncertainty

M.G. Rodio[1], R. Abgrall[2] and P.M. Congedo[3]

[1] Den-Service de thermo-hydraulique et de mécanique des fluides (STMF), CEA, Université Paris-Saclay, F-91191, Gif-sur-Yvette, France
[2] Universität Zürich, Winterthurerstrasse 190, CH-8057 Zürich, Switzerland
[3] Team CARDAMOM, INRIA Bordeaux Sud-Ouest, Talence, 33405 Cedex, France

E-mail: `mariagiovanna.rodio@cea.fr`

Abstract. Cavitation is characterized by vapor bubbles creation in the liquid phase as a consequence of a pressure drop. This phenomenon can be reproduced by means of several two-phase models. An equation of state is commonly used in order to define the thermophysical properties of the two fluids and to close the model. The aim of this work is to study how the uncertain parameters of the equation of state (EOS) can influence the prediction of the cavitation structures.

These uncertainties are propagated through a two-phase numerical solver for evaluating the impact on the predictive character of the numerical solution.

The variability of the mixture velocity and the mixture pressure are analyzed.

1. Introduction

The cavitation is a phenomenon affecting several mechanical devices. Classical examples are the propellers, the hypervelocity projectiles and all devices in which a section decrease can generate a very fast liquid acceleration, as in the pump or in the rocket engines. Since the nucleation of a new phase appears, cavitation can yields dramatic consequences, such as material erosion [21], noise [22] and performance decrease [23].

Several numerical models have been proposed to simulate this phenomenon. Principally two classes can be identified: (1) methods considering interfaces as true discontinuities and (2) methods allowing a numerical diffusion at the interface.

Methods belonging to the second class treat the interface like an artificial transition region where the thermodynamic conditions are unknown and they are known as *Interface diffusive Models*[5, 4, 3].

In this work, a five equation model, with heat and mass transfer terms, is used for reproducing cavitating flows, supposing that the two phases have the same velocity and pressure. The numerical approach proposed in [6] has been followed here. It is based on a splitting procedure based on three steps for assuring a good accuracy and a reduced computational cost. The step associated to definition of the heat and mass transfer terms is strongly connected to the EOS chosen in order to close the system. This means that the mass transfer prediction is sensitive to the choice of the EOS parameters [2, 7, 6].

Despite the use of stochastic methods applied to the numerical simulation in fluid mechanics being more and more diffused, only few studies exist concerning the application of uncertainty quantification tools to cavitating flows. Li *et al.* [9] proposed a Markov stochastic model to reproduce the random behaviour of cavitation bubble(s) near compliant walls. Fariborza *et al.* [8] proposed an empirical model for the time-discrete stochastic nucleation of intergranular creep cavities. They assumed nucleation to occur randomly in time, with the temporal behavior being governed by an inhomogeneous Poisson process. Giannadakis *et al.* [10] described the bubble breakup in lagrangian models using a stochastic Monte-Carlo approximation. This study was oriented on the particular topic of cavitation in the Diesel nozzle holes. Wilczynski [11] and Goel *et al.* [12] performed an uncertainties-based study on some hydrodynamic cavitation model parameters. In particular, Wilczynski [11] applied a stochastic model to capture the interaction of turbulent pressure field on cavitation nuclei population. Moreover, Goel *et al.* [12] performed a sensitivity analysis on several empirical parameters used typically in two-phase models. This study was performed using a finite differences method. In this case, input data uncertainty characterization is not required for the sensitivity analysis, that can be performed basing only on the mathematical form of the model. A recent work [13] presents a sensitivity study of the cavitation and turbulence models, where different combinations of empirical coefficients are considered. Recently, Rodio & Congedo [14, 26, 25] have proposed a study about the impact of various sources of uncertainty (on the cavitation model and on the inlet conditions) on the prediction of cavitating flows by coupling a non-intrusive Polynomial Chaos stochastic method with a cavitating CFD solver.

The aim of this paper is to evaluate the influence of experimental and EOS uncertainties on the cavitation model prediction. Note that a Stiffened Gas EOS is used here.

First, the uncertainties and their variation are estimated. Then a forward problem is applied, *i.e.* the uncertainties are propagated through a two-phase numerical solver for evaluating the impact on the predictive character of the numerical solution. In particular, the variability of some quantities of interest, such as the cavitation length, mixture pressure, is computed, thus permitting to evaluate the robustness of the physical model.

The paper is organized as follows. Section 2 presents the five-equation model with heat and mass transfer. Section 3 illustrates the main ingredients of the splitting method used for solving the cavitation model. In Section 4, the Stiffened Gas equation of state is be described. In section 5, some details concerning the Uncertainty Quantification method are provided. Finally, section 6 illustrate numerical results obtained on the most used numerical configuration in literature.

2. The cavitation model

Supposing a mechanical relaxation between the two phases, *i.e.* supposing that the two phases have the same pressure and velocity, a five equations model can be used for reproducing a cavitation problem. It is composed by a transport equation for the vapor volume fraction, the mass conservative equations for each phase and, finally, the momentum and the energy conservative equation for the liquid/vapor mixture, as follows:

$$\begin{aligned}
&\frac{\partial \alpha_1}{\partial t} + v\frac{\partial \alpha_1}{\partial x} = K\frac{\partial v}{\partial x} + \eta Q + \frac{\rho}{\rho_I}\dot{Y} \\
&\frac{\partial \alpha_1\rho_1}{\partial t} + \frac{\partial(\alpha_1\rho_1 v)}{\partial x} = \rho\dot{Y} \\
&\frac{\partial \alpha_2\rho_2}{\partial t} + \frac{\partial(\alpha_2\rho_2 v)}{\partial x} = -\rho\dot{Y} \\
&\frac{\partial \rho v}{\partial t} + \frac{\partial(\rho v^2 + P)}{\partial x} = 0 \\
&\frac{\partial \rho E}{\partial t} + \frac{\partial(\rho E + P)v}{\partial x} = 0,
\end{aligned} \tag{1}$$

where α_1, ρ_k, v, P, e_k, Q and $\dot{Y}$ are the gas volume fraction, the phase density, the mixture velocity, the mixture pressure, the specific phase internal energy , the heat term and the mass term, respectively. $E = e + \frac{1}{2}v^2$ is the mixture total energy and $e = (\alpha_1\rho_1 e_1 + \alpha_2\rho_2 e_2)/\rho$ is the mixture internal energy. The parameters $K = \frac{\alpha_1\alpha_2(\rho_2 c_2^2 - \rho_1 c_1^2)}{(\alpha_1\rho_2 c_2^2 + \alpha_2\rho_1 c_1^2)}$ and $\eta = \frac{\alpha_1\alpha_2}{\alpha_2\rho_1 c_1^2 + \alpha_1\rho_2 c_2^2}\left(\frac{\Gamma_1}{\alpha_1} + \frac{\Gamma_2}{\alpha_2}\right)$. For closing the system, an equation of state is required, in order to define e_k as a function of pressure and density for both phases, and the definition of Q and $\dot{Y}$. This is detailed in Section 3.

3. A splitting method

The system (1) is solved by applying a splitting method proposed in [6]. Assuming that the characteristic time of a mechanical relaxation, $1/\mu$, is much smaller than the characteristic time scales $1/\theta$ and $1/\nu$ of heat and mass transfer terms, respectively, the solution at each time can be found by performing the following three steps :

- *Step 1*: compute the numerical solution of the hyperbolic part of system (1) without heat and mass transfer term source.
- *Step 2*: update the solution of system (1), by solving the temporal ODEs system with the heat and mass transfer terms. This consists in applying a thermo-chemical relaxation that allows to attain a new equilibrium solution on the saturation curve.
- *Step 3*: the thermo-chemical relaxation does not guarantee that the positivity of the solution be preserved, so a positivity check of the solution is performed.

Since the originality of this work is focused on the uncertainty quantification analysis, we will describe, in the follow-up of this section, the main ingredients of the splitting method. We suggest, anyway, to read [6] for a more accurate description of the numerical method.

3.1. Step 1: Numerical solution of a five-equation model without heat and mass transfer

This step is necessary for solving the system (1) without heat and mass transfer terms. The problem is, thus, reduced to apply any numerical method able to solve a classical five equation model, as for example in [7, 3]. In this work, we applied the discrete equation method (DEM)[16], since it allows to treat the approximation of a non-conservative term, when the velocity $\vec{v}$ and K are simultaneously discontinuous [15, 16, 4].

3.2. Step 2: Numerical solution of the temporal ODEs with heat and mass transfer terms

This step consists in solving the following temporal ODEs, considering the new source terms:

$$
\begin{aligned}
\frac{\partial \alpha_1}{\partial t} &= \eta Q + \frac{\rho}{\rho_I}\dot{Y} := S_{\alpha_1} \\
\frac{\partial \alpha_1\rho_1}{\partial t} &= \rho\dot{Y} := S_{Y_1} \\
\frac{\partial \alpha_2\rho_2}{\partial t} &= -\rho\dot{Y} := -S_{Y_1} \\
\frac{\partial \rho v}{\partial t} &= 0 \\
\frac{\partial \rho E}{\partial t} &= 0.
\end{aligned}
\tag{2}
$$

The mass and heat transfer are two unknowns of the system. Applying a thermochemical relaxation, *i.e.* imposing temperature and Gibbs free energy equilibrium between the phases, a non trivial manipulation of the system 2 allows to find the final expressions for Q and $\dot{Y}$:

$$\begin{cases} Q = -\dfrac{B'}{AB' - A'B}\dfrac{(\Delta T)^n}{\Delta t} + \dfrac{B}{AB' - A'B}\dfrac{(\Delta g)^n}{\Delta t} \\ \dot{Y} = \dfrac{A'}{AB' - A'B}\dfrac{(\Delta T)^n}{\Delta t} + \dfrac{A}{AB' - A'B}\dfrac{(\Delta g)^n}{\Delta t} \end{cases} \tag{3}$$

where:

$$\begin{aligned} A &= -\xi\left[\left(\frac{T_2 D_2 Y_1}{T_1 D_1 Y_2} - 1\right)\left(Z_1\eta - \frac{T_2 C_{v_2}}{\rho}C_1\eta\right) + \frac{T_2}{\alpha_2\rho_2}C_1\eta\right] \\ B &= -\xi\left(\frac{T_2 D_2 Y_1}{T_1 D_1 Y_2} - 1\right)\left[\theta_1 + \frac{T_2 C v_2}{\rho}\left(C_2 - \frac{\rho}{\rho_I}\right)\right] + \frac{T_2}{\alpha_2\rho_2}\left(C_2 - C_1\frac{\rho}{\rho_I}\right) \\ A' &= -\left(\frac{1}{\rho_2} - \frac{1}{\rho_1}\right)\frac{c_1^2}{\alpha_1\gamma_1}\rho_1\eta + \frac{S_2 T_2}{\alpha_2\rho_2}C_1\eta - \xi\delta_1\left(Z_1\eta - \frac{T_2 C_{v_2}}{\rho}C_1\eta\right) \\ B' &= \left(\frac{1}{\rho_2} - \frac{1}{\rho_1}\right)\frac{c_1^2}{\alpha_1\gamma_1}\left(\rho - \rho_1\frac{\rho}{\rho_I}\right) - \frac{S_2 T_2}{\alpha_2\rho_2}\left(C_2 - C_1\frac{\rho}{\rho_I}\right) - \xi\delta_1\left[\theta_1 + \frac{T_2 C_{v_2}}{\rho}\left(C_2 - \frac{\rho}{\rho_I}\right)\right] \end{aligned} \tag{4}$$

and $\xi = (T_1 D1)/(Y_1(C_{v_1}T_1 D_1 + C_{v_2}T_2 D_2))$, the interface density has been defined in [2] as follows: $\rho_I = (\frac{\rho_1 c_1^2}{\alpha_1} + \frac{\rho_2 c_2^2}{\alpha_2})/(\frac{c_1^2}{\alpha_1} + \frac{c_2^2}{\alpha_2})$. Moreover, $D_1 = \alpha_1\gamma_1 c_2^2$, $D_2 = \alpha_2\gamma_2 c_1^2$, $C_2 = \frac{D_2 + D_1}{D_1}\rho$, $\theta_1 = (e_1 - e_2) + \rho\left[\frac{P_{\infty,2}Y_2}{\rho_2^2}\left(\frac{1}{\alpha_2} - \frac{\rho_2}{\alpha_2\rho_I}\right) - \frac{P_{\infty,1}Y_1}{\rho_1^2}\left(\frac{1}{\alpha_1} - \frac{\rho_1}{\alpha_1\rho_I}\right)\right]$, $Z_1 = Y_1\frac{P_{\infty,1}}{\rho_1\alpha_1} - Y_2\frac{P_{\infty,2}}{\rho_2\alpha_2}$, $Y_k = \alpha_k\rho_k$ and finally, $\delta_1 = \left(\frac{1}{\rho_2} - \frac{1}{\rho_1}\right)\frac{c_1^2\rho_1}{T_1\gamma_1} + S_1 - S_2\frac{T_2 D_2 Y_1}{T_1 D_1 Y_2}$. In all equations T_k, S_k, Cv_k, c_k and ρ_k are the phase temperature, entropy, heat capacity at constant volume, speed of sound and density respectively.

Note that the expressions of Q and $\dot{Y}$ have been obtained supposing the use of a Stiffened Gas as equation of state. So the terms 4 are clearly dependent on the EOS parameters. Their influence on the simulation is shown in section 4.

3.3. Step 3: Mass fraction and density positivity

The previous step allows to compute the heat and mass transfer terms, but it does not guarantee a consistent solution. So, in order to preserve the positivity of the solution, a limitation is put on the source terms, by determining the maximum admissible values, as follows:

$$S_{max,\alpha_1} = \begin{cases} \frac{1-\alpha_1}{\Delta t} & \text{if } S_{max} > 0 \\ \frac{-\alpha_1}{\Delta t} & \text{otherwise,} \end{cases} \qquad S_{max,Y_1} = \begin{cases} \frac{(1-\alpha_1)\rho_2}{\Delta t} & \text{if } S_{max} > 0 \\ \frac{-\alpha_1\rho_1}{\Delta t} & \text{otherwise.} \end{cases}$$

Thus, if $|S_{max,\alpha_1}| > |S_{\alpha_1}|$ and $|S_{max,Y_1}| > |S_{Y_1}|$, the source terms is used.

Otherwise, if the limit value is not respected, the gas volume fraction $\alpha_1 = 1$ is imposed if $S_1 > S_2$, otherwise $\alpha_1 = 0$. Knowing α_1, we can find the new pressure, P, by solving a single quadratic equation (see [6] for more details):

$$0 = (P^{\star\star})^2 + bP^{\star\star} + d \tag{5}$$

where:

$$\begin{aligned} b &= \frac{F_1}{F_2} + \frac{q_1(\alpha_1^{\star\star}C_{v_2}\Gamma_2) + q_2(\alpha_2^{\star\star}C_{v_1}\Gamma_1)}{Z_2} + F_2\frac{P_{\infty,1}(\alpha_1^{\star\star}C_{v_2}\Gamma_2) + P_{\infty,2}(\alpha_2^{\star\star}C_{v_1}\Gamma_1)}{\rho^\star\Gamma_1\Gamma_2} \\ d &= \frac{F_1}{Z_2}\frac{P_{\infty,1}(\alpha_1^{\star\star}C_{v_2}\Gamma_2) + P_{\infty,2}(\alpha_2^{\star\star}C_{v_1}\Gamma_1)}{\rho^\star\Gamma_1\Gamma_2} + \frac{P_{\infty,1}q_1(\alpha_1^{\star\star}C_{v_2}\Gamma_2) + P_{\infty,2}q_2(\alpha_2^{\star\star}C_{v_1}\Gamma_1)}{Z2} \end{aligned} \tag{6}$$

$$F_1 = P_{\infty,1}\gamma_1\frac{(\alpha_1^{\star\star}C_{v_2}\Gamma_2)}{C_{v_2}} + P_{\infty,2}\gamma_2\frac{(\alpha_2^{\star\star}C_{v_1}\Gamma_1)}{C_{v_1}} - (\rho e)^{\star}\Gamma_1\Gamma_2,$$

$$F_2 = \alpha_1^{\star\star}\Gamma_2 + \alpha_2^{\star\star}\Gamma_1 \qquad and \qquad Z_2 = F_2\frac{(\alpha_1^{\star\star}C_{v_2}\Gamma_2) + (\alpha_2^{\star\star}C_{v_1}\Gamma_1)}{\rho^{\star\star}\Gamma_1\Gamma_2}.$$

Then, by solving the single quadratic equation of $P^{\star\star}$, we select the physically admissible solution of the quadratic equation that maximizes the total entropy $s^{\star\star} = Y_1^{\star\star}s_1^{\star\star} + Y_2^{\star\star}s_2^{\star\star}$, where $Y_k^{\star\star} = (\alpha_k^{\star\star}\rho_k^{\star\star})/\rho^{\star\star}$

This procedure allows to reduce the computational cost compared to [2, 17]: it is not necessary to integrate over a fractional hydrodynamic time step.

4. Thermodynamic closure

The equation of state chosen for this analysis is the Stiffened Gas (SG). This EOS has been largely used for simulating two-phase flows problems. On the contrary of more complex EOSs, the expressions for each thermodynamic variables are simple and defined as follows:

$$\begin{aligned} P_k(\rho_k, e_k) &= (\gamma_k - 1)(e_k - q_k)\rho_k - \gamma_k P_{\infty,k}, \\ e_k(\rho_k, T_k) &= T_k c_{v,k} + \frac{P_{\infty,k}}{\rho_k} + q_k, \\ \rho_k &= \frac{P + P_{\infty,k}}{(\gamma-1)Cv_k T_k} \\ s_k(P_k, T_k) &= c_{v,k} ln\frac{T_k^{\gamma_k}}{(P_k + P_{\infty,k})^{(\gamma_k - 1)}} + q_k'. \end{aligned} \tag{7}$$

where P_k, ρ_k and e_k are the pressure, the density and the specific internal energy of the phase, respectively. The polytropic coefficient γ_k is the constant ratio of specific heat capacities $\gamma_k = c_{p,k}/c_{v,k}$, $P_{\infty,k}$ is a constant reference pressure,q_k is the specific creation energy of the fluid at a given reference state and q' is a fluid specific constant. Moreover, T_k and $c_{v,k}$ are the temperature, the specific heat at constant volume and the enthalpy, respectively. Moreover, the sound speed remains strictly positive, ensuring the hyperbolicity of the system. As explained in section 3.2, the mass term is dependent on the EOS parameters.

Let focus the attention on the parameter q'. It is usually estimated by means the procedure described in [2]. However, for the same fluid and the same test case, several values are used in literature. We focus the attention for example on the water and its vapor parameters used for the two-phase expansion tube test (see [2]). In [2, 17], $q' = -23.0 \times 10^3$, in [7] $q' = -23.4 \times 10^3$ and in [6] $q' = -23.2 \times 10^3$.

The aim of this paper is, thus, to evaluate and quantify the influence of this parameter on the model prediction.

5. Forward Uncertainty Quantification method

For the uncertainty propagation, a truncated Polynomial Chaos expansion (see [20]) is computed. Using this non-intrusive uncertainty quantification method means that a single deterministic computation is replaced with a whole set of computations, each one of those being run for specific values of the uncertain conditions.

Polynomial Chaos (PC) expansions are derived from the original theory of Wiener on spectral representation of stochastic processes using Gaussian random variables. Let ξ be a vector of standard independent random variables $\xi_i, i = 1, 2, ..., n_\xi$. Any well-behaved process u (i.e. a second-order process, then with a finite variance) can be expanded in a convergent (in the mean square sense) series of the form

$$u(\mathbf{x}, t, \xi) = \sum_{\alpha} u_\alpha(\mathbf{x}, t)\Psi_\alpha(\xi), \tag{8}$$

where α are multi-indices, $\alpha = (\alpha_1, \alpha_2, ..., \alpha_n)$, with each component $\alpha_i = 0, 1, ...,$ and Ψ_α are multivariate polynomial functions orthogonal with respect to the probability distribution function of the vector ξ. Each Ψ_α is defined by a product of orthogonal polynomials $\Phi_i^{\alpha_i}(\xi_i)$, that is, $\Psi_\alpha(\xi) = \prod_{i=1}^{n_\xi} \Phi_i^{\alpha_i}(\xi_i)$, where each $\Phi_i^{\alpha_i}$ is a polynomial of degree α_i, so that the degree of Ψ_α is $|\alpha|_1 = \sum_{i=1}^{n_\xi} \alpha_i$. A one-to-one correspondence exists between the choice of stochastic variable ξ_i and the polynomials $\Phi_i^{\alpha_i}(\xi_i)$. For instance, if ξ_i is a normal/uniform variable, the corresponding $\Phi_i^{\alpha_i}(\xi_i)$ are Hermite/Legendre polynomials of degree α_i. Coefficients $u_\alpha(x, t)$ are called the PC coefficients of the random process u and are obtained by

$$u_\alpha(\mathbf{x}, t) = \langle u(\mathbf{x}, t), \Psi_\alpha \rangle \|\Psi_\alpha\|^{-2}, \tag{9}$$

where the scalar product is defined by the expectation operator. For practical use, the PC expansions are truncated to degree No

$$u(\mathbf{x}, t, \boldsymbol{\xi}) = \sum_{|\alpha|_1 \leq No} u_\alpha(\mathbf{x}, t)\Psi_\alpha(\boldsymbol{\xi}). \tag{10}$$

The number of multivariate polynomials Ψ_α , that is, the dimension of the expansion basis, is related to the stochastic dimension n_ξ and the degree No of polynomials ; it is given by the formula $(n_\xi + No)!/(n_\xi!No!)$.

Several approaches can be used to estimate PC coefficients. The approach used in this study is based on Gauss-Legendre quadrature nodes. From the PC expansion of the random process, it is then easy to derive its mean and variance and to estimate sensitivity information using the analysis of variance (ANOVA) decomposition[18].

6. Results

In this section, we present some results. The test case reproduced for the validation of the cavitation model and the uncertainties propagation is a two-phase expansion tube.

Physical parameters are summarized in Table 1. A second order scheme has been performed for all the numerical test-cases by means of a MUSCL scheme[15], coupled with a Relaxation solver[16] and a Van Leer limiter. Note also that a mesh-convergence study has been done, even if it is not reported here for brevity.

Table 1. EOS coefficients for liquid and gas phases.

TEST CASE	FLUID	SG EOS					
		γ	P_∞ $[Pa]$	C_p $[\frac{J}{KgK}]$	C_v $[\frac{J}{KgK}]$	q	q'
Expansion tube	Liq. Water	2.35	10^9	4267	1816	-1167×10^3	0
	Vap. Water	1.43	0	1487	1040	2030×10^3	-23.2×10^3

6.1. Two-phase expansion tube test

Now, let us consider a tube of unit length, filled with water at atmospheric pressure $P = 10^5$, density $\rho_2 = 1150$ kg$/m^3$ and temperature $T_2 = 354.728$ K. A small amount of vapor, $\alpha_1 = 10^{-2}$, is uniformly distributed in the whole domain. By assuming the flow in thermal and pressure initial equilibrium $T_2 = T_1$, the vapor density can be computed by means of the equations (7)

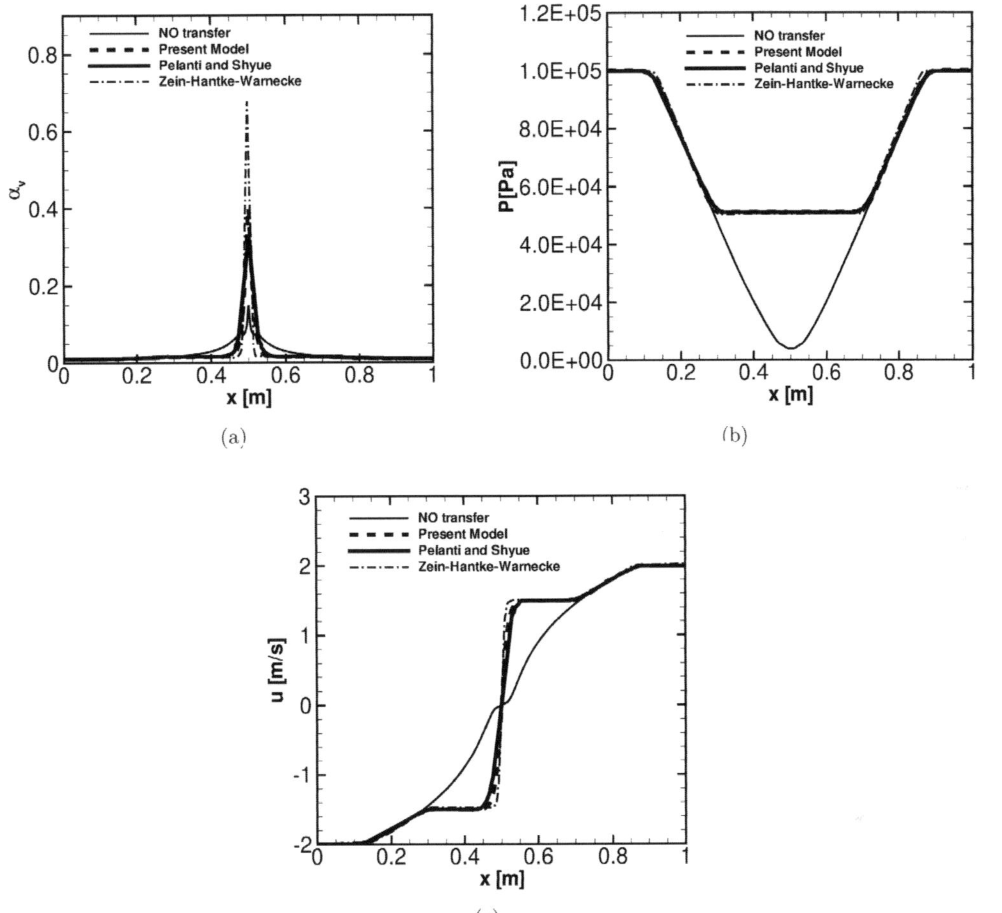

(a) (b) (c)

Figure 1. (a) Vapor volume fraction; (b) mixture pressure; (c) mixture velocity.

and it is equal to $\rho_1 = 0.63$ kg/m^3. The solution is shown at a time t=3.2 ms in Fig. 1. The computation is performed by using a mesh of 5000 cells. A velocity discontinuity ($v = -2$ m/s on the left and $v = 2$ m/s on the right) is set at x=0.5 m.

Supposing theoretically that the liquid phase can not evaporate, because of the incremental velocity, the pressure drops quite to zero in the center of the tube, as we can observe in figure 1 (solid line). In real conditions, the pressure can drops until it attains the saturation value corresponding to the flow temperature. So, between $x = 0.3$ m and $x = 0.7$ m, the pressure reaches its saturation value of about 5×10^4 Pa.

The comparison with the results obtained by Pelanti and Shyue [7] and by Zein *et al.* [17], shows a very good agreement in terms of pressure and velocity (Fig. 1b-c), but some differences in terms of vapor volume fraction (see Fig. 1a) can be observed in particular with [17]. Note that this difference could be explained by the different value used for the equation of state's parameter q'.

6.2. Uncertainty assessment

In this case, three uncertainties are considered: the EOS parameter, q', the initial left velocity and the vapour volume fraction. Because of the lack of knowledge concerning the parameter q', it is assumed as an epistemic uncertainty. Its variation range is chosen relying on a calibration with respect to the National Institute of Standards and Technology (NIST [24]) experimental data. In particular, the saturation curves data obtained in terms of pressure against temperature and in terms of the liquid volume against temperature are considered. By varying q' several curves are obtained (Figures 2(a) and 2(b)). Considering the experimental uncertainty provided by NIST, q' is assumed to vary between -23300 and -23500. s The initial left velocity is assumed as an epistemic uncertainty, with a arbitrary variation of $\pm 10\%$ with respect to the initial value of -2 m/s (treated with an uniform probability density function).

The variation of the initial vapor volume fraction is established by considering the existing literature about the existence of non-condensable gas in the liquid phase. In particular, relying on experimental observations for water at room temperature[1], the initial gas volume fraction is assumed to vary between $5 \cdot 10^{-3}$ and $1 \cdot 10^{-2}$.

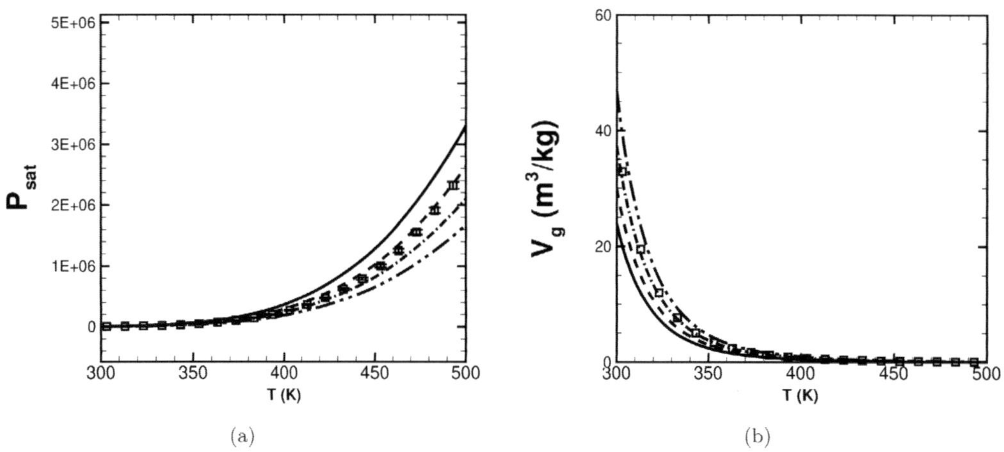

Figure 2. (a) Variability of the saturation curve ((a) pressure-temperature, (b) liquid volume-temperature) w.r.t. q': $q' = -23200$ (solid), $q' = -23300$ (dashed), $q' = -23400$ (dashdot) and $q' = -23500$ (dashdot dotted).

6.3. Forward propagation

First, the uncertainties propagation study has permitted to compute the mean and the variance of the pressure and the velocity, obtained at the final time (t=3.2 ms) along the x coordinate.

The mean profiles of the pressure is very similar to the deterministic solution, shown in figure 1b. However the numerical errors bars show a large variability of the pressure. In order to identify which is the most important uncertainty that produces this variability, an ANOVA analysis is then performed. Figure 3(b) shows the pressure variance and all the contributions to the variance related with the three uncertainties. The pressure variation is only dependent on the uncertainties on α_{GAS} and q'. In particular the initial α variation produces a non-negligible pressure variation, but it does not affect the transition phase region ($0.3 < x < 0.7$ m), which, on the contrary, is strongly influenced by the q' variation.

In terms of velocity, the variation is less important respect to the pressure one (see figure 4(a)). Anyway, we remember that the velocity variation has been applied to the left part of the tube. It explains the variance behavior for $x < 0.5$ m. In fact, the contribution of initial velocity contribution, v_{IN}, is more important than the others, but remains of the same order of magnitude. For $x > 0.5$ m, on the contrary, the v_{IN} contribution drops quite to zero and the q' contribution remains the most important in the cavitation region.

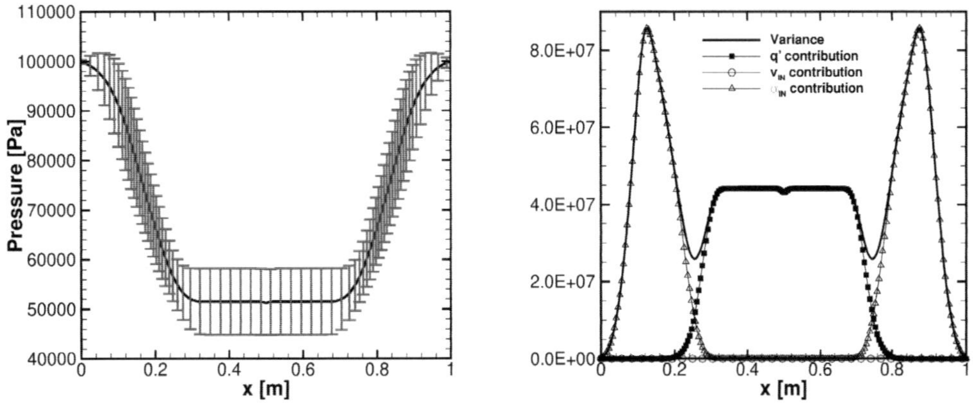

(a) (b)

Figure 3. (a) Mean pressure profile. The numerical error bars represents the pressure standard deviation for each x location. (b) ANOVA analysis: contributions of each source of uncertainty to the global variance of the pressure for each x location.

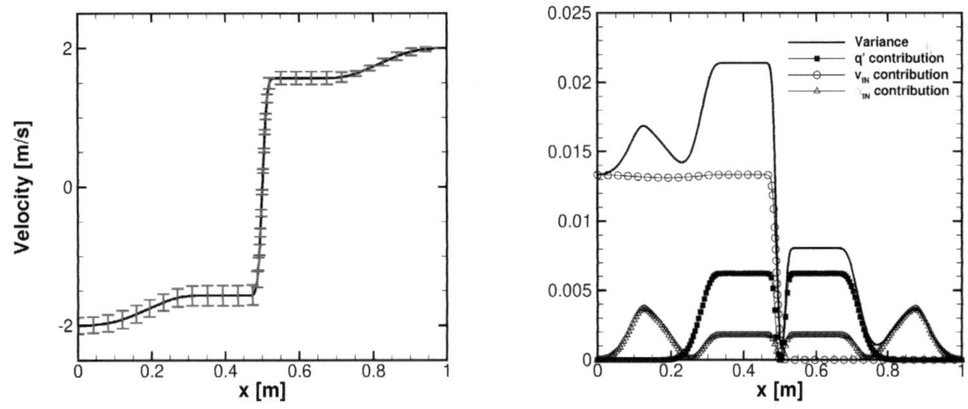

(a) (b)

Figure 4. (a) Velocity mean profile. The numerical error bars represents the pressure standard deviation for each x location. (b) ANOVA analysis: contributions of each source of uncertainty to the global variance of the pressure for each x location.

7. Conclusions and Perspectives

This paper is focused on an uncertainty propagation study aimed to estimate the influence of experimental uncertainties on the cavitation-driven phenomena. The influence of uncertainties on boundaries condition and EOS uncertainties have been assessed on the mixture pressure and velocity. The study showed that the cavitation region is strongly influenced by the EOS parameter uncertainties. This clearly questions the quality of the prediction in this kind of configuration, considering the ranges of variation.

References

[1] Brennen. C.E. 1995 *Cavitation and bubble dynamics* (Oxford University Press)
[2] Saurel R., Petitpas F. and Abgrall R. 2008 *J. Fluid Mech.* **607** 313-350
[3] Murrone A. and Guillard H. 2005 *J. Comput. Phys.* **202** 664-698
[4] Abgrall R. and Rodio M.G. 2014 *Comp. Fluids* **91** 164-181
[5] Saurel R. and LeMetayer O. 2001 *J. Fluid Mech.* **431** 239-271
[6] Rodio M.G. and Abgrall R. 20015 Int. J. Heat Mass Tran. **89** 1386-1401
[7] Pelanti M. and Shyue K.M. 2014 *J. Comput. Phys.* **259** 331-357
[8] Fariborza S.J., Harlowa D.G. and Delpha T. J. 1986 *Acta Metall.* **34** 1433-1441
[9] Li S., Zuo Z.G. and Li S.C. 2006 *J. Hydrodyn* **18**, 487-491
[10] Giannadakis E., Papoulias D., Gavaises M. *et al.* 2007 *Proc SAE Int. Conf.* (Detroit)
[11] Wilczynski L. 2000 *Advances in Fluid Mechanics III* **29** p 1
[12] Goel T., Thakur S., Haftka R.T. *et al.* 2008 *Int. J. Numer. Meth. Fluids* **58** 969-1007
[13] Tseng C.C. and Wang L.J. 2014 *Comp. Fluids* **103** 262-274
[14] Rodio M.G. and Congedo P.M. 2014 *Europ. J. Mech.-B/Fluids* **44** 88-99
[15] Abgrall R. and Saurel R. 2003 *J. Comput. Phys.* **186** 361-396
[16] Abgrall R. and Perrier V. 2006 *SIAM J. Sci Comput* **5** 84-115
[17] Zein A., Hantke M. and Warnecke G. 2010 *J. Comput. Phys.* **229** 2964-2998
[18] Crestaux T., Le Maître O. and Martinez J-M. 2009 *Relia. Eng. System. Safety* **94** 1161-1172
[19] Rodio M.G., Congedo P.M. and Abgrall R. 2014 *Europ. J. Mech.-B/Fluids* **45** 20-35
[20] Congedo P.M., Corre C. and Martinez J-M. 2011 *Comp. Meth. Appl. Mech. Engineering* **200** 216-232
[21] Franc J-P, Riondet M., Karimi A. and Chahine G.L. 2012 *Wear* **274-275** 248-259
[22] Wittekind D. and Schuster M. 2016 *Ocean Engineering* **120** 116-121
[23] Ju J., Jang J., Choi M. and Baek J.H. 2016 *Adv. Mech. Eng* **8** 1-9
[24] http://webbook.nist.gov/chemistry/fluid/
[25] Abgrall R., Congedo P.M., Geraci G. and Rodio M.G. 2014 *J. Comput. Phys.* **299** 281-306
[26] Congedo P.M., Goncalves E. and Rodio M.G., 2015 *Europ. J. Mech.-B/Fluids* **53** 190-204

A flamelet model for transcritical LOx/GCH$_4$ flames

Hagen Müller & Michael Pfitzner

Institut für Thermodynamik, Universität der Bundeswehr München, Werner-Heisenberg-Weg 39, 85577 Neubiberg, Germany

E-mail: hagen.mueller@unibw.de

Abstract. This work presents a numerical framework to efficiently simulate methane combustion at supercritical pressures. A LES flamelet approach is adapted to account for real-gas thermodynamics effects which are a prominent feature of flames at near-critical injection conditions. The thermodynamics model is based on the Peng-Robinson equation of state (PR-EoS) in conjunction with a novel volume-translation method to correct deficiencies in the transcritical regime. The resulting formulation is more accurate than standard cubic EoSs without deteriorating their good computational performance. To consistently account for pressure and strain fluctuations in the flamelet model, an additional enthalpy equation is solved along with the transport equations for mixture fraction and mixture fraction variance. The method is validated against available experimental data for a laboratory scale LOx/GCH$_4$ flame at conditions that resemble those in liquid-propellant rocket engines. The LES result is in good agreement with the measured OH* radiation.

Introduction

Todays main stage liquid-propellant rocket engines (LRE) typically operate at supercritical pressures, i.e., at chamber pressures that exceed the critical pressure of the propellants, and at cryogenic injection temperatures. One or both of the propellants are thus injected at near-critical conditions and mixing, ignition and combustion are affected by non-ideal thermodynamic effects. In particular, the thermodynamic- and transport properties, e.g., density, enthalpy, viscosity, are highly non-linear functions of temperature and pressure. Moreover, the experiments of Mayer *et al.* [1, 2] showed that the surface tension between liquid and vapor is diminished at sufficiently high pressures and mixing is characterized by continuous-phase diffusion rather than by two-phase spray atomization. In these diffusion mixing layers, the fluid properties change drastically and the density may vary by two orders of magnitude within a few micrometers.

Such configurations pose a serious challenge for numerical as well as for experimental studies. At the same time, their investigation is important to better understand the involved processes and to develop tools that help in the design of LREs, but also of other high pressure combustion devices, such as novel Diesel motors or gas turbines. The topic received considerable attention in the last decade and several valuable experiments were carried out. Thorough overviews are given, for instance, by Oschwald *et al.* [3] and by Chehroudi [4] focusing on the joint work of the U.S. Air Force Research Laboratory (AFRL) and the German Aerospace Center (DLR) as well as by Habiballah *et al.* [5] focusing on the Mascotte testing facility. Along with the better understanding that has been generated by the experimental efforts, several groups developed models and numerical tools to perform simulations of near-critical mixing and combustion allowing for a detailed view on the flow. Among the first to conduct large-eddy simulations (LES) of supercritical injection was Oefelein and Yang [6] and later Zong *et al.* [7] as well as Oefelein [8]. These studies showed that the accurate treatment of thermodynamic non-idealities are key to obtain realistic numerical representations. However, accuracy comes at the cost of numerical effort and a compromise has to be made in order to keep the computational effort manageable. Matheis *et al.* [9] identified and compared various volume-translation methods [10, 11] for cubic equations of state (EoS), such as Peng-Robinson (PR) [12] or Soave-Redlich-Kwong (SRK) [13]. These methods have the advantage to improve the predictive quality of cubic EoSs for near-critical conditions without a significant degradation of their good computational performance. In the present work, we employ the volume-translation method of Abudour *et al.* [14] which is based on the work of Chou and Prausnitz [15]. This method has already been used in our previous LES studies of supercritical nitrogen injection [16] as well as of inert coaxial LN_2/GH_2 injection [17].

In the present work, we focus on simulating the injection and combustion of LOx/GCH_4, at the experimental operating conditions of Singla *et al.* [18]. An additional subject that needs to be addressed is the choice of an appropriate combustion model for this configuration. In previous LES studies of this test case, Guézennec *et al.* [19] employed a reduced reaction mechanism accounting for finite-rate chemistry, while Schmitt *et al.* [20] made the assumption of infinitely fast chemistry. This assumption was also made by Cutrone *et al.* [21] as well as by Kim *et al.* [22] who simulated the test case of Singla *et al.* [18] using a Reynolds-averaged Navier-Stokes (RANS) flamelet model. In the present work, we employ the steady laminar flamelet model for LES [23, 24] using a presumed β-shape probability density function (PDF) for the subgrid-scale (SGS) fluctuations. The choice is supported by the comparative study of Zong *et al.* [25], who showed that the flamelet approach performed better than alternative combustion models for a splitter plate configuration at conditions that were similar to those in Singla *et al.'s* experiments. Furthermore, Ribert *et al.* [26] showed that the chemical consumption rates scale as the square-root of pressure. Thus, the assumption that chemical reactions are considerable faster than mixing time scales and the reaction zone is thin , which is a precondition for the flamelet approach, appears to be appropriate for the present high-pressure configuration. This is also supported by Lacaze and Oefelein [27] who investigated the applicability of the flamelet approach for the LES of supercritical LOx/H_2 combustion.

Numerical and physical modeling

LES equations

In LES, only the large-scale turbulent fluctuations are computed while fluctuations on subgrid-scale (SGS) level have to be modeled. The threshold between unresolved and resolved scale is determined by a certain filter width which is the local cell size in the present work. Filtering the Navier-Stokes equations leads to the filtered LES equations for mass, momentum and energy, cf. Garnier *et al.* [28].

As mentioned before, we employ the flamelet concept [24] to avoid the computational cost of solving a complex chemistry kinetics mechanism. The species composition in the flow field is represented by three parameters: the scalar dissipation rate $\tilde{\chi}$, the resolved mixture fraction $\tilde{f}$ and its subgrid variance $\widetilde{f''^2}$. The scalar dissipation is modeled with an algebraic equation, whereas additional transport equations are introduced for the mixture fraction and its variance that are solved along with the filtered Navier-Stokes equations:

$$\frac{\partial\left(\bar{\rho}\tilde{f}\right)}{\partial t} + \frac{\partial\left(\bar{\rho}\tilde{u}_i\tilde{f}\right)}{\partial x_i} = \frac{\partial}{\partial x_i}\left(\left(\frac{\bar{\mu}}{Sc} + \frac{\mu_e}{Sc_t}\right)\frac{\partial\tilde{f}}{\partial x_i}\right), \tag{1}$$

$$\frac{\partial\left(\bar{\rho}\widetilde{f''^2}\right)}{\partial t} + \frac{\partial\left(\bar{\rho}\tilde{u}_i\widetilde{f''^2}\right)}{\partial x_i} = \frac{\partial}{\partial x_i}\left(\left(\frac{\bar{\mu}}{Sc} + \frac{\mu_e}{Sc_t}\right)\frac{\partial\widetilde{f''^2}}{\partial x_i}\right) - 2\bar{\rho}\tilde{\chi} + 2\left(\frac{\bar{\mu}}{Sc} + \frac{\mu_e}{Sc_t}\right)\left(\frac{\partial\tilde{f}}{\partial x_i}\right)^2, \tag{2}$$

where a bar $\bar{\star}$ denotes the finite-volume filter and a tilde denotes Favre-filtering, i.e., $\tilde{\star} = \overline{\rho\star}/\bar{\rho}$. x_i are Cartesian coordinates, t is the time, $\bar{\rho}$ is the density, $\tilde{u}_i$ is the velocity component in direction i. μ and μ_e are the molecular and the eddy viscosity, respectively.

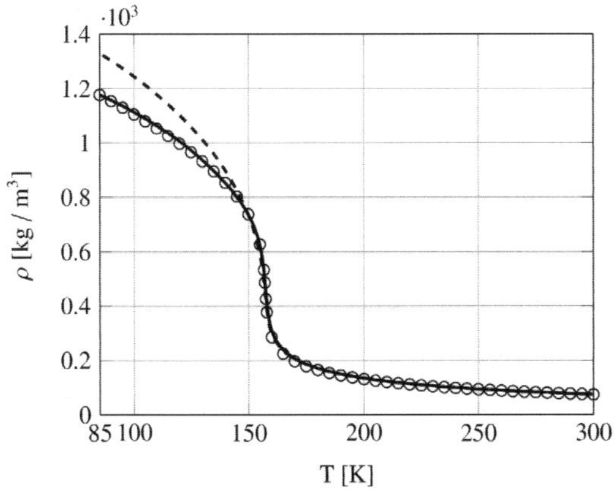

(a) Density

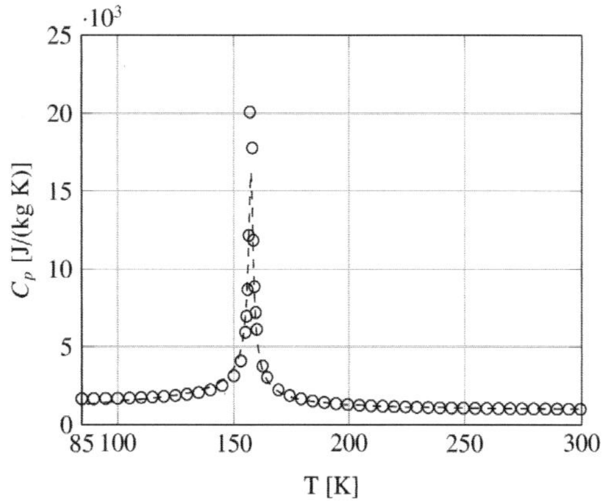

(b) Isobaric heat capacity

Figure 1. Verification of the real-gas thermodynamics model for O_2 at p = 5.61 MPa: — PR-VTA EoS, -- PR EoS, ∘ NIST chemistry webbook [33].

The eddy viscosity μ_e is calculated using the SGS turbulence model of Vreman [29]. The diffusion of mixture fraction and of mixture fraction variance is modeled using the standard gradient diffusion assumption with the diffusion coefficient being connected with the viscosity through the molecular and turbulent Schmidt number. These are set constant to $Sc = 1.0$ and $Sc_t = 0.7$, respectively. The scalar dissipation rate in Eq. 2 is decomposed into a resolved and a SGS contribution.

$$2\bar{\rho}\tilde{\chi} = 2\frac{\tilde{\mu}}{Sc}\left(\frac{\partial \tilde{f}}{\partial x_i}\right)^2 + C_\chi \frac{\mu_e}{Sc_t}\frac{\widetilde{f''^2}}{\Delta^2} \tag{3}$$

Here, Δ is the local filter size and the model constant C_χ is set to 2.

In the present work, we employ a *Pressure Implicit with Splitting of Operators* (PISO) algorithm [30, 31] for solution of the Navier-Stokes equations. Instead of solving the continuity equation directly, the discretized momentum equation is used to derive a pressure evolution equation, which guarantees mass conservation. This has the advantage that stiff equation systems, especially for low Mach number flows, are avoided and the computation can be run using larger time steps. However, several modifications are necessary to consistently incorporate real-gas thermodynamics models into the standard PISO approach. For details, please see the work of Jarczyk and Pfitzner [32].

Real-gas thermodynamics

The thermodynamic model is based on the mixture PR EoS [12]:

$$p = \frac{RT}{v_{PR} - b_m} - \frac{a_m}{v_{PR}^2 + 2v_{PR}b_m - b_m^2}, \tag{4}$$

where v_{PR} is the molar volume of the mixture, T is the temperature and R is the universal gas constant. Intermolecular attractive forces are described by the temperature dependent function a_m and the reduction of free volume due to the finite volume of the molecules is taken into account by b_m. The mixing rules for these parameters are taken from the work of Harstad *et al.* [10]. The PR EoS is an adequate choice for temperatures above and close to the critical temperature, however, in the transcritical regime considerable deviations from experimental data can be observed [11]. In the present work, we therefore employ the volume-translation method of Abudour *et al.* [14] (PR-VTA), who proposed a generalized form of the

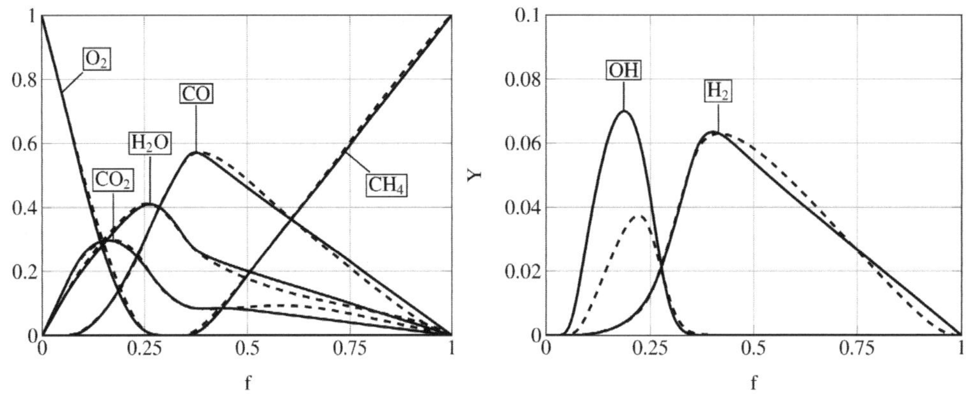

(a) Major species (b) Minor species

Figure 2. CH_4/O_2 flamelet solution for $\chi_{st} \approx 0$ and $p = 5.61$ MPa: —full mechanism (GRI 3.0), -- reduced mechanism of Frassoldati *et al.* [37].

method of Chou and Prausnitz [15] that considerably improves the accuracy of the density prediction at minimal extra computational cost. The basic principle is to shift the thermodynamic state of the fluid along its volume axis keeping pressure and temperature constant.

The caloric properties, e.g. heat capacities and enthalpy, are calculated using the departure function formalism. The quantity of interest is decomposed into an ideal gas contribution, which is calculated at a low reference pressure using NASA polynomials [34], and a pressure-dependent departure function. More details are given, for instance, by Poling *et al.* [35]. The viscosity and the thermal conductivity is modeled using the empirical correlation for dense fluids of Chung *et al.* [36].

The left plot in Figure 1 demonstrates the effect of the volume-translation method on the density prediction for O_2 at $p = 5.61$ MPa. While the uncorrected density deviates from the reference data of the NIST [33] for temperatures below 150 K, the density obtained with PR-VTA is in good agreement with the reference data over the entire temperature range considered in the present work. In particular, for the oxidizer injection condition ($T_{O2} = 85$ K, $p = 5.61$ MPa) of the test case presented later on, the deviation from NIST is 10.33% for the uncorrected PR EoS and only 0.005% for the PR-VTA EoS. The right side of Figure 1 shows the constant pressure heat capacity c_p. Note that only the PR EoS result is shown, since the volume translation of Abudour *et al.* could not be considered in the evaluation of the caloric properties for the reasons mentioned above, cf. Matheis *et al.* [9] for details.

Flamelet model for real fluids

A flamelet model for turbulent non-premixed combustion [23, 24] is employed to cut the cost of solving the chemistry kinetics along with the transport equations. The concept is based on the view of the turbulent flame as an ensemble of thin laminar diffusion flames, generally referred to as flamelets. The underlying assumption is that the characteristic chemical time scales are small compared to the turbulent time scales (high Damköhler number). Another condition for the validity of the flamelet approach is that turbulent vortices do not perturb the inner structure of the flame, i.e. the flame thickness is sufficiently small compared to the turbulent length scales (small turbulent Reynolds number). The main advantage of the flamelet approach is that the flamelets, which describe the local structure of the turbulent flame, are coupled to the turbulent flow by only a few parameters, i.e., the mixture fraction, its variance and the scalar dissipation rate. The chemistry kinetics can thus be handled separately from the turbulent flow. This feature is used to calculate the flamelets in a pre-processing step and store them in look-up tables which are accessed during the LES to retrieve the local species composition.

Table 1. Boundary conditions of Singla *et al.* [18].

LOx			CH_4		
$\dot{m}$ [g/s]	T_{in} [K]	ρ_{in} [kg/m^3]	$\dot{m}$ [g/s]	T_{in} [K]	ρ_{in} [kg/m^3]
44.4	85	1177	143.1	288	42

Flamelet models have often been used for flows at atmospheric pressure where broad experience has been gathered for different flame types. However, the approach needs to be revised when used to simulate combustion at supercritical pressures. The influence of real-gas thermodynamics on the inner structure of the supercritical hydrogen flames has been investigated by several groups [38, 26, 39, 40]. They show that thermodynamic non-idealities have no significant effect on the inner structure of the flame and an ideal gas law is sufficient to calculate the flamelets. The reason is the strong temperature increase near the flame front that limits the influence of real-gas effects to regions close to $f = 0$ and $f = 1$. A further aspect is the influence of pressure on the reaction kinetics. Since the reaction speed increases with the pressure, supercritical flames are shown to be thin and resistant to strain [38, 26]. This supports the flamelet assumptions of high Damköhler number and small turbulent Reynolds number. However, as pointed out by Lacaze and Oefelein [27], an additional energy equation is needed in the LES to account for pressure and strain variations. In particular, the effect of pressure on the flamelet solution is typically neglected in the flamelet calculation. Considering that the caloric properties are pressure-dependent when using the PR EoS, energy conservation can only be guaranteed by solving an additional energy equation in the LES.

In the present work, we use the Flamemaster software of Pitsch [41] to solve the flamelet equations for a strained counterflow diffusion flame configuration under the assumption that the fluid is an ideal gas. The reduced methane oxidation mechanism of Jones and Lindstedt [42] with additional reactions that account for dissociation of water and oxygen as proposed by Frassoldati *et al.* [37] is used to reduce the number of species. The flamelet solution is filtered using a β-shape PDF for the mixture fraction and a Dirac delta function for the PDF of the scalar dissipation rate to account for SGS fluctuations. The filtered species composition is accessed during the LES using the filtered values of mixture fraction, mixture fraction variance and scalar dissipation. The effect of pressure fluctuations and real-gas thermodynamics is thus not reflected in the species composition of the flame structure. However, we solve an enthalpy transport equation and calculate the thermodynamic properties in the LES using the real-gas thermodynamics models based on the volume correction of Abudour *et al.* [14].

Figure 2 shows the result of a CH_4/O_2 flamelet simulation at $\chi_{st} \approx 0$ and $p = 5.61$ MPa. To verify that the simplified reaction mechanism of Frassoldati *et al.* [37] is appropriate for the present configuration, the results are compared with a flamelet simulation using the full CH_4/O_2 reaction mechanism (GRI 3.0 [43]). Only minor differences can be observed for the major species.

Experimental and numerical setup

Reference experiment

The present numerical approach to simulate real-gas combustion is tested against the measurements of Singla *et al.* [18] who performed a series of experiments for trans- and supercritical LOx/CH_4 flames at the Mascotte testing facility. The test bench has previously been used to study high-pressure hydrogen flames [5, 44] and has been extended to allow for the use of methane as fuel. The combustion chamber is a 50×50 mm^2 square duct with a length of 400 mm and a converging-diverging nozzle at the chamber end. The propellants are injected through a single coaxial injector element that is mounted on the faceplate. All four side walls of the chamber can be equipped with rectangular windows to allow for optical measurements. Singla *et al.* [18] have used a intensified (CCD) camera to collect the spontaneous emission of OH* at a rate of 15 Hz. In addition the CH* signal has been detected with an ICCD camera at

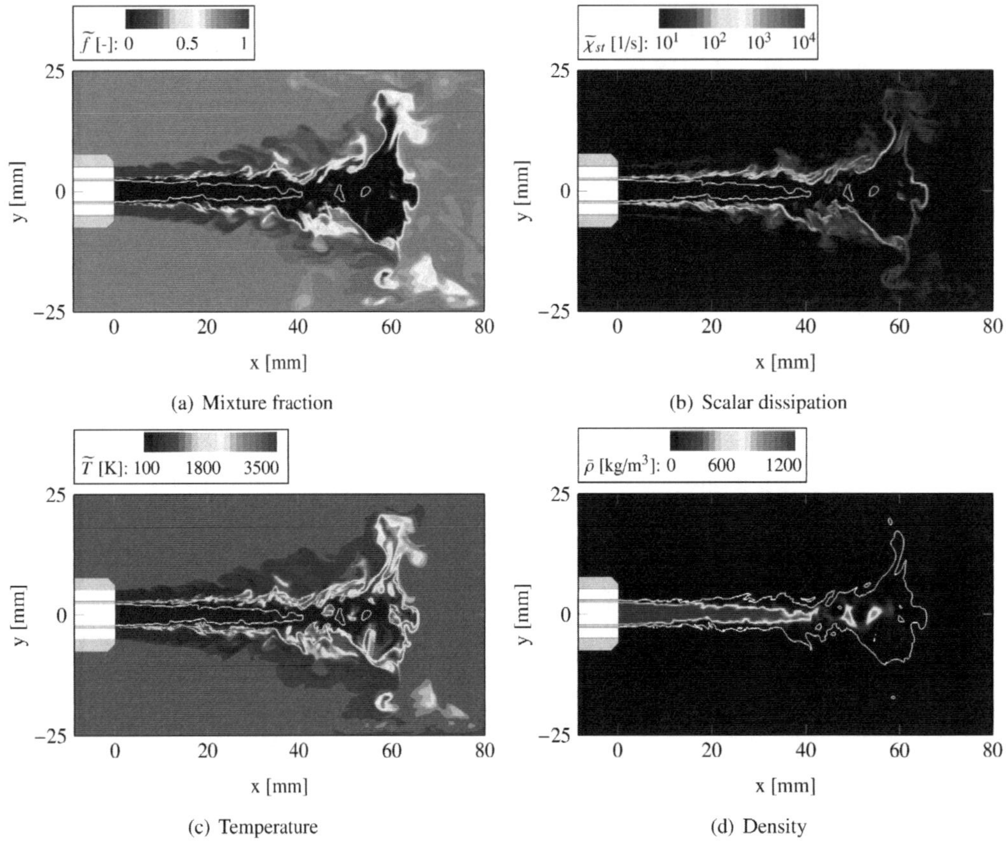

Figure 3. Instantaneous fields of mixture fraction, scalar dissipation rate, temperature and density in a plane perpendicular to the inflow plane. The white line in a), b) and c) denotes the pseudo-critical temperature of oxygen $T_{pb,O_2} = 157.4$ K. The white line in d) denotes the stoichiometric mixture $f = 0.2$.

a rate of 2.5 Hz. After averaging the instantaneous images, an Abel transformation has been carried out to obtain slices through the flame which are used for the present comparison with the numerical results.

Singla *et al.* [18] have performed several experiments varying both the chamber pressure and the injection temperature. For the present study, we simulate the operating point G2 for which the injection conditions are summarized in Table 1. The chamber pressure $p = 5.61$ MPa is supercritical with respect to the pure species ($p_{c,O2} = 5.04$ MPa, $p_{c,CH4} = 4.6$ MPa). The temperature of the injected oxygen $T_{O_2} = 85$ K is smaller than the pseudo-critical temperature at these conditions ($T_{pb,O_2} = 157.4$ K) and the fluid is thus in a transcritical state which is characterized by a high liquid-like density and viscosity. However, methane is injected in a purely supercritical state, i.e., both the pressure and the temperature exceed their critical value, and the fluid features a much smaller, gas-like density. The heat release at these conditions is 0.55 MW given that the oxidizer is fully consumed. This is a valid assumption considering that the mass flux ratio $\dot{m}_{O_2}/\dot{m}_{CH_4} = 0.31$ is well below the mass stoichiometric value of 4.

Computational setup

The LES is performed with the open-source CFD software OpenFOAM which has been extended by the real-gas flamelet model described above. The grid comprises 16×10^6 hexahedral cells and is block-structured. The minimal axial and radial cell size at the injector is 0.2 mm and 0.025 mm, respectively.

To reduce the numerical cost, the numerical domain is truncated after 150 mm. This is sufficient to accommodate the flame and avoid interference with the outflow boundary. The inner tube of the coaxial injector is $D = 4.4$ mm and the width of the annulus and the post is $\Delta D_{CH_4} = 2.2$ mm and $\Delta D_p = 0.6$ mm, respectively.

The turbulent inflow data for both the O_2 and the CH_4 inlet are generated with a separate precursor incompressible LES using cyclic boundary conditions in axial direction. The domain length for these simulations is $2\pi D_h$, where D_h is the hydraulic diameter of the respective inflow. Slices of the turbulent velocity fields are extracted from this simulation and accumulated in a database which is then used to interpolate the turbulent velocity field onto the coarser grid of the flame simulation.

For spatial discretization we use a second-order central differences scheme with a van Leer limiter. To avoid unphysical oscillations, it was necessary to use localized artificial dissipation μ^* in regions of high density gradients.

$$\mu^* = C_\mu \bar{\rho} a_s \Delta^2 \left| \frac{\partial Z}{\partial x_j} \right| \tag{5}$$

a_s and Δ are the speed of sound and the local filter width, respectively. The constant C_μ is set to 0.01. The method is similar to that proposed by Terashima *et al.* [45], however, as a sensor we use the local gradient of the compressibility factor ($Z = pv/RT$), which is zero in regions where real gas effects are negligible.

Results

General flame features

To give a qualitative impression of the general features of the flame, Figure 3 shows the instantaneous fields of mixture fraction, scalar dissipation rate, temperature and density. The white line in Figure 3 a), b) and c) denotes the pseudo-critical temperature of oxygen and marks the region of high density gradient. The fluid which is enclosed by this line features high liquid-like densities and a low velocity. The white line in Figure 3 d) denotes the stoichiometric mixture fraction and marks the flame front.

Close to the injector at $x < 15$ mm, a thin diffusion flame can be observed. The flame surface is perturbed by small scale turbulent structures and it exerts a relatively high level of strain. The maximum scalar dissipation rates in this region are in the order of 10^4 s^{-1}. Further downstream the boundary layer instability grows and large scale eddies evolve leading to a higher degree of flame wrinkling and a thickening of the mean flame front. However, the continuous dense oxygen core persists until $x \approx 40$ mm. Thereafter dense pockets of O_2 detach from the core. They are transported further downstream

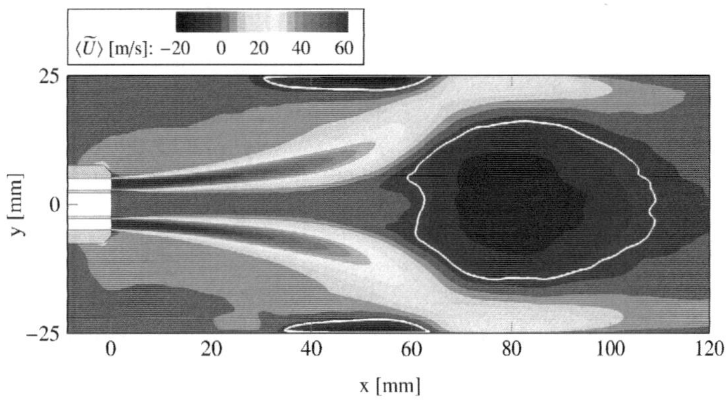

Figure 4. Averaged axial velocity field in a plane perpendicular to the inflow plane. The white line denotes $\langle U \rangle = 0$ m/s.

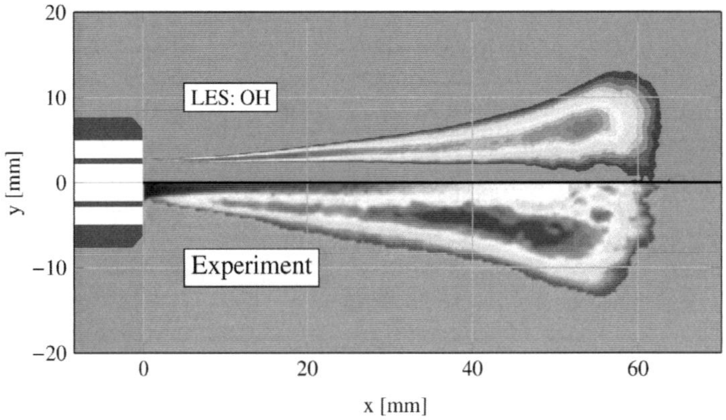

Figure 5. Averaged OH mass fraction in comparison with the measured OH* radiation of Singla *et al.* [18].

where they react with the surrounding fuel. In this region, i.e., at $x \approx 50$ mm, a strong radial expansion can be observed. The flow is redirected towards the chamber wall and reactions stop shortly thereafter at $x \approx 60$ mm.

This characteristic flame shape can be explained considering the averaged velocity field which is shown in Figure 4. The white lines denote zero axial velocity and mark the recirculation zones. In addition to the backflow region that forms close to the chamber wall between the flame and the face plate, a secondary recirculation can be observed downstream of the reaction zone. This was also observed in the LES result of Schmitt *et al.* [20]. Due to the interaction with the chamber wall in the rear part of the flame, the flow of hot combustion products is redirected towards the chamber axis where is flows back towards the flame. The momentum of the backflow further increases the expansion of the flame leading to a stable flow condition.

Comparison with experimental data

Temporal averaging was started after the flow field was considered fully developed in the region of interest and then continued for $\Delta t = 0.08s \approx 45D/\bar{U}_{O_2}$. In addition, we make use of the symmetry and spatially average the flow field using the 4 half planes that are perpendicular to the injector plane and the chamber wall. The averaged OH mass fraction is compared with the experimental result of Singla *et al.* [18] in Figure 5.

The characteristic flame shape, which has been discussed in the previous section, can also be seen in the experimental result. The first section of the flame is characterized by a constant spreading angle. At $x \approx 40$ mm a sudden expansion can be observed, which is followed by an abrupt end of reaction. This is well reproduced by the LES. Also the calculated length of the flame as indicated by the OH mass fraction is in excellent agreement with the measurement.

Conclusion

The current contribution presents a real-gas flamelet model for large-eddy simulations (LES) to simulate non-premixed combustion at supercritical pressures. The main challenges are the accurate modeling of thermodynamic non-idealities as well as their consistent coupling with the flamelet approach. The thermodynamic model used herein consists of a cubic Peng-Robinson equation of state (PR EoS) and a volume-translation method that corrects deficiencies of the PR EoS at low temperatures. This method yields a highly accurate thermal equation of state at reasonable computational effort. Pressure and strain fluctuations are taken into account by solving the energy equation along with the mixture fraction transport equation which represents the local species composition. Subgrid mixture fraction fluctuations are taken into account with a β-shape probability density function and a reduced methane reaction mechanism that consists of 9 species and 6 reactions is used to generate the flamelets.

Simulations have been performed for the experimental test case of Singla *et al.* [18], in which liquid oxygen and gaseous methane are injected into a rectangular combustion chamber. The pressure is supercritical. A comparison with the available OH^* radiation measurements shows that the predicted flame length is in excellent agreement with the experiment and characteristic flame features are well reproduced. The result indicates that the real-gas adapted flamelet approach for combustion at supercritical pressures that is presented herein is well suited to simulate this flame.

Acknowledgments
The authors gratefully acknowledge the German Research Foundation (Deutsche Forschungsgemeinschaft (DFG)) for providing financial support in the framework of SFB/TRR 40 *Fundamental Technologies for the Development of Future Space-Transport-System Components under High Thermal and Mechanical Loads* and the Gauss Centre for Supercomputing e.V. (www.gauss-centre.eu) for funding this project by providing computing time on the GCS Supercomputer SuperMUC at Leibniz Supercomputing Centre (LRZ, http://www.lrz.de).

References

[1] W. Mayer and H. Tamura, "Propellant injection in a liquid oxygen/gaseous hydrogen rocket engine," *Journal of Propulsion and Power*, vol. 12, no. 6, pp. 1137–1147, 1996.

[2] W. Mayer, A. Schik, B. Vielle, C. Chaveau, I. Gökalp, and D. Talley, "Atomization and breakup of cryogenic propellants under high pressure subcritical and supercritical conditions," *Journal of Propulsion and Power*, vol. 14, no. 5, pp. 835–842, 1998.

[3] M. Oschwald, J. J. Smith, R. Branam, J. Hussong, A. Schik, B. Chehroudi, and D. Talley, "Injection of fluids into supercritical environments," *Combustion Science and Technology*, vol. 178, no. 1-3, pp. 49–100, 2006.

[4] B. Chehroudi, "Recent experimental efforts on high-pressure supercritical injection for liquid rockets and their implications," *International Journal of Aerospace Engineering*, vol. 2012, 2012.

[5] M. Habiballah, M. Orain, F. Grisch, L. Vingert, and P. Gicquel, "Experimental studies of high-pressure cryogenic flames on the Mascotte facility," *Combustion Science and Technology*, vol. 178, no. 1-3, pp. 101–128, 2006.

[6] J. Oefelein and V. Yang, "Modeling high-pressure mixing and combustion processes in liquid rocket engines," *Journal of Propulsion and Power*, vol. 14, no. 5, pp. 843–857, 1998.

[7] N. Zong, H. Meng, S. Y. Hsieh, and V. Yang, "A numerical study of cryogenic fluid injection and mixing under supercritical conditions," *Physics of Fluids (1994-present)*, vol. 16, no. 12, pp. 4248–4261, 2004.

[8] J. Oefelein, "Mixing and combustion of cryogenic oxygen-hydrogen shear-coaxial jet flames at supercritical pressure," *Combustion Science and Technology*, vol. 178, no. 1-3, pp. 229–252, 2006.

[9] J. Matheis, H. Müller, C. Lenz, M. Pfitzner, and S. Hickel, "Volume translation methods for real-gas computational fluid dynamics simulations," *Journal of Supercritical Fluids*, vol. 107, pp. 422–432, 2016.

[10] K. G. Harstad, R. S. Miller, and J. Bellan, "Efficient high-pressure state equations," *AIChE Journal*, vol. 43, no. 6, pp. 1605–1610, 1997.

[11] K. Frey, M. Modell, and J. W. Tester, "Density-and-temperature-dependent volume translation for the SRK EOS: 1. Pure fluid," *Fluid Phase Equilibria*, vol. 279, no. 1, pp. 56–63, 2009.

[12] D.-Y. Peng and D. B. Robinson, "A new two-constant equation of state," *Industrial & Engineering Chemistry Fundamentals*, vol. 15, no. 1, pp. 59–64, 1976.

[13] G. Soave, "Equilibrium constants from a modified Redlich-Kwong equation of state," *Chemical Engineering Science*, vol. 27, no. 6, pp. 1197–1203, 1972.

[14] A. M. Abudour, S. A. Mohammad, R. L. Robinson, and K. A. M. Gasem, "Volume-translated Peng-Robinson equation of state for liquid densities of diverse binary mixtures," *Fluid Phase Equilibria*, vol. 349, pp. 37–55, 2013.

[15] G. F. Chou and J. M. Prausnitz, "A phenomenological correction to an equation of state for the critical region," *AIChE Journal*, vol. 35, no. 9, pp. 1487–1496, 1989.

[16] H. Müller, C. A. Niedermeier, J. Matheis, M. Pfitzner, and S. Hickel, "Large-eddy simulation of nitrogen injection at trans- and supercritical conditions," *Physics of Fluids (1994-present)*, vol. 28, no. 1, p. 15102, 2016.

[17] H. Müller, M. Pfitzner, J. Matheis, and S. Hickel, "Large-Eddy Simulation of coaxial LN2/GH2 injection at trans- and supercritical conditions," *Journal of Propulsion and Power*, vol. 32, no. 1, pp. 46–56, 2016.

[18] G. Singla, P. Scouflaire, C. Rolon, and S. Candel, "Transcritical oxygen/transcritical or supercritical methane combustion.," in *Proceedings of the Combustion Institute*, vol. 30, pp. 2921–2928, 2005.

[19] N. Guezennec, M. Masquelet, and S. Menon, "Large Eddy Simulation of flame-turbulence interactions in a LOX-CH4 shear coaxial injector," in *50th AIAA Aerospace Sciences Meeting including the New Horizons Forum and Aerospace Exposition: AIAA paper 2012-1267*, 2012.

[20] T. Schmitt, Y. Méry, M. Boileau, and S. Candel, "Large-eddy simulation of oxygen/methane flames under transcritical conditions," in *Proceedings of the Combustion Institute*, vol. 33, pp. 1383–1390, 2011.

[21] L. Cutrone, P. De Palma, G. Pascazio, and M. Napolitano, "A RANS flamelet-progress-variable method for computing reacting flows of real-gas mixtures," *Computers and Fluids*, vol. 39, no. 3, pp. 485–498, 2010.

[22] T. Kim, Y. Kim, and S. K. Kim, "Effects of pressure and inlet temperature on coaxial gaseous methane/liquid oxygen turbulent jet flame under transcritical conditions," *Journal of Supercritical Fluids*, vol. 81, pp. 164–174, 2013.

[23] N. Peters, "Laminar diffusion flamelet models in non-premixed turbulent combustion," *Progress in Energy and Combustion Science*, vol. 10, no. 3, pp. 319–339, 1984.

[24] N. Peters, *Turbulent combustion*. Cambridge University Press, 2000.

[25] N. Zong, G. Ribert, and V. Yang, "A flamelet approach for modeling of liquid oxygen (LOX)/methane flames at supercritical pressures," in *46th AIAA Aerospace Sciences Meeting and Exhibit: AIAA paper 2008-946*, 2008.

[26] G. Ribert, N. Zong, V. Yang, L. Pons, N. Darabiha, and S. Candel, "Counterflow diffusion flames of general fluids: Oxygen/hydrogen mixtures," *Combustion and Flame*, vol. 154, no. 3, pp. 319–330, 2008.

[27] G. Lacaze and J. C. Oefelein, "A non-premixed combustion model based on flame structure analysis at supercritical pressures," *Combustion and Flame*, vol. 159, no. 6, pp. 2087–2103, 2012.

[28] E. Garnier, N. A. Adams, and P. Sagaut, *Large eddy simulation for compressible flows*. Springer, 2009.

[29] A. W. Vreman, "An eddy-viscosity subgrid-scale model for turbulent shear flow: Algebraic theory and applications," *Physics of Fluids (1994 - present)*, vol. 16, no. 10, pp. 3670–3681, 2004.

[30] R. I. Issa, "Solution of the implicitly discretised fluid flow equations by operator-splitting," *Journal of Computational Physics*, vol. 62, no. 1, pp. 40–65, 1985.

[31] R. I. Issa, "Solution of the implicitly discretised reacting flow equations by operator-splitting," *Journal of Computational Physics*, vol. 93, no. 2, pp. 388–410, 1991.

[32] M. Jarczyk and M. Pfitzner, "Large Eddy Simulation of supercritical nitrogen jets," in *50th AIAA Aerospace Sciences Meeting including the New Horizons Forum and Aerospace Exposition: AIAA paper 2012-1270*, 2012.

[33] "National Institute of Standards and Technology."

[34] E. Goos, A. Burcat, and B. Ruscic, "Extended third millennium ideal gas and condensed phase thermochemical database for combustion with updates from active thermochemical tables," 2009.

[35] B. E. Poling, J. M. Prausnitz, and J. P. O'Connell, *The properties of gases and liquids*. Mc Graw Hill, 2001.

[36] T.-H. Chung, M. Ajlan, L. L. Lee, and K. E. Starling, "Generalized multiparameter correlation for nonpolar and polar fluid transport properties," *Industrial & Engineering Chemistry Research*, vol. 27, no. 4, pp. 671–679, 1988.

[37] A. Frassoldati, A. Cuoci, T. Faravelli, E. Ranzi, C. Candusso, and D. Tolazzi, "Simplified kinetic schemes for oxy-fuel combustion," in *1st International Conference on Sustainable Fossil Fuels for Future Energy*, pp. 6–10, 2009.

[38] L. Pons, N. Darabiha, S. Candel, G. Ribert, and V. Yang, "Mass transfer and combustion in transcritical non-premixed counterflows," *Combustion Theory and Modelling*, vol. 13, no. 1, pp. 57–81, 2009.

[39] S. Pohl, M. Jarczyk, M. Pfitzner, and B. Rogg, "Real gas CFD simulations of hydrogen/oxygen supercritical combustion," in *Progress in Propulsion Physics*, vol. 4, pp. 583–614, EDP Sciences, 2013.

[40] T. Kim, Y. Kim, and S.-K. Kim, "Real-fluid flamelet modeling for gaseous hydrogen/cryogenic liquid oxygen jet flames at supercritical pressure," *Journal of Supercritical Fluids*, vol. 58, no. 2, pp. 254–262, 2011.

[41] H. Pitsch, "Flamemaster."

[42] W. Jones and R. Lindstedt, "Global reaction schemes for hydrocarbon combustion," *Combustion and Flame*, vol. 73, no. 3, pp. 233–249, 1988.

[43] G. P. Smith, D. M. Golden, M. Frenklach, N. W. Moriarty, B. Eiteneer, M. Goldenberg, C. T. Bowman, R. K. Hanson, S. Song, W. C. Gardiner, V. V. Lissianski, and Z. Qin, "Gas Research Institute."

[44] S. Candel, G. Herding, R. Snyder, P. Scouflaire, C. Rolon, L. Vingert, M. Habiballah, F. Grisch, M. Péalat, P. Bouchardy, D. Stepowski, A. Cessou, and P. Colin, "Experimental investigation of shear coaxial cryogenic jet flames," *Journal of Propulsion and Power*, vol. 14, no. 5, pp. 826–834, 1998.

[45] H. Terashima, S. Kawai, and N. Yamanishi, "High-resolution numerical method for supercritical flows with large density variations," *AIAA Journal*, vol. 49, no. 12, pp. 2658–2672, 2011.

Rocket engine high-enthalpy flow simulation using heated CO_2 gas to verify the development of a rocket nozzle and combustion tests

K. Takeishi[1], K. Ishizaka[2], J. Okamoto[3] and Y. Watanabe[4]

[1]Tokushima Bunri University, Sanuki, Kagawa, Japan
[2]Mitsubishi Heavy Industries Ltd., Takasago, Hyogo, Japan
[3]Mitsubishi Heavy Industries Ltd., Komaki, Aichi, Japan
[4]Japan Aerospace Exploration Agency (JAXA), Tsukuba, Japan
Corresponding author: takeishi@fst.bunri-u.ac.jp

Abstract. The LE-7A engine is the first-stage engine of the Japanese-made H-IIA launch vehicle. This engine has been developed by improving and reducing the price of the LE-7 engine used in the H-II launch vehicle. In the qualification combustion tests, the original designed LE-7A (LE-7A-OR) engine experienced two major problems, a large side load in the transient state of engine start and stop and melt on nozzle generative cooling tubes. The reason for the troubles of the LE-7A-OR engine was investigated by conducting experimental and numerical studies. In actual engine conditions, the main hot gas stream is a heated steam. Furthermore, the main stream temperature in the nozzle changes from approximately 3500 K at the throat to 500 K at the exit. In such a case, the specific heat ratio changes depending on the temperature. A similarity of the Mach number should be considered when conducting a model flow test with a similar flow condition of the Mach number between an actual engine combustion test and a model flow test. High-speed flow tests were conducted using CO_2 gas heated up to 673 K as a working fluid and a 1:12 sub-scaled model nozzle of the LE-7A-OR engine configuration. The problems of the side force and the conducted form of the shock waves generated in the nozzle of the LE-7A-OR engine during engine start and stop were reproduced by the model tests of experimental and numerical investigations. This study presented that the model flow test using heated CO_2 gas is useful and effective in verifying the numerical analysis and the design verification before actual engine combustion tests.

1. Introduction

The H-IIA launch vehicle was developed based on an H-II launch vehicle while planning for the improvement of reliability and attaining a cost reduction of approximately 50% to acquire international competitiveness [1]. Table 1 shows the specifications of the LE-7A engine used for the H-IIA launch vehicle in comparison with that of the LE-7 engine for the H-II launch vehicle. The nozzle skirt of the LE-7A-OR engine comprised the upper regenerative tubes cooled by liquid hydrogen and the lower sheet metal made by a forging process. A step was established between the lower- and upper-part skirts. In addition, film cooling with hydrogen gas was adopted on the lower-part nozzle's inner surface to keep the metal temperature of the lower-part nozzle under the allowable metal temperature. However, an unexpected high side force of more than the permission level of the actuator was obtained when the qualification combustion tests of the LE-7A-OR engine were conducted using the ground facilities at the liquid engine examination test stand of the Tanegashima Space Center. The side force value was over 300 kN. The improvement of the LE-7A-OR engine should be forced to reduce the side force within the actuator's permission level [2, 3].

Table 1 Comparison of the engine specifications of LE-7 and LE-7A

	LE-7A	LE-7
Thrust [kN]	1098	1078
Specific Impulse [s]	440	446
Combustion Chamber Pressure [MPa]	12.2	12.7
Total Length [mm]	3680	3243
Maximum Diameter [mm]	1965	2570
Expansion Ratio	51.7	51.4
Weight [kg]	1830	1720
Burn Time [s]	Max 400	350

Experiments were conducted to investigate the reason of the LE-7A-OR troubles by using dried pure nitrogen gas as a working fluid. However, the tests could not reproduce the troubles. Accordingly, we determined that the reason why the high-speed flow test of the LE-7A-OR engine by using dry nitrogen gas could not reproduce the trouble is the specific heat ratio of the working fluid. The specific heat ratio of the combustion gas (mainly steam) changed from 1.16 at the combustion chamber to 1.28 at the nozzle exit. The use of pure dried nitrogen gas, which is a well-known and established high-speed experimental method, keeps the constant specific heat ratio of 1.4 and could not simulate the actual LE-7A-OR engine states. We examined the specific heat ratio of three atomic gases to fit with steam under a rocket engine combustion state. Consequently, we found that CO_2 is the best.

As part of the development of an improved LE-7A(LE-7A-IM) engine nozzle, we experimentally conducted an evaluation of the influence of the nozzle shape on the heat flux distribution on the nozzle wall surfaces and thrusts using 1/12 sub-scaled nozzle models of the LE-7 and the LE-7A-OR engine with heated CO_2 gas. Furthermore, we conducted experiments by using a 1/12 sub-scaled nozzle of the LE-7A-IM engine. We also performed an analysis to verify the re-designed improved nozzle within a permitted small thrust.

This study describes the novel experimental technique to simulate the real rocket engine combustion flow by using heated CO_2 gas. The test results were obtained for the purpose mentioned above and compared with the CFD analysis results.

2. Phenomenon with the original LE-7A engine

The LE-7 engine nozzle had a truncated perfect (TP) shape. The LE-7A-OR engine nozzle shape chose a compressed truncated perfect (CTP) nozzle shape dissimilar to the LE-7 engine. The CTP nozzle shape is the design method used to axially shorten the nozzle head by compressing a TP nozzle. The CTP nozzle was adopted as the design technique for the LE-7A-OR nozzle because the CTP nozzle had splendid characteristics of freely selecting the expansion ratio depending on the compression of TP nozzle and generating a small performance loss with shape compression [4]. The initial expansion angle of the LE-7A-OR engine is approximately 3° bigger than that of the LE-7 engine. Figure 1 shows the comparison of the nozzle contour of the LE-7A-OR engine with that of the LE-7 engine.

Two deficient phenomena on the nozzle skirt of the LE-7A-OR engine were generated in the combustion tests in the initial stage of the LE-7A engine development. One phenomenon is the large side thrust in the transient state of engine start and stop. The other is the damage of the nozzle generative cooling tubes of the upper-part nozzle skirt.

The side load origin was estimated because of the transition of the separation pattern from shock separation to restricted shock separation (RSS) [5, 6]. Figure 2 is referred from reference [6], in which M. Frey and G. Hagemann described the detailed flow structure of the FSS and the RSS. The RSS is attributed to a reattachment of the separated flow to the nozzle wall, which induces shocks and expansion waves. The shocks generated high pressure on the wall, and the reattached flow caused a high heat flux region.

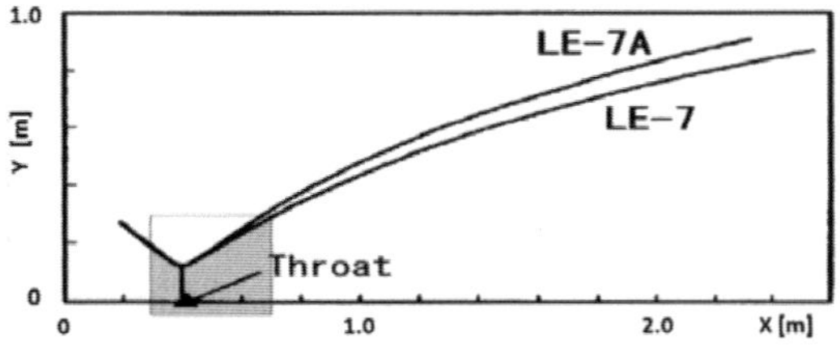

Fig. 1 Nozzle contour comparison of the LE-7A and LE-7 engines

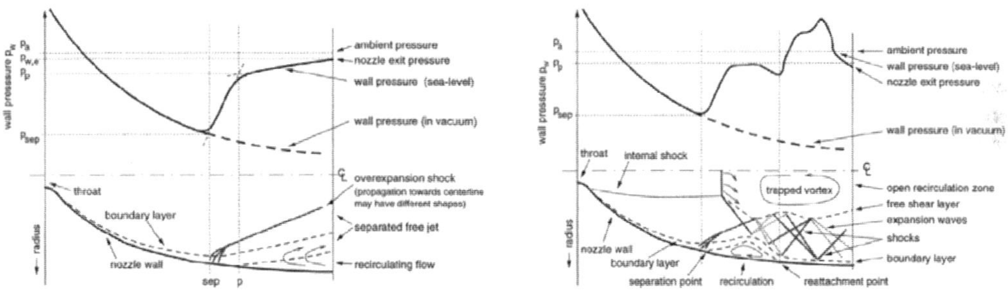

(a) Free shock separation (FSS) (b) Restricted shock separation (RSS)

Fig. 2 Phenomenological sketch of the FSS and the RSS [6]

3. Experimental test facility and measuring techniques

The combustion temperature of the LE-7A engine under the actual engine operating conditions was approximately 3500 K. It was thought and created to simulate the high-temperature steam flow of the rocket engine nozzle by using the heated CO_2 gas as working fluid to obtain the flow resemblance at a laboratory level. Figure 3 clearly shows that the steady-state specific heat ratio of the original LE-7A engine in the actual engine operating conditions can be almost simulated if CO_2 is used as the actuating experimentally working fluid and the stagnation temperature of the CO_2 gas can be heated up to 1200 K.

The Takasago R&D Center of Mitsubishi Heavy Industries, Ltd. has an experimental test facility, including the pebble heater heating arrangement that can increase gas by possibly up to a maximum of 400 °C (673 K). A rocket engine nozzle flow test can be simulated and the resemblance between the combustion gases of 3500 K flow through the nozzle and CO_2 gas in the experiment is possible by performing the experimental test utilizing CO_2 heated up to 673 K as the working fluid using this device. We revised the contour of the sub-scaled model nozzle to fit 673 K, instead of heating up the CO_2 gas to 1200 K as mentioned earlier. We re-designed the sub-scaled test nozzle contour to agree with the Mach number distribution of the actual LE-7A-OR engine operation of 3500 K flowing through the test nozzle contour in a stagnation point temperature of 673 K. The design was implemented by re-designing the original nozzle shape with a throat radius of 0.398 times of the throat curvature with the non-viscosity method. The nozzle length/throat diameter was fixed at the same value with the actual engine configuration. Furthermore, the correction of the boundary layer thickness by the difference of the Reynolds number between the CO_2 test and the actual engine operation was not revised. The maximum expansion corner was designed such that the position where the internal shock intersected at the sub-

IOP Conf. Series: Journal of Physics: Conf. Series **821** (2017) 012011 doi:10.1088/1742-6596/821/1/012011

scaled model nozzle center agreed with that of the LE-7A-OR engine. Furthermore, the scaled test nozzle contour was revised such that the Mach number distribution of the radial direction at the nozzle exit agreed with that of the LE-7A-OR nozzle.

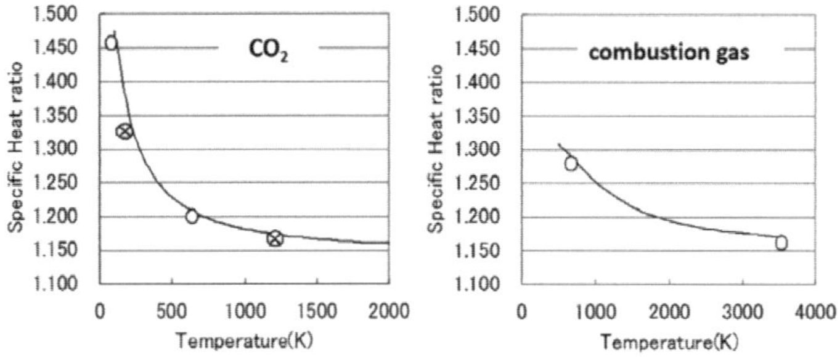

Fig. 3 Comparison of the specific heat ratio between the combustion gas and the heated CO_2

The 1/12 sub-scaled nozzle models were manufactured with the abovementioned concepts. High-speed flow tests were also performed. Figure 4 shows a comparison of the numerical results of the CO_2 high-speed flow test of the LE-7A-OR sub-scaled model with the actual combustion condition.

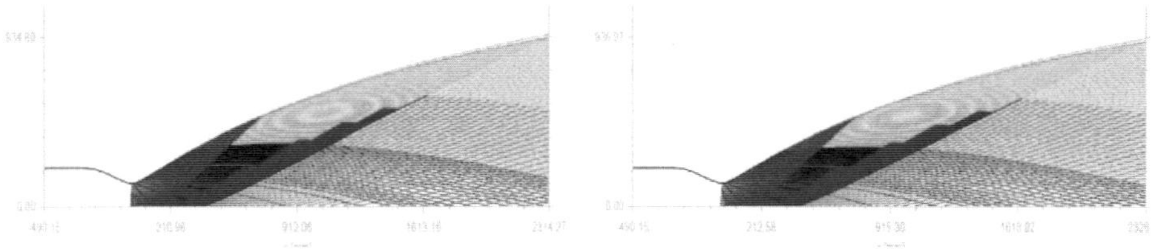

((a) Steady Combustion condition ((a) CO_2 high speed flow test condition at 673K

Fig. 4 Computed characteristic curve contours of the LE-7A-OR nozzle

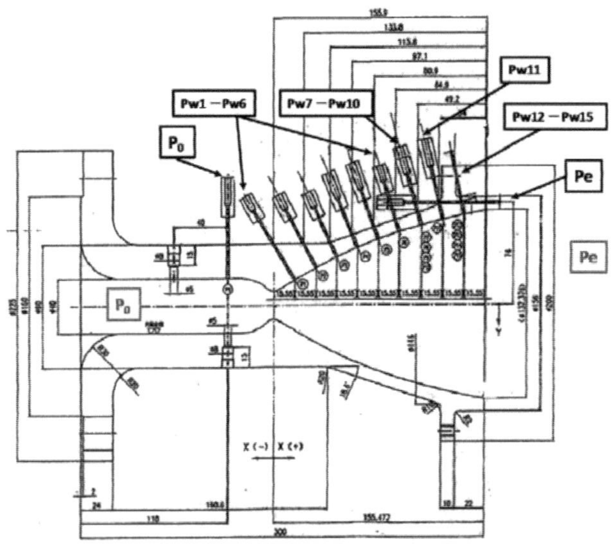

Fig. 5 Scaling nozzle model and static pressure taps

The test facilities comprised a pebble heater cumber, a heated gas reservoir tank, a high-speed opening and shutting valve, a 20 MPa reservoir tank with a high pressure, an ejector, and a silencer. The sub-scaled test nozzle is mounted downstream of the high-speed opening and shutting valve.

The test nozzles represent the LE-7, LE-7A-OR, and LE-7A-IM nozzles. Figure 5 shows the LE-7A-IM model nozzle shape, dimensions, and static pressure tap-mounting locations. Pressure monitoring in the nozzle wall surface of 1.2 mm was performed by piercing the nozzle wall with a stainless steel pipe having a 1.2 mm outside diameter and approximately 30 mm length. Pressure gauges of Kulite XT-190-15G and Kulite XT-190-25G were also attached to the pipe ends. The pressure gauges had an accuracy of $\pm 0.5\%$ of full scale. The pressure gauge was a semiconductor gauge type having a response frequency of 100 kHz and approximately 2 kHz, including a 30 mm-long pressure pipe.

4. Experimental results and discussion

The experiments on the sub-scaled nozzle models of the LE-7, LE-7A-OR, and LE-7A-IM engines were conducted by processing with the test section of the ramjet test facility. Accordingly, CO_2 gas was heated using a pebble heater up to 673 K and 2 MPa. The carbon dioxide concentration was examined and confirmed by measuring the oxygen concentration before and after the examinations.

Two test cases were considered. In case A, the high-speed opening and shutting valve was opened, and the nozzle upper stagnation pressure was maintained at approximately 2 MPa. Furthermore, the transient test was started by operating an ejector and changing the pressure back from atmospheric pressure to 0.01 MPa or from 0.01 MPa to atmospheric pressure. The pressure-changing time t of the back pressure was approximately 10 s in the start and stop simulation tests.

Figure 6 shows the test result at the time of shut down and start-up of the LE-7 sub-scaled nozzle test. Figure 6 also plots each pressure ratio Po/Pe in a transverse axis versus each wall pressure/back pressure (Pw/Pe) in a vertical axis. Po is the chamber pressure; Pe is the nozzle exit pressure; and Pwi is the pressure at number i static pressure taps. Figures 7 and 8 show the test result of the LE-7A-OR and LE-7A-IM sub-scaled nozzles, respectively, in the same manner as that of the LE-7 sub-scaled nozzle. The Pw/Pe of shutdown in Figure 7 clearly shows that the RSS[7] was estimated to occur on the LE-7A-OR sub-scaled nozzle surface because of $Pw/Pe > 1$. Meanwhile, the pressure ratio Pw/Pe is less than 1.0 in the case of the LE-7 and the LE-7A-IM. Moreover, no big pressure change in Pw/Pe of the LE-7 and the LE-7A-IM occurred (Figures 6 and 7). The test results of the LE-7 and the LE-7A-OR sub-scaled nozzles clearly simulated the results of the actual LE-7 and the LE-7A-OR engine combustion tests.

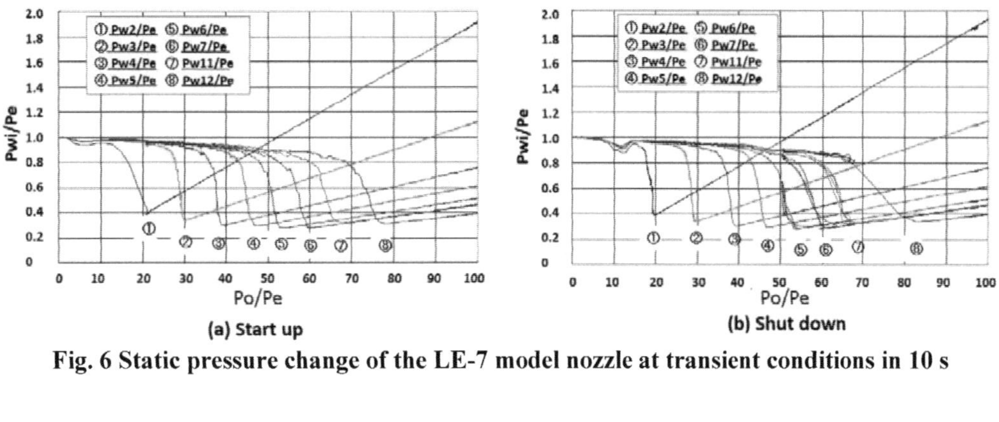

Fig. 6 Static pressure change of the LE-7 model nozzle at transient conditions in 10 s

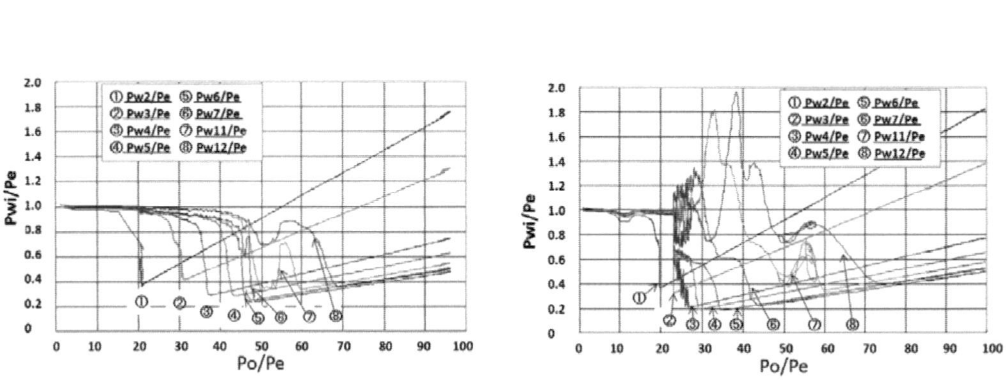

(a) Start up (b) Shut down

Fig. 7 Static pressure change of the LE-7A-OR model nozzle at transient conditions in 10 s

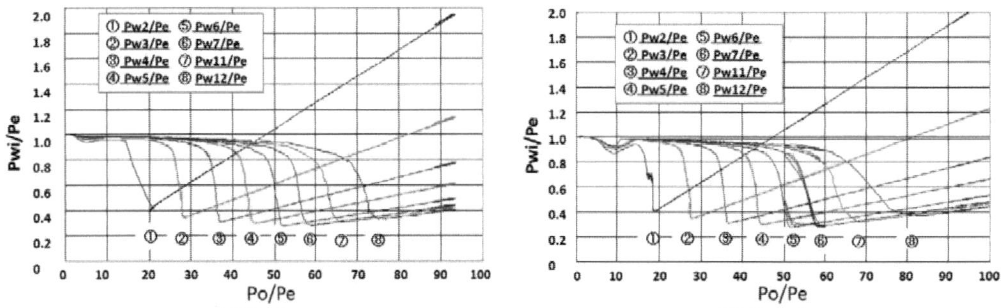

(a) Start up (b) Shut down

Fig. 8 Static pressure change of the LE-7A-IM model nozzle at transient conditions in 10 s

In case B, an ejector to control the back pressure was operated to maintain the back pressure *Pe* at 0.01 MPa. The high speed opening and shutting valve was then opened or closed to raise or reduce the nozzle entrance pressure *P*o to 2 MPa or 0.01 MPa in 0.2 s. Figure 9 shows the test result of case B using the sub-scale model nozzle of the LE-7 engine.

The RSS for the LE-7 sub-scale model test in case B is estimated to occur at a period where the wall surface pressure *P*w was higher than the nozzle back pressure *P*e (*P*w/*P*e > 1) at the shut down tests shown in Figure 9. In other words, the wall surface pressure of the LE-7 sub-scaled nozzle showed an unstable flow condition in the transient test of case B at a very rapid shut down in 0.2 s.

The facts obtained in this experiment meant that the RSS occurred even with the LE-7 sub-scaled nozzle by the short shut down time of 0.2 s. However, the RSS did not occur in the actual combustion examination of the LE-7 engine by the shut down time of 4.0 s.

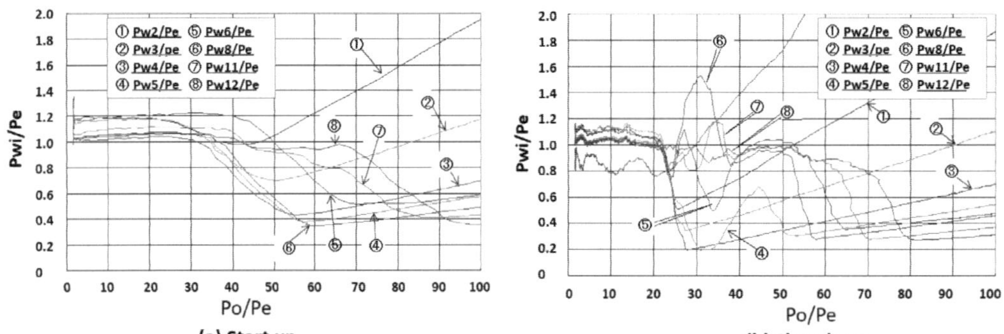

(a) Start up (b) Shut down

Fig. 9 Static pressure change of the LE-7 model nozzle at transient conditions in 0.2 s

In contrast to the test results of the LE-7A-OR sub-scaled nozzle, the wall surface pressure ratio did not exceed the nozzle back pressure ($Pw/Pe < 1$) in all the examinations in the LE-7A-IM sub-scaled nozzle, even with the hard simulation test of shutdown in 0.2 s in case B.

The test results show that the LE-7A-OR nozzle shape causes a more unstable flow than the LE-7 nozzle shape. The case B tests clearly illustrate that the LE-7A-IM is stable with the same level of the LE-7 nozzle. The simulation test clearly demonstrates that using the heated CO_2 gas is very useful in verifying the development of a rocket nozzle and combustion tests. In addition, the severe shut down test operated in 0.2 s is useful in checking and assuring the stability of the flow of the newly designed nozzle in transient conditions.

5. Analytical method and results

The analysis was conducted using the second-order upwind method, the positive elucidation at a two-dimensional axis symmetry analysis, and the second space precision with a TVD scheme. In addition, a two-equation turbulence model (i.e., $k - \varepsilon$ model) is used. The numerical analysis was conducted according to the transient CO_2 sub-scaled model tests under the boundary conditions that the circumference side and the nozzle upper-reach circumference are assumed as isolated walls. The number of grid points was 401 (axial direction) × 71 (radial direction) in the nozzle section. The total number of grids was about 8×10^4. Meanwhile, the minimum grid spacing in the r direction satisfies $\boldsymbol{y^+ < 1}$ at a grid point adjacent to the nozzle wall. The thermal properties of the CO_2 were modeled in the simulations by the models published by McBride et al.[11].

Figures 9–11 illustrate the Mach number distributions and the characteristic curves in the nozzle at the steady-state actual engine conditions of the LE-7, LE-7A-OR, and LE-7A-IM nozzles, respectively.

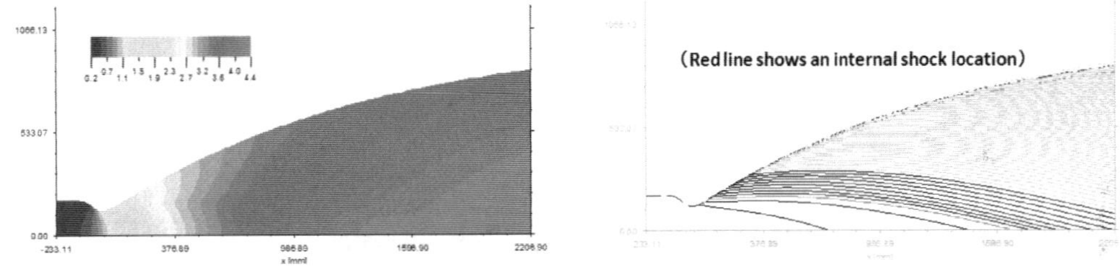

(a) Mach number contour of LE-7 CO_2 test **(b) Characteristic curve of LE-7 CO_2 test**

Fig. 10 Computed z-plane Mach number contours of the LE-7 nozzle

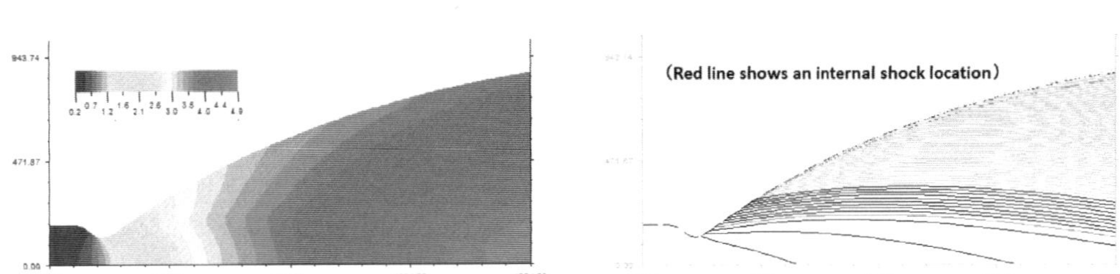

(a) Mach number contour of LE-7A-OR at CO_2 test **(b) Characteristic curve of LE-7A-OR at CO_2 test**

Fig. 11 Computed z-plane Mach number contours of the LE-7A-OR nozzle

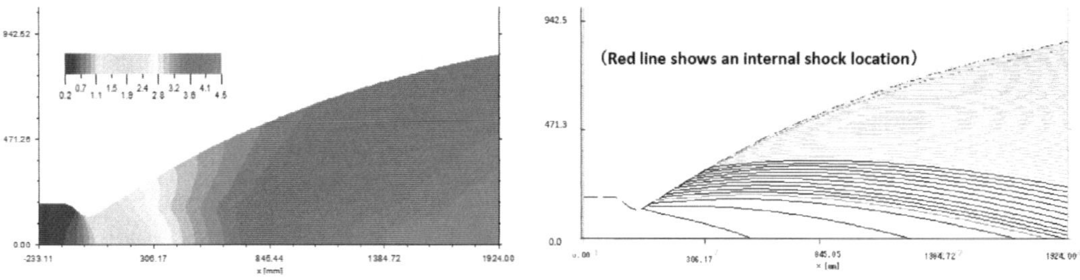

(a) Mach number contour of the LE-7A-IM at CO_2 test **(b) Characteristic curve of the LE-7A-IM at CO_2 test**

Fig. 12 Computed z-plane Mach number contours of the LE-7A-IM nozzle

Figures 13–15 show the numerical results of the simulation of the shutdown tests in 0.2 s by using CO_2 as a working fluid of the LE-7, LE-7A-OR, and LE-7A-IM sub-scaled nozzles, respectively.

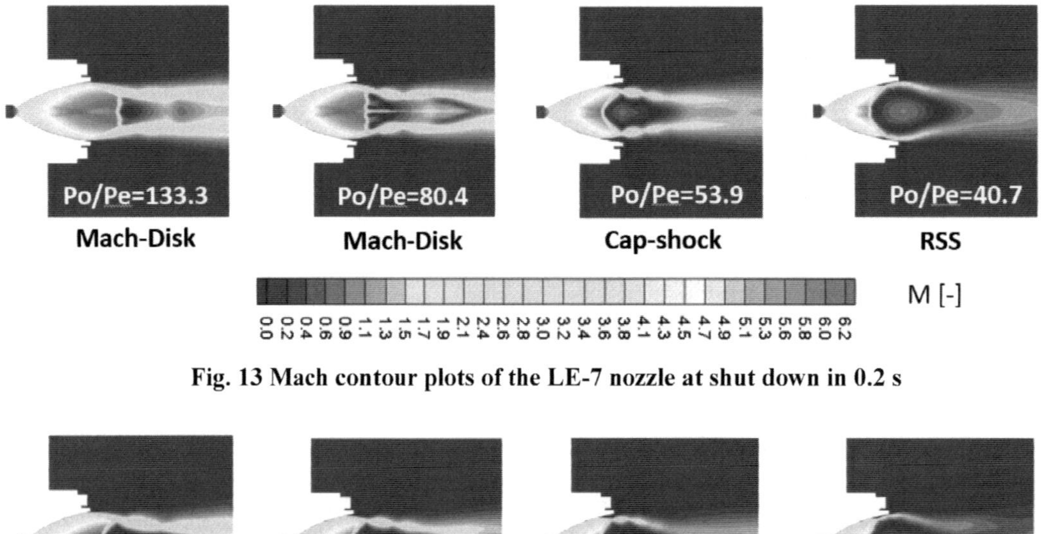

Fig. 13 Mach contour plots of the LE-7 nozzle at shut down in 0.2 s

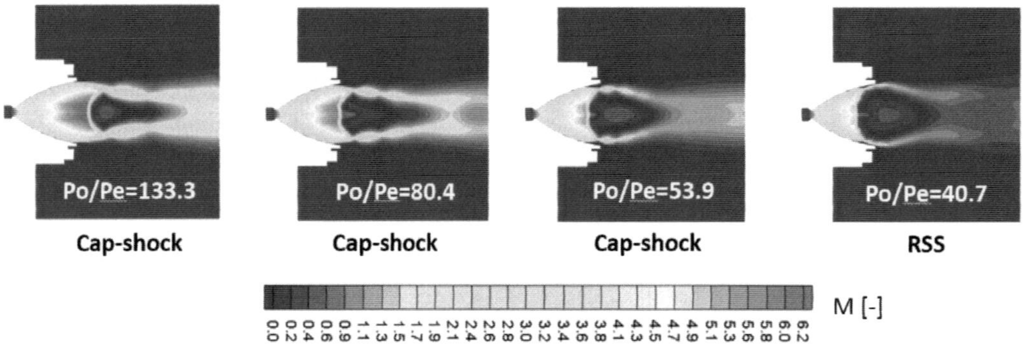

Fig. 14 Mach contour plots of the LE-7A-OR nozzle at shut down in 0.2 s

The strong side force tends to occur in shutdown at a pressure ratio *Po*/*Pe* around 40 and 80 at the actual LE-7A-OR engine combustion tests. The notes of Figures 13–15 show the flow patterns estimated from the Mach number distribution [8–10]. The CFD results of the LE-7 and the LE-7A-OR sub-scaled nozzles shown in Figures 13 and 14 clearly reproduce the RSS outbreak at around *Po*/*Pe* = 40. The CFD can estimate the wall surface pressure decreases just before the RSS outbreak. The RSS outbreak phenomenon is clearly captured by the CFD analysis. The CFD analysis in Figure 15 shows that the RSS does not occur at the shutdown test in 0.2 s by using the LE-7A-IM sub-scaled nozzle at a pressure ratio *Po*/*Pe* around 40. This result guarantees the CFD analysis accuracy that can be used for the rocket nozzle design.

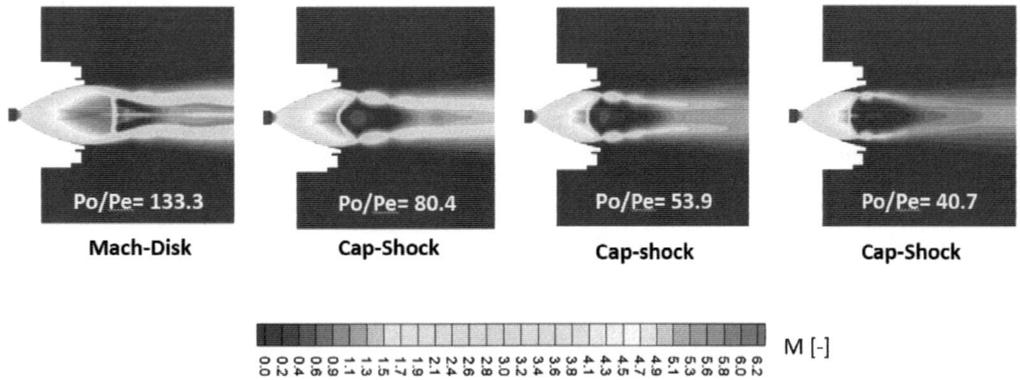

Fig. 15 Mach contour plots of the LE-7A-IM nozzle at shut down in 0.2 s

Figure 16 shows the comparison between the experimental measurement of the LE-7A-OR sub-scaled nozzle test at the shutdown test in 0.2 s and the numerical analysis results just under the same conditions with the experiments. Figure **16** illustrates that the CFD analysis can estimate the RSS outbreak even though it occurs at a little lower pressure ratio of *Po*/*Pe*.

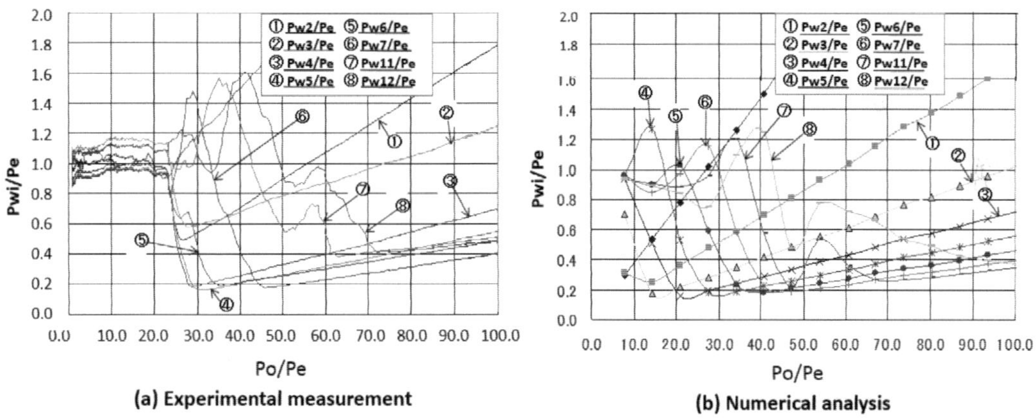

Fig. 16 Comparison of the static pressure change between the numerical analysis and the experimental measurement of the LE-7A-OR nozzle at the shut down condition

6. Conclusion

This study presented a novel experimental technique to simulate the real rocket engine combustion flow by using heated CO_2 gas. The experiments and the numerical analysis of the sub-scaled nozzle model tests using CO_2 gas of the LE-7 engine, original LE-7A engine, and re-designed LE-7A engine lead to the following conclusions:

1. The new verification method of a rocket nozzle design and a rocket engine combustion test is presented by using the heated CO_2 gas instead of the high-enthalpy flow of a rocket engine combustion gas.
2. The specific heat ratio **κ** depending on the temperature can be exactly adjusted between the CO_2 gas and the rocket engine combustion gas.
3. The flow visualization and detailed flow measurement can be conducted by utilizing the heated CO_2 gas to simulate the real rocket engine combustion gas flow. The model flow test using the heated CO_2 gas is useful and effective for the accuracy verification of the numerical analysis and the design verification before the actual engine combustion tests.
4. The rapid shut down test in 0.2 s made the LE-7 sub-scale nozzle flow into the RSS. This severe CO_2 test condition can be used to confirm the nozzle flow stability and is useful for the design criteria of the rocket engine operation and nozzle contour.

Nomenclature

FSS: free shock separation
k: turbulent kinetic energy
*P*e: back pressure at a nozzle exit
*P*w: static wall pressure
*P*o: stagnation pressure
RSS: restricted shock separation
r: *r* coordinate
t: time
z: *z* coordinate
ε: turbulent dissipation rate

References

[1] Fukushima, Y. and Nakatsuji, H., "Development status of LE-7A and LE-5B engines for H-2A family," Acta Astronautica, 50, pp.275-284, 2002.
[2] Watanabe, Y., Sakazume, N. and Tsuboi, M., "LE-7A Engine Nozzle Problems during the Transient Operations," AIAA Paper 2002-3841, 2002.
[3] Matsuyama, K., Ohigashi, H., Ito, T. Yasui, M. and Manako, H., "H-IIA Rocket Engine Development," Mitsubishi Heavy Industries, Ltd., Technical Review, 39, pp.51-56, 2002.
[4] Hoffman, J. D., "Design of Compressed Truncated Perfect Nozzle," AIAA Paper 85-1171, 1985.
[5] Baars, W. J. and Tinney, C. E., "Wall pressure unsteadiness and side loads in over expanded rocket nozzles," AIAA Journal, 50, pp61-73, 2012.
[6] Frey, M and Hagemann, G., "Flow Separation and Side-Loads in Rocket Nozzles," 35th AIAA/ASME/SAE/ASEE Joint Propulsion Conference & Exhibit, Los Angeles, USA, 1999.
[7] Hagemann, G., Alting, J. and Preclik, D., "Scalability for Rocket Nozzle Flows Based on Subscale and Full-Scale Testing," Journal of Propulsion and Power, 19, pp.321-331, 2003.
[8] Frey, M. and Hagemann, G., "Restricted shock separation in rocket nozzles," AIAA J. of Propulsion and Power, Vol.16, No.3, pp. 478-484, 2000.
[9] Hagemann, G. and Frey, M., "Shock pattern in the plume of rocket nozzles: needs for design consideration," Shock Waves J., Vol. 17, No.2, pp. 387-395, 2008.
[10] Wang, T. S., "Transient Three-Dimensional Analysis of Nozzle Side Load in Regeneratively Cooled Engines," 41st AIAA/ASME/SAEYASEE Joint Propulsion Conference and Exhibit, Tucson, Arizona, 2005.
[11] McBride, B. J., Heimel, S., Ehlers, J. G. and Gordon, S., "Thermodynamic Properties to 6000 K for 210 Substances Involving the First 18 Elements", NASA SP-3001, 1963

IOP Conf. Series: Journal of Physics: Conf. Series **821** (2017) 012012 doi:10.1088/1742-6596/821/1/012012

Numerical characterization under uncertainties of a piston expander for exhaust heat recovery on heavy commercial vehicles

P.M. Congedo[1], J. Melis[2], R. Daccord[2]

[1] INRIA Bordeaux Sud-Ouest, 200 Rue de la Vieille Tour, 33405 Talence, France

[2] EXOES, 6 Avenue de la Grande Lande, 33170 Gradignan, France

E-mail: pietro.congedo@inria.fr,julien.melis@exoes.com,remi.daccord@exoes.com

Abstract. While nearly 30 percent of the fuel energy is lost as waste heat in the form of hot exhaust gases, exhaust heat recovery promises one of the biggest fuel economy potential regarding the technologies available in the next decade. Applied to heavy commercial vehicles (HCVs), buses or off road vehicles, a bottoming Rankine Cycle (RC) on exhaust heat shows a great potential in recovering the exhaust gases energy, even for part loads. The objective of this paper is to illustrates the interest in assessing the uncertainty of this kind of systems for getting a robust prediction of the associated numerical model. In particular, the focus here is on the simulation of a piston expander for exhaust heat recovery.

Uncertainties associated to the experimental measurements are propagated through the numerical code by means of uncertainty quantification techniques. Several sources of uncertainties are taken into account at the same time, thus yielding various indications concerning the most predominant parameters, and their influence on several quantities of interest, such as the mechanical power, the mass flow and the exhaust temperature.

1. Introduction

To date, emission regulations focusing on local pollutants have limited the potential efficiency gains of internal combustion engine (ICE). However, future regulations will focus on CO2 emissions, requiring high efficiency increase of the whole drivetrain. The best efficiency of a modern ICE will remain below 42%. Electrification of ancillaries and hybridization seem to lead to little fuel cuts on Heavy Commercial Vehicles (HCVs) and, at least, would be too expensive to reach future CO2 emissions regulation compared to air drag reduction and waste heat recovery (WHR). While HCVs aerodynamic is mostly constrained by regulation and also depends on trailer manufacturers, WHR appears as essential in the future innovation panel for HCVs. Though the potential of WHR is clear, the technology has to prove the business, durability and safety cases to be launched in mass production.

A great interest has been developed in the last years concerning the potential new technologies for exploiting the waste heat recovery (WHR) from IC engines. Several options have been investigated [1]: Organic Rankine Cycles [2], six-stroke cycle IC engine [3], thermoelectric generator [4], turbocharger technology [5, 6]. Among several technologies, Rankine cycles (RC) seem promising to win the competition regarding their efficiency, cost and maturity. The key components of a RC are shown hereafter (see Figure 1). Key choices when designing such a system are the selection of the hot source, the working fluid, the expander and the cold sink.

ORC technology for recover low grade heat sources, such as in solar or geothermal plant, feature the use of turbines, scroll expander, screw expander [7]. Nevertheless, these solutions are not ideal when considering the WHR in IC engines because of the space and weight constraints. For this reason, some studies are focused on the development of an ORC design presenting a reciprocating machine, permitting an optimal recovery of the waste heat energy, based on a low working fluid flow, high values in expansion ratios and space restrictions. Some works [8] have been devoted to the design of an ORC facility for waste heat recovery in IC engines. Several studies [9, 10, 11, 13, 14, 15] suggested the use of a piston expander for exhaust heat recovery (EHR). In [12], they tested an axial piston expander and calibrated a semi-empirical model.

Uncertainty quantification (UQ) is an emerging field in scientific computing. UQ methods compute the impact of inevitable uncertainties in boundary and initial conditions, and in model parameter inputs on the output quantities of interest. One of the objectives of UQ is to enable a more rigorous comparison between numerical predictions and experimental measurements with their respective uncertainty margins. Few works exist on the application of these techniques in ORC-based framework (see for example [17, 18, 19]).

In this paper, we focus on a piston expander (designed by the Exoès engineering company) devoted to exhaust heat recovery, and how uncertainty quantification techniques could be exploited in order to characterize the global performances of the system and to drive further improvements. After providing a general description of the expander in Section 2, the numerical model and an analysis over the main uncertainties of the system are presented in Section 3. Some results are provided in Section 4, and conclusions are drawn in Section 5.

2. Description of the expander

While using a RC, the majority of the tailpipe exhaust heat will be rejected in the cooling fluid under hood. A trade-off due to weight, packaging, fan engagement and aero complexities must be found. Designing a compact cooling system is also a main challenge to reach the expected fuel consumption reduction [16]. Exoès decided to use the existing ICE cooling loop to avoid an additional front radiator and to be compatible with all the different truck cooling architectures. Exoès assume then a cooling temperature between 75 and 95°C according to a typical front radiator management of a HCV.

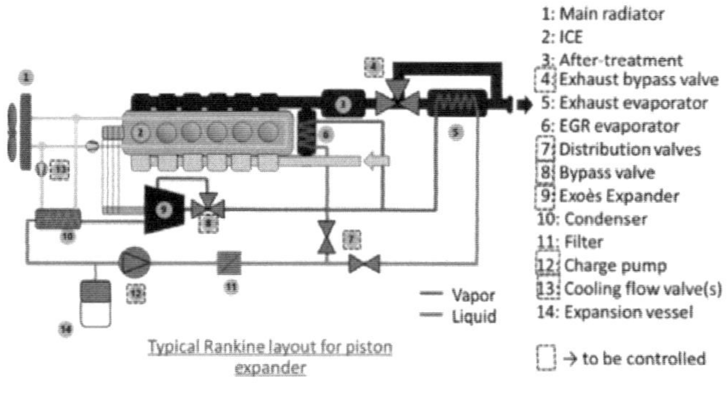

Figure 1. Bottoming cycle on exhaust line.

As we focus on the expander, what matters are the expander boundaries: the connections to the RC with vapor state points and the connection to the ICE with its speed. These boundaries are summed up in table 1. The main performance objectives for this prototype are summed up in table 2.

Table 1. Expander boundaries.

	Nominal ranges	Design Point	Full Load
Inlet pressure [Bar]	10-35	20	35
Inlet temperature [°C]	180-240	210	240
Outlet pressure [Bar]	1-2.3	1.2	2.3
Engine Speed [Rpm]	600-1800	1200	1200
Massflow [g/s]	20-80	40	80

Table 2. Expander performance target (The maximal effective isentropic efficiency is obtained at an outlet pressure of 1.2 Bar).

	Target	Achieved to date
Max effective isentropic efficiency	>60%	54%
Weight (without the coupling to the ICE)	12	18
Size (without the coupling to the ICE)	D180xL300	D215xL350

The expander is built on three double acting crosshead pistons arranged around a swash plate (see Figure 2). The machine has been designed to be mechanically coupled to an ICE on the front end accessory drive with a pulley, or on the back of the ICE through the gear of a power-take-off. The expander can be declutched or not. This feature may be granted by a free wheel, a clutch or a gearbox. The inlet system has been developed and patented to work at 250°C and to endure the heavy accelerations required to supply this two-stroke high speed steam engine. Together with the piston rings, the poppet valves of the inlet system also work with the scarce lubrication provided by the circulating oil not viscous enough at these high temperatures. The lubrication of the expander is controlled through an internal oil pump directly mounted on the shaft tip. The oil is pumped in the oil tank at the bottom of the expander. The oil passes through an oil filter integrated in the expander and is ducted through the shaft and/or through the crankcase to distribute lubricant in the swashplate chamber including the bearings, in the inlet valves chambers and on the shaft seal. An integrated oil separator at the outlet port recovers the possible leaks of oil into the vapor to fill them back to the tank.

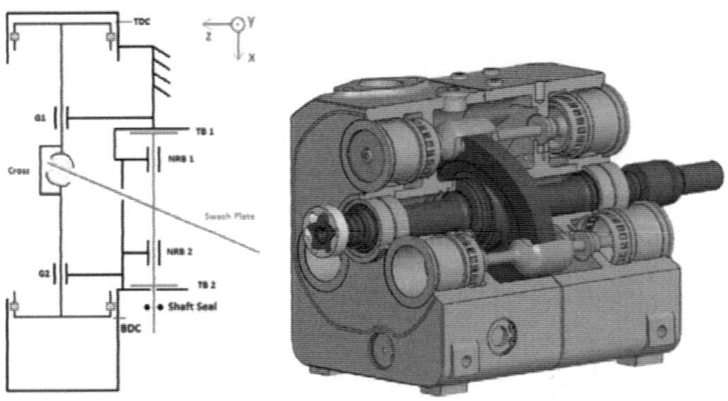

Figure 2. Expander architecture.

3. Numerical model

A Matlab 1D expander model has been built. The model consists in calculating the in-cylinder pressure during a complete rotation and ensuring the convergence of the computed mass-flow.

The model carries out the balances of energy and mass in the inlet and outlet collectors as well as in the piston chamber discretized per each half shaft angle of rotation (denoted with a subscript i in the scheme reported in Figure 3). Note that the software Refprop [20] is used to compute the thermodynamic behavior of the working fluid. In addition, several losses are taken into account as described in the scheme presented in Figure 3. As a result, a typical Pressure-Volume diagram is computed (see Figure 4) as well as all the macroscopic values, such as massflow ($\dot{M}$), mechanical power ($\dot{W}$), exhaust temperature (T_{out}) :

$$\left(\dot{M}, \dot{W}, T_{out}\right) = f\left(RPM, P_{in}, h_{in}, P_{out}, X_j\right) \tag{1}$$

where P_{in} and h_{in} are the inlet pressure and enthalpy, P_{out} is the outlet pressure, and X_j are some parameters calibrated with respect to the experimental data.

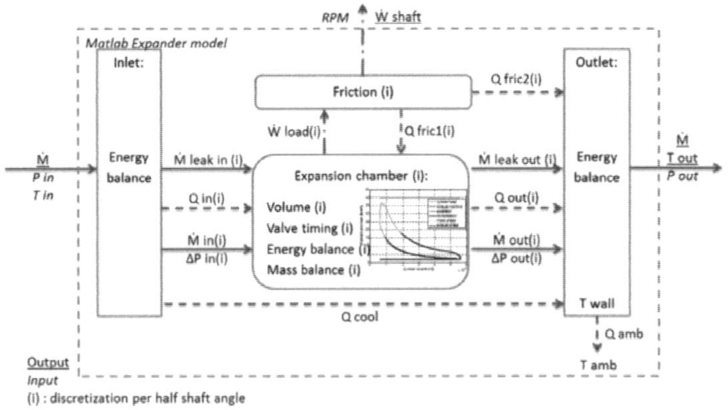

Figure 3. Expander model.

The model f is formulated according to the following hypothesis :

- The pressure losses during the intake and the exhaust phases are calculated through a calibrated discharge coefficient applied to theoretical flow crossing the throat.
- The steam properties and velocity at the throat are computed assuming that the flow is isentropic before the throat and isenthalpic on the total transformation.
- Leaks are modeled through the piston rings, the inlet valve seat and the inlet valve guide with Couette-Poiseuille equations.
- Convective heat transfers is implemented with a fixed convective heat transfer coefficient that has to be calibrated. Heat transfer occurs between inlet and outlet, inlet and cylinder chamber, cylinder chamber and outlet and through cylinder walls.
- Friction models use a simple calibrated friction coefficient, except for the sliding shoe on the plate. For this contact a dedicated 2D model is used to calculate the oil film thickness and then the equivalent friction coefficient.

The working fluid is a mixture constituted by water 4.4%wg, ethanol 92.6%wg and lubricant 3%wg. For calculations hereafter we simplify the mixture neglecting the lubricant percentage.

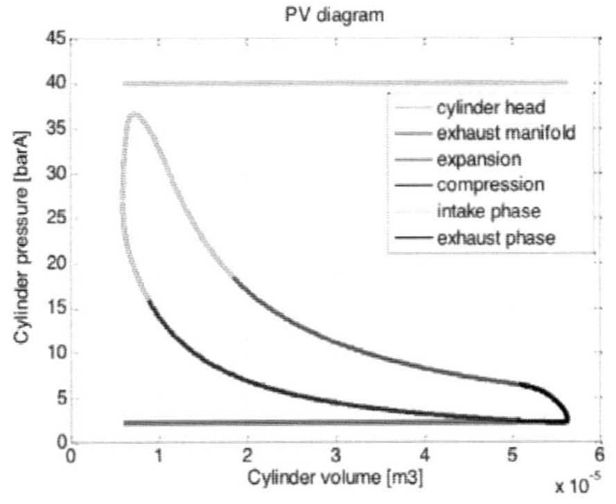

Figure 4. PV diagram.

3.1. Uncertainty assessment and design variables

We have identified three different sets of parameters. First group is related to some physical parameters, which are characterized in terms of uncertainties, but they are not treated as design variables. In practice, only certain nominal conditions are taken into account, without exploring the design space. These variables are the following : Cylinder Head Enthalpy, Bore, Stroke, Implantation radius, Inlet diameter, Outlet width, Half ball radius.

Second group is related to the parameters which are characterized in terms of uncertainties and also their variability in a larger design space is explored. These parameters are the Rotational speed, Cylinder Head Pressure, Exhaust pressure, Dead volume, Inlet lift, Outlet lift.

The third group of parameters is associated to the set of calibration parameters. In this case, since these parameters have been obtained by performing a calibration starting from experimental data, it is harder to fully characterize their variability. For this reason, concerning the uncertainty on calibration parameters, we consider a ± 0.01 variation based on some experimental evidence [14], which should provide a first estimation about the potential influence of the uncertainty on these parameters. Future studies should consider a larger variation, or solving the whole problem by means of Bayesian-based approaches. Finally, the following calibration parameters are considered: Inlet Leak coefficient (denoted with X1 in the following), Piston Leak coefficient (X2), Inlet Silent coefficient (X3), Outlet Silent coefficient (X5), Intern heat transfer coefficients (X6), Extern heat transfer coefficients (X7), Dead Volume coefficient (X8), Inlet Lash coefficient (X9), Discharge coefficient (X10), Relative Pressure coefficient (X13), PS Friction Coefficient (X14), SSP Friction Coefficient (X15).

The variation of the whole set of parameters are then resumed in Figure 3.

3.2. Numerical method for uncertainty quantification

Let us consider a stochastic differential equation of the form:

$$L(\mathbf{x}, \boldsymbol{\xi}, \phi) = f(\mathbf{x}, \boldsymbol{\xi}) \tag{2}$$

where L is a non-linear spatial differential operator (for instance, L is the steady Navier-Stokes operator) depending on a set of uncertainties, defined with a random vector $\boldsymbol{\xi}$ (whose dimension depends on the number of uncertain parameters in the problem) and $f(\mathbf{x}, \boldsymbol{\xi})$ is a source term depending on the position vector $\mathbf{x}$ and on $\boldsymbol{\xi}$. The solution of the stochastic equation (2) is the unknown dependent variable $\phi(\mathbf{x}, \boldsymbol{\xi})$, and is a function of the space variable $\mathbf{x} \epsilon \mathbf{R}^d$ and of $\boldsymbol{\xi}$. One of the objective of Uncertainty Quantification is to compute the statistics of the quantity of interest, *i.e.* $\phi(\mathbf{x}, \boldsymbol{\xi})$ with respect to the system uncertainties $\boldsymbol{\xi}$.

Table 3. Uncertainty assessment.

Parameter	Nominal Point	Uncertainty
Rotational Speed (tr/min)	3000	0.30%
Cylinder Head Pressure (barA)	25	0.36 bar
Cylinder Head Temperature (°C)	210	5°C
Cylinder Head Enthalpy (kJ/Kg)	1094.27	1.30%
Exhaust pressure (barA)	1.3	0.06bar
Bore (mm)	40	0.12mm
Stroke (mm)	40.04	0.10%
Implantation radius (mm)	55	0.40%
Dead volume (cc)	6.13	6.0%
Inlet diameter (mm)	10	0.40%
Inlet lift (mm)	2	0.05mm%
Inlet duration (°)	82	2°
Inlet top angle (° after TDC)	15.5	2°
Outlet lift (mm)	4.5	0.33mm
Outlet width (%)	65	0.30%
Inlet leak coefficient (X1)	0.036	0.01
Piston leak coefficient (X2)	1.94	0.01
Inlet silent coefficient (X3)	0.1	0.01
Outlet silent coefficient (X4)	0.0	0.01
Inlet heat transfer coefficient (X5)	2.22	0.01
External heat transfer coefficient (X6)	1.62	0.01
Inlet valve stem leak coefficient (X7)	0.969	0.01
Dead volume coefficient (X8)	1.03	0.01
Inlet lash coefficient (X9)	0.978	0.01
Discharge coefficient (X10)	1.36	0.01
Relative pressure coefficient (X13)	0.911	0.01
PS friction coefficient (X14)	0.145	0.01
SSP friction coefficient (X15)	0.235	0.01

Using a non-intrusive uncertainty quantification tool means that a single deterministic computation (used to solve for example the differential operator L) is replaced with a whole set of such computations, each one of those being run for specific values of the uncertain conditions. In this work, a non-intrusive techniques is used to propagate physical uncertainties through the system under consideration. An anchored-ANOVA [21, 22] approach is applied in this paper, since a large set of uncertainties is analyzed here (the method can treat until hundreds of uncertainties). This analysis permits to detect the hierarchy of the most predominant uncertainties and sensibly reduced the number of deterministic simulations needed to perform the stochastic analysis.

4. Results

This section is devoted to present the results coming from different sensitivity analysis, which are performed considering several sets of conditions. Note that the sensitivity analysis is performed for each of the four quantities of interest in this study: the mechanical power, the mass flow rate, the exhaust temperature and the isentropic efficiency, defined as $\Delta h/\Delta h_{ideal}$ where Δh

represents the variation of the static enthalpy between inlet and outlet boundaries, and Δh_{ideal} is the variation of enthalpy when an ideal isentropic transformation with the same initial conditions and pressure ratio as the real flow.

First, the uncertainties on the physical parameters are taken into account, then a problem with 14 (fourteen) uncertainties should be solved. Secondly, the sensitivity analysis is extended by including also the uncertainties coming from the calibration parameters. The interest is into a deeper understanding of the most predominant set of parameters. In this case, a problem with 26 (twenty-six) parameters has to be treated.

Let us present the result coming from the sensitivity analysis applied to the problem featuring 14 (fourteen) uncertainties on the physical parameter. First, a sensitivity analysis study is applied to the variability of the Mechanical Power, which has been reported in Figure 5. As it can be easily observed, there are three predominant uncertainties (Dead volume, rotational speed, exhaust pressure, in order of importance), and a fourth (inlet lift) with a small contribution. Concerning the variability associated to the Mass flow rate, plotted in Figure 6, the uncertainty on the Dead Volume contributes nearly to the 60% of the variance. Three other uncertainties, *i.e.* on the cylinder head pressure, on the cylinder head enthalpy and on the exhaust pressure contribute nearly to the 36% of the variance. Note also that another uncertainty, *i.e.* on the inlet lift, contribute nearly to the 2%, while all the other uncertainties can be neglected. Finally, the results of the sensitivity analysis is reported in Figure 7 for the exhaust temperature and in Figure 8 for the isentropic efficiency, respectively. Concerning the variability of the exhaust temperature, the uncertainty on the cylinder head enthalpy contributes the most, attaining nearly the 97% of the total variance. Concerning the isentropic efficiency, there are two predominant uncertainties, on the exhaust pressure and on the dead volume, which contribute nearly to the 90% of the variance.

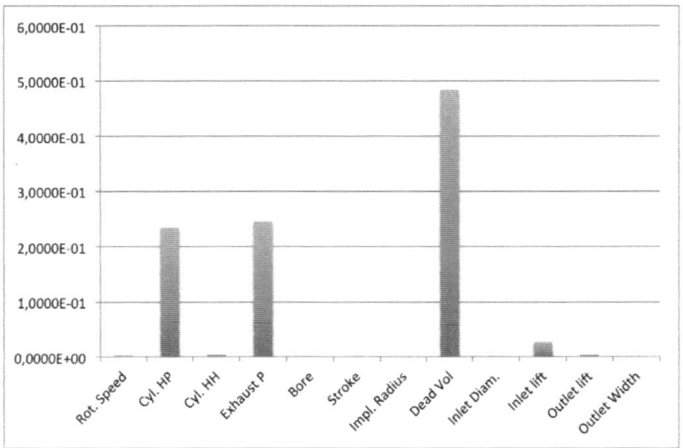

Figure 5. Hierarchy of the most predominant parameters for the Mechanical Power.

Let us now introduce the results of the sensitivity study when also the uncertainties on the calibration parameters are introduced. Note that in this case, a fixed variation interval has been imposed for each uncertainty. Results for the mechanical power (only this figure is reported in this paper, i.e. Figure 9, for sake of brevity), mass flow, exhaust pressure and isentropic efficiency, show that the hierarchy computed by considering only the uncertainties on the physical parameters remains nearly the same for each quantity of interest. Then, the influence of the calibration parameter can generally be considered as negligible with respect to the uncertainty of the physical parameters. Note then that the only parameter that provides a certain contribution to the global variance is the Dead volume coefficient, attaining nearly the 1%, in the context of the sensitivity analysis for the mechanical power, the mass flow, and the

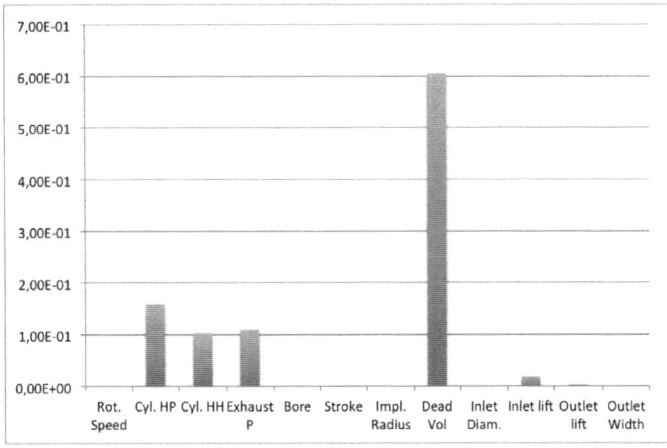

Figure 6. Hierarchy of the most predominant parameters for the Mass Flow.

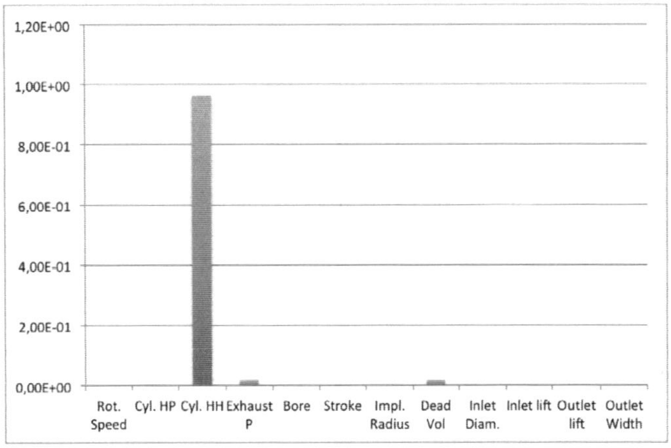

Figure 7. Hierarchy of the most predominant parameters for the Exhaust Temperature.

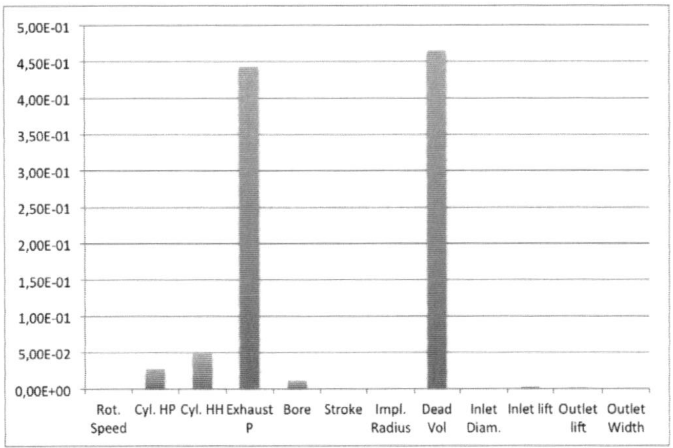

Figure 8. Hierarchy of the most predominant parameters for the Isentropic efficiency.

isentropic efficiency.

Note that anyway this analysis could be in theory biased by the assumption about the interval variation, which has been imposed to each calibration parameter, in absence of a more

Table 4. Some statistics for the Mechanical Power, the Mass Flow, the exhaust temperature, and the isentropic efficiency.

	Mech. Pow. [kW]	Massflow [kg/s]	Exh. Temp. [^{o}C]	Net Is. Eff. [%]
μ	8.541781	67.29476	100.6077	61.36836
σ	0.05593562	0.5739631	1.584921	0.07429255
COV	0.006548473	0.00852909	0.01575348	0.0012106
90% conf. interv.	(8.4510 ; 8.6362)	(66.473 ; 68.321)	(98.289 ; 103.265)	(61.220 ; 61.463)

precise characterization. Clearly, if the interval variation is increased, the expected behavior could feature also an increasing importance of the uncertainty on the calibration parameter. Nevertheless, this study provides an useful indication about the non-influence of these parameters when a reduced interval variation is considered. An optimal way to further investigate this problem could be the calibration of these parameters by means of a bayesian approach starting directly from the experimental data.

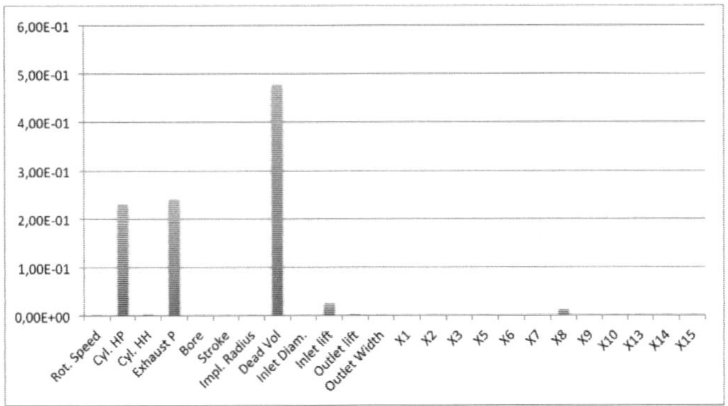

Figure 9. Hierarchy of the most predominant parameters for the Mechanical Power, considering the uncertainties both on the physical and calibration parameters.

Moreover, in Table 4, several statistical quantities are computed, in particular the mean, the standard deviation, the coefficient of variation (standard deviation to mean ration), the 90% confidence interval. Note that the PDF of the mechanical power is truly of gaussian-type (see Figure 10), with a coefficient of variation of 0.6%. The PDF of the exhaust temperature (see Figure 11) displays a quasi multi-modal behavior with two peaks at 99°C and 102°C. In this case, the coefficient of variation is much higher, attaining 1.5%.

5. Conclusions

This paper illustrates a study concerning the uncertainty characterization and assessment of the simulation of a piston expander for waste heat recovery. Experimental and model parameters uncertainties are propagated through a simple numerical model in order to get the sensitivity analysis for some important quantities of interest, such as the mechanical power, the mass flow and the exhaust pressure. Results show that the uncertainty on the dead volume contributes largely to the variance of each quantity of interest except for the exhaust temperature. This last one is mostly dependent on the uncertainty of the cylinder head enthalpy. Other quantities to control (since they explain a large contribution to the variance) are the exhaust pressure and the rotational speed, which affect the variance of the mechanical power and of the isentropic

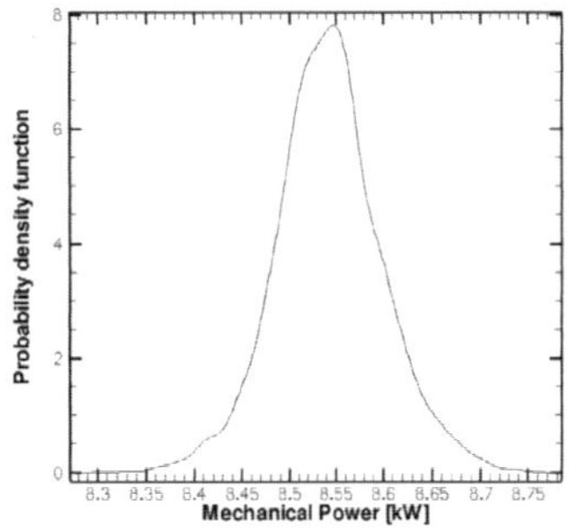

Figure 10. PDF of the mechanical power.

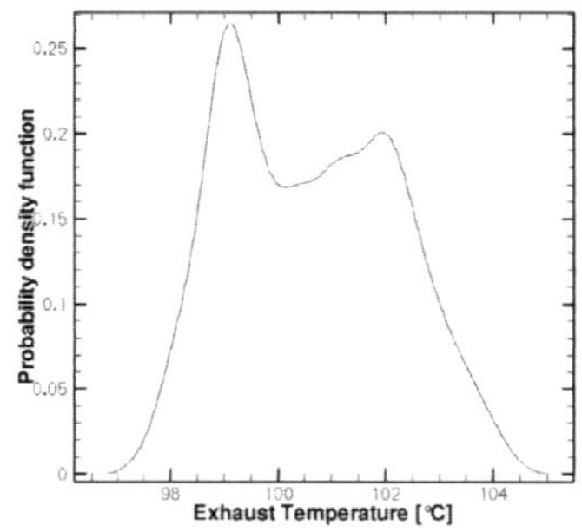

Figure 11. PDF of exhaust temperature.

efficiency as well. On the contrary, uncertainties on the calibration parameters are negligible. Globally, the min/max variation for each quantity of interest is around 3%, then, they are quite stable w.r.t. the system uncertainties.

This analysis has then permitted to identify four important parameters (on twenty-six) to control in order to reduce the global amount of uncertainty associated to the quantity of interest, thus providing interesting insights for the model improvement. Future works should be directed to extend this study including a larger set of operating conditions and to the calibration of the model parameters, exploiting directly the experimental data.

References

[1] Saidur R and Rezaei M and Muzammil W and Hassan M and Paria S and Hasanuzzaman M 2015 *Renewable and Sustainable Energy Reviews* **16** 1260-1268.

[2] Apostol V and Pop H and Dobrovicescu A and Prisecaru T and Alexandru A and Prisecaru M 2015 *Procedia Engineering* **100** 549-558.

[3] Conklin J and Szybist J 2014 *Energy Procedia* **45** 121-130.

[4] Yang J 2005 *24th International Conference on Thermoelectrics* 1-5.

[5] Dolz V and Novella R and Garcia A and Sanchez J 2012 *Applied Thermal Engineering* **36** 269-278.

[6] Serrano J and Dolz V and Novella R and Garcia A 2012 *Applied Thermal Engineering* **36** 279-287.

[7] Qiu G and Liu H and Riffat S 2011 *Applied Thermal Engineering* **31** 3301-3307.

[8] Macian V and Serrano J and Dolz V and Sanchez J 2013 *Applied Energy* **104** 758-771.

[9] Demler R 1976 *SAE Int. Tech. Paper* 1-15.

[10] Badami M and Mura M 2009 *Energy* **34** 1315-1324.

[11] Endo T and Kawajiri S and Kojima Y and Takashi K 2007 *SAE World Congress* 1-10.

[12] Glavatskaya Y and Podevin P and Lemort V and Shonda O and Decombes G 2012 *Energies* **5** 1751-1765.

[13] Daccord R and Kientz T and Melis J 2013 *ASME ORC 2013.*

[14] Daccord R and Darmedru A and Melis J 2014 *SAE 2014 World Congress & Exhibition* **2014-01-0675** 12.

[15] Galindo J and Doltz V and Royo L and Haller R and Melis J 2015 *3rd International Seminar on ORC Power Systems* **85** 17.

[16] Dickson J and Ellis M and Rousseau T 2014 *SAE International.* **2014-01-2339** 99-120.

[17] Abgrall R and Congedo PM and Geraci G and Iaccarino G 2016 *CMAME.* **301** 80-115.

[18] Congedo PM and Geraci G and Abgrall R and Pediroda V and Parussini L 2013 *Engineering Computations.* **30** 1032-20153.

[19] Congedo PM and Corre C and Martinez JM 2011 *CMAME.* **200** 216-232.

[20] Lemmon EW and Huber ML and McLinden MO 2013 *NIST Standard Reference Database 23: Reference Fluid Thermodynamic and Transport Properties-REFPROP, Version 9.1.* NIST.

[21] Tang K and Congedo PM and Abgrall R 2015 *IJNME.* **102** 1584.

[22] Tang K and Congedo PM and Abgrall R 2016 *Journal of Computational Physics.* **314** 589.

1st International Seminar on Non-Ideal Compressible-Fluid Dynamics for Propulsion & Power IOP Publishing
IOP Conf. Series: Journal of Physics: Conf. Series **821** (2017) 012013 doi:10.1088/1742-6596/821/1/012013

SU2: the Open-Source Software for Non-ideal Compressible Flows

M. Pini[1], S. Vitale[1], P. Colonna[1], G. Gori[2], A. Guardone[2], T.Economon[3], J.J.Alonso[3], F.Palacios[4]

[1] Propulsion & Power group, Faculty of Aerospace Engineering, Delft University of Technology
[2]Department of Aerospace Science and Technology, Politecnico di Milano
[3]Aerospace Design Laboratory, Stanford University
[4]Advanced Concepts, Boeing Commercial Airplanes, The Boeing Company, Everett, United States

E-mail: m.pini@tudelft.nl

Abstract.
The capabilities of the open-source SU2 software suite for the numerical simulation of viscous flows over unstructured grid are extended to non-ideal compressible-fluid dynamics (NICFD). A built-in thermodynamic library is incorporated to account for the non-ideal thermodynamic characteristics of fluid flows evolving in the close proximity of the liquid-vapour saturation curve and critical point. The numerical methods, namely the Approximate Riemann Solvers (ARS), viscous fluxes and boundary conditions are generalised to non-ideal fluid properties. Quantities of interest for turbomachinery cascades, as loss coefficients and flow angles, can be automatically determined and used for design optimization.

A variety of test cases are carried out to assess the performance of the solver. At first, numerical methods are verified against analytical solution of reference NICFD test cases, including steady shock reflection and unsteady shock tube. Then, non-ideal gas effects in planar nozzles and past turbine cascades, typically encountered in Organic Rankine Cycle applications, are investigated and debated. The obtained results demonstrate that SU2 is highly suited for the analysis and the automatic design of internal flow devices operating in the non-ideal compressible-fluid regime.

Introduction

Non-ideal compressible fluid dynamics (NICFD) is the discipline devoted to the study of the thermo-physical characteristics of fluid flows departing from gas ideality, namely flows not obeying to the perfect gas law. Supercritical flows, dense vapors, and two-phase flows belong to this category. Due to the ever more stringent environmental regulations, the interest around non-ideal fluid flows has peaked in the propulsion and power field. For example, they are by far the essential element in Organic Rankine Cycle (ORC) turbogenerators, which are energy conversion systems renowned for the efficient exploitation of renewable energy sources [1, 2, 3, 4]. Supercritical CO_2 have become an attractive working medium for next-generation solar and nuclear power generation [5, 6] [7], as well as the key-enabler of highly-efficient refrigeration devices [8]. The successful deployment of these sustainable technologies is primarily dictated by the performance maximization of their components (e.g. turbomachinery, heat exchangers, ejectors) which can be only driven by CFD, as non-ideal flows usually exhibit gas-dynamic phenomena largely unpredictable with simplified methods [9]. To the authors knowledge, there is currently no computational infrastructure providing analysis and design capability for non-ideal fluid flows. Robust and accurate simulations of non-ideal fluid flows is still a challenge, and the quasi-absence of experimental data in the thermodynamic regions of interest renders uncertain the reliability of the physical models embedded in CFD tools. i.e. the thermo-physical and turbulence models. Presently, several research programs are underway [10] to conduct experiments for non-ideal flow fluids aimed at providing sets of data for the validation of NICFD tools.

The SU2 software suite [11] has recently gained great interest as open-source platform for solving multi-physics PDE problems and PDE-constrained optimization problems on general unstructured meshes. The code resolves steady and unsteady Reynolds-averaged Navier-Stokes (RANS) equations for incompressible and compressible, laminar and fully turbulent, flows. As unique feature, SU2 accommodates built-in design functionalities through a continuous and a discrete adjoint solver. Therefore, equipped with accurate thermo-physical models, SU2 can eventually meet the research and industrial needs of NICFD community paving the way to unsteady design for all those applications involving non-ideal flows.

The first extension of SU2 to NICFD was thoroughly addressed in [12]. This paper documents the latest efforts undertaken by the authors to expand the capabilities towards analysis and design under non-ideal flow conditions. Especially, the thermo-physical library has been enriched by a direct coupling with the FluidProp database [13].

These new features are tested on a variety of model problems and practical applications comprising a rarefaction shock-wave over a wedge, an unsteady shock-tube, and supersonic flow within a converging-diverging nozzle and turbine vanes typical of ORC blade passages. The collection of test cases not only provides evidence of the capability of the tool but can be also considered as benchmarks for developers and users of NICFD tools.

The paper is organized as follows: section 1 reminds the governing equations for an arbitrary equilibrium real gas. Section 2 briefly recalls the adopted numerical methods. Section 3 outlines the thermo-physical models currently available in SU2, while in section 4 the obtained results are shown and discussed.

1. Governing Equations

SU2 has been designed to solve the compressible Reynolds-Averaged Navier-Stokes (RANS) equations [14], as high Mach number flows are of concern in NICFD. The system of PDE equations including the inviscid and viscous terms is usually written as

$$\partial_t \vec{U} + \nabla \cdot \vec{F}^c - \nabla \cdot \vec{F}^v = \vec{Q} \quad \text{in } \Omega,\, t > 0. \tag{1}$$

Equation (1) describes how mass, momentum and energy evolve in a control domain. $\vec{U}$ symbolizes the vector of conservative variables, i.e. $\vec{U} = (\rho, \rho v_1, \rho v_2, \rho v_3, \rho E)^{\mathsf{T}}$, where ρ is the fluid density, E is the total energy per unit mass, and $\vec{v} = (v_1, v_2, v_3) \in \mathbb{R}^3$ is the flow velocity in a Cartesian coordinate system. A thorough description of the model can be found in [12] and [15].

2. Numerical Algorithms

The convective fluxes are calculated by means of the Roe's approximate Riemaann solver (ARS) generalised to arbitrary fluids. As reported in [16], the averaged Jacobian matrix resulting from a local linearization of the Riemann problem (2) is no longer uniquely determined for equilibrium non-ideal flows and a new condition (3) must be satisfied.

$$\left(\vec{F}_{\mathrm{i}}^{c} - \vec{F}_{j}^{c}\right) = \bar{A}\left(\vec{U}_{\mathrm{i}} - \vec{U}_{\mathrm{j}}\right), \quad \bar{A} = A(\bar{\vec{U}}), \tag{2}$$

$$\bar{\chi}(\rho_\mathrm{i}-\rho_\mathrm{j})+\bar{\kappa}(\rho_\mathrm{i}e_\mathrm{i}-\rho_\mathrm{i}e_\mathrm{j})=(P_\mathrm{i}-P_\mathrm{j}), \tag{3}$$

where $\bar{\chi}$ and $\bar{\kappa}$ are averaged secondary thermodynamic properties defined as:

$$\chi=\left(\frac{\partial P}{\partial \rho}\right)_{\rho e}=\left(\frac{\partial P}{\partial \rho}\right)_{e}-\frac{e}{\rho}\left(\frac{\partial P}{\partial e}\right)_{\rho}\quad ,\kappa=\left(\frac{\partial P}{\partial \rho e}\right)_{\rho}=\frac{1}{\rho}\left(\frac{\partial P}{\partial e}\right)_{\rho}. \tag{4}$$

In SU2 the intermediate state is retrieved through the well-established Vinokur-Montagne' [17] approach. Second-order accuracy is achieved using a Monotone Upstream-centered Schemes for Conservation Laws (MUSCL) approach [18] with gradient limitation. The numerical viscous fluxes are conversely evaluated by averaging the flow variables, flow derivatives, and transport properties at two neighbouring cells. Characteristics based boundary conditions are enforced to the inflow/outflow boundaries. To automatically detect the incoming and outgoing waves, the approach proposed in [19] is employed. By means of an eigenvalue analysis the number of enforceable unknowns is determined and variables that can be specified at the boundary are automatically selected. For more details the reader is referred to [12].

To enhance portability, SU2 relies on widely popular open-source software for parallelization (MPI and ParMetis). More details on the parallelization paradigm can be found in [15].

3. Computation of Thermo-physical Properties

Unlike standard CFD solvers, the numerical schemes for non-ideal fluid flows require the calculation of primary and secondary, i.e. partial derivatives of primary thermodynamic variables, thermodynamic properties as a function of density ρ and internal energy e, which are in turn a recombination of the conservative variables as follows

$$\rho=U_1,\quad e=\frac{U_5}{U_1}-\frac{(U_2+U_3+U_4)^2}{2U_1^2}=E-\frac{\|v\|^2}{2}. \tag{5}$$

The stable equilibrium state principle assures that any other generic thermodynamic property X (e.g. pressure, temperature, etc.) is determined through an arbitrary thermodynamic model in the form

$$X=X(\rho,e)=X(\vec{U}). \tag{6}$$

Equation 6 is explicit for polytropic models, like the ideal gas (PIG), the Van der Waals (PVdW), and the Peng-Robinson (PR) available in the built-in thermodynamic library. These models ensure high computational efficiency but are not accurate approaching the critical point. However, they are useful to initialize simulations where strong non-ideal flow effects are of concern. For these latter cases, SU2 can exploit the advantages offered by the direct coupling with a general purpose thermophysical library [13] originally developed at Delft University of Technology.

4. Results and Discussion

As opposed to classic aerodynamics, no canonical test cases exist for assessing the accuracy and performance of non-ideal compressible flow solvers. Consequently, a series of numerical applications are presented thereafter, in which SU2 is first verified against known analytical solutions and then compared to results obtained by a commercial package.

4.1. Rarefaction shock-wave

In thermodynamic conditions close to the critical point fluids of complex molecules are supposed, by theory, to produce non-classical gas dynamic phenomena such as rarefaction shock waves or compression fans. The non-classical region encloses all the possible states of a thermodynamic

	Density	Pressure	Temperature	Mach
S_A	0.79	1.06	1.01	1.70
S_B	0.45	0.93	1.00	1.19

Table 1: Fluid state before and after the rarefaction shock-wave computed with SU2.

mesh tag	A	B	C	D	E	F
N elements	2301	5251	14751	49551	104351	179151
Iter time [s]	0.0145	0.03	0.086	0.4	0.6	1.3

Table 2: Number of elements composing each mesh used to discretise the domain and computational time for a single solver iteration.

system, defined as a combination of pressure, temperature and density, that produce a negative value of the *fundamental derivative of gas-dynamics* $\Gamma < 0$ [9]. For instance, linear siloxanes, synthetically denoted as MD_nM, are supposed to be endowed with this very particular region. For this family of fluids, cubic equations of state, like the Van der Waals model, admit the onset of the non-classical phenomena mentioned earlier. The fluid considered in this section is a MDM siloxane and its properties are listed in Table 3, the value of the specific heat ratio is assumed to be $\gamma = 1.0125$ for this particular test case.

The geometry consists of a simple square domain containing an edge with a design slope $\theta = 15.945°$, and the fluid flows through the domain from the left to the right boundary. Symmetry condition is applied at the upper border, while at the bottom surface an inviscid wall boundary condition is prescribed. From now on, we identify the fluid state S before the shock with the subscript A and the state S after the shock with the subscript B. Results here presented were obtained assuming fluid conditions close to the critical point to show and highlight non-classical phenomena. Convective fluxes were evaluated using an implicit, second-order accurate in space, scheme based on a standard Roe numerical scheme. Dimensionless state variables, with respect to the fluid critical values, are reported in Tab. 1 for state A and B. If the Van der Waals equation of state is used in place of the PIG law, a non-classical rarefaction shock wave occurs. Results, reported in Tab. 1, show that values of pressure and density drop across the shock. The deviation between the numerical and the theoretical value for p_B, computed using the Rankine-Hugoniot condition for a Van der Waals fluid, is approximately $5.3163 \cdot 10^{-5}\%$.

Different structured meshes, with a different level of resolution, were used to simulate the phenomenon. Tab. 2 lists the total number of elements in each mesh and the computational cost, in terms of time, of each single iteration carried out by the SU2 solver. The overall computational time is not reported here since convergence history changes significantly from mesh to mesh due to the intervention of the automatic CFL (Courant-Friedrichs-Lewy) adaptation routine. By the way, just to give an idea, the total computational time spans from roughly two hundreds seconds, for mesh A, to tens of thousands of seconds for the finer grid F.

Flow solution, reported in Fig. 1 for grid D, clearly show the presence of a non-classical rarefaction shock wave. Fig. 2 depicts Mach profiles at $y = 0.4$ along the x axis, for all the different level of space dicretisation (see Tab. 2). For grid A numerical dissipation, bound to a non adequate level of refinement, causes the shock to be smeared out over a wide portion of the domain. Trends reported in Fig. 2 show that the solution tends to converge to a shock discontinuity by increasing the level of mesh resolution. Though very small differences can still be found, Mach trends for grid E and F suggest that the solution is by now independent from the domain dicretisation. Finally, the value of the angle β, i.e. the angle between the shock and

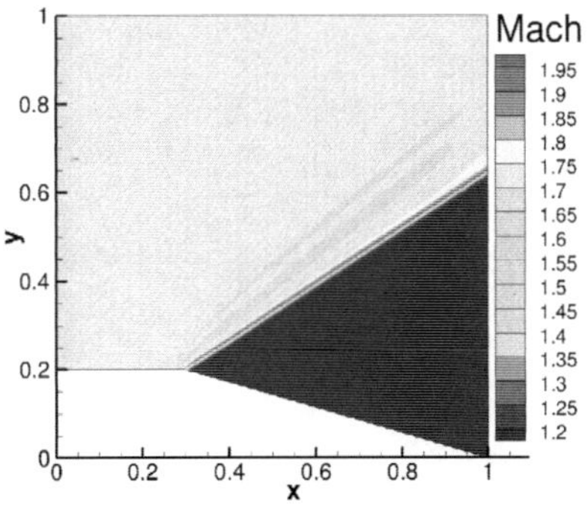

Figure 1: Van der Waals, mesh D, second order: Mach field.

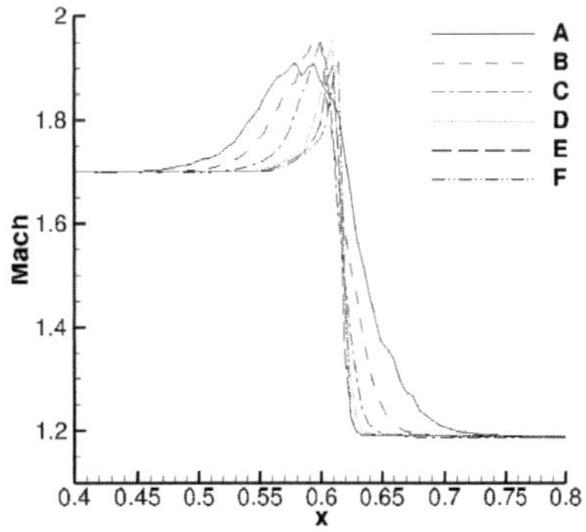

Figure 2: Mach profile along $y = 0.4$ for solution computed over different meshes.

Fluid	R_{gas} [J/KgK]	γ	P_{cr} [Pa]	T_{cr} [K]	ρ_{cr} [Kg/m^3]
Air	287.058	1.4	-	-	-
MDM	35.152	1.018	1415000	564.09	256.82

Table 3: Fluid parametes for air and MDM specified for numerical simulation of unsteady shock tube.

the direction of the x-axis, is compared against its theoretical value. The analytical relation, which depends only on the thermodynamic state of the fluid before and after the shock reads

$$\beta = \arcsin \sqrt{\frac{p_B - p_A}{\rho_B - \rho_A} \frac{\rho_B}{\rho_A} \frac{1}{c_A^2 M_A^2}}. \tag{7}$$

In the present case the analytical value is $\beta = 32.8602°$, while the computed shock angle is $\beta = 32.735°$, fairly close to the theoretical one.

4.2. 2D unsteady shock tube

The motion of a shock-wave travelling inside a pipe is presented in this section. The problem is assumed to be two-dimensional and viscous phenomena are neglected without any loss of generality. Unsteady simulation were carried out using a first order accurate dual-time stepping method. In the following, comparisons between the analytical and the numerical solution obtained by SU2 are shown. The problem is studied using both ideal and non-ideal fluids, namely air and MDM. Tab. 3 reports fluid parameters such as gas constants, critical values and specific heat ratio specified for the numerical simulations. The pipe is represented by a rectangular domain of length $L = 10$ m, along x axis, and height $h = 0.5$ m, on y axis. The shock moves from left to right and it is generated by applying a pressure and a temperature jump at the left boundary. The upper and lower boundaries are assumed to be free-slip walls. The pipe has an opening on the right boundary which allows the fluid to be discharged into an infinite-acting reservoir: a constant pressure condition is imposed here, $p_{out} = 101325$ [Pa] for air and $p_{out} = 800000$ [Pa] for MDM, is applied.

Initially, the left-to-right movement of the shock creates two uniform regions, namely 1 (upstream state with respect to the shock) and 2 (downstream state with respect to the shock). It is possible to retrieve the analytical solution of these kind of problem considering both an ideal gas or a complex fluid whose thermodynamic behaviour is characterize by means of the

	S_1	S_2	S_3	$\bar{S}_1$	$\bar{S}_2$	$\bar{S}_3$
P [Pa]	101325	130795	101325	101325	130795	101324
T [K]	303.15	326.32	303.15	303.15	326.35	303.37
ρ [Kg/m^3]	1.1644	1.3963	1.6641	1.1644	1.3961	1.1635

Table 4: Thermodynamic states retrieved through the analytical solution, and by means of numerical simulations (barred values) for air.

	S_1	S_2	S_3	$\bar{S}_1$	$\bar{S}_2$	$\bar{S}_3$
P [Pa]	800000	838120	800000	800000	838120	800000
T [K]	570.5	571.17	570.5	570.5	571.17	570.5
ρ [Kg/m^3]	49.52	52.53	49.52	49.52	52.53	49.52

Table 5: Thermodynamic states retrieved through the analytical solution, and by means of numerical simulations (barred values) for MDM.

simple, but yet qualitatively sound, Van der Waals equation of state. The so called *Rankine-Hugoniot* condition impose a relation between the pre- and the post- shock states. See [20] for the definition the Rankine-Hugoniot condition for a van der Waals fluid. Since state 1 is fully known and the value of the static pressure p_2 is chosen arbitrarily as a boundary condition, it is possible to solve Rankine-Hugoniot equation numerically to retrieve the value of the density in the post-shock region. Once ρ_2 and p_2 are available, the post-shock state is uniquely defined and the speed of the shock can be reconstructed.

After the shock is discharged through the open end, at $x = 10$ m, region 1 does no longer exist and a third region 3 arise in the domain. At the discharge section the fluid may be supersonic, transonic or subsonic leading to three different analytical solutions: only the subsonic discharge is considered hereinafter. For a subsonic discharge, a rarefaction fan is produced at the open section. Rarefaction waves travel upstream, from right to left, and generate an isentropic expansion which allows to match the outer pressure and brings to a new thermodynamic state which we will define as S_3 hereinafter. The analytical solution for this particular configuration may be retrieved by exploiting properties of Riemann invariants from characteristic theory.

The physics of the rarefaction fan for a van der Waals flow is governed by more complex equations which must account for non-ideal effects. A detailed procedure to retrieve the exact solution for a van der Waals fluid can be found in [20]. Tab. 4 and Tab. 5 report the thermodynamic states in region 1, 2, 3, respectively for air and MDM, reconstructed analytically together with their values computed through SU2 (pointed by subscript $\bar{\ }$). Due to the very weak shock intensity the entropy loss across the shock is negligible thus total quantities are approximately conserved, for this reason state S_3 failry matches state S_1. Figure 3 shows the analytical solution for the shock-wave travelling rightward, both for a dilute gas and for a non-ideal flow. The exact solution is compared with first and second order space accurate calculations proving that numerical results from SU2 are in good agreement with the analytical solution. The shock is correctly captured in terms of speed and intensity but the discontinuity is slightly smeared out due to numerical dissipation effects arising from space discretization and from first order time approximation.

Fig. 4 reports the same comparison for a left-ward travelling rarefaction fan. Numerical solutions, first and second order accurate in space, first order in time, are plotted against the analytical solution. The analytical solution is again correctly represented by the numerical solution.

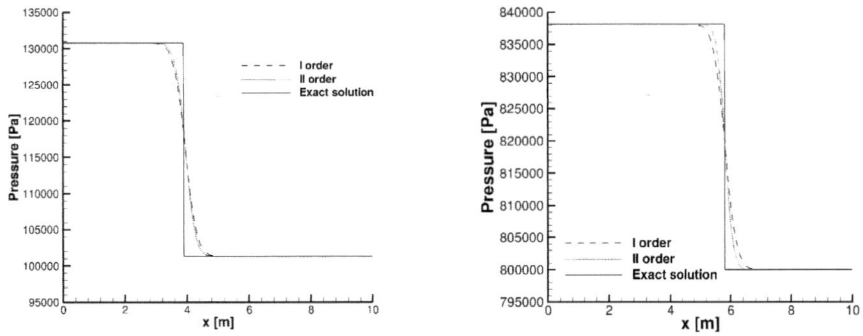

Figure 3: Comparison of the numerical against the analytical solution (continuous line). First order solution in space is represented by dashed line, second order by a dotted curve. Comparison is made at $t = 0.01$ sec for air (left) and $t = 0.05$ sec for MDM (right). The shock is moving from left to right.

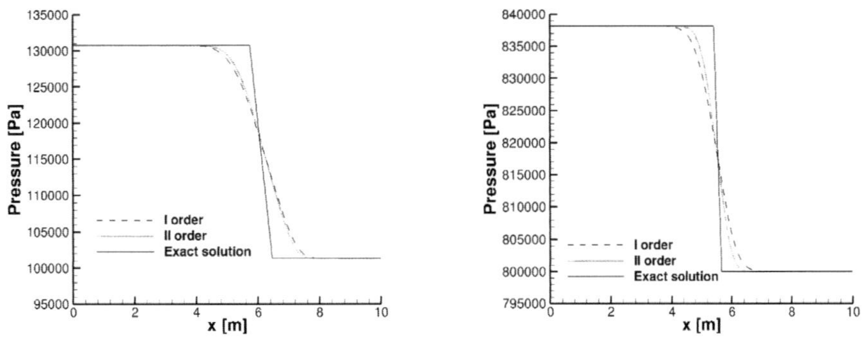

Figure 4: Comparison of the numerical against the analytical solution (continuous line). First order solution in space is represented by dashed line, second order by a dotted curve. Comparison is made at $t = 0.04$ sec for air (left) and $t = 0.13$ sec for MDM (right). The rarefaction fan is moving from left to right.

4.3. Supersonic Nozzle for Laboratory Experiments

The flow-field within a supersonic nozzle is discussed as first practical example. The nozzle is one of the test sections of the ORCHID facility, a closed-loop research rig currently under construction at Delft University of Technology [21] whose purpose is to perform fundamental studies on dense organic flows and validate NICFD tools. The test-bench is operated with siloxane MM, here modelled through the multi-parameter Span-Wagner EoS available in FluidProp. Table 6 summarizes the input parameters for this simulation. Results provided by the commercial package Ansys-CFX [22] are used for verification purposes. In both cases the Navier-Stokes equations were solved using an implicit Euler algorithm based on CFL adaptation. Note that in CFX the MM properties are available in tabulated form, which makes the computational cost of the single iteration of CFX about 50% lower than that of SU2 on four Intel i7-3630QM CPU 2.4 GHz physical cores. Figures 5 and 6 display the Mach contour and the pressure distribution along the nozzle mid section respectively. The results fairly well correlate, suggesting that, though missing experimental confirmation, SU2 can be adopted for simulating NICFD flows in practical applications.

Total inlet temperature	257.0 $°C$
Total inlet pressure	18.423 bar
Static back-pressure	2.0 bar
Turbulence model	SST-kω
Inlet turbulence intensity	0.05
Space discretization (SU2)	Upwind generalized Roe 2nd order
Space discretization (CFX)	High resolution method

Table 6: Input parameters for the supersonic nozzle test case.

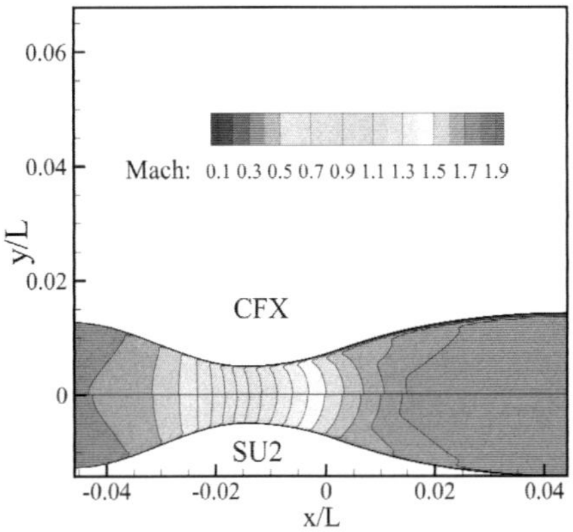

Figure 5: Mach contour at the mid section of the nozzle.

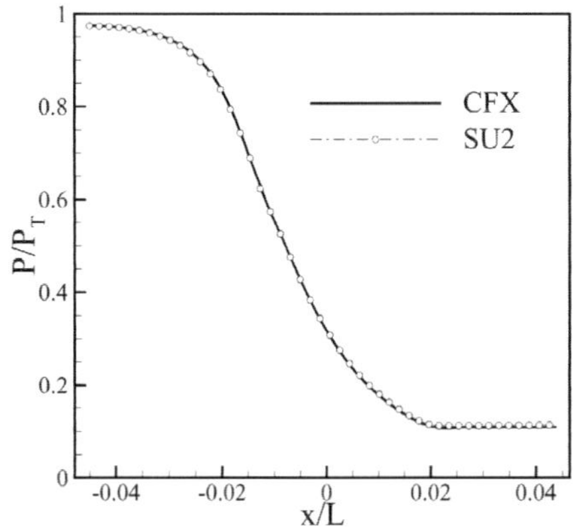

Figure 6: Streamwise distribution of the static pressure at the mid section of the nozzle.

4.4. Supersonic ORC Blade Passage

The last test case is representative of stator vanes of single-stage axial ORC expanders. As pointed out in [23], the passage was designed by resorting to oversimplified flow models and was recently used to assess the effectiveness of shape optimization for highly non-ideal flows [24]. The turbine is operated with the siloxane MDM modelled by the polytropic improved Peng-Robinson-Stryjek-Vera equation of state. Three structured grids of increasing density level (with two cells in span-wise direction) are used for the calculations. The SST-kω turbulence model is used ensuring wall $y+$ well below the unity all along the blade surface. As only blade-to-blade flow features are of interest, free-slip conditions are imposed on the endwalls, and periodicity is set on the upper and the lower boundaries. A summary of the input parameters for the simulation can be found in Table 7. As in the previous case, SU2 simulations are verified against the results provided by the Ansys-CFX solver. Again, in both cases the equations were solved using an implicit Euler algorithm based on CFL adaptation. At first, a mesh sensitivity study is carried out to gain insight of the convergence rate of the two solvers. The mass-averaged entropy loss coefficient $Y = \frac{S_{in}-S_{out}}{S_{in}}$ is considered as merit function for the study. It is worth mentioning that this quantity is automatically computed by SU2 at runtime. Figure 9 shows the variation of such coefficient for increasing mesh sizes. As can be observed, the two solvers exhibit similar convergence properties, the solution being mesh-independent for grids finer than 200k cells. The calculation with SU2 on the coarsest grid reached full convergence after 200

Total inlet temperature	272 °C
Total inlet pressure	8 bar
Static back-pressure	1.0 bar
Inlet turbulence intensity	0.05
Space discretization (SU2)	Upwind generalized Roe 2nd order
Space discretization (CFX)	High resolution method

Table 7: Input parameters for the supersonic ORC blade passage.

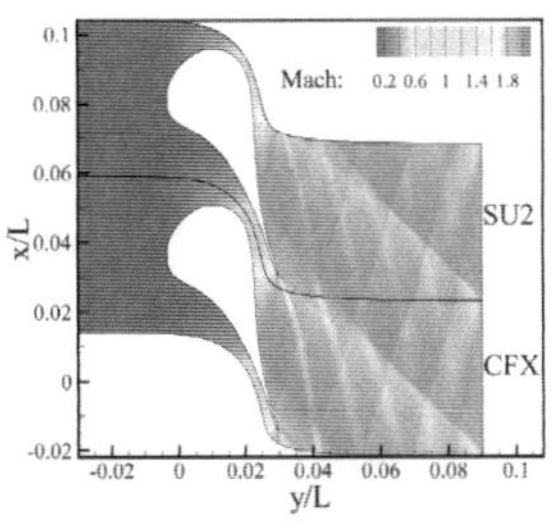

Figure 7: Mach contour.

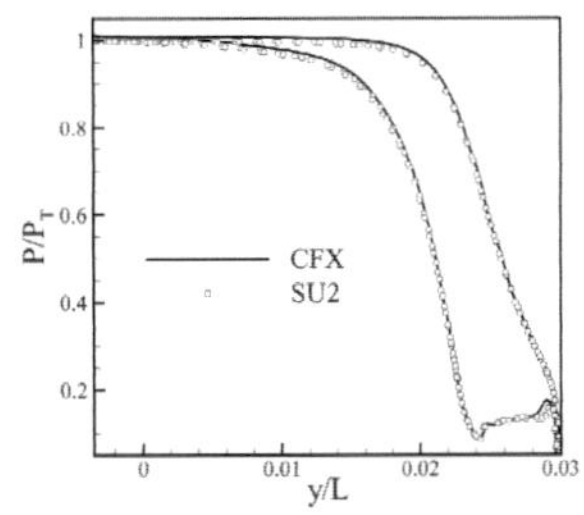

Figure 8: Pressure distribution along the blade surface.

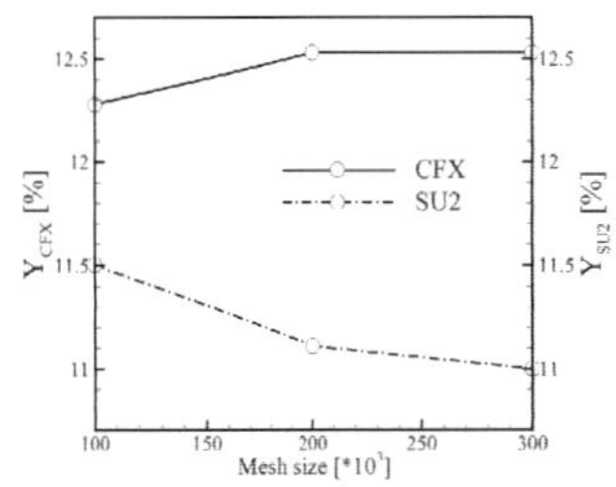

Figure 9: Mesh sensitivity study.

iterations (about 250 seconds on four Intel i7 - 2.2 GHz physical cores). Figure 7 displays the Mach number contour obtained by the two solvers for the finest grid. As expected, two shock waves are generated at the blade trailing edge, forming the so-called fishtail shock configuration. The trailing-edge shock from the pressure side is reflected on the suction side of the successive blade, while the suction-side trailing shock propagates downstream towards the outlet. At approximately $y/L = -0.06$ the two shocks coalesce into a single, stronger shock wave traveling downstream. Such interaction is somehow more visible in the SU2 solution, which is therefore found to capture these flow features more accurately.

5. Conclusions

In this paper, the capability of SU2 for simulating non-ideal compressible flows were illustrated and demonstrated through model problems and practical examples.

The results showed that, equipped with complex equations of state and appropriate numerical schemes, SU2 can be successfully used to predict non-classical gas dynamic phenomena such as rarefaction shock waves. The numerical outcomes furthermore indicated that SU2 is highly suited for the analysis of both steady and unsteady flows in the non-ideal compressible flow regime. Moreover, the novel integrated turbomachinery features makes the tool attractive for automated design of two and three-dimensional ORC turbine cascades and, more in general, of internal flow devices operating in the vicinity of the critical point.

Current developments are devoted to implement look-up tables to further enhance computational efficiency and to multi-row turbomachinery calculations for analysis and design purposes.

6. Acknowledgements

Matteo Pini and Salvatore Vitale would like to acknowledge the STW and Dana Spicer on the CC-Powertrain project. Alberto Guardone and Giulio Gori would like to acknowledge the

European Research Council (ERC) of the European Research Council Executive Agency.

References

[1] Drescher, U., and Bruggeman, D., 2007. "Fluid selection for the Organic Rankine Cycle (ORC) in biomass power and heat plants". *Applied Thermal Engineering,* ***27***(1), pp. 223–228.

[2] Lang, W., Almbauer, R., and Colonna, P., 2013. "Assessment of Waste Heat Recovery for A Heavy-duty Truck Engine Using An ORC Turbogenerator". *Journal of Engineering for Gas Turbines and Power-Transactions of the ASME,* ***135***(4), pp. 042313–1–10.

[3] Quoilin, S., Broek, M. V. D., Declaye, S., Dewallef, P., and Lemort, V., 2013. "Techno-economic survey of Organic Rankine Cycle (ORC) systems". *Renewable and Sustainable Energy Reviews,* ***22***(0), pp. 168–186.

[4] Casati, E., Vitale, S., Pini, M., Persico, G., and Colonna, P., 2014. "Centrifugal Turbines for Mini-Organic Rankine Cycle Power Systems". *Journal of Engineering for Gas Turbines and Power,* ***136***, pp. 122607–1–11.

[5] Lettieri, C., Baltadjiev, N., Casey, M., and Spakovszky, Z., 2014. "Low-Flow-Coefficient Centrifugal Compressor Design for Supercritical CO_2". *Journal of Turbomachinery,* ***136***, pp. 081008–1–9.

[6] Lettieri, C., Yang, D., and Spakovszky, Z., 2014. "An Investigation of Condensation Effects in Supercritical Carbon Dioxide Compressors". In The 4th International Symposium - Supercritical CO_2 Power Cycles.

[7] Dostal, V., Hejzlar, P., and Driscoll, M., 283-301. "The supercritical carbon dioxide power cycle: comparison to other advanced power cycles". *Nuclear technology,* ***154***(3).

[8] Elbel, S., 2011. "Historical and present developments of ejector refrigeration systems with emphasis on transcritical carbon dioxide air-conditioning applications". *International Journal of Refrigeration,* ***34***(7), 11, pp. 1545–1561.

[9] Colonna, P., and Guardone, A., 2006. "Molecular interpretation of nonclassical gas dynamics of dense vapors under the van der Waals model". *Physics of Fluids,* ***18***, pp. 056101–14.

[10] Guardone, A., Spinelli, A., and Dossena, V., 2013. "Influence of Molecular Complexity on Nozzle Design for an Organic Vapor Wind Tunnel". *ASME Journal of Engineering for Gas Turbines and Power,* ***135***.

[11] Palacios, F., Colonno, M. R., Aranake, A. C., Campos, A., Copeland, S. R., Economon, T. D., Lonkar, A. K., Lukaczyk, T. W., Taylor, T. W. R., and Alonso, J. J., 2013. "Stanford University Unstructured (SU^2): An open-source integrated computational environment for multi-physics simulation and design". *AIAA Paper 2013-0287,* ***51st AIAA Aerospace Sciences Meeting and Exhibit***, January.

[12] Vitale, S., Gori, G., Pini, M., Guardone, A., Economon, T. D., Palacios, F., Alonso, J. J., and Colonna, P., 2015. "Extension of the su2 open source cfd code to the simulation of turbulent flows of fluids modelled with complex thermophysical laws". *AIAA Paper,* ***2760***, p. 2015.

[13] Colonna, P., and van der Stelt, T., 2005. FluidProp: A program for the estimation of thermo-physical properties of fluids. Tech. rep.

[14] Landau, L., and Lifshitz, E., 1993. *Fluid Mechanics (2nd Edition).* Pergamon Press.

[15] Economon, T. D., Mudigere, D., Bansal, G., Heinecke, A., Palacios, F., Park, J., Smelyanskiy, M., Alonso, J. J., and Dubey, P., 2016. "Performance optimizations for scalable implicit {RANS} calculations with {SU2}". *Computers & Fluids,* ***129***, pp. 146 – 158.

[16] Cinnella, P., 2006. "Roe-type schemes for dense gas flow computations". *Computers & Fluids,* ***35***(10), pp. 1264–1281.

[17] Montagne, J., and Vinokur, M., 1990. "Generalized flux-vector splitting and Roe average for an equilibrium real gas". *Journal of Computational Physics,* ***89***(2), pp. 276–300.

[18] van Leer, B., 1979. "Towards the ultimate conservative difference scheme V. a second-order sequel to Godunov's method". *Journal of Computational Physics,* ***32***(1), July, pp. 101–136.

[19] Guardone, A., Isola, D., and Quaranta, G., 2011. "Arbitrary Lagrangian Eulerian formulation for two-dimensional flows using dynamic meshes with edge swapping". *Journal of Computational Physics,* ***230***(20), pp. 7706–7722.

[20] Quartapelle, L. P., Castelletti, Guardone, A., and Quaranta, G., 2003. "Solution of the Riemann problem of classical gasdynamics". *Journal of Coputational Physics,* ***190***, p. 2003.

[21] Head, A. J., De Servi, C., Casati, E., Pini, M., and Colonna, P., 2016. "Preliminary design of the ORCHID: a facility for studying non-ideal compressible fluid dynamics and testing ORC expanders". In ASME Turbo Expo 2016, no. GT2016-56103.

[22] ANSYS Inc. Academic Research, R. ., 2012. *CFX.*

[23] Colonna, P., Harinck, J., Rebay, S., and Guardone, A., 2008. "Real-gas effects in organic rankine cycle turbine nozzles". *Journal of Propulsion and Power,* ***24***(2), pp. 282–294.

[24] Pini, M., Persico, G., Pasquale, D., and Rebay, S., 2015. "Adjoint method for shape optimization in real-gas flow applications". *Journal of Engineering for Gas Turbines and Power,* ***137***(3), p. 032604.

Robust optimization of supersonic ORC nozzle guide vanes

Elio A. Bufi
Laboratoire DynFluid Arts et Metiers ParisTech, Paris (France)
Politecnico di Bari, Bari (Italy)
E-mail: elio-antonio.bufi@ensam.eu

Paola Cinnella
Laboratoire DynFluid Arts et Metiers ParisTech, Paris (France)
E-mail: paola.cinnella@ensam.eu

Abstract. An efficient Robust Optimization (RO) strategy is developed for the design of 2D supersonic Organic Rankine Cycle turbine expanders. The dense gas effects are not-negligible for this application and they are taken into account describing the thermodynamics by means of the Peng-Robinson-Stryjek-Vera equation of state. The design methodology combines an Uncertainty Quantification (UQ) loop based on a Bayesian kriging model of the system response to the uncertain parameters, used to approximate statistics (mean and variance) of the uncertain system output, a CFD solver, and a multi-objective non-dominated sorting algorithm (NSGA), also based on a Kriging surrogate of the multi-objective fitness function, along with an adaptive infill strategy for surrogate enrichment at each generation of the NSGA. The objective functions are the average and variance of the isentropic efficiency. The blade shape is parametrized by means of a Free Form Deformation (FFD) approach. The robust optimal blades are compared to the baseline design (based on the Method of Characteristics) and to a blade obtained by means of a deterministic CFD-based optimization.

1. Introduction

The use of dense gases as working media in turbomachinery, referred to as Organic Rankine Cycle (ORC) turbines, is proposed as a method of recovery of variable energy sources such as waste heat from industrial processes. Proposed heat sources for ORC turbines typically include variable energy sources such as solar thermal collectors or waste heat from industrial processes. In such conditions, traditional design at nominal operating conditions may lead to poor performance in practice. Optimization under uncertainties is called Robust Optimization (RO). The robustness is determined by a measure of insensitivity of the design with respect to variations of the design parameters, like geometrical tolerances or fluctuations of the operating conditions. To measure the robustness of a new design, statistics such as mean and variance (or standard deviation) of a response are calculated in the RO process. In this work a RO of a supersonic ORC nozzle guide vane is performed with the aim to minimize the variance and maximize the mean value of the isentropic efficiency. An original global optimization methodology has been developed in order to overcome the numerical issues typical of CFD dense gas calculations, which affect convergence and time calculation. A Non-dominated Sorted Genetic Algorithm (NSGA) is coupled with a non-intrusive Uncertainty Quantification (UQ) model which provides the statistics (mean and variance). The UQ subloop is based on a Bayesian-Kriging surrogate trained with the system response to the uncertain variables. For ORC applications, three main sources of uncertainty need to be considered: thermodynamic model, geometry and operational conditions [1]. The number of parameters in each category has a strong influence on the feasibility of the UQ analysis and, then, of the RO process. One way to reduce the space parameters dimension is to carry out a sensitivity analysis and show that it is possible to consider only operational (total pressure and temperature) and geometrical (blade thickness) parameters as the main contributors to the total variance of the system output. The nozzle guide vane baseline is designed by means of a Method Of Characteristics (MOC) generalized to accurate Equations Of State (EOS) [2, 3]. A Free Form Deformation (FFD) approach, which allows to easily handle any kind of shape independently from the geometrical complexity, is applied to generate a variety of nozzle designs. The resulting parametrization of the nozzle guide vane shape depends on a relatively low number of parameters. The evaluation of the isentropic efficiency is carried out by means of 2D Reynolds Average Navier-Stokes (RANS) simulations. Given the computational cost of viscous CFD simulations, we propose to speed-up the RO by training a second surrogate Kriging model for the multi-objective fitness function. An adaptive sample infill strategy based on the Multi Objective Expected Improvement (MOEI) is implemented in order to improve the surrogate accuracy during the genetic optimization, with a minimal additional cost.

2. Governing equations and CFD solver

The compressible Reynolds-Averaged Navier-Stokes (RANS) equations are solved, supplemented by a dense-gas EOS, and specifically the Peng-Robinson-Stryjek-Vera (PRSV) cubic equation [4], which has been proven to represent a good compromise between accuracy and computational cost. The caloric behaviour is modelled by assuming that the isocoric specific heat in the dense gas limit follows a simple power law. The governing equations are discretized by using a cell-centered finite volume scheme for structured multi-block meshes of third-order accuracy [5]. To preserve the high accuracy on non-Cartesian grids, the numerical fluxes are constructed by using weighted discretization formulas, which take into account the mesh deformations [6]. The equations are then integrated in time using a four-stage Runge-Kutta scheme. Local time stepping, implicit residual smoothing and multi-grid acceleration are used in order to drive the solution to the steady state. The turbulence is taken into account by means of the Spalart-Allmaras transport equation. The accuracy of the numerical solver has been already demonstrated in previous works [5, 7].

3. Uncertainty Quantification

The evaluation of mean and variance of a Quantity of Interest (QoI), in our case the isentropic efficiency of the ORC expander, is carried out by means of a surrogate model of the output response and a Bayesian-Kriging approach is used. In the Bayesian framework, the Kriging surrogate predicts a set of M values $\mathbf{y}$, conditional on N data $\mathbf{y}^*$, where the data $\mathbf{y}^*$ are a subset of $\mathbf{y}$ selected by the observation matrix $\mathbf{H}$. Assuming that $\mathbf{y}$ and $\mathbf{y}^*$ are normalized such that $\mathbf{y}^*$ has zero mean and unit variance, the covariance matrix $\mathbf{P}$ of the prior distribution $p(\mathbf{y})$ is defined such that its elements are generated by the Gaussian function:

$$\mathbf{P} = [P_{ij}(\theta)] = \left[exp\left(-\frac{h_{ij}^2}{2\theta^2} \right) \right] \tag{1}$$

where $h_{ij} = \xi_i - \xi_j$ is the correlation range, ξ_i the generic coordinate of the prediction $\mathbf{y}$ and θ an hyperparameter which need to be estimated. By considering a uniform and uncorrelated observation error ϵ, the covariance matrix of the likelihood can be also defined as:

$$\mathbf{R} = \epsilon^2 \mathbf{I} \tag{2}$$

where $\mathbf{I}$ is the identity matrix. By means of the Bayes theorem it is possible to evaluate the posterior distribution of prediction conditioned on the observation:

$$p(\mathbf{y}|\mathbf{y}^*) = \frac{p(\mathbf{y}^*|\mathbf{y})p(\mathbf{y})}{p(\mathbf{y}^*)} \tag{3}$$

Then, by following Wikle and Berliner [8], the Kriging predictor mean and variance are:

$$E(\mathbf{y}|\mathbf{y}^*) = \mathbf{PH}^T(\mathbf{R}+\mathbf{HPH}^T)^{-1}\mathbf{y}^* \tag{4}$$

$$var(\mathbf{y}|\mathbf{y}^*) = [\mathbf{I}-\mathbf{PH}^T(\mathbf{R}+\mathbf{HPH}^T)^{-1}]\mathbf{P} \tag{5}$$

The advantage of this methodology is to obtain directly an estimation of the surrogate model accuracy by means of the Kriging variance. Because we deal with an ordinary Kriging model where the hyperparameters need to be estimated, the Maximum Likelihood Estimation (MLE) approach is implemented in order to evaluate the vector θ as the solution of the optimization problem:

$$max_\theta \left\{ -\left(ln\,|\mathbf{A}| + \mathbf{y}^{*T}\mathbf{A}^{-1}\mathbf{y}^* \right) \right\} \tag{6}$$

where $\mathbf{A} = (\mathbf{R}+\mathbf{HPH}^T)$. Since the problem (6) is performed over a multidimensional space whose cardinality is given by the number of uncertain variables and the cost varies as $O(n_s N^3)$, where n_s is the number of optimization steps and N is the number of observed samples, it represents a bottleneck for the above methodology. Here the MLE problem is solved by means of the Nelder-Mead downhill simplex method and the inversion of the matrix $\mathbf{A}$ is improved with a Cholesky decomposition. Finally, once the Kriging surrogate has been trained and the hyperparameters estimated, the statistics are calculated by means of a Montecarlo simulation of the model by randomly sampling the uncertain variable distributions.

4. Shape parametrization: FFD method

The FFD methods are widely used in computer graphics to model 3D objects. The term "free-form" designates: *"whatever the object is, whatever its description and topology, we are able to deform it"* [9]. In this sense, these methods are useful when the geometrical complexity is high and a low number of control parameters are required to deform the object. Besides, the use of a combination of Bernstein polynomials allows to take easily into account geometrical singularities, control smoothness and represent the baseline exactly. This approach has been successfully applied in the past to airfoil optimization in external Aerodynamics [10]. The FFD is performed by parametrizing the space surrounding the object of interest (the blade, in our case). A lattice is created, whose node coordinates represent the control points and a deformation space is defined by means of a trivariate (bivariate for 2D case) tensor product Bernstein polynomial. For the 2D case, a baseline point $\mathbf{X}$ can be described in the lattice reference system as [11]:

$$\mathbf{X} = \mathbf{X}_0 + s\mathbf{S} + t\mathbf{T} \tag{7}$$

where the *(s,t)* lattice coordinates are limited in the [0,1] interval and can be related to the baseline reference system as $s = \frac{\mathbf{T}\times\mathbf{U}(\mathbf{X}-\mathbf{X_0})}{\mathbf{T}\times\mathbf{U}\cdot\mathbf{S}}$, $t = \frac{\mathbf{S}\times\mathbf{U}(\mathbf{X}-\mathbf{X_0})}{\mathbf{S}\times\mathbf{U}\cdot\mathbf{T}}$, where $\mathbf{U}$ is the direction normal to the object plane. Given a $(l \times m)$ lattice (where l and m are the number of interval along the horizontal and vertical direction, respectively) the node locations (i.e. the control points) are calculated as $\mathbf{P_{ij}} = \mathbf{X_0} + \frac{i}{l}\mathbf{S} + \frac{j}{m}\mathbf{T}$. Then, once the *(s,t)* coordinates are calculated and the deformation specified by the new $\mathbf{P_{ij}}$ position, the deformed location of a point $\mathbf{X}$ will be given by:

$$\mathbf{X_{FFD}} = \sum_{i=0}^{l} \binom{l}{i} (1-s)^{l-i} s^i \left[\sum_{j=0}^{m} \binom{m}{j} (1-t)^{m-j} t^j \mathbf{P_{ij}} \right] \tag{8}$$

An application of the above method to the parametrization of an ORC nozzle guide vane is shown in Fig. 1 for a lattice with 12 nodes. The four corners are fixed and only vertical shifts are allowed, leading 8 control parameters, which are the design variables of the subsequent optimization step.

IOP Conf. Series: Journal of Physics: Conf. Series **821** (2017) 012014 doi:10.1088/1742-6596/821/1/012014

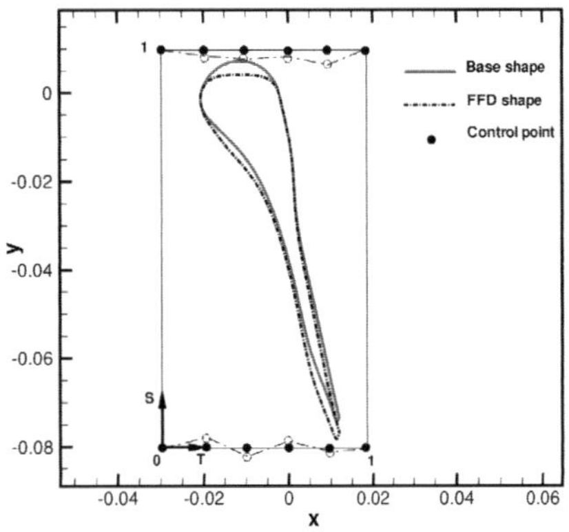

Figure 1. Application of FFD to a supersonic nozzle blade for a $l \times m$ lattice ($l = 6, m = 1$) with 8 control parameters.

5. Robust Optimization strategy

The RO is carried out by coupling a NSGA with the Kriging-based UQ method. The objective functions are the statistical expectancy and variance of a QoI, here the isentropic efficiency. The NSGA is then used to construct a Pareto front of optimal solutions to the problem *"find a shape for which the expectation of the QoI is maximal and its variance is minimal"*. A straightforward coupling of the NSGA with UQ solver is overly costly for the present application, due to the high number of function evaluations, each one involving a CFD simulation. If we consider a Kriging surrogate trained with $10n_{dim}$ samples, where n_{dim} is the number of uncertain variables, and a population of 30 individuals in each genetic generation, it results that $300n_{dim}$ CFD calculations have to be performed for a single genetic step. To reduce the computational cost, a second Kriging surrogate has been developed to predict the response of the multi-objective fitness function, i.e. the mean and variance of the QoI, to the design variables. Since a single output Kriging model is used, the multi-output response is obtained by training two independent surrogates, one for each fitness component. Fig. 2 presents a flow chart of the RO algorithm. The

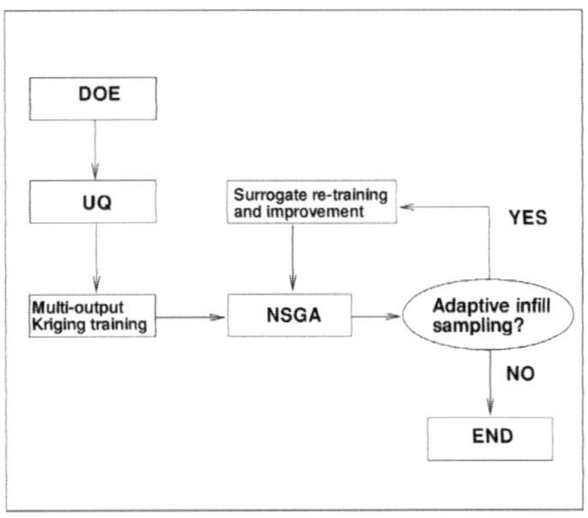

Figure 2. Flow chart of the RO algorithm.

first step consists of a Design Of Experiments (DOE) plan based on Latin Hypercube Sampling (LHS) of the design space. A large training set is sampled with $N > 10n_{dim}$ samples and, after the evaluation of mean and variance through the UQ, the multi-output Kriging surrogate is trained. This is then used in the genetic algorithm to evaluate the bi-objective fitness function. The cost of the surrogate RO is reduced to $10n_{par}10n_{dim}$ CFD calculations, with respect to the ($n_{gen} n_{individuals} 10 n_{dim}$) required by the straightforward RO, where n_{gen}, $n_{individuals}$ and n_{par} are the number of genetic generations, individuals and design parameters, respectively. Since the Kriging surrogate automatically provides a surrogate accuracy criterion through the Kriging error estimate, a sample infill strategy can be used to improve the model during the NSGA loop. However, an approach based only on the mean value of the Kriging error can mislead the improvement procedure. This is avoided through a more sophisticated method, described in the next Section.

5.1. Adaptive sample infill strategy

The adaptive infill sampling consists of adding to the initial DOE new samples selected during the NSGA iteration, and to retrain the Kriging model in order to improve its accuracy (see Fig. 2). A possible approach consists in selecting some individuals from the Pareto front according to performances criteria. Unfortunately, this method does not ensure to explore the design parameter space in such a way that the direction of the optimization process is preserved. Besides, the number of samples chosen among the individuals of the Pareto front influences the Kriging accuracy. To overcome these issues, a more efficient approach is to exploit the probabilistic nature of the surrogate error estimate to adapt it in regions where the *expected improvement* (EI) of the global minimum is maximised [12, 13]. The EI quantifies the probability to improve the surrogate on the design space parameters and, then, a global optimization process of the EI function provides the most suitable location for the new surrogate training. Since we deal with a multi-objective optimization, the EI function will be a surface on a hyper-dimensional parameter space, then a Multi Objective Expected Improvement (MOEI) approach is carried out [13]. The global optimization is performed by means of a differential evolution algorithm. Once the best MOEI is found in the design space, its fitness is calculated using the UQ solver and added to the training set. The accuracy of the surrogate is rapidly improved with a few MOEI evaluations, then it is possible to have accurate predictions of the Pareto fronts by using a number of MOEI (i.e. UQ) evaluations $n_{MOEI} << n_{gen}$, as shown in the next Section. The cost of the RO based on the adaptive infill strategy is of $(10 n_{par} + n_{MOEI}) 10 n_{dim}$ CFD calculations.

6. Numerical results

In this Section the MOEI procedure is first tested on the analytic multidimensional Kursawe function in order to show the feasibility and accuracy of the adaptive infill strategy. Then, the surrogate-based RO algorithm is applied to a supersonic 1D inviscid nozzle with real gas flow, for which it is possible to carry out a reference CFD-based RO. Finally, the RO methodology is applied to a supersonic ORC nozzle guide vane.

6.1. Surrogate-based optimization of the 2D Kursawe function

The adaptive sampling strategy through MOEI is validated against the analytic Kursawe function. This is a multidimensional function with two outputs, widely used for global optimization tests in the form of multi-objective minimization problem. Here the two-dimensional form is used to perform the cycle shown in Fig. 2, without the UQ loop. The Kriging surrogate is trained on an initial LHS training set of 20 samples and used to perform the NSGA. Two calculations, with and without the activation of the MOEI, are carried-out and compared with the exact Pareto front. The maximum number of individuals and generations are set to 50 and 60, respectively, and the MOEI is activated every 20 generations. As a consequence, the Kriging surrogate is updated with only 3 new samples during the genetic loop. Figure 3 compares the exact Pareto front with those provided by the Kriging surrogates with and without MOEI. In the first case the discrepancy is high, whereas the MOEI front is well predicted. To verify the exact position of the two Kriging surrogates respect to the exact Pareto front, the fitness of individuals laying on the approximate Pareto fronts is recalculated with the Kursawe function. Fig. 4 shows that the MOEI strategy greatly improves accuracy compared to Kriging without adaptive sampling. Other tests have been performed by increasing the space parameter up to 8 dimensions, showing that the MOEI suffers of a curse of dimensionality problem. Indeed the higher is the cardinality, the higher are the number of MOEI calculations required to reach good results by providing more samples to the updated training set. However, the total cost of the algorithm in terms of function evaluations remains much lower than a high-fidelity RO.

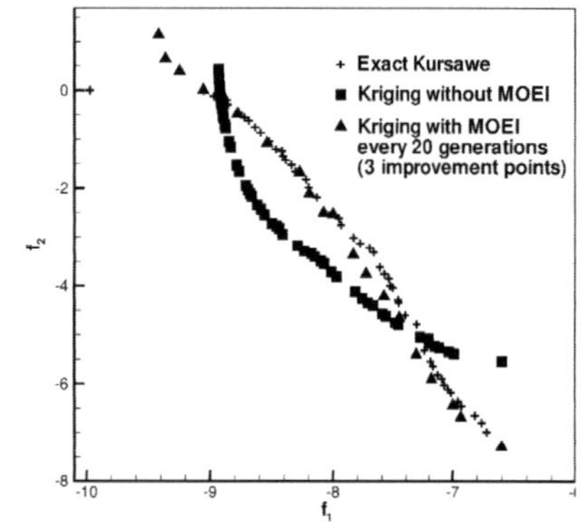

Figure 3. Pareto fronts comparison among exact Kursawe function and Kriging surrogates with and without MOEI.

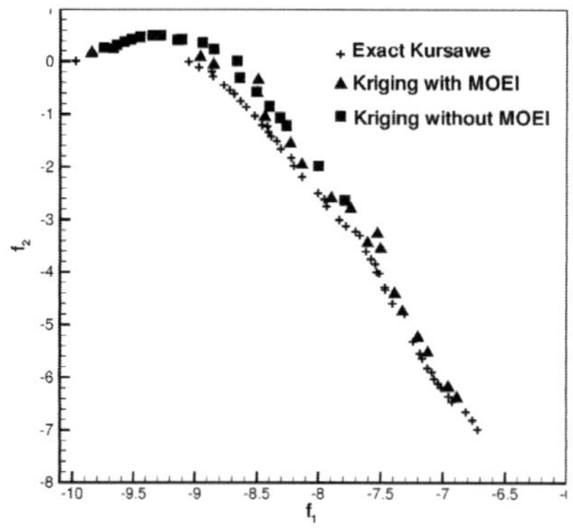

Figure 4. Exact Pareto front and comparison with the Kriging surrogate best individuals recalculated with the exact Kursawe function.

6.2. RO of a quasi-1D supersonic nozzle

The RO methodology described in the previous section is first applied to a supersonic quasi-1D inviscid nozzle. The flow is modelled by the Euler equations supplemented by the PRSV EOS. More details about the solver can be found in [14]. The working fluid is MDM and the uncertain parameters in terms of mean, percent coefficient of variation (CoV) and probability density function (pdf) are shown in Tab. 1. Three uncertainties are taken into account: two operational (total pressure and temperature p_0, T_0) and one geometrical (wall nozzle displacement ε) parameters. The nominal operating conditions are normalized with the critical values and are chosen such that a normal shock is set in the nozzle divergent. The chosen CoV ensures that a shock is always present in the convergent, while allowing a significant variation of the shock position. The latter is treated a random QoI, whose mean and variance are estimated through UQ.

Table 1. Uncertain variables for the 1D supersonic nozzle. Total pressure and temperature are normalised respect to the critical values, ε is the wall nozzle displacement.

Parameter	Mean	CoV%	pdf
p_0	0.212	15%	UNIFORM
T_0	0.0709	15%	UNIFORM
ε	1	1%	UNIFORM

The RO problem is defined as follows:

$$\min_{\alpha} |\mu(x_s) - \mu(x_s)_{target}| \, , \min_{\alpha} \left|\sigma(x_s)^2 - \sigma(x_s)^2_{target}\right| \quad (9)$$

where α is the vector of design parameters, x_s the shock position normalised with the nozzle length, $\mu_{target} = 0.7$ and $\sigma^2_{target} = 1.0 \times 10^{-4}$ two target values to which to converge during the optimization. The nozzle geometry is deformed by means of the FFD and $n_{par} = 8$ control parameters are defined by a *6* $\times$ *1* lattice around the nozzle with the four corners blocked and only vertical shifts are allowed. The accuracy of the Kriging-based UQ solver is first assessed by means of a convergence study. The QoI is the shock location and the maximum Kriging error estimate $\varepsilon_{K,max}$ is computed as a function of the number of samples N. The results, displayed in Figs. 5,6, show that after 30 samples the Kriging error is below 6×10^{-5} and small variations of mean and and variance are observed. Then, the UQ-Kriging with $N = 30$ is coupled to the NSGA. The results are compared with those of a surrogate-based optimization, with and without MOEI. The initial surrogate is based on a LHS training set of $10n_{par} = 80$ samples. The results of the RO are shown in Figs. 7,8, where $f1 = \left|\sigma(x_s)^2 - \sigma(x_s)^2_{target}\right|$ and

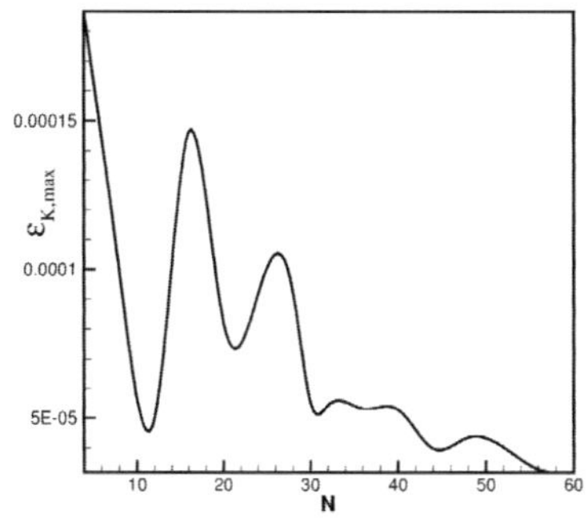

Figure 5. Maximum Kriging error estimate convergence plot as function of the number of samples.

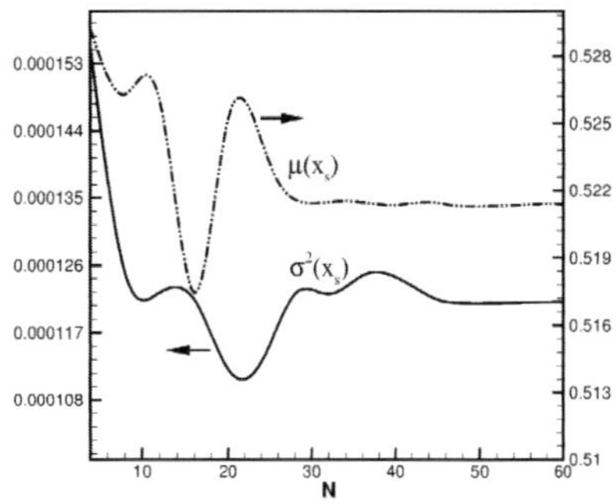

Figure 6. Variance and mean convergence plot as function of the number of samples.

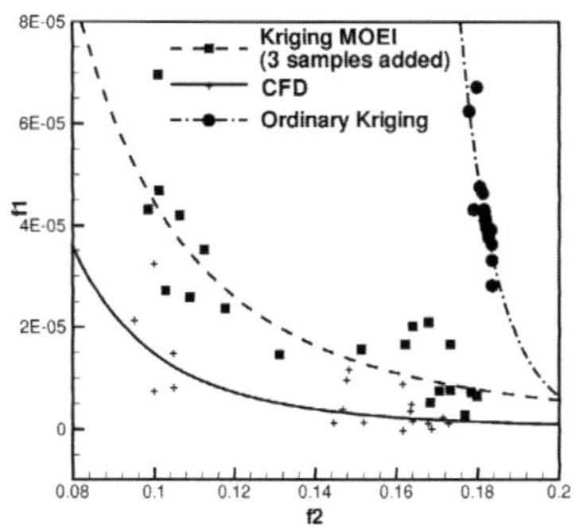

Figure 7. Pareto fronts comparison between high-fidelity CFD RO and Kriging surrogate (with and without MOEI) RO.

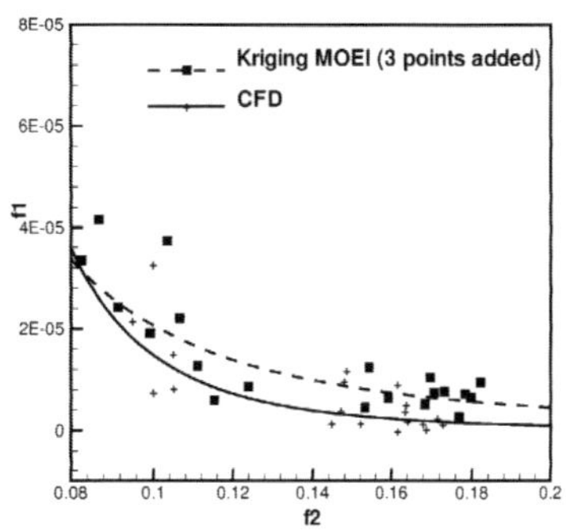

Figure 8. Pareto individuals recalculation for the Kriging surrogate with MOEI and comparison with CFD.

$f2 = |\mu(x_s) - \mu(x_s)_{target}|$. The first figure compares CFD-based and surrogate-based Pareto fronts. The ordinary Kriging surrogate, without MOEI, fails to predicts adequately the high-fidelity Pareto front. On the other hand, by adaptively adding only 3 samples with MOEI (i.e. one every 5 generations of the NSGA), the prediction is more accurate. To verify if the individuals of the surrogate-based Pareto front with MOEI lie on the high-fidelity one, the fitness function of the non-dominated individuals is recalculated with the CFD solver. Fig. 8 shows an even better agreement between the two fronts, which proves the accuracy and efficiency of the MOEI strategy for the present test case.

6.3. RO of ORC nozzle guide vanes

A deterministic optimization is first performed for the baseline geometry of Fig. 1, obtained by applying an accurate MOC based on a complex EOS for R245FA working fluid (Bufi et al. 2015). The viscous CFD calculations are carried out on a structured 384×128 mesh, for which $y^+ < 1$. Grid convergence is checked for the wall pressure distribution and isentropic efficiency, showing a maximum difference below 2% and 8% respectively, by referring to a 768×192 mesh. These convergence levels are acceptable for RO applications. The maximum deformations allowed with the FFD are set to 20% in the lattice reference system, and 8 control points are taken into account, as shown in Fig. 1. In this way the axial chord is kept constant whereas the shape can change freely. In Tab. 2 the nominal design parameters are shown. In Figs. 9,10 the

Table 2. Nominal nozzle parameters. Total pressure and temperature are normalised with respect to the critical values whereas β is the throat-to-exit static pressure ratio.

Fluid	p_0	T_0	β
R245FA	1.1	0.98	5

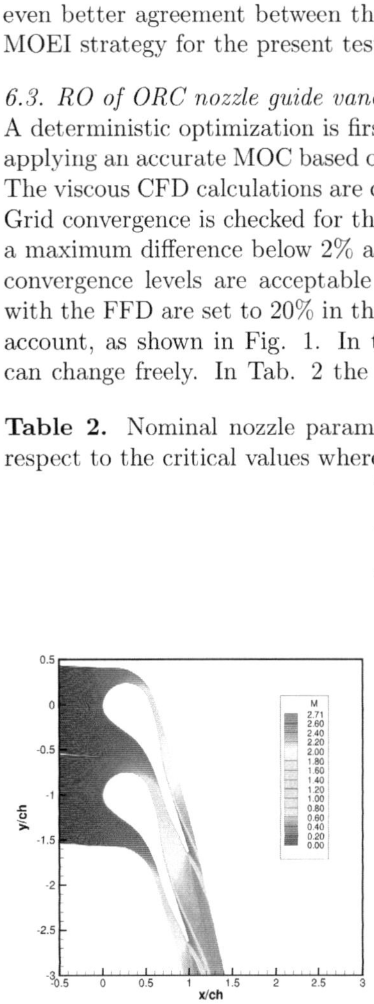

Figure 9. Baseline Mach contour plots.

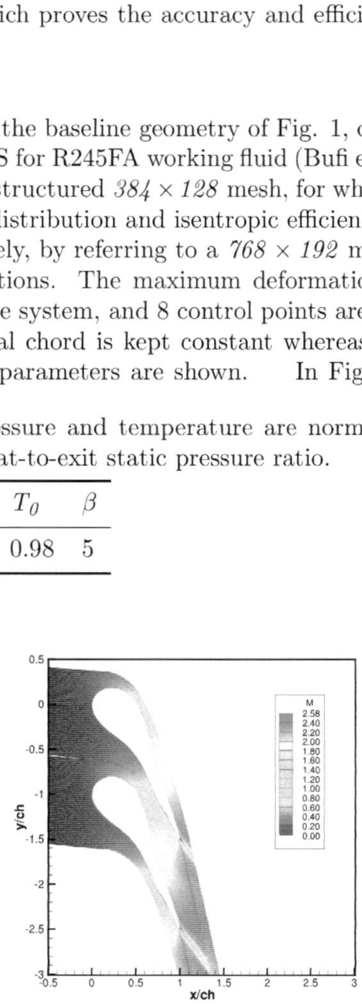

Figure 10. Deterministic optimized shape Mach contour plots.

Mach contour plots are shown for baseline and optimized geometry, respectively. In the first case, the presence of a weak oblique shock at the trailing edge lead the Mach number from 2.37 to 2.31. The optimized blade is narrower and exhibits significantly different shapes at the leading edge. This shows that the MOC procedure provides already a satisfactory design for the divergent part, whereas the convergent benefits from CFD optimization. The throat area per unit depth is 9% higher, leading to a slightly higher massflow rate. The expansion pressure ratio through the nozzle is unchanged, and the exit Mach number is 2.1 for both cases. The optimized shape still exhibits oblique shocks on the trailing edge, but with lower upstream Mach number, so that shocks are weaker. The stagger angle is slightly changed from 72° to 71.4° and the viscous wake and the separation zone on the suction side, are reduced. These improves the isentropic efficiency from 0.879 to 0.962 (8%). Afterwards, a RO of the baseline blade is also carried out by maximizing mean μ_{μ_s} and minimizing variance $\sigma^2_{\mu_s}$ of the isentropic efficiency and using the multi-objective Kriging surrogate with MOEI adaptive sampling. To evaluate the objective function, the UQ Kriging with 24 samples is used. Total inlet conditions (p_0, T_0) and blade thickness ε are set as uncertain variables, with the distribution shown in Tab. 4. The choice of beta pdfs, with a proper calculation of the shape parameters based on the known mean and standard deviation, allows to avoid expansions in the liquid-vapor region. The optimization converges after 60 generations with 3 samples adaptively added each 20 generations. The Fig. 11 shows the Pareto front. The recalculated samples show the goodness of the NSGA-Kriging with MOEI surface method. The stochastic performance of deterministically-optimized blade is also represented for comparison. For this blade the average value of the isentropic efficiency is $\mu_{\mu_{s,det}} = 0.946$, i.e. slightly lower than the deterministic value $\mu_{s,det} = 0.962$ with a variation of 1.5%. The higher variance compared to the robust individuals shows that the deterministic optimization provides a less robust blade, in spite of the high performances. The stochastic performance of the baseline blade shows that the RO provides a 6% average increase of the isentropic efficiency and a 1% decrease of the standard deviation. In Fig. 12 the robustly optimized shape, selected on the Pareto front (black arrow in Fig. 11), is compared with the baseline and deterministic optimized ones. Once again, the more significant geometrical variations are observed in the subsonic zone.

Table 3. Uncertain variables for the ORC supersonic nozzle. Total pressure and temperature are normalised respect to the critical values, ε is the blade thickness.

Parameter	Mean	CoV%	pdf
p_0	0.98	8%	BETA
T_0	1.13	8%	BETA
ε	1	1%	UNIFORM

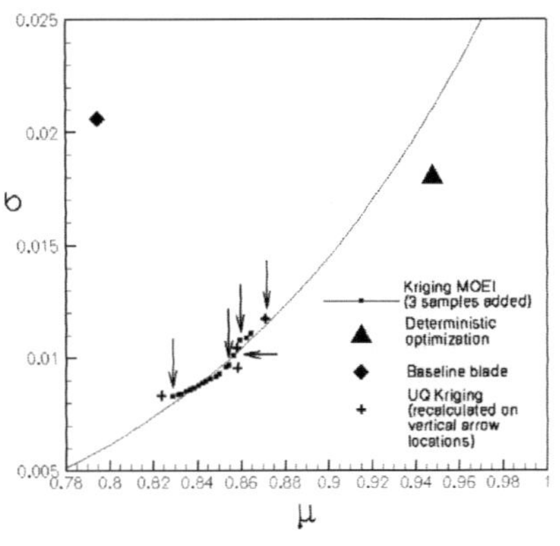

Figure 11. Pareto front of the ORC nozzle RO. Horizontal arrow: design compared with the deterministic optimization; vertical arrows: recalculated samples.

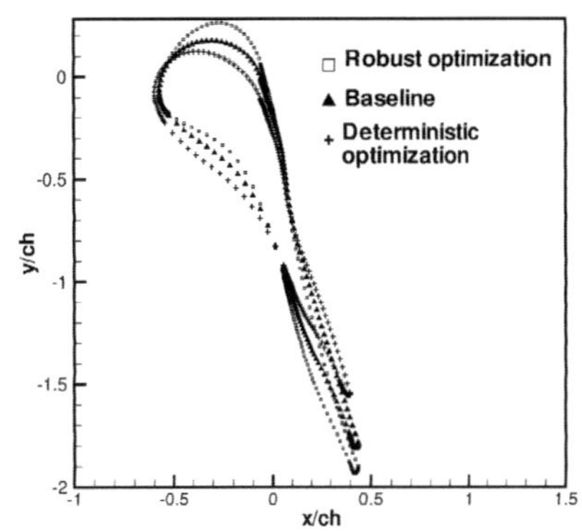

Figure 12. Comparison among baseline and deterministically and robustly optimized shapes. The robust shape is selected from the center zone of the pareto front.

7. Conclusions

A RO has been carried out for an ORC supersonic nozzle with isentropic efficiency as QoI. The efficiency of the optimization procedure with the NSGA has been improved by means of a Kriging

Table 4. Uncertain variables for the ORC supersonic nozzle. Total pressure and temperature are normalised respect to the critical values, ε is the blade thickness.

Parameter	Mean	CoV%	pdf
p_0	0.98	8%	BETA
T_0	1.13	8%	BETA
ε	1	1%	UNIFORM

surrogate model coupled with the MOEI technique for an adaptive sampling strategy. The results have been compared with those obtained from a deterministic optimization, showing that the deterministic shape provides a less robust shape. Indeed, the deterministic design exhibits a not negligible sensitivity to the operational and geometrical uncertain variable variations. The RO provides a good compromise between average performance and stability. Besides, the calculation performances of the RO methodology are promising and able to handle the intrinsic high computational cost of dense gas calculations by reducing the number of CFD simulations. Finally, results show that the initial design of the nozzle divergent, based on MOC along with boundary layer corrections, is little modified by both the robust and deterministic optimization, which affect essentially the design of subsonic part.

References

[1] Congedo, P. M., Cinnella, P., Hercus, S., and Corre, C. E. (2011, May). Efficient robust optimization techniques for uncertain dense gas flows. In CFD and Optimization-ECCOMAS Thematic Conference.

[2] Bufi, E. A., Cinnella, P., and Merle, X. (2015, June). Sensitivity of supersonic orc turbine injector designs to fluctuating operating conditions. In Proceedings of the ASME 2015 Turbo Expo Turbine Technical Conference, ASME TURBO EXPO 2015 June 15-19, 2015, Montreal, Canada.

[3] Bufi, E. A., Obert, B., and Cinnella, P. (2015, October). Fast design methodology for supersonic rotor blades with dense gas effects. In 3 International Seminar on ORC Power Systems, October 12-14, 2015, Brussels, Belgium.

[4] Stryjek, R., and J. H. Vera. "PRSV: An improved PengRobinson equation of state for pure compounds and mixtures." The canadian journal of chemical engineering 64.2 (1986): 323-333.

[5] Cinnella, Paola, and Pietro M. Congedo. "Numerical solver for dense gas flows." AIAA journal 43.11 (2005): 2458-2461.

[6] Rezgui, Ali, Paola Cinnella, and Alain Lerat. "Third-order accurate finite volume schemes for Euler computations on curvilinear meshes." Computers and fluids 30.7 (2001): 875-901.

[7] Cinnella, Paola, and Pietro M. Congedo. "Aerodynamic Performance of Transonic Bethe-Zal'dovich-Thompson Flows past an Airfoil." AIAA journal 43.2 (2005): 370-378.

[8] Wikle, Christopher K., and L. Mark Berliner. "A Bayesian tutorial for data assimilation." Physica D: Nonlinear Phenomena 230.1 (2007): 1-16.

[9] Raffin, Romain. "Free Form Deformations or Deformations Non-Constrained by Geometries or Topologies." *Deformation Models.* Lecture Notes in Computational Vision and Biomechanics, Volume 7, Springer Netherlands, 2013. pp 49-74.

[10] Duvigneau, Rgis. Aerodynamic shape optimization with uncertain operating conditions using metamodels. Doctoral dissertation INRIA, 2007.

[11] Sederberg, Thomas W., and Scott R. Parry. "Free-form deformation of solid geometric models." ACM SIGGRAPH computer graphics 20.4 (1986): 151-160.

[12] Dwight, R., J. De Baar, and I. Azijli. "A tutorial on adaptive surrogate modeling." (2012).

[13] Keane, Andy J. "Statistical improvement criteria for use in multiobjective design optimization." AIAA journal 44.4 (2006): 879-891.

[14] Pini, M., and P. Cinnella. "Hybrid adjoint-based robust optimization approach for fluid-dynamics problems." 15th AIAA Non-Deterministic Approaches Conference. 2012.

IOP Conf. Series: Journal of Physics: Conf. Series **821** (2017) 012015 doi:10.1088/1742-6596/821/1/012015

Evolutionary Optimization of Centrifugal Nozzles for Organic Vapours

Giacomo Persico

Laboratorio di Fluidodinamica delle Macchine
Dipartimento di Energia
Politecnico di Milano
Via Lambruschini 4, I-20156 Milano, Italy

E-mail: giacomo.persico@polimi.it

Abstract. This paper discusses the shape-optimization of non-conventional centrifugal turbine nozzles for Organic Rankine Cycle applications. The optimal aerodynamic design is supported by the use of a non-intrusive, gradient-free technique specifically developed for shape optimization of turbomachinery profiles. The method is constructed as a combination of a geometrical parametrization technique based on B-Splines, a high-fidelity and experimentally validated Computational Fluid Dynamic solver, and a surrogate-based evolutionary algorithm. The non-ideal gas behaviour featuring the flow of organic fluids in the cascades of interest is introduced via a look-up-table approach, which is rigorously applied throughout the whole optimization process. Two transonic centrifugal nozzles are considered, featuring very different loading and radial extension. The use of a systematic and automatic design method to such a non-conventional configuration highlights the character of centrifugal cascades; the blades require a specific and non-trivial definition of the shape, especially in the rear part, to avoid the onset of shock waves. It is shown that the optimization acts in similar way for the two cascades, identifying an optimal curvature of the blade that both provides a relevant increase of cascade performance and a reduction of downstream gradients.

1 Introduction

Among the several industrial application of organic fluids, Organic Rankine Cycle (ORC) power systems represent one of the most attractive technology for the exploitation of energy sources featuring medium-low enthalpy level [1]. The performance of the entire ORC power system is crucially dependent on the efficiency of the turbine [2, 3], whose optimization is complicated by the character of organic fluids, which combine low enthalpy drops with high expansion ratios. The design of centripetal or axial turbo-expanders for organic fluids usually leads to machines with a single or a few stages (up to 3), featuring converging-diverging channels and strong shocks at blade outlet regions [4, 5]; such configurations induce limitations in the performance and in the degree of control of the turbine. These inherent difficulties are further amplified by the severe non-ideal effect exhibited by organic vapours in their flow within the machine.

To overcome the aforementioned limitations the novel radial-outflow, or centrifugal, turbine configuration has been recently proposed in [6, 7], receiving scientific recognition and industrial exploitation. The centrifugal turbine can better accomplish the large volumetric flow ratio, thanks to the natural increase of passage area along the flow path, thus limiting the flaring angle without significant increase in meridional velocity; furthermore many stages can be disposed in a relatively compact machine, reducing the cascade expansion ratios, potentially avoiding supersonic flow conditions. For these reasons, the centrifugal architecture allows to increase the aerodynamic performance as well as to enlarge the power control capability.

A research program on the development of novel centrifugal turbines for ORC applications is presently ongoing at Politecnico di Milano, with the aim of proposing specific design guidelines for this technology. To this end, a design methodology has been conceived with a hierarchical approach, following the path laid-down in [8]. The preliminary design of the whole machine was first studied by means of mean-line and throughflow codes [7], then the basic criteria for the aerodynamic design of centrifugal turbine profiles were set-up [9] and the three-dimensional aerodynamics of centrifugal turbine stators and rotors was studied [10].

The present work provides a step forward in the design procedure of centrifugal turbines, by discussing the shape optimization of centrifugal turbine blades. As a matter of fact, the selection of the most suitable profile such non-conventional architecture working with organic fluids cannot rely on empirical methods, but it demands the set-up of a completely new technique based on the combination of aerodynamic indications, high-fidelity computational tools, and systematic optimization strategies. This new procedure is constructed by combining a generalized geometrical parametrization technique, a high-fidelity Computational Fluid Dynamic (CFD) model, and a surrogate-based gradient-free optimization method.

This paper discusses the application of the design procedure to the shape optimization of two centrifugal turbine stators, characterized by different radial effect and aerodynamic loading. The comparison between the two cases allows to highlight the specific design features of centrifugal turbine blades and, eventually, to derive some design guidelines for the aerodynamic design of this class of machines. The paper is organized as follows: at first the procedure for the construction of a first-guess shape for the turbine profile is recalled; then the shape optimization tool for turbomachinery cascades operating with non-ideal fluids is introduced; finally the optimal design of two centrifugal nozzle cascades is extensively discussed.

2 Preliminary Cascade Design

The centrifugal nozzles studied in this work are conceived for implementation in the six-stage centrifugal turbine proposed in [7] for application in high-temperature ORC systems, and using the Siloxane MDM as working fluid. Operating conditions, size, power release, and total-total efficiency of the turbine, as resulting from the preliminary design, are listed in Table 1.

Fluid	$P_{T,in}$	$T_{T,in}$	P_{out}	Ω	*Diameter*	*Power*	*Efficiency*
MDM	10 bar	274 C	0.17 bar	3000 rpm	1.0 m	1.27 MW	86.9%

Table 1. 6-stage turbine conditions, size and performance

The layout of the turbine consists in six stages which share the reaction degree (equal to 0.5), the discharge flow angle (equal to 66.7 deg, opposite in sign between stators and rotors), and the radial chord (equal to 26 mm). With these choices, a smooth shape of the meridional channel is achieved. The throughflow analysis, performed with the in-house code TzFlow [11] and extensively reported in [7], indicated that the flow remains mostly subsonic.

The present work discusses the optimal aerodynamic design of the first and sixth nozzle cascades of the turbine, re-named CN-I and CN-VI in the following. As usual in turbomachinery applications, a preliminary (or baseline) shape for the blades is first defined using simple criteria, then the initial configuration is optimized using a systematic technique. In the present study, the baseline shape of the two profiles was defined by applying the concepts proposed in [9], in which aft-loaded profiles featuring a smooth acceleration are shown to be suitable centrifugal turbine cascades (at least for flow turning within 70 deg). The mean-line of the profiles is then assigned as elliptic arc, whose aspect ratio is determined after a parametric study. Then, a conventional distribution of thickness is applied to the mean-line to construct the blade. The blade geometric angles are assigned imposing null incidence and deviation estimated by resorting to the correlation of Ainley & Mathieson. The resulting profiles, originally constructed in the Cartesian coordinate system, are finally mapped on the polar coordinate system using a conformal transformation, in order to maintain the geometric angles of the blade.

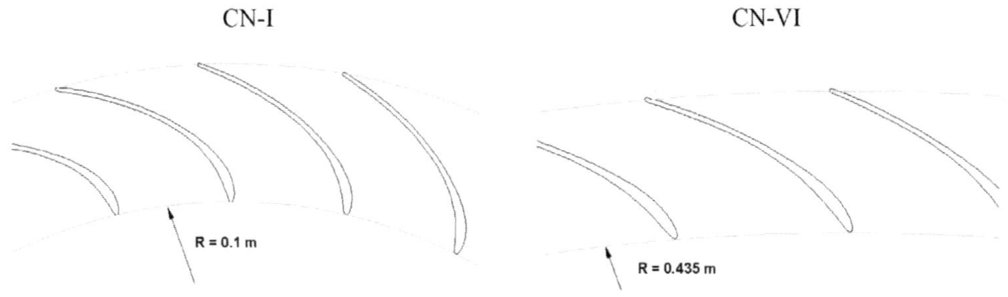

Figure 1. Preliminary design of the CN-I and CN-VI cascades of the centrifugal turbine

The CN-I cascade was constructed in order to deflect the flow from purely meridional (radial-outward) flow to 66.7 deg; the blade number, selected using the Zweifel criterion, resulted equal to 28. The distribution of thickness was taken from that of a lightly loaded low-pressure axial turbine. The CN-VI cascade was constructed to deflect the flow of 40 deg, using the same thickness distribution of the first nozzle; the blade number resulted 82. Figure 1 reports the layout of the two nozzle cascades under consideration, constructed by applying the aforementioned procedure. These blades were the starting point for the application of the systematic optimization technique, which is described in the next Section.

3 Optimization Technique

The progress in computational capability, the increased fidelity of CFD models, and the improving effectiveness of optimization algorithms have triggered the development of novel design tools to determine the optimal shape of aerofoils and turbomachinery blades. Among the techniques presently available, deterministic adjoint-based methods [12, 13] and heuristic evolutionary methods [14, 15] are nowadays object of intense application and research in Aerodynamics. Evolutionary methods, in particular, are of interest as they allow to explore a wide range of feasible solutions [16] without being 'intrusive' with respect to the source flow model, as they only require the use of direct calculation tools.

In this work, the shape-optimization of the centrifugal nozzles is carried out by applying the in-house design package FORMA (Fluid-dynamic OptimizeR for turbo-Machinery Aerofoils) recently set-up at Politecnico di Milano. The optimization strategy, briefly recalled in the following, is constructed by combining three main bricks, namely a geometry-parameterization code, a high-fidelity and experimentally validated CFD solver, and a surrogate-based evolutionary algorithm.

3.1 Geometry Parameterization Method

In shape optimization methods, it is common practice to parametrize the geometry with piecewise functions, so to control the shape by moving a limited number of so-called Control Points (CPs). In the FORMA package, B-Spline curves are used to parameterize the line of the blade profile; the global and local control of the shape provided by the use of B-Splines makes them a powerful tool for aerodynamic designs [18]. The position of the CPs their range of variation define the design space of the optimization problem. Piecewise lines of order 3 are used in this work. More details on the parametrization technique can be found in [17].

To set-up the optimization, at first an approximate representation of the baseline blade shape is constructed using a B-Spline curve, defined and manipulated by the position of the CPs; once the number and the spacing of the CPs is prescribed, the coordinates of the CPs are found via a least squares interpolation method. In the FORMA package, the pressure and suction sides of the blade are generated as a unique B-Spline curve, filleted by a circular-arc trailing edge.

3.2 Computational Flow Model

To evaluate the performance of the blade configurations progressively identified by the optimization algorithm, high-fidelity CFD simulations are performed using the ANSYS-CFX solver. Turbulence effects are introduced using the $k - \omega$ SST model, ensuring wall y^+ below unity all along the blade profile. To properly introduce the non-ideal thermodynamic behavior of MDM, a look-up table approach was used. The look-up table was constructed by sampling the Span-Wagner Equations of State [19] though the thermodynamic library *FluidProp* [20]; tabulated transport properties were also introduced. Since the object of this study is the optimization of the profile, a quasi-3D flow model is used, considering a straight stream-tube.

Calculations were performed on structured grids composed by hexahedral elements. Grids composed by 50 kcells in the blade-to-blade surface were found to provide the optimal trade-off between computational cost (10 minutes for a CFD simulation on a 16-processor Linux cluster) and fidelity of result (5% overestimate in the entropy production with respect to the grid independent value [9]). The mesh was re-generated for each CFD run performed throughout the optimization process. The performance of the optimized blades, as well as the computed flow fields reported in this paper, were evaluated by using a grid independent mesh.

The reliability of the flow model here used was assessed against experiments performed by the authors themselves on a cold-flow gas turbine stage installed at Politecnico di Milano [21].

3.3 Surrogate-Based Optimization Strategy

The optimization technique applied in this work is based on Genetic Algorithms (GAs). GAs are attractive as they allow to deal with oscillating and non-smooth objective functions, as well as to easily handle constrained and multi-objective optimization problems [22]. Furthermore, GAs are global optimization methods and, hence, are best suited in presence of a multiplicity of local optima (a situation that cannot be excluded a priori in aerodynamic design).

Nevertheless, GAs require a massive application of the direct computational tool, that in case of CFD often result in an unacceptable computational burden. To tackle this cost, a surrogate evolutionary strategy was applied in this study. This technique is conceived so that the GA operates only on a surrogate model of the objective function. To be effective, the surrogate model has to be a reliable representation of the objective function; in surrogate-based strategies the model is initialized and progressively updated during the optimization, so that the shape is optimized while the reliability of the surrogate improves. Among the surrogate models available [23], the analytical Kriging formulation was used in this work to approximate the objective function. The Kriging formulation is based on a set of interpolation methods and allows to properly approximate complex surfaces featuring sharp curvature changes, even in presence of highly irregular distributions of interpolation points. This makes the Kriging model suitable for complex optimization problems such as the design of transonic turbomachinery.

The surrogate model is initialized by interpolating a database of configurations tested with the CFD model. To improve the reliability of the surrogate model, a Surrogate-Based Global Optimization (SBGO) is applied. In SBGO, the model is 'trained' by adding progressively the optima found in the previous iterations to the initial database. By virtue of this strategy, the global strategy results computationally efficient, even though the convergence is not guaranteed. However, a previous study on the optimization of an axial supersonic turbine [17], as well as the preliminary trials made on the present centrifugal configuration, did not show convergence issues if a sufficiently large initial data-base is assigned.

The FORMA package was assembled making use of the object-oriented framework Dakota [24]. A single-objective GA was used as optimization tool, by resorting to the JEGA library (Java Engine for Genetic Algorithms). After a preliminary parametric study, the GA was set-up with a population size of 300 generations, a crossover rate of 0.8, and a mutation rate of 0.02; elitism was also used as a selection technique. To determine the Kriging parameters, the Surfpack library was employed; universal Kriging formulation was used, with non-zero covariance based on a Gaussian correlation function. In order to initialize the surrogate model, a Latin Hypercube technique was used, with a population equal to 10 times the number of design variables.

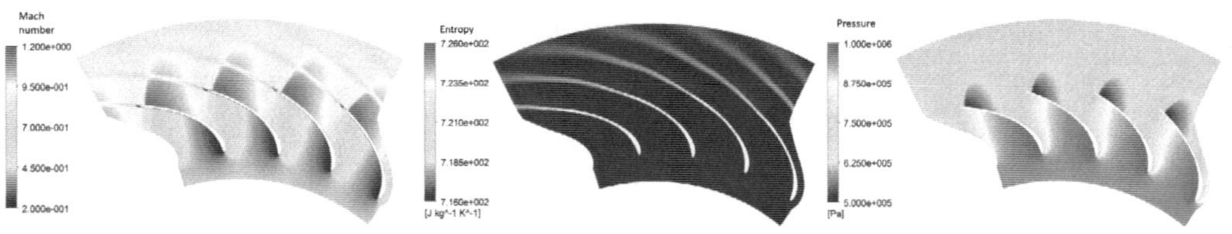

Figure 2. CN-I flow field for the baseline shape: Mach number distribution (left), entropy field (center), pressure field (right)

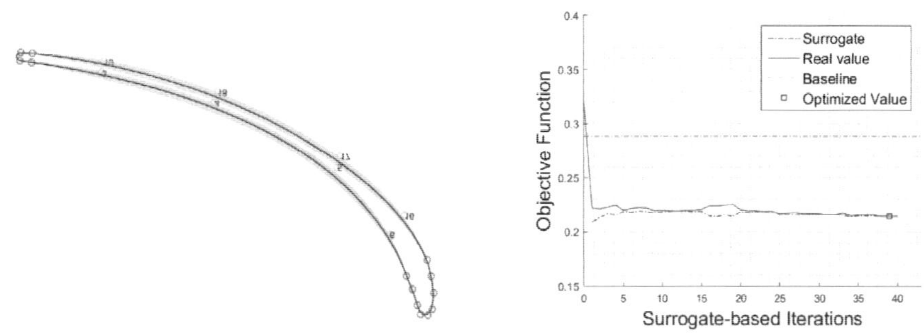

Figure 3. CN-I optimization process. Left: Design space. Right: Convergence

4 Results

In this Section, the FORMA package is applied to the configurations of interest, with the aim to both obtain an optimal shape and derive design remarks. The optimization of CN-I is first considered, then optimal design of the CN-VI is presented.

4.1 First-Stage Centrifugal Nozzle

The CN-I has to provide conventional flow deflection for turbine applications, but it exhibits a very significant radial effect, as a 25% change in radius and then of meridional cross-section occurs along the blade radial extension. In such a configuration the application of conformal mapping to profiles with a rear curvature leads to blades with high backbone length with a large and curved semi-bladed region downstream of the throat. This makes critical the design of the rear suction side of the blade, especially considering the transonic flow conditions.

These critical aspects are well visible in Figure 2, which reports about the Mach number, entropy and pressure distributions for the CN-I in the baseline configuration. Even though the target cascade-exit Mach number is below 1, a significant over-speed is established in the rear part of the profile downstream of the throat, with local Mach number exceeding 1.2. This feature results in a relatively strong shock just upstream of the trailing edge. This is detrimental for the turbine efficiency for both direct and indirect reasons, as such a shock reduces both the performance of the nozzle itself and that of the subsequent rotor, and also induces a severe aerodynamic forcing on the rotor blades.

Aiming at maximizing the turbine performance, the shape optimization of CN-I has to focus on the rear part of the profile. To this end, the baseline geometry was parametrized as reported in Figure 3-left, namely with 21 CPs in total and 8 CPs movable (of $\pm 0.4mm$ in vertical direction), resulting in the design space highlighted in green in the figure. Both the leading edge and trailing edge regions of the profile are kept unchanged during the design process, for different reasons. The four CPs on the trailing edge are crucial to constraint the thickness (for structural reasons) and to guarantee the required regularity in a region where the B-spline of the profile is merged with the circular-arc of the trailing edge. The leading edge region is

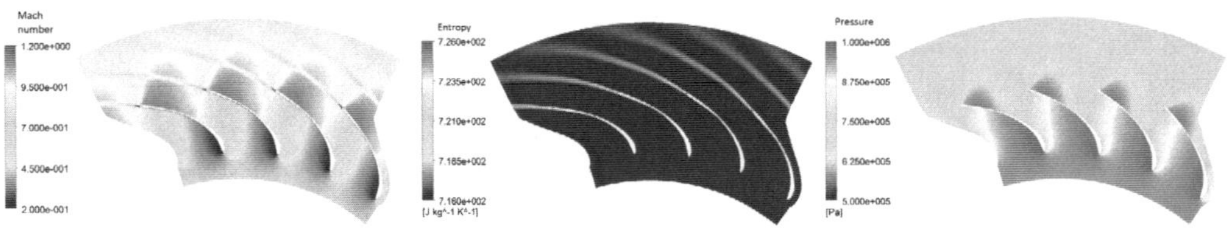

Figure 4. CN-I flow field for the optimal shape: Mach number distribution (left), entropy field (center), pressure field (right)

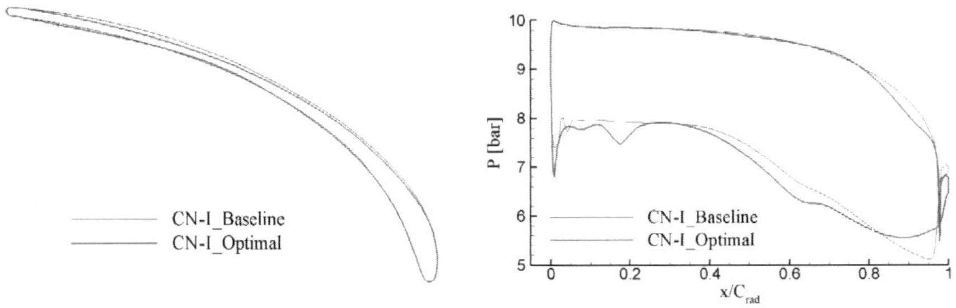

Figure 5. CN-I baseline / optimization comparison. Left: Blade profiles. Right: Pressure distribution.

kept fixed as, apparently, it has a weak effect on the performance and a reduction of movable CPs results in a reduction of design variables and, eventually, in a significant limitation of the computational burden. The mass-averaged entropy production between the inlet section and a section placed 0.2 C_{rad} away from the blade trailing edge is used as Objective Function.

Figure 3-right reports the convergence process of the surrogate-based optimization, and indicates that after the initialization the surrogate function already exhibits a satisfactory reliability. A marginal improvement is, however, achieved in the following 50 iterations, resulting in a reduction of entropy production of about 30% with respect to the baseline configuration. Considering the cost of each CFD run, the whole optimization required about 1 day of calculation (using the aforementioned 16-processor cluster machine).

Figure 4, which reports the flow field for the optimal configuration with the same scheme of Figure 2, clarifies that the improvement in performance is achieved by a reduction of the over-speed on the rear suction side of the blade, which results in a much weaker shock. The weakening the shock induces a reduction of both shock losses and wake losses, as the effects of shock-boundary layer interaction are also minimized. As a result, the total pressure loss coefficient, defined as $Y = (P_{T,in} - P_{T,out})/(P_{T,in} - P_{out})$, drops from 4.4% of the baseline configuration to 3.3% of the optimal configuration (Y was evaluated on a section placed half a C_{rad} downstream of the trailing edge, where most of the mixing has already occurred).

By considering the detailed shape of the baseline and optimal blades, reported in Figure 5-left, it can be observed that both the mean line and the thickness distribution change for the optimal blade. In particular, the optimization drives the design towards a profile with an increased curvature and a reduction of thickness in the central region of the blade, followed by a more flat mean line and a newly increased thickness in the rear section. Figure 5-right, which provides the pressure distribution on the profiles, highlights that the shape modification allows to reduce the over-speed in the rear suction side as well as to eliminate the abrupt pressure rise induced by the shock at the trailing edge. The pressure side, instead, exhibits almost the same trend of the baseline profile. This is a fairly expected result as in this kind of transonic profiles the flow mechanisms occurring on the suction side are the major sources of loss.

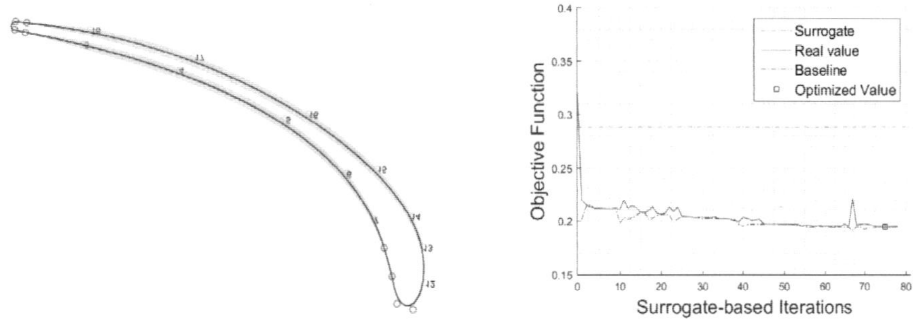

Figure 6. CN-I advanced optimization process. Left: Design space. Right: Convergence

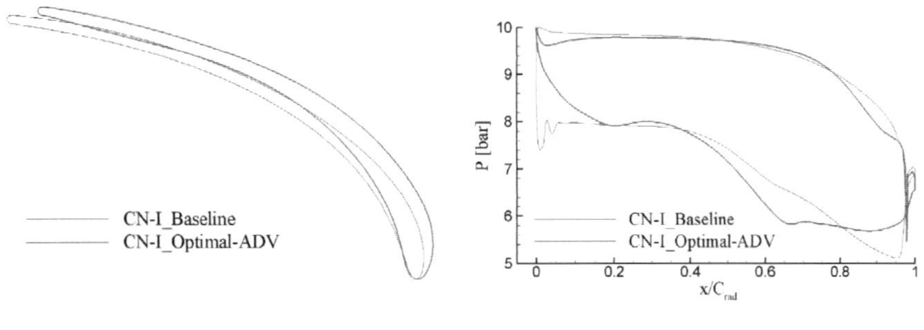

Figure 7. CN-I advanced optimization result. Left: Mach number distribution. Right: entropy field.

The pressure distribution provided in Figure 5 also shows that significant adverse pressure gradients affect the front area of the profile. Even though the calculations suggest that these pressure gradients do not induce local separations, in an effort of further improving the cascade performance it would be beneficial to smooth the pressure distribution in the front area by a proper re-design of the blade. However, the geometry modifications required to improve the front area of the blade are trivial and, hence, they do not need the application of an automatic technique. For this reason, a new baseline profile was constructed by modifying the front part of the blade both in terms of mean-line inclination (which is now inclined by 10 deg at leading edge, instead of being radial as in the first design) and in terms of thickness (which is now 50% larger than in the previous case up to mid-chord length, and is smoothly reduced in the rear half of the profile to become equal to the original one at the trailing edge). The newly-constructed baseline blade was then parametrized by using 21 CPs, 12 of them are movable (by $\pm 0.4mm$) in such a way that the entire suction side of the blade can be modified. The new baseline blade and the corresponding design space are provided in Figure 6-left. The convergence process of this second 'advanced' optimization, reported in Figure 6-right, confirms the smooth and regular trend observed previously, and shows a further minimization with respect for the first optimization

The resulting optimal blade differs significantly from the original baseline configuration. The curvature of the mean-line is increased in the front and central part of the blade, so to impose a higher local flow turning; conversely, the blade becomes almost straight in the semi-bladed region, so to avoid severe over-speed in the rear suction side. The impact of these features on the pressure distribution on the blade can be appreciated in Figure 6-right, which shows that the advanced optimization has both eliminated the severe adverse pressure gradients in the front area and further flattened the pressure distribution in the rear suction side.

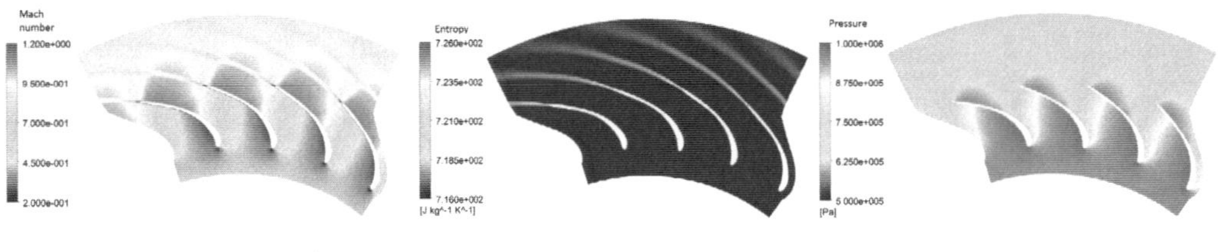

Figure 8. CN-I flow field after the advanced optimization: Mach number distribution (left), entropy field (center), pressure field (right)

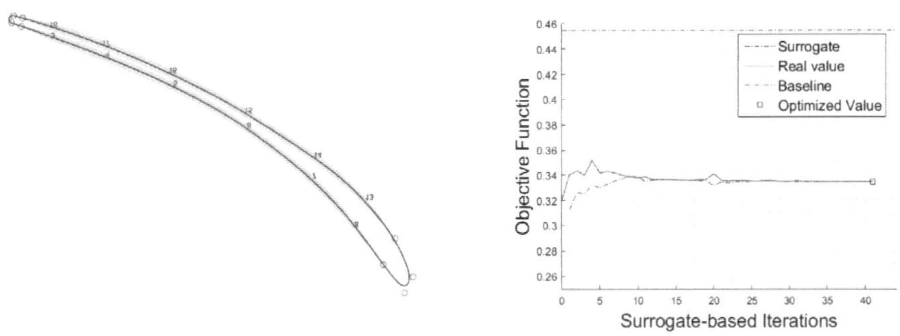

Figure 9. CN-VI optimization process. Left: Design space. Right: Convergence

The aerodynamics of the cascade benefit from these improvements, as visible in Figure 8. The pressure and Mach number distributions highlight the reduced gradients in the leading edge region as well as the further weakening of the rear shock, which eventually result in a weaker wake and reduced losses. The corresponding loss coefficient Y further reduces to 3.1%.

On the whole, the systematic optimization has allowed to improve the pressure distribution along the entire blade profile resulting in a reduction from 4.4% to 3.1% in the loss coefficient, which meas an improvement of 30% in relative terms. The dramatic weakening of the shock in the rear part of the profile is expected to also improve the aerodynamics of the subsequent rotor, which is placed 0.2 C_{rad} away from the CN-I trailing edge.

4.2 Sixth-stage centrifugal nozzle

The optimization strategy was applied for the optimal design of the CN-VI cascade, which features a low aerodynamic loading (flow deflection of 40 deg) and transonic condition; full details on the preliminary design of the blade and the three-dimensional aerodynamics of the cascade can be found in [10].

The baseline shape of the CN-VI blade was still constructed using an elliptic-arc mean-line and the same thickness distribution used for the baseline CN-I blade. The blade was parametrized using 20 CPs, 12 of them can be moved by $\pm 0.4mm$ in vertical direction, as shown in Figure 9-left. The convergence process, reported in Figure 9-right exhibits the usual smooth trend, in which it is apparent that the surrogate achieves a satisfactory reliability 10 iterations after the initialization.

The effectiveness of the optimization can be appreciated by comparing the flow fields reported in Figures 10 and 11, which refer to the baseline and optimal configurations respectively. As observed for the CN-I cascade, the use of an elliptic-arc mean line allows to control the expansion process in the bladed channel, but not in the semi-bladed rear suction side, resulting in a relatively strong shock close to the trailing edge. The optimization acts on the blade so to increase the mean-line curvature in the central section of the blade, while making more straight the rear suction side of the blade. The result is an anticipated acceleration of the flow and a

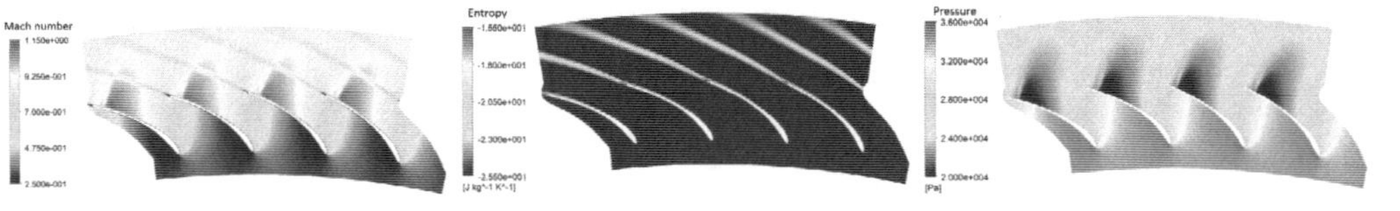

Figure 10. CN-VI flow field for the baseline shape: Mach number distribution (left), entropy field (center), pressure field (right)

Figure 11. CN-VI flow field for the optimal shape: Mach number distribution (left), entropy field (center), pressure field (right)

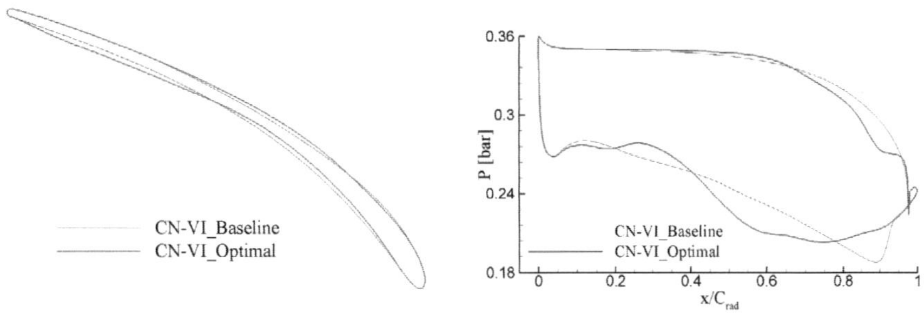

Figure 12. CN-I advanced optimization result. Left: Mach number distribution. Right: entropy field.

much lower over-speed, so that the shock close to the trailing edge is completely eliminated. These features are well visible in Figure 12, which provides a direct comparison between the blades and pressure distribution on the surface.

As a result of these shape modifications, the total pressure loss coefficient drops from 3.1% of the baseline configuration to 2.8% of the optimal cascade; furthermore, the pressure and Mach number gradients downstream of the CN-VI cascade are drastically reduced, with beneficial effects on the subsequent rotor, which is placed 0.2 C_{rad} downstream of the trailing edge.

5 CONCLUSIONS

This paper has presented the application of an evolutionary shape-optimization technique to the design of non-conventional centrifugal nozzles for application in ORC turbines. The technique, implemented in the in-house optimization package FORMA, makes use of a geometry-parametrization tool, an high-fidelity flow solver, and a surrogate-based evolutionary algorithm.

Two cascades have been considered, featuring different flow turning and radial extension, namely the first and the sixth nozzle of a six-stage radial-outflow turbine. The baseline shape of the blades, constructed by means of analytical tools, are parametrized via B-Splines, whose local control capability allowed a detailed shape reconstruction while preserving surface smoothness. The generalized thermodynamic treatment required to deal with the non-ideal gases featuring ORC turbines is achieved via a look-up-table approach, and is introduced straightforwardly into the optimization thanks to the non-intrusive character of the evolutionary strategy. To tackle the computational cost of the optimization, the genetic algorithm is coupled to a Kriging surrogate model. The selection of the design space and the convergence process have been highlighted to show the path of the optimization and the related computational cost.

The application to a novel cascade configuration has shown that the systematic optimization is able to improve significantly the performance of the nozzles and to reduce the pressure and velocity gradients at the inlet of the subsequent rotors. If the reduction of loss coefficient achieved for the first and last stator were extended to the remaining ones, a gain in 0.5% in the turbine efficiency would be achieved. This is a relevant result, especially because the estimated efficiency of the baseline turbine is relatively high (87%). Moreover, application of the same technique to the rotor blades would provide a further direct performance increase. Finally, all the stages would achieve a secondary benefit from optimization, especially from the structural point of view, thanks to the much more uniform flow released by optimized blade rows. These considerations clearly demonstrate the potential of a systematic blade optimization for multi-stage centrifugal turbines.

The analysis of the optimal blades has shown that centrifugal cascades are highly sensitive to the distribution of blade curvature. In particular, the aerodynamics of centrifugal nozzles is complicated by the inherent enlargement of meridional cross-section along the blade, which couples with the local flow deflection and with the non-ideal volumetric behaviour of the fluid. Simple assumptions on the functional form of the blade mean-line do not allow to control properly such effects, resulting in a relevant over-speed and in the onset of shocks in the rear suction side of the blade. Thanks to the flexibility and the generality of the systematic optimization technique here applied, optimal distributions for the mean-line curvature and the blade thickness are determined along the profiles. The trends obtained for the two different cascades exhibit similar features, which might be used to set-up generalized design remarks for centrifugal nozzles.

References

[1] Gaia M 2011 *First International Seminar on ORC Power Systems, Keynote Lecture*
[2] Macchi E 2013 *Second International Seminar on ORC Power Systems, Keynote Lecture*
[3] La Seta A, Meroni A, Andreasen J G, Pierobon L, Persico G and Haglind F 2016 *Energies* **9**
[4] Hoffren J T T, Larjola J and Siikonen T 2002 *ASME J. Eng. Gas Turb. Power* **124** 395–403
[5] Harinck J, Colonna P, Guardone A and Rebay S 2010 *ASME J. Turbomach.* **132** pp. 011001–17
[6] Spadacini C, Centemeri L, Xodo L, Astolfi M, Romano M and Macchi E 2011 *First International Seminar on ORC Power Systems*
[7] Pini M, Persico G, Casati E and Dossena V 2013 *ASME J. Eng. Gas Turb. Power* **135**
[8] Persico G and Pini M 2016 *Organic Rankine Cycle (ORC) Power Systems: Technologies and Applications* (Woodhead Publishing - Elsevier) chap Fluid dynamic design of Organic Rankine Cycle turbines
[9] Persico G, Pini M, Dossena V and Gaetani P 2013 *ASME Turbo Expo 2013* GT2013-95770
[10] Persico G, Pini M, Dossena V and Gaetani P 2015 *ASME J. Eng. Gas Turb. Power* **137** 1–11
[11] Persico G and Rebay S 2012 *Comput. Fluids* **60** 86–98
[12] Peter J and Dwight R 2010 *Comput. Fluids* 373–391
[13] Pini M, Pasquale D, Persico G and Rebay S 2014 *ASME J. Eng. Gas Turb. Power* **137** 032604
[14] Coello C 2000 *ACM Comput. Surv.* **32** 109–143 ISSN 03600300
[15] Verstraete T, Alsalihi Z and Van Den Braembussche R 2010 *ASME J. Turbomach.* **132** 031004
[16] Pierret S, Coelho R and Kato H 2006 *Struct. Multidiscip. O.* **33** 61–70 ISSN 1615147X
[17] Rodriguez-Fernandez P and Persico G 2015 *3rd International Seminar on ORC Power Systems* 133
[18] Farin G 2002 *Curves and Surfaces for CAGD: A Practical Guide* 5th ed (San Francisco, CA, USA: Morgan Kaufmann Publishers Inc.) ISBN 1-55860-737-4
[19] Colonna P, Nannan N and Guardone A 2008 *Fluid Ph. Eq.* **263** 115–130
[20] Colonna P and van der Stelt T P 2004 FluidProp: a program for the estimation of thermo physical properties of fluids Software URL `http://www.fluidprop.com`
[21] Persico G, Mora A, Gaetani P and Savini M 2012 *ASME J. Turbomach.* **134** 061030 (10 pages)
[22] Reeves C R and Rowe J E 2002 *Genetic Algorithms: Principles and Perspectives: A Guide to GA Theory* (Norwell, MA, USA: Kluwer Academic Publishers) ISBN 1402072406
[23] Simpson T, Poplinski J, Koch P N and Allen J 2001 *Eng. Comput.* **17** 129–150 ISSN 0177-0667
[24] Adams B, Bauman L, Bohnhoff W, Dalbey K, Ebeida M, Eddy J, Eldred M, Hough P, Hu K, Jakeman J, Swiler L and Vigil D 2013 *Sandia Technical Report SAND2010-2183*

IOP Conf. Series: Journal of Physics: Conf. Series **821** (2017) 012016 doi:10.1088/1742-6596/821/1/012016

CFD analysis of thermally induced thermodynamic losses in the reciprocating compression and expansion of real gases

Aly I. Taleb[1], Paul Sapin[1], Christoph Barfuß[1,2], Drazen Fabris[1,3], Christos N. Markides[1]

[1]Clean Energy Processes (CEP) Laboratory, Deparment of Chemical Engineering, Imperial College London, London SW7 2AZ, UK

[2]Lehrstuhl für Energiesysteme, Fakultät für Maschinenwesen, Technische Universität München, 85478 Garching b. München, Germany

[3]Department of Mechanical Engineering, Santa Clara University, Santa Clara, CA 95053, USA

E-mail: a.taleb@imperial.ac.uk

Abstract. The efficiency of expanders is of prime importance in determining the overall performance of a variety of thermodynamic power systems, with reciprocating-piston expanders favoured at intermediate-scales of application (typically 10–100 kW). Once the mechanical losses in reciprocating machines are minimized (e.g. through careful valve design and operation), losses due to the unsteady thermal-energy exchange between the working fluid and the solid walls of the containing device can become the dominant loss mechanism. In this work, gas-spring devices are investigated numerically in order to focus explicitly on the thermodynamic losses that arise due to this unsteady heat transfer. The specific aim of the study is to investigate the behaviour of real gases in gas springs and to compare this to that of ideal gases in order to attain a better understanding of the impact of real-gas effects on the thermally induced losses in reciprocating expanders and compressors. A CFD-model of a gas spring is developed in OpenFOAM. Three different fluid models are compared: (1) an ideal-gas model with constant thermodynamic and transport properties; (2) an ideal-gas model with temperature-dependent properties; and (3) a real-gas model using the Peng-Robinson equation-of-state with temperature and pressure-dependent properties. Results indicate that, for simple, mono- and diatomic gases, like helium or nitrogen, there is a negligible difference in the pressure and temperature oscillations over a cycle between the ideal and real-gas models. However, when considering heavier (organic) molecules, such as propane, the ideal-gas model tends to overestimate the pressure compared to the real-gas model, especially if the temperature and pressure dependency of the thermodynamic properties is not taken into account. In fact, the ideal-gas model predicts higher pressures by as much as 25% (compared to the real-gas model). Additionally, both ideal-gas models underestimate the thermally induced loss compared to the real-gas model for heavier gases. This discrepancy is most pronounced at rotational speeds where the losses are highest. The real-gas model predicts a peak loss of 8.9% of the compression work, while the ideal-gas model predicts a peak loss of 5.7%. These differences in the work loss are due to the fact that the gas behaves less ideally during expansion than during compression, with the compressibility factor being lower during compression. This behaviour cannot be captured with the ideal-gas law. It is concluded that real-gas effects must be taken into account in order to predict accurately the thermally induced loss mechanism when using heavy fluid molecules in such devices.

1. Introduction

Reciprocating machines have potential applications in the efficient conversion of external low and medium-grade heat (e.g. solar, geothermal, waste) to useful work. Piston expanders, for instance, are a viable alternative to turbomachines and other positive displacement (volumetric) expanders in organic Rankine cycles (ORCs) at intermediate-scales of application (typically 10–100 kW). They provide higher efficiencies at these lower power outputs and at off-design operating speeds which may occur due to the intermittent nature of the aforementioned heat sources. Other examples of reciprocating external-heat conversion devices include two-phase thermofluidic oscillators [1–3] and liquid-piston Fluidyne engines [4, 5], both of which rely on liquid pistons to transfer the hydraulic power generated by periodic oscillations of pressure and volume of a suitable working fluid within an enclosed space.

In reciprocating machines, thermodynamic losses occur due to mechanical friction between moving parts, flow losses through valves and due to leakage, and irreversibilities due to heat transfer across finite temperature differences. In reciprocating compressors, for example, the majority of thermodynamic losses do not occur due to leakage or during the discharge or suction processes but during superheating [6]. Once mechanical losses are minimized, thermally induced losses can begin to dominate and must be taken into account. In reciprocating expanders and compressors, they can reach similar values as valve losses [7, 8]. Thermally induced losses tend to peak around intermediate rotational speeds with Péclet numbers (defined in Sec. 2.1) on the order of 10. Reciprocating piston expanders in ORCs have speeds ranging from 50 to 4000 RPM [9–11] which translates into a Péclet number from somewhere around 500 to 50000 (assuiming a bore diamter of 10 cm). Thermofluidic oscillators have frequencies on the order of 0.01 to 1 Hz [1–3] which is equivalent to Péclet numbers from 0.1 to 100 (assuming a cylinder diameter of 2 cm).

To understand the basic irreversibility due to the periodic heat transfer without considering inlet and exit flows, we consider a gas spring—in which a fixed mass undergoes sequential compression and expansion within a closed space. In the present paper, a gas spring is simulated numerically using computational fluid dynamics (CFD). Gas springs do not include the complex effects of valve timing and gas exchange of reciprocating-piston devices while retaining the fluctuating heat exchange phenomena between the gas and the cylinder walls. Performance losses arising due to unsteady heat transfer occur even if the overall process is globally adiabatic (in the mean), since heat exchange between the gas and the cylinder wall across a finite temperature difference is an inherently irreversible process. Heat is transferred from the gas to the wall during compression when the gas temperature is higher than the wall temperature. During expansion, heat is returned back to the gas as its temperature falls below that of the wall. Even if the net fluctuating heat transfer is zero, a finite exergetic penalty is paid. Therefore, it is of primary importance, to quantify and characterize the heat exchange fluctuations between the fluid and the solid.

Early studies in this area were undertaken by Pfriem [12], who described analytically a phase difference between the heat flux and the temperature difference between the wall and the bulk gas. Lee [13] described the oscillating heat flux by developing a model for a complex Nusselt number at low Péclet numbers (or low frequencies). Experiments were conducted by Kornhauser and Smith [14, 15] from which the thermal loss was evaluated for a wide range of Péclet numbers and different compression ratios. Additionally, these authors proposed empirical *Nu*-*Re* correlations which accounted for the aforementioned phase difference with imaginary components. The correlation was only developed for low compression ratios and helium as a working fluid. Later on, Lekić and Kok [16] simulated Kornhauser and Smith's experiments with CFD simulations and examined local effects of temperature and pressure in the compression space.

The heat-conversion technologies of interest here commonly use more complex organic working fluids than those classically investigated in gas spring systems. This is especially valid for ORC engines and thermofluidic oscillators, which rely on refrigerants and hydrocarbons such as pentane, HFC-245fa or HFC-134a [17]. Lee [13] considered ideal gases while Kornhauser and Smith [14, 15] experimented with light or noble gases such as hydrogen, helium and argon for which the ideal-gas law is an accurate approximation. However, in the temperature and pressure ranges relevant to this study (200–700 K, 0.5–10 bar) and for organic fluids, real-gas effects need to be considered. In this work, the irreversibility due to heat transfer is quantified while accounting for real-gas effects, specifically by employing different equations-of-state and by accounting for thermodynamic property variations due to pressure and temperature.

2. Methodology

2.1. Problem setup

The geometry of the gas-spring arrangement is defined by the following parameters: bore diameter: $d = 2r = 105$ mm, stroke length: $L_\text{s} = 2r_\text{s} = 78$ mm, connecting rod length: $L_\text{r} = 149$ mm, dead volume height: $L_\text{dv} = 14$ mm. The geometry was chosen to match an experimental gas-spring apparatus [18]. The oscillatory Péclet number, Pe, is used as a dimensionless measure for the rotational speed of the gas spring:

$$Pe = \frac{\omega D_\text{h}^2}{4\alpha_0}, \tag{1}$$

where ω is the angular frequency, $\alpha_0 = k/\rho c_p$ is the time and spatially-averaged thermal diffusivity and $D_\text{h} = 4V_\text{md}/A_\text{md}$ is the hydraulic diameter at mid-stroke (V_md and A_md are the cylinder volume and surface area at mid-stroke). The thermal diffusivity is calculated with the thermal conductivity k, the density ρ and the specific heat capacity c_p.

To quantify the thermal loss over one complete oscillation cycle, we define the thermal loss parameter ψ as the ratio of net work done on the gas during one cycle $W_\text{com} - W_\text{exp}$ to the work done onto the gas just during compression W_com:

$$\psi = \frac{W_\text{com} - W_\text{exp}}{W_\text{com}} = \frac{\oint P\text{d}V}{\int_\text{BDC}^\text{TDC} P\text{d}V}, \tag{2}$$

where TDC stands for top dead centre and BDC for bottom dead centre. The pressure inside the gas spring (assumed uniform) is represented by P and the volume by V.

2.2. Geometry and mesh

The CFD model was developed in the open-source code OpenFOAM (v. 2.1.1). As the geometry is axisymmetric (Fig. 1(a)), the gas spring is modelled in two spatial dimensions. The simulation is performed over 10 cycles in order to reach a cyclic steady state, i.e. a state where the cycle-to-cycle variation is below 0.5%. The geometry consists of a wedge with a width of 0.02 rad and there is only one layer of cells in circumferential direction.

The changing volume of the compression space is modelled with a dynamic mesh, with the moving lower boundary representing the face of the piston. Every node in the mesh compresses vertically relative to the stationary cylinder head (top boundary) and the moving piston head (bottom boundary). Nodes closer to the piston move to a greater extent than nodes closer to the cylinder head. Thus, the mesh is distorted during compression/expansion as in Fig. 1(b). The mesh is modelled to have a cell aspect ratio of unity at mid-stroke so that the distortion at bottom dead center (BDC) is approximately equal to that at top dead center (TDC) in terms of the ratio of a cell's larger dimension to smaller dimension (if grading is not considered). The size of the cells decreases gradually towards the boundaries to increase the mesh resolution inside the boundary layers near the walls. At mid-stroke, the smallest cell radial dimension is one fourth of the largest cell radial dimension. The same is valid in axial direction. The mesh resolution used depends on the rotational speed where the total number of cells increases for higher speeds. A

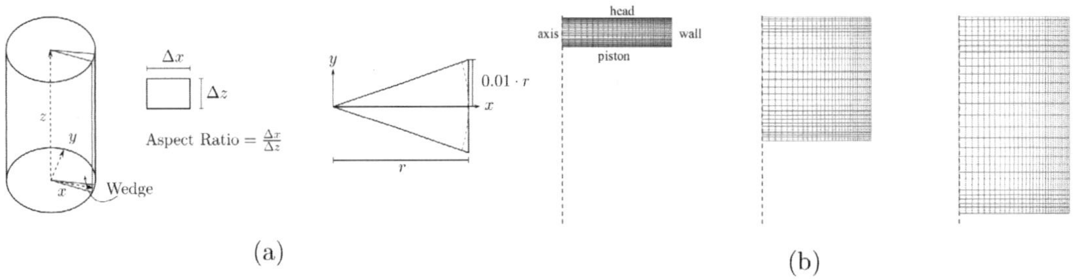

Figure 1. (a) Modelled wedge geometry and definition of the aspect ratio. Left-hand diagram shows the wedge in a perspective view. Right-hand diagram shows a top view of the geometry (not to scale). (b) Mesh movement during an expansion stroke of a 30 × 30 mesh. A grading factor increases resolution near the walls.

convergence study was carried out to investigate the independence of the results from the mesh resolution and the Courant number. For frequencies (speeds) around 100 RPM, for example, the mesh is sufficiently resolved at 90 × 90 [19].

The time step is controlled by a user-specified maximum Courant number Co_{max}. For a given cell, the Courant number is defined as:

$$Co = \frac{|u|\Delta t}{\Delta x}, \tag{3}$$

where $|u|$ is the magnitude of the flow velocity (flow speed) through a cell, Δx is the cell size in the direction of the velocity vector and Δt is the size of the time step. The time step is chosen such that the pre-set Co_{max} value is never exceeded. Due to the dynamic mesh, the time step not only adapts to the flow velocity but also to the change in cell size. The maximum Courant number is set to $Co_{\text{max}} = 0.1$ for speeds around 100 RPM.

The compressible solver uses a modified version of *coldEngineFOAM*, provided by OpenFOAM for piston engines without combustion. This solver is based on the PIMPLE algorithm which is a combination of the SIMPLE (semi-implicit method for pressure linked equations) and PISO (pressure-implicit split operator).

2.3. Turbulence model

The simulations have shown that during the upstroke and the downstroke the velocity profile is primarily axial and uniform. There is, however, a significant velocity gradient in radial direction near the wall due to the no-slip boundary condition. Turbulence is generated in regions of high perpendicular velocity gradients to the main flow direction (shear) which require a sufficiently high inertial force to overcome the damping effect of the viscosity. This can occur in the boundary layer during the stroke. The boundary-layer or local Reynolds number Re_δ can provide some insight into the transition behaviour of the boundary layer.

To the authors' knowledge, no critical Reynolds number is currently available to predict the transition from laminar to turbulent flow in such a (gas-spring) flow configuration. Merkli and Thomann [20] investigated experimentally the flow transition of the Stokes boundary layer in oscillating pipe flow where no significant compression of the fluid occurs. They observed a critical boundary Reynolds number of $Re_{\delta,\text{c}} = 2u_{\text{max}}/\sqrt{\nu\omega} \approx 400$.

It is proposed to use this empirical result to determine whether turbulence must be accounted for in our simulations, with an instantaneous $Re_\delta(\phi)$ defined as:

$$Re_\delta(\phi) = \frac{2u(\phi)}{\sqrt{\omega\nu(\phi)}}, \tag{4}$$

where ϕ is the crank angle (equivalent to piston position), $u(\phi)$ is the instantaneous piston velocity equivalent to the average flow velocity and $\nu(\phi)$ is the instantaneous kinematic viscosity. The critical Reynolds number of $Re_{\delta,c} = 400$ occurs for propane gas springs at $Pe > 11000$ or at speeds over 280 RPM and for nitrogen gas springs at $Pe > 8500$ or at speeds over 680 RPM. The critical Reynolds number was not attained for gas springs using helium for the speeds considered in this study below 1000 RPM.

Furthermore, Lekić and Kok [16] performed direct numerical simulations of compression and expansion of helium in a gas spring and found that turbulence has a negligible impact on the governing physical phenomena. For these reasons, the momentum balance equation solved in the current model is a laminar Navier-Stokes equation.

2.4. Real-gas effects

Both real and ideal-gas equations-of-state are compared in this study. The ideal-gas equation-of-state is replaced by the Peng-Robinson equation-of-state for real gases [21]. Additionally, the temperature-dependent mole-specific heat capacity $c_p^0(T)$, enthalpy $h^0(T)$ and entropy $s^0(T)$ for ideal gases are calculated with temperature-dependent polynomials often referred to as JANAF or NASA heat capacity polynomials [22]. For brevity, only the heat capacity is shown here:

$$c_p^0(T) = R\left(a_1 T^{-2} + a_2 T^{-1} + a_3 + a_4 T + a_5 T^2 + a_6 T^3 + a_7 T^4\right) . \tag{5}$$

With $c_v^0 = c_p^0 - R$, a correction of the thermodynamic properties for the density (or pressure) can be calculated from [21]:

$$c_v(T, v) = c_v^0(T) + T \int_v^{\infty} \left[\left(\frac{\partial^2 P}{\partial T^2}\right)_v\right]_T \mathrm{d}v \, , \tag{6}$$

and:

$$c_p(T, v) = c_v(T, v) - T \frac{\left(\frac{\partial P}{\partial T}\right)_v^2}{\left(\frac{\partial P}{\partial v}\right)_T} . \tag{7}$$

2.5. Initial and boundary conditions

The initial conditions of the pressure and temperature within the gas-spring space are set at BDC (crank angle $\phi = -180$ °). For all cases, the initial pressure and temperature are set to 3 bar and 330 K. The wall temperatures are uniform and constant at 350 K. Adair et al. [23] observed in their experiments that the cylinder wall temperatures in reciprocating compressors oscillate by less than 1 K. The same temperature was selected for all wall boundaries (cylinder head, piston head and cylinder wall). The CFD simulations of Lekić and Kok [16] made the same assumption of constant and uniform wall temperature.

3. Results

3.1. Influence of real-gas effects on temperature and pressure

Three gases were considered in this study: helium, nitrogen and propane. A similar Péclet number ranging from $Pe \approx 360$ to $Pe \approx 540$ was adopted for all three gases. Even for the same gas there are different Pe at identical speeds because of variations in the mean thermal diffusivity. The rotational speeds considered were 250 RPM for helium, 25 RPM for nitrogen and 10 RPM for propane. Figures 2 and 3 show: (a) T–V and (b) P–V curves for the three substances when modelled as ideal gases with constant (solid blue) and temperature-dependent thermodynamic/transport properties (dashed red) and as real gases with density and temperature-dependent properties (dotted green).

As expected, nitrogen and helium behave similarly regardless of the model, although, the temperatures are slightly overestimated compared to the real-gas model when assuming ideal-gas properties. The maximum deviations between the varying-property ideal-gas and the real-gas model for helium and nitrogen are 5 K and 4 K (or $\approx$ 1% of the average temperature) for the temperature and 0.14 bar and 0.60 bar (< 4% of average pressure) for the pressure respectively. The maximum temperature-deviation for propane is also very small: 6 K or 2%. However, propane shows non-negligible deviations for the pressure. One can see in the the P–V plot of Fig. 3(b) that both ideal-gas models – although they are in good agreement with each other – deviate significantly from the real-gas simulation. At BDC where the pressure and temperature are at a minimum, the constant-property ideal-gas (blue in Figs. 2/3), varying-property ideal-gas (red) and the real-gas model (green) calculate similar pressure values of 2.9 bar, 2.9 bar and 2.8 bar respectively. However, when the pressure and temperature are high, the deviation between the pressure values of the different models is high: 24.2 bar, 23.4 bar and 18.0 bar for the constant-property model, varying property model and real gas model respectively. Because both ideal-gas models do not have a correction for the pressure (or density) dependency of the gas properties, they both deviate in the pressure by as much as 25% at TDC from the real-gas model.

Discrepancies between ideal and real gases can be understood via the compressibility factor $Z = Pv/RT$, which is plotted in Fig. 4. By definition, $Z = 1$ for ideal gases and values different (and deviating further) from unity provide a measure for the magnitude of the real-gas effects. These tend to increase when one is close to the critical point or saturation, which for the cases considered, corresponds to low temperatures and high pressures. Figure 4 shows that Z is almost always equal to 1 at all crank angles for nitrogen. When propane is used, Z moves away from 1, especially at TDC where it can go as low as 0.75. The largest differences between the ideal and real-gas models are observed at this position.

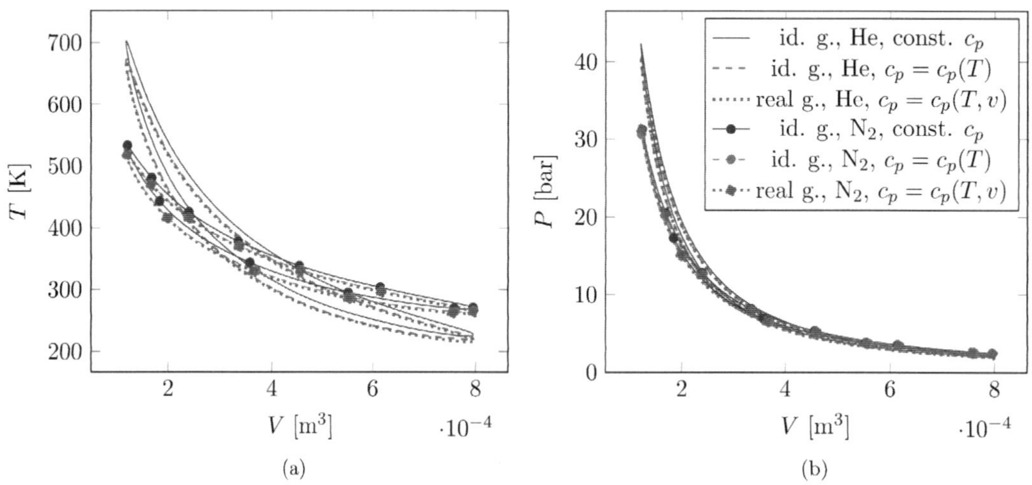

Figure 2. (a): T–V diagram and (b): P–V diagram for helium (no markers) at 250 RPM and nitrogen (with markers) for the ideal-gas model with constant thermophysical properties, $Pe = 446$ for helium and $Pe = 371$ for nitrogen (solid blue); ideal-gas model with temperature-dependent properties, $Pe = 502$ for helium and $Pe = 355$ for nitrogen (dashed red); and real-gas model with temperature and density dependent properties $Pe = 457$ and $Pe = 355$ for nitrogen (dotted green).

3.2. Thermally induced losses

Figure 5 shows the loss parameter ψ for helium, nitrogen and propane for varying Péclet numbers. The general trend which has been reported in previous work [13, 14, 16, 24] is also observed here:

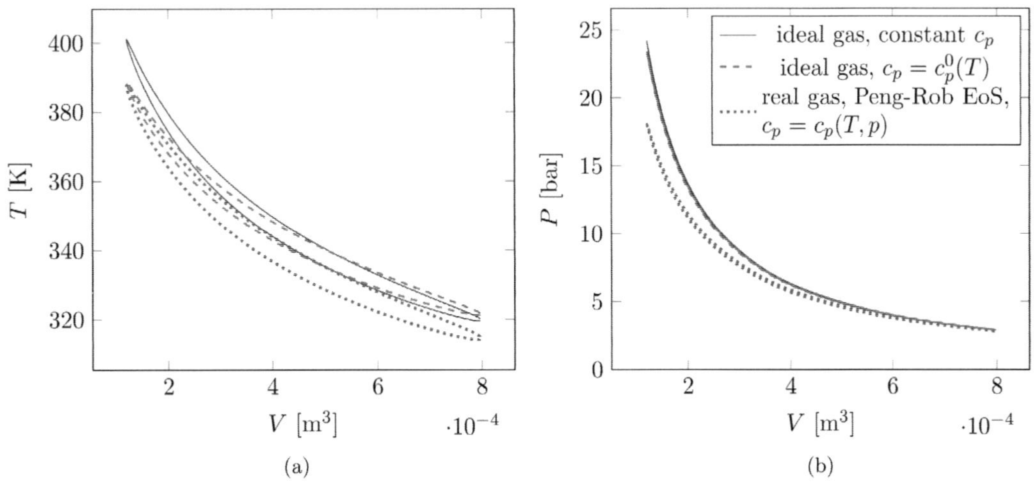

Figure 3. (a): T–V diagram and (b): P–V diagram for propane at 10 RPM for the ideal-gas model with constant thermophysical properties, $Pe = 545$ (solid blue); ideal-gas model with temperature-dependent properties, $Pe = 442$ (dashed red); and real-gas model with temperature and density dependent properties, $Pe = 382$ (dotted green).

at low and high Pe the loss is low, and a maximum is observed at an intermediate value which is around $Pe \approx 10$. The losses are as high as 28.0% of the compression work for helium, 18.6% for nitrogen and 8.9% for propane. At low Péclet numbers, the oscillation frequency is small, and there is enough time for any work done by the gas to be exchanged in the form of heat with the solid walls. The gas temperature remains constant and equal to the wall temperature, and therefore the system undergoes isothermal compression. On the other hand, the oscillation frequency is high at elevated Péclet numbers. The variation between compression and expansion occurs too quickly for significant heat transfer to occur, and the system experiences adiabatic compression and expansion. At intermediate Péclet numbers, a (net) work results from the interplay between both the non-zero internal energy of the gas and the non-zero heat transfer to the solid, leading to a net loss of work during the cycle.

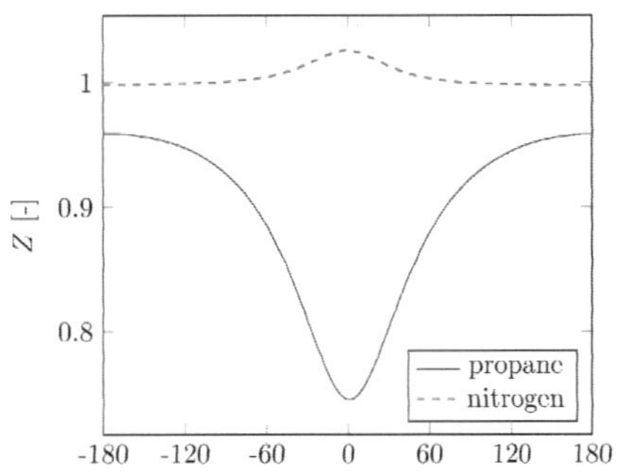

Figure 4. Compressibility Z over a cycle for nitrogen and propane.

At low Pe, all three models predict similar loss values for helium and nitrogen, as shown in Fig. 5(a). When moving to $Pe > 10$, one can observe that the constant property model slightly

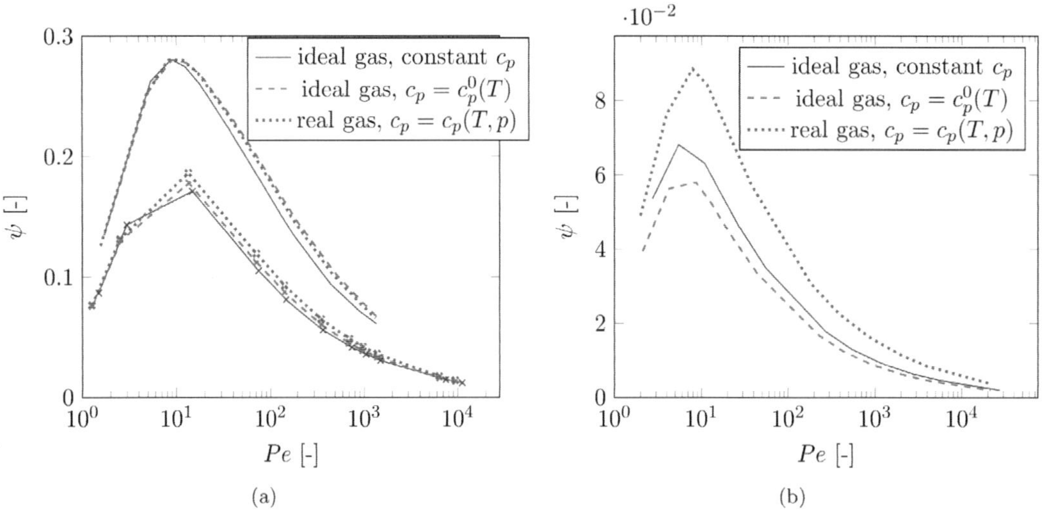

Figure 5. Loss ψ as a function of *Pe* numbers with: (a) helium (no markers) and nitrogen (markers x) and (b) propane.

underestimates the loss compared to the other two alternatives. As for propane in Fig. 5(b), both ideal-gas models underestimate the thermal loss compared to the real-gas model across the whole Péclet number range. At $Pe \approx 10$ (where the losses peak), the differences between the models is greatest. The temperature-dependent ideal-gas model predicts a peak loss of 0.057 and the constant-property ideal-gas model predicts a peak loss of 0.068. On the other hand, the real-gas model shows a peak loss of 0.089. Real-gas effects increase the thermally induced losses significantly, which is not captured with the ideal-gas models.

We now define parameter λ as:

$$\lambda = 1 - Z \tag{8}$$

as a measure of how non-ideal a gas behaves. It is higher during the expansion stroke than during compression meaning that the gas is less ideal during expansion. This is shown in Fig. 6(a) where λ is plotted against the cylinder volume in one oscillation cycle. This "asymmetry" between expansion and compression for real gases leads to an increased loss compared to ideal gases. For the case from Fig. 6(a) ($Pe \approx 10$), for example, the mean value for λ during compression is 0.09 while during expansion it is 0.11. The gas is less ideal during expansion than during compressions. Integrating λ with respect to the instantaneous volume of the cylinder provides a good measure of the asymmetry in gas behaviour between expansion and compression:

$$\Lambda = \frac{1}{V_{\text{TDC}}} \oint \lambda \mathrm{d}V \,. \tag{9}$$

The integral Λ gives the (non-dimensionalized) area enclosed within the plot of Fig. 6(a). If $\Lambda > 0$, then the mean compressibility factor is lower during expansion than compression (the gas is more non-ideal during expansion). The higher the value of Λ, the larger the "asymmetry" in compressibility factor between the expansion and compression strokes. This is plotted for different *Pe* numbers in Fig. 6(b) using propane. The integral Λ is always positive for all Péclet numbers, and it has a maximum around $Pe \approx 10$ of $\Lambda_{\text{max}} = 0.096$ where the loss (see Fig. 5(b)) also has its maximum. The difference between the real gas and ideal-gas loss $\Delta\psi$ is also shown in the same figure. It is defined as:

$$\Delta\psi = \psi_{\text{RG}} - \psi_{\text{IG}} \,, \tag{10}$$

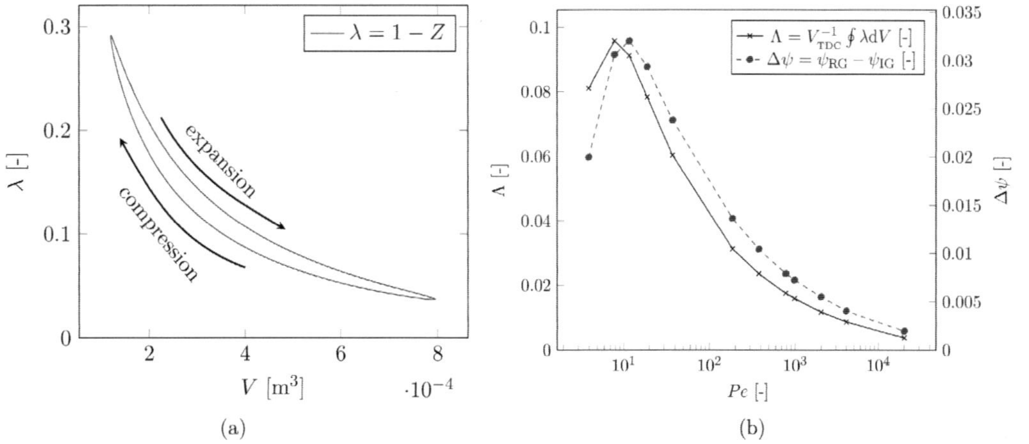

Figure 6. (a) Plot of $\lambda = 1 - Z$ against volume over a cycle with propane and $Pe \approx 10$, 0.2 RPM. (b) Black (x) symbols/solid line: Integral of $\lambda = 1 - Z$ with respect to the cylinder volume over a cycle as a measure of the asymmetry in gas behaviour between compression and expansion. Blue (●) symbols/dashed line: Difference in thermally induced losses between real and ideal-gas models $\Delta\psi$ for different Péclet numbers. For both curves, propane is considered.

where ψ_{RG} and ψ_{IG} are the thermally induced losses for the real and temperature-dependent ideal-gas model respectively. Figure 6(b) shows that $\Delta\psi$ behaves in accordance with Λ: a high asymmetry in $\lambda = 1 - Z$ (high Λ) between expansion and compression leads to a large discrepancy in the losses between the models. At $Pe \approx 10$, the difference between the loss terms $\Delta\psi$ peaks at $\Delta\psi_{max} = 0.032$ corresponding to $\Lambda_{max} = 0.096$. This corresponds to a difference in efficiencies (ratio of expansion work to compression work, also $1 - \psi$) between 91.1% for the real gases and 94.3% for the ideal gas. Since the ideal-gas model has a constant compressibility factor $Z = 1$, strong changes in Z in the real-gas model will lead to large deviations between the two models. The increase in real-gas behaviour during expansion means that less work can be recovered. Therefore, the losses are higher when taking real-gas effects into account, however these deviations diminish as one moves away from $Pe \approx 10$.

4. Conclusion

A two dimensional CFD model of a gas-spring wedge has been developed to simulate the compression/expansion processes and thermally induced thermodynamic losses in gas springs. The model can also take real-gas effects into account by calculating pressure and temperature-dependent thermodynamic and transport properties and by using the Peng-Robinson equation-of-state. Due to the low velocity of the flow and piston, the Reynolds numbers were well below the critical Reynolds numbers for the transition to turbulence in the boundary layer thereby allowing a laminar model to be used. The wall temperatures surrounding the gas were set to be uniform and constant. Three different gas models were compared: (1) an ideal-gas model with constant thermodynamic and transport properties, (2) an ideal model with temperature-dependent properties, and (3) a real-gas model based on the Peng-Robinson equation-of-state with pressure and temperature-dependent properties.

When comparing helium and nitrogen at Péclet numbers between 350 and 550, all three models show similar variations in pressure and temperature over the cycle. For propane, the ideal-gas models deviate from real-gas models especially at the higher pressures occurring at and near TDC where the deviation was as high as 25% for pressure. When comparing to the results of the real-gas model, the ideal-gas models overestimate the pressures and to a lesser extent the temperatures in these conditions. The compressibility factor Z, which is used here to quantify deviations from ideality, is almost always 1, but for propane Z falls down to 0.75 at TDC. This explains the discrepancies between the ideal and real-gas models of propane.

Thermally induced losses were evaluated and compared for the gases over a range of Péclet numbers. Significant differences between the models were only observed for propane: both ideal-gas models underestimate the thermal loss compared to the real-gas models of this gas. This occurs over the whole Péclet number range investigated indicating that real-gas effects can increase the thermally induced losses in gas springs. The highest difference between loss values was observed at peak loss ($Pe \approx 10$) where the real-gas loss was 8.9% of the compression work and the ideal-gas loss was 5.7% of the compression work. The reason behind the difference in the predicted losses between is the different gas behaviour during compression and expansion. Propane has a lower compressibility factor during expansion than compression, making it behave more like a real gas during the expansion stroke. For ideal-gas models, the compressibility factor is by definition always equal to unity. This increase in real-gas behaviour during expansion means less work is recovered and therefore higher losses over the cycle.

References

[1] Solanki R, Mathie R, Galindo A and Markides C N 2013 *Applied Energy* **106** 337–354
[2] Taleb A I, Timmer M A G, El-Shazly M Y, Samoilov A, Kirillov V A and Markides C N 2016 *Energy* **104** 250–265
[3] Kirmse C J W, Oyewunmi O A, Taleb A I, Haslam A J and Markides C N 2016 *Applied Energy* In press
[4] Stammers C W 1979 *Journal of Sound and Vibration* **63** 507–516
[5] West C D 1983 Dynamic analysis of the Fluidyne *Proceedings of the 18th Intersociety Energy, Conversion Engineering Conference* pp 779–784
[6] Ribas F A, Deschamps C J, Fagotti F, Morriesen A and Dutra T 2008 Thermal analysis of reciprocating compressors – a critical review *International Compressor Engineering Conference*
[7] Yu L 2015 Dynamic modelling of a reciprocating piston expander literature Review
[8] Stouffs P, Tazerout M and Wauters P 2001 *International Journal of Thermal Sciences* **40** 52–66
[9] Curran H M 1981 *Journal of Energy* **5** 218–223
[10] Zahoransky R, Knöringer P, Schelling U and Wittig S 1983 *Wärme* **91** 43–53
[11] Bao J and Zhao L 2013 *Renewable and Sustainable Energy Reviews* **24** 325–342
[12] Pfriem H 1943 Periodic heat transfer at small pressure fluctuations Tech. rep. National Advisory Committee for Aeronautics
[13] Lee K P 1983 A simplistic model of cyclic heat transfer phenomena in closed spaces *Proceedings of the 18th Intersociety Energy Conversion Engineering Conference* pp 720–723
[14] Kornhauser A A and Smith J L 1993 *Journal of Energy Resources Technology* **115** 70–75
[15] Kornhauser A A and Smith J L 1994 *Transactions of the American Society of Mechanical Engineers, Journal of Heat Transfer* **116** 536–542
[16] Lekić U and Kok J 2010 *Open Thermodynamics Journal* **4** 13–26
[17] Chen H, Goswami Y G and Stefanakos E K 2010 *Renewable and Sustainable Energy Reviews* **14** 3059–3067
[18] Sapin P, Taleb A I, Barfuß C, White A J, Fabris D and Markides C N 2016 Thermodynamic losses in a gas spring: Comparison of experimental and numerical results *12th International Conference on Heat Transfer, Fluid Mechanics and Thermodynamics*
[19] Barfuß C 2016 *Numerical Simulation of Heat Transfer Effects during Compression and Expansion in Gas Springs* Master's thesis Technische Universität München URL `https://www.researchgate.net/publication/308404171_Numerical_Simulation_of_Compression_and_Expansion_Processes_in_Gas_Springs_for_Innovative_Electricity_Storage`
[20] Merkli P and Thomann H 1975 *Journal of Fluid Mechanics* **68** 567–575
[21] Poling B E, Prausnitz J M and O'Connell J P 2001 *The Properties of Gases and Liquids* (McGraw Hill)
[22] Burcat A and Ruscic B 2005 Third millennium ideal gas and condensed phase thermochemical database for combustion with updates from active thermochemical tables Tech. Rep. ANL-05/20 Argonne National Laboratory
[23] Adair R P, Qvale E B and Pearson J T 1972 Instantaneous heat transfer to the cylinder wall in reciprocating compressors *International Compressor Engineering Conference*
[24] Mathie R, Markides C N and White A J 2014 *Heat Transfer Engineering* **35** 1435–1449

IOP Conf. Series: Journal of Physics: Conf. Series **821** (2017) 012017 doi:10.1088/1742-6596/821/1/012017

Direct numerical simulations of homogeneous isotropic turbulence in a dense gas

A Giauque[1], C Corre[1] and M Menghetti[1]

[1] LMFA - Laboratoire de Mecanique des Fluides et d'Acoustique
Ecole Centrale de Lyon, 36 avenue Guy de Collongue, 69134 Ecully Cedex, France

E-mail: alexis.giauque@ec-lyon.fr

Abstract. A study of turbulence in BZT dense gas flows is performed using DNS. It is shown that for a large but realistic intensity, the turbulence in dense gas flows behaves in a highly compressible manner when the average thermodynamic state lies within the inversion region in which the gas fundamental derivative is negative. A close similarity is observed in the evolution of the kinetic energy when the initial turbulent Mach number and the Taylor Reynolds number are matched regardless of the Equation of State (EoS) considered. A large turbulent Mach number is yet more easily attained in dense gas flows lying in the inversion region because of the low speed of sound associated with it. In this case the turbulence shows a highly compressible evolution with periodic exchanges between the internal and kinetic energies. In order to assess the capabilities of currently available Large Eddy Simulation (LES) subgrid-scale models, a-posteriori tests are performed using the dynamic Smagorinsky model. Coherently with the hypothesis it relies on, the model perfectly captures the evolution of the kinetic energy when the turbulent Mach number is low enough. When using the perfect gas EoS at a higher turbulent Mach number the agreement is reasonable. Yet, when the average thermodynamic state lies within the inversion region and when using the thermal and caloric Martin&Hou EoS, the model is not able to capture the correct evolution of the kinetic energy. The results presented in this study call for a specific research effort directed towards the assessment and possibly the development of advanced subgrid-scale models for LES of turbulent dense gas flows.

1. Introduction

Engineering systems harvesting dense gas thermodynamic cycles for the recovery of fatal heat (i.e. heat that is lost in the output gases during any transformation process) experience an accelerated development under the combined effect of the increase of the energy price and of the public awareness for climate issues. Companies developing this technology face numerous challenges among which some come from a lack of basic knowledge regarding the turbulent flows of dense gases in expanders. Indeed, to obtain efficient turbomachinery systems, numerous prototypes are designed on the basis of Computational Fluid Dynamics (CFD) predictions before being tested. Today, Reynolds Averaged Navier-Stokes (RANS) simulations have become a usual tool in the design stage of these machines. Large Eddy Simulation (LES) is an additional useful tool for the understanding of intrinsically unsteady phenomena occurring in Organic Rankine Cycle (ORC) turbines. Yet, for both these methods, the turbulence models upon which they rely were all developed in the context of perfect gases.

Among dense gases, Bethe-Zel'dovitch-Thompson (BZT) fluids [1] have raised a particular interest because their fundamental derivative defined as $\Gamma = 1 + \frac{\rho}{a}\left[\frac{\partial a}{\partial \rho}\right]_s$, where ρ is the density and a the speed of sound, can be lower than one and even become negative in which case shock waves are dampened along with the associated losses. It is also expected that for appropriate inflow conditions, the turbulent Mach number will be large enough to modify the behavior of turbulence down to the smallest scales, which will significantly impact viscous losses. In order to assess the need for accurate and reliable models tailored for dense gases, it is first necessary to study turbulence in dense gas flows without resorting to a turbulence model. In this study, Direct Numerical Simulation (DNS) is used to investigate decaying homogeneous isotropic turbulence (HIT) and understand the influence of thermodynamics on turbulence in dense gas flows. The article is organized as follows. In the first section, the numerical tool used to perform DNS is introduced along with the test cases performed in the study to compare perfect and dense gas behavior. Section 2 focuses on the DNS results and compares the properties of turbulence between the different test cases. Section 3 presents, for the different test cases, a comparison between results obtained by LES using the dynamic Smagorinsky model and DNS.

2. Numerical Setup and Test cases

For this study, FC-70 is used as the working fluid. FC-70 is suspected to be dense enough to possess an inversion region in which its fundamental derivative is negative [2]. All along this study, the initial pressure, temperature and specific volume correspond to the thermodynamic state 4 for the first MAH case in the article of Fergason *et al* [3] in which the reduced (normalized by the critical values) pressure and specific volume are respectively equal to 0.9866 and 1.5733. All decaying HIT cases proposed in this study are performed for an initial Taylor Reynolds number equal to 75 and simulations are initiated using an incompressible homogeneous isotropic turbulent velocity field following the Passot-Pouquet spectrum with k_e the wavelength of maximum initial kinetic energy defined by the relationship $Re_\lambda = 2u_{rms}/(\nu k_e)$. Given the low speed of sound obtained using the thermal and caloric Martin&Hou [4] Equation of State (EoS) for the chosen thermodynamic point, a 20 m/s turbulent velocity translates into a turbulent Mach number equal to 0.8. This case, referred to as case MAH1 from now on, is the main physical case of interest in the present study. In order to assess the influence of the average thermodynamic state, an additional test case is performed using the Martin&Hou EoS for the same turbulent velocity and Taylor Reynolds number but for a lower reduced pressure equal to 0.1. In this thermodynamic region much further away from the saturation curve, the speed of sound is larger and this translates into a smaller turbulent Mach number equal in this case to 0.3. This test case will be referred to as MAH2. Additionally, in order to assess the influence of the EoS, two cases using the perfect gas EoS are performed in which for the first one, the turbulent velocity is the same as for MAH1 and for the second one the turbulent Mach number is the same as for MAH1. These cases will be referred to respectively as PG1 and PG2. For cases MAH1 and MAH2, the Equation of state is the one proposed by Martin and Hou [4]. For these cases, the dynamic viscosity as well as the thermal conductivity follow the model proposed by Chung *et al* [5]. For cases PG1 and PG2, the dynamic viscosity follows Sutherland's law and a constant Prandtl number equal to 0.71 is used. Table1 summarizes the characteristics of each of the test cases performed in this study. Note that even if the reduced pressure and temperature are close to unity for both PG1 and PG2 cases, it does not influence the results since the Perfect Gas EoS is used.

DNS are performed using the numerical solver AVBP [6]. The parallel LES and DNS code solves the full compressible Navier-Stokes equations using a two-step time-explicit Taylor-Galerkin scheme (TTGC or TTG4A) for the hyperbolic terms based on a cell-vertex formulation [7]. The schemes provide high spectral resolution as well as low numerical dissipation and dispersion. They also ensure third-order accuracy in space and third-order (TTGC) or

Table 1. Physical characteristics of the DNS test cases

Case name	Turbulent velocity (m/s)	Turbulent Mach Number	Reduced Pressure	Reduced Temperature
MAH1	20	0.8	0.98	1.0015
MAH2	20	0.3	0.1	1.0015
PG1	20	0.32	0.98	1.0015
PG2	63	0.8	0.98	1.0015

fourth-order (TTG4A) accuracy in time [7]. The main difference between TTGC and TTGA4 is linked to the amplification factors of the schemes that show a different behavior on small spatial wave length structures: TTG4A has a more dissipative behavior on small wave lengths compared to TTGC. Nevertheless, TTG4A is less prone to numerical instabilities issued by the centered spatial discretization. These schemes have been extensively validated in the context of Large Eddy Simulation [8, 9, 10] but their precision can be thought of as being in the low limit for performing Decaying HIT Direct Numerical Simulations (DNS). In order to make sure the results are reliable, for each case three simulations are performed using different spatial resolutions. The first computation uses 240^3 cells, the second one 320^3 and the most resolved 400^3. Given the common initial Taylor Reynolds number, the respective resolutions correspond to discretizations of the Kolmogorov length scale with 2, 2.5 and 3 points respectively. Figure 1 displays the temporal evolution of the kinetic energy and enstrophy. The kinetic energy is normalized by its initial value, the enstrophy is normalized by the maximum value reached in the most resolved case and the time is normalized by the characteristic turn-over time which is defined as $\tau_0 = L_i/u_{rms}$ and depends on the case considered. Figure 1 shows that all computations provide exactly the same evolution both for the kinetic energy and for the enstrophy although this last quantity is of a higher order. This result is coherent with the fact that in order to provide that evolution, the Kolmogorov length scale does not need to be perfectly discretized given the fact that only a very small percentage of the kinetic energy remains at this scale [11].

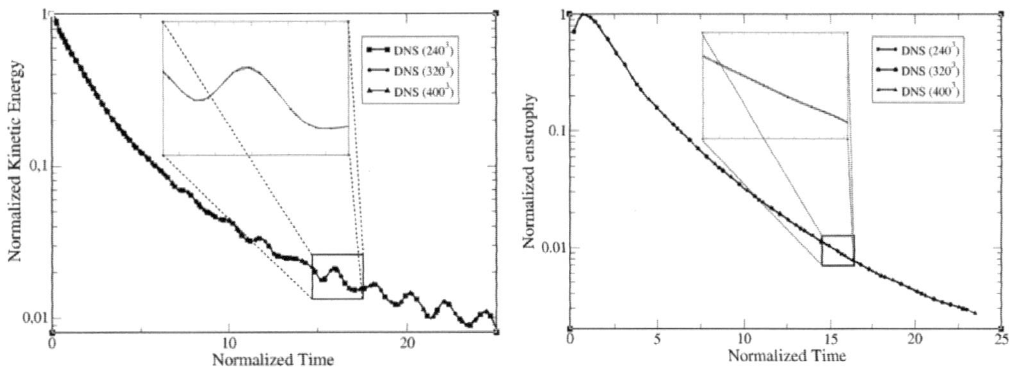

Figure 1. Evolution of the normalized kinetic energy (left) and normalized enstrophy (right) with time in the MAH1 case for three different spatial resolutions (240^3,320^3 and 400^3).

3. Results

3.1. Kinetic Energy and Skewness

The cases corresponding to large initial turbulent Mach number show a transition period before they adapt to the initial incompressible condition as already described in [12, 13]. During this period, shocklets appear in cases MAH1 and PG2. Figure 2 compares the dilatation between the two cases. One can see that because the average thermodynamic state lies within the inversion region for the MAH1 case, regions of very large dilatation are present together with compression shocklets. In the PG2 case, only compression shocklets are visible. This result is coherent with recent findings from Sciacovelli & Cinnella [14].

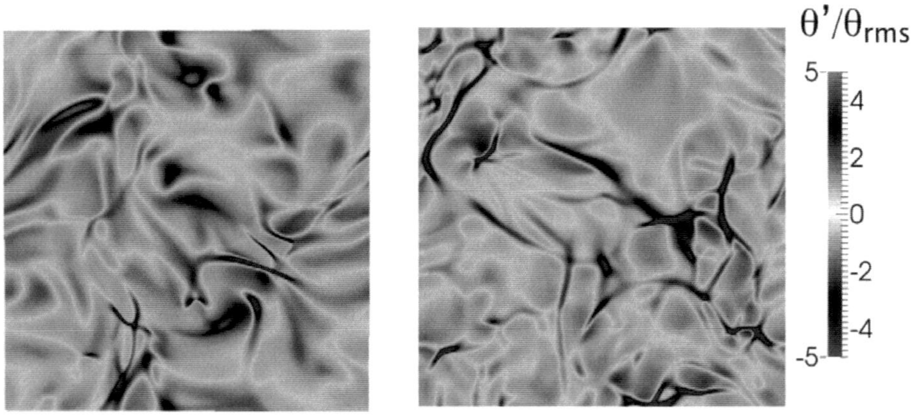

Figure 2. Sectional cut of the flow colored by the local divergence divided by its spatial r-m-s value at time $\tau = 1$. Left:MAH1, Right:PG2.

Figure 3 compares for the four cases of table 1 the evolution of the kinetic energy with respect to time. The first remark that can be made is that the temporal evolution of the kinetic energy depends more on the common value of the turbulent Mach number than on the type of EoS given the fact that the Taylor Reynolds number is identical for all cases. Indeed, cases MAH2 and PG1 show almost identical evolutions of their kinetic energy. In cases MAH1 and PG2, the evolutions are rather similar until $t/\tau = 15$, with a slightly smaller slope afterwards for MAH1. An important aspect for cases MAH1 and PG2 is the observed oscillation of the kinetic energy starting around $t/\tau = 8$ for both cases. This has already been observed by Kida & Orszag [12] and is related to a periodic exchange between the internal and kinetic energies through the pressure gradient term.

Also of interest is the evolution of the skewness of the velocity gradient as this quantity is expected to be statistically constant in decaying HIT. Figure 4 compares the evolution of this statistical quantity in the cases using the Martin&Hou EoS. One can see on the left that in the case MAH2, the isotropy is well preserved and that skewness factors closely match the value of -0.4 experimentally obtained and also predicted by the Heisenberg's transfer theory [15, 16]. In cases MAH1 and PG2, the use of the Passot-Pouquet spectrum, derived from incompressible turbulence studies, induces an adaptation period during which the skewness rapidly evolves.

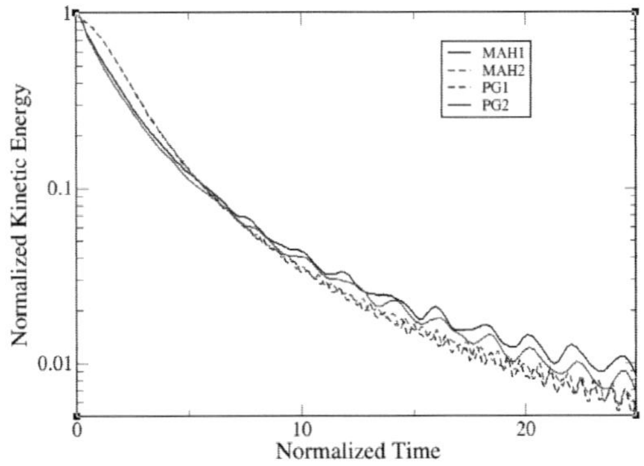

Figure 3. Evolution of the normalized kinetic energy with time.

Once the numerical computation has adapted to the initial solution, the values tend to remain close to the theoretical value of -0.4. One can also note that for $t/\tau > 8$, some departure from isotropy occurs as the three directional skewness factors do not perfectly follow the same temporal evolution. Although no definitive proof is provided here, it is reasonable to assume that this behavior is linked with the observed periodic exchange between the internal and kinetic energies shown in Fig 3, that starts at the same instant in time for case MAH1. Similar results are obtained for cases PG1 and PG2. Regardless of the case considered, it is interesting to note that some stationarity is achieved for this statistical quantity. Together with the already demonstrated grid convergence, it enables the analysis and comparisons of Turbulent Kinetic Energy (TKE) spectra with a fair degree of confidence.

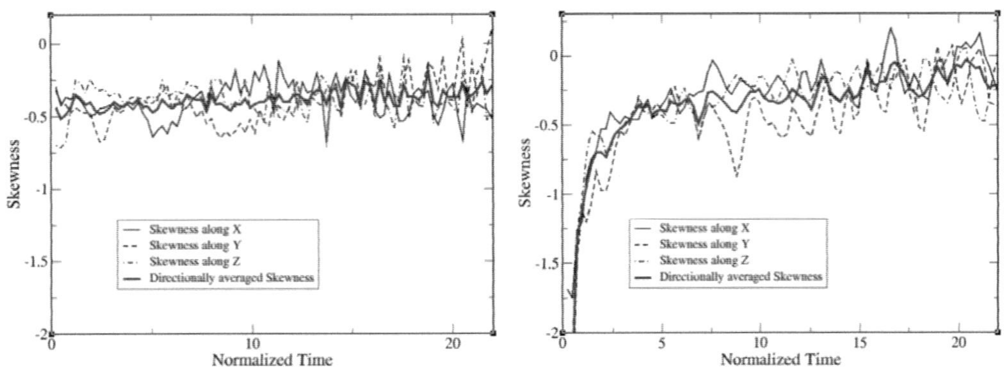

Figure 4. Evolution of the skewness of the velocity gradient with time. Left:MAH2, Right:MAH1

3.2. Spectrum and Thermodynamics

The temporal evolution of the spectra of normalized TKE (i.e $\widehat{E_n} = (\widehat{E_c/\rho}) * \epsilon^{-2/3}k^{5/3}$) is first analyzed for cases PG1 and MAH2. In these cases, the low initial turbulent Mach number

IOP Conf. Series: Journal of Physics: Conf. Series **821** (2017) 012017 doi:10.1088/1742-6596/821/1/012017

leads to an almost incompressible evolution of the turbulence. Therefore, as verified in Figure 5 the non-dimensionalization provides spectra that collapse on each other at all times. Because of the low Taylor Reynolds number used in this study the inertial region of the spectrum is quite narrow. In this region, the retrieved constant is equal to 1.2 which is smaller than the Kolmogorov constant ($\approx 1.5 - 1.6$). From this comparison, one can note that at low turbulent Mach number FC-70, though being a highly compressible fluid, behaves in an incompressible manner as far as turbulence is concerned. This behavior goes together with a monotonic decrease of the kinetic energy as observed in Fig 3.

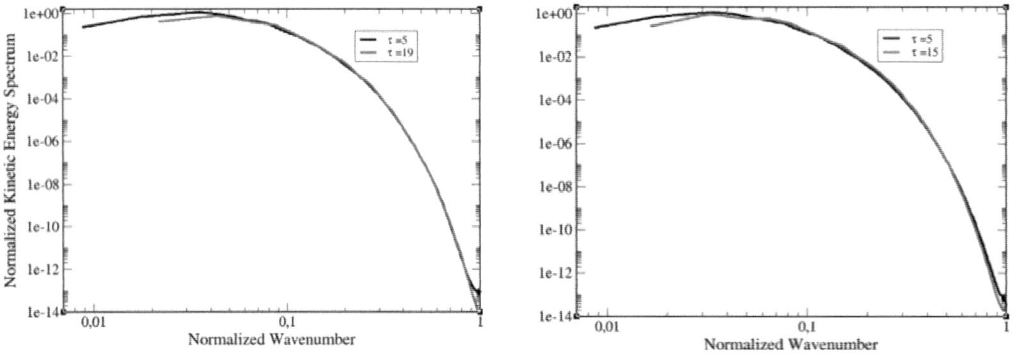

Figure 5. Normalized spectrum of TKE at different instants in time for the case MAH2 (left) and PG1 (right).

Cases MAH1 and PG2 are much more challenging in terms of physical interpretation. Figure 6 presents a comparison of the temporal evolution of the normalized TKE spectra for the two cases. First, one can observe that the overall levels are quite different between MAH1 and PG2. The PG2 case shows a much more energetic content at small scales. For example, the non dimensional TKE level of 10^{-4} is reached by structures only 1.5 times larger than the Kolmogorov length whereas such levels in the MAH1 case correspond to turbulent structures two times larger. Comparing Figure 6 with Figure 5 results, it is interesting to see that as far as the overall TKE spectrum level is concerned, the MAH1 case is much closer to the MAH2 and PG1 cases than the PG2 case.
Another important observation can be made from analyzing Figure 6. For both MAH1 and PG2 cases, the normalized spectra do not collapse on each other for length scales smaller than approximately 5 Kolmogorov length scales. Depending on the time at which the normalized spectrum of TKE is observed, relative levels fluctuate in the region of small but yet well resolved length scales. Grid convergence analysis applied to the TKE spectra (not shown here for brevity reasons) clearly demonstrate that these fluctuations are not produced by some numerical dissipation issue but are most probably connected to physical phenomena taking place at these scales. Because this behavior occurs for the two cases also showing temporal oscillations of the kinetic energy (see Fig 3) one can reasonably suppose that there is a connection between the two features of the turbulence. Yet, the fact that the normalized spectra do not collapse over time in the region of small length-scales ($l < 5l_\eta$) also shows that an additional parameter should be considered during the nondimensionalization process in addition to the wavenumber k and the dissipation ϵ.

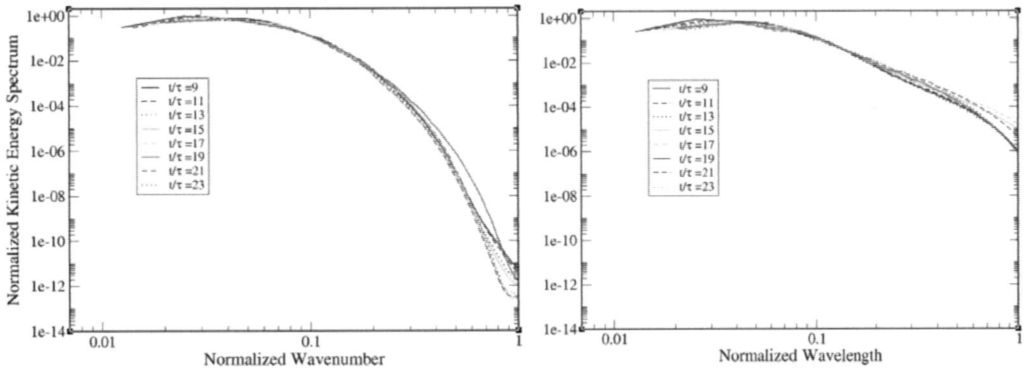

Figure 6. Normalized spectrum of TKE at different instants in time for the case MAH1 (left) and PG2 (right).

Also of interest for case MAH1 is the temporal evolution of the distribution of thermodynamic states. One can see in Figure 7 the evolution of the curves formed, at different reduced times in the simulation, by the maxima of the reduced pressure distribution for each reduced volume. Most represented thermodynamic states closely follow the adiabatic curve passing through the initial conditions. Because static thermodynamic conditions are of course influenced by the kinetic energy, they lie inside the inversion region but also cross the limit $\Gamma = 0$. For a large part of the total temporal extension of the simulation ($t/\tau < 30$) there is a coexistence of both types of thermodynamic conditions. Although it is not further analyzed here, it is interesting to note that such a distribution might include very close points in physical space being in highly different thermodynamic states, for instance one being in the inversion region and the other outside. In this case, judging from previous inviscid results obtained by Fergason *et al* [3], one would expect to detect the presence of expansion shocklets as seemingly observed in Fig 2. Finally, as time goes by, the initially spread distribution slowly collapses back to the initial single mean thermodynamic point though with a slightly larger temperature corresponding to the dissipation of the entire kinetic energy into internal energy.

3.3. A-posteriori Large eddy simulation

The previous sections show that given a realistic intensity, the behavior of turbulence in dense gas flows is significantly modified by the thermodynamic state of the fluid. It is therefore interesting to assess the present capability of usual LES subgrid-scale models to capture the evolution of the kinetic energy in dense gas flows. To do so, the dynamic Smagorinsky model is used to simulate cases MAH1, MAH2, PG1 and PG2 with two different levels of spatial resolution. A first resolution corresponding to 32^3 elements is used. In this case, the grid spacing corresponds to 4 Kolmogorov length scales. This resolution, being already quite fine, leads to grid filtering lying in the dissipation range which is not representative of LES requirements. Therefore a second resolution is considered comprising 16^3 elements. For this last resolution, the grid spacing corresponds to 8 Kolmogorov length scales which although still not in the inertial range makes it more representative of LES requirements. Figure 8 shows that for cases MAH2 and PG1, LES is able to correctly capture the temporal evolution of the kinetic energy with a good accuracy. This confirms that when the turbulent Mach number is low enough, thermodynamics does not

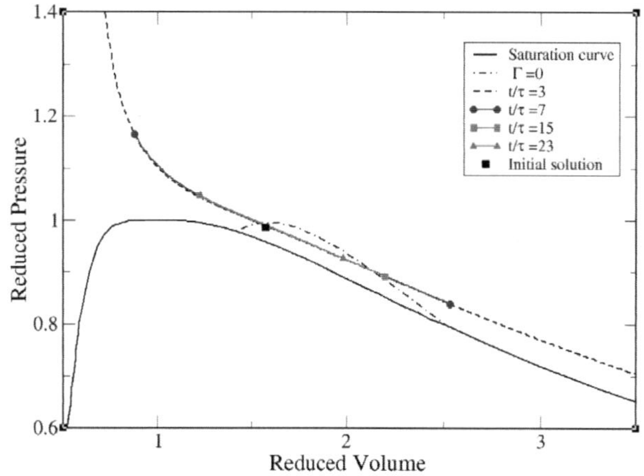

Figure 7. Distribution of the thermodynamic states in the $p_r - v_r$ diagram in the case MAH1 at different instants in time.

significantly influence turbulence. In these cases, the standard dynamic Smagorinsky subgrid-scale model is good enough to study turbulent dense gas flows using LES. For this condition to be verified though, the turbulent velocity needs to be small enough compared to the speed of sound. This is where the thermodynamic inversion region plays a key role. Indeed, when the average thermodynamic state lies within this inversion region, the speed of sound is drastically decreased and the same turbulent speed (here 20 m/s) leads to a highly compressible evolution of the turbulence. As it can be seen in Figure 9 for case MAH1, both LES simulations cannot capture the right evolution of the kinetic energy. One can first note that no oscillations of the kinetic energy are visible for the realistic resolution of 16^3. This is coherent with the fact that these fluctuations relate to length scales smaller than 5 Kolmogorov length scales as suspected from the time evolution of the spectra in Figure 6. More importantly, even for the 32^3 resolution, LES fails to obtain the right slope of the curve of the kinetic energy in the case MAH1. Figure 9 also presents the same comparison for the case PG2. One can see that kinetic energy oscillations also disappear for the resolution comprising 16^3 elements. Yet, in this case LES results are much closer to DNS once the initial highly compressible phase of the simulation is over. After this transitional phase, LES seems to correctly predict the slope of the curve of the kinetic energy in contrast with what is observed for case MAH1.

4. Concluding remarks

A study of turbulence in BZT dense gas flows is performed using DNS. It is shown that for a large but realistic intensity (≈10% of a 200m/s average bulk velocity), the turbulence behaves in a highly compressible manner when the average thermodynamic state lies within the inversion region. As a consequence, similarities between the temporal evolutions of the kinetic energy are found between the case using the Martin & Hou Equation of State (EoS) and the perfect gas EoS when in addition of the Taylor Reynolds number, the turbulent Mach number is kept constant. Yet, in order to ensure such matching values, the turbulence intensity has to be increased in the perfect gas case to unrealistic levels (that might only be of interest for

IOP Conf. Series: Journal of Physics: Conf. Series **821** (2017) 012017 doi:10.1088/1742-6596/821/1/012017

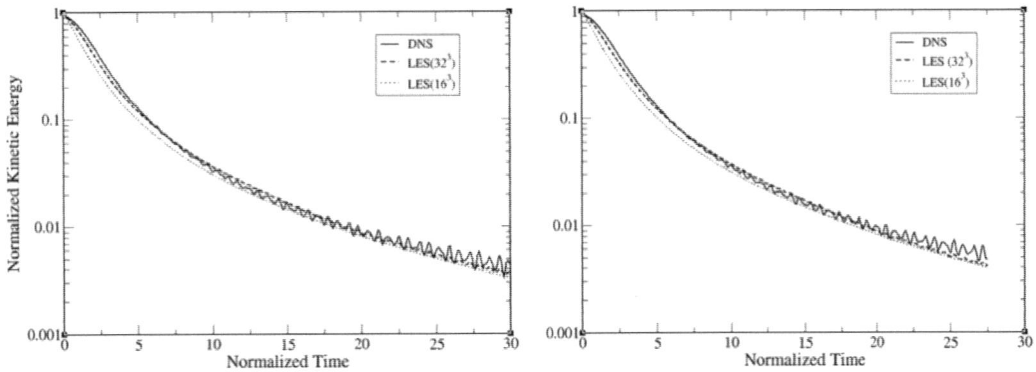

Figure 8. Evolution of the normalized kinetic energy with time with comparison between LES and DNS in the case MAH2(left) and PG1 (right).

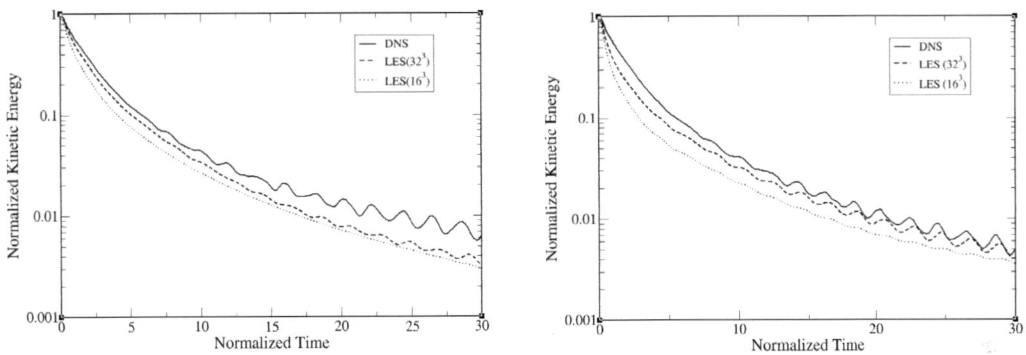

Figure 9. Evolution of the normalized kinetic energy with time with comparison between LES and DNS in the case MAH1(left) and PG2 (right).

interstellar flows). It is therefore expected that turbulent flows of dense gases will more easily present a characteristic compressible behavior that needs to be addressed through modeling if accurate Large Eddy Simulations (LES) of those flows are to be performed. In order to assess the capabilities of present LES subgrid-scale models, a-posteriori tests are performed using the dynamic Smagorinsky model. Coherently with the hypothesis it relies on, the model perfectly captures the evolution of the kinetic energy when the turbulent Mach number is low enough. When using the perfect gas EoS at a higher turbulent Mach number, the agreement is reasonable. Yet, for the dense gas case, when the average thermodynamic state lies within the inversion region, the model is not able to capture the correct evolution of the kinetic energy. Because of the renewed interest in this thermodynamic region, such turbulent flows will certainly be encountered in the future when using dense gases. Therefore, the results of this study call for a specific research effort assessing available subgrid-scale models and possibly developing advanced models for LES of turbulent dense gas flows (as for example the one proposed by Kosovic *et al* [17]). This research should also enable a deeper understanding of the phenomena at play in dense gas turbulence (such as for example the possible occurring of backscatter [18]).

Acknowledgement
The authors would like to thank Pr. Laadhari for insightful discussions on the theory of homogeneous isotropic turbulence. Simulation are carried out using HPC ressources at CINES under the project grant #c2016a7564.

References

[1] P. A. Thompson, "A Fundamental Derivative in Gasdynamics," *Physics of Fluids*, vol. 14, no. 9, p. 1843, 1971.
[2] A. Guardone and B. M. Argrow, "Nonclassical gasdynamic region of selected fluorocarbons," *Physics of Fluids*, vol. 17, no. 11, pp. 1–17, 2005.
[3] S. Fergason, T. Ho, B. M. Argrow, and G. Emanuel, "Theory for producing a single-phase rarefaction shock wave in a shock tube," *Journal of Fluid Mechanics*, vol. 445, pp. 37–54, 2001.
[4] J. J. Martin and Y.-c. Hou, "Development of an Equation of State for Gases," *AIChE journal*, vol. 2, no. 4, pp. 142–151, 1955.
[5] T. H. Chung, M. Ajlan, L. L. Lee, and K. E. Starling, "Generalized multiparameter correlation for nonpolar and polar fluid transport properties," *Industrial & Engineering Chemistry Research*, vol. 27, pp. 671–679, apr 1988.
[6] T. Schönfeld and M. Rudgyard, "Steady and Unsteady Flow Simulations Using the Hybrid Flow Solver AVBP," *AIAA Journal*, vol. 37, pp. 1378–1385, nov 1999.
[7] O. Colin and M. Rudgyard, "Development of High-Order TaylorGalerkin Schemes for LES," *Journal of Computational Physics*, vol. 162, pp. 338–371, aug 2000.
[8] G. Staffelbach, L. Gicquel, G. Boudier, and T. Poinsot, "Large Eddy Simulation of self excited azimuthal modes in annular combustors," *Proceedings of the Combustion Institute*, vol. 32, no. 2, pp. 2909–2916, 2009.
[9] L. Gicquel, G. Staffelbach, and T. Poinsot, "Large Eddy Simulations of gaseous flames in gas turbine combustion chambers," *Progress in Energy and Combustion Science*, vol. 38, pp. 782–817, dec 2012.
[10] A. Giauque, L. Selle, L. Gicquel, T. Poinsot, H. Buechner, P. Kaufmann, and W. Krebs, "System identification of a large-scale swirled partially premixed combustor using LES and measurements," *Journal of Turbulence*, vol. 6, p. N21, jan 2005.
[11] P. Moin and K. Mahesh, "DIRECT NUMERICAL SIMULATION: A Tool in Turbulence Research," *Annual Review of Fluid Mechanics*, vol. 30, no. 1, pp. 539–578, 1998.
[12] S. Kida and S. A. Orszag, "Energy and Spectral Dynamics in Decaying Compressible Turbulence," *Journal of Scientific Computing*, vol. 7, no. 1, 1992.
[13] R. Samtaney, D. I. Pullin, and B. Kosović, "Direct numerical simulation of decaying compressible turbulence and shocklet statistics," *Physics of Fluids*, vol. 13, no. 5, pp. 1415–1430, 2001.
[14] L. Sciacovelli and P. Cinnella, "Numerical Simulation of Dense Gas Compressible Homogeneous Isotropic Turbulence," in *15th European Turbulence Conference*, (Delft), 2015.
[15] W. H. Reid and D. L. Harris, "Similarity Spectra in Isotropic Turbulence," *Physics of Fluids*, vol. 2, p. 139, nov 1959.
[16] G. K. Batchelor, *The Theory of Homogeneous Turbulence*. Cambridge University Press, 1953.
[17] B. Kosović, D. I. Pullin, and R. Samtaney, "Subgrid-scale modeling for large-eddy simulations of compressible turbulence," *Physics of Fluids*, vol. 14, no. 4, pp. 1511–1522, 2002.
[18] J. O'Brien, J. Urzay, M. Ihme, P. Moin, and a. Saghafian, "Subgrid-scale backscatter in reacting and inert supersonic hydrogen-air turbulent mixing layers," *Journal of Fluid Mechanics*, vol. 743, pp. 554–584, 2014.

DNS of turbulent flows of dense gases

L Sciacovelli[1,2], P Cinnella[1], X Gloerfelt[1] and F Grasso[1,3]

[1]Laboratoire DynFluid, Arts et Métiers ParisTech, 75013 Paris, FR
[2]Dipartimento di Meccanica, Matematica e Management, Politecnico di Bari, 70125 Bari, IT
[3]Laboratoire DynFluid, Conservatoire National des Arts et Métiers, 75003 Paris, FR

E-mail: luca.sciacovelli@ensam.eu, paola.cinnella@ensam.eu, xavier.gloerfelt@ensam.eu, francesco.grasso@cnam.fr

Abstract. The influence of dense gas effects on compressible turbulence is investigated by means of numerical simulations of the decay of compressible homogeneous isotropic turbulence (CHIT) and of supersonic turbulent flows through a plane channel (TCF). For both configurations, a parametric study on the Mach and Reynolds numbers is carried out. The dense gas considered in these parametric studies is PP11, a heavy fluorocarbon. The results are systematically compared to those obtained for a diatomic perfect gas (air). In our computations, the thermodynamic behaviour of the dense gases is modelled by means of the Martin-Hou equation of state. For CHIT cases, initial turbulent Mach numbers up to 1 are analyzed using mesh resolutions up to 512^3. For TCF, bulk Mach numbers up to 3 and bulk Reynolds numbers up to 12000 are investigated. Average profiles of the thermodynamic quantities exhibit significant differences with respect to perfect-gas solutions for both of the configurations. For high-Mach CHIT, compressible structures are modified with respect to air, with weaker eddy shocklets and stronger expansions. In TCF, the velocity profiles of dense gas flows are much less sensitive to the Mach number and collapse reasonably well in the logarithmic region without any special need for compressible scalings, unlike the case of air, and the overall flow behaviour is midway between that of a variable-property liquid and that of a gas.

1. Introduction

The physical understanding of the turbulence dynamics of dense gas flows is relevant for several engineering systems, like high-Reynolds wind tunnels [1], chemical transport and processing [2], energy conversion cycles [3] and refrigeration [4]. Dense gases (DG) are single-phase fluids with a molecular complexity such that the fundamental derivative of gas dynamics [5] $\Gamma := 1 + \frac{\rho}{c} \left.\frac{\partial c}{\partial \rho}\right|_s$ (where ρ is the density, p the pressure, s the entropy, and c the sound speed), which measures the rate of change of the sound speed in isentropic transformations, is less than one or even negative in a range of thermodynamic conditions close to the saturation curve. In such conditions, the speed of sound increases in isentropic expansions and decreases in isentropic compressions, unlike the case of perfect gases. Dense gas effects are expected to give the most interesting phenomena for a family of heavy polyatomic compounds, named Bethe–Zel'dovich–Thompson (BZT) fluids [5], which exhibit a region of negative Γ values (called the "inversion zone" [6]) in the vapour phase close to the liquid/vapour coexistence curve. This is theoretically predicted to result in non-classical compressibility effects in the transonic and supersonic flow regimes, like rarefaction shock waves, mixed shock/fan waves, and shock splitting [6, 7]. Furthermore, the entropy change across a weak shock is much weaker than usual for dense gases with $\Gamma << 1$, leading to reduced shock losses. Numerical simulations of turbulent dense gas flows of engineering interest are based on the well-known RANS equations, which need to be supplemented by a model for the Reynolds stress

	$\mathcal{M}$ g mole^{-1}	T_c K	ρ_c kg m^{-3}	p_c MPa	Z_c -	ω -	d.m. D	T_b K	$c_v(T_c)/R$ -	n -
PP11	624.11	650.2	627.14	1.46	0.2688	0.4963	0.0	488.15	97.3	0.5776

Table 1: Thermophysical properties of PP11: molecular weight ($\mathcal{M}$), critical temperature (T_c), density (ρ_c) and pressure (p_c), critical compressibility factor (Z_c), acentric factor (ω), dipole moment (d.m.) of the gas phase, boiling temperature (T_b), ratio of ideal-gas specific heat at constant volume over the gas constant ($c_v(T_c)/R$) at the critical point and power-law exponent for the low-density specific heat (n).

tensor. The accuracy of RANS models for dense-gas flows has not been properly assessed up to date, due to the lack of both experimental and numerical reference data. The creation of reliable and thorough DNS databases is then needed to quantify the deficiencies of existing turbulence models and to develop and calibrate improved models. Additionally, DNS may improve our understanding of modifications to compressible turbulence due to the fluid in use. Indeed, for dense gases, the perfect gas (PFG) model is no longer valid, and more complex equations of state (EoS) must be used to account for their peculiar thermodynamic behaviour. Moreover, in the dense gas regime, the dynamic viscosity μ and the thermal conductivity κ depend on temperature and pressure through complex relationships. Similarly, the approximation of nearly constant Prandtl number ($Pr = \mu c_p/\kappa \approx$ const) is no longer valid. In the following, we present recent DNS results for the decay of compressible homogeneous isotropic turbulence (CHIT), and of turbulent plane channel flow (TCF), representative of free and wall-bounded turbulence, respectively, and we highlight the most striking differences with respect to perfect-gas turbulence.

2. Governing Equations and Numerical Method

In the present study we consider flows of gases in the single-phase regime, governed by the compressible Navier-Stokes equations. The gas behaviour is modelled through the Martin–Hou (MAH) thermal equation of state [8], which provides a realistic description of the gas behaviour and of the extent of the inversion zone. This equation, involving five virial terms and satisfying ten thermodynamic constraints, is reasonably accurate for the fluid of interest and requires a minimum amount of experimental information. A power law of the form $c_{v_\infty} = c_{v_\infty}(T_c)(T/T_c)^n$ is used to model variations of the low-density specific heat with temperature, where n is a material-dependent parameter. In addition to the EoS, thermodynamic models relating the dynamic viscosity μ and thermal conductivity κ to the gas temperature and pressure have to be specified. In the following calculations, the transport properties follow the Chung-Lee-Starling model [9], which incorporates a correction term in the dense-gas region. For air cases, the classical polytropic perfect gas model is used along with a temperature-dependent power law for the viscosity, $\mu = \mu_{ref}(T/T_{ref})^{0.7}$, and a constant Prandtl assumption, $Pr = 0.7$. The working fluid is perfluoro-perhydrophenanthrene (chemical formula $C_{14}F_{24}$), called hereafter with its commercial name PP11; its thermodynamic properties are provided in table 1. This fluid has been often used in the dense-gas literature since it exhibits a relatively wide inversion zone and, as a consequence, may lead to the appearance of non-classical effects.

The space derivatives in the governing equations are approximated by means of optimized finite difference schemes. Specifically, the convective terms are approximated by fourth-order optimized finite differences with an eleven-point stencil in each mesh direction [10], and the viscous terms by standard fourth-order finite differences. As a part of the algorithm, an optimized selective sixth-order filter also using an eleven-point stencil is applied in each direction to damp grid-to-grid oscillations in smooth flow regions. In order to ensure computational robustness for compressible flows with shocks, an adaptive nonlinear selective filtering strategy is employed [10].

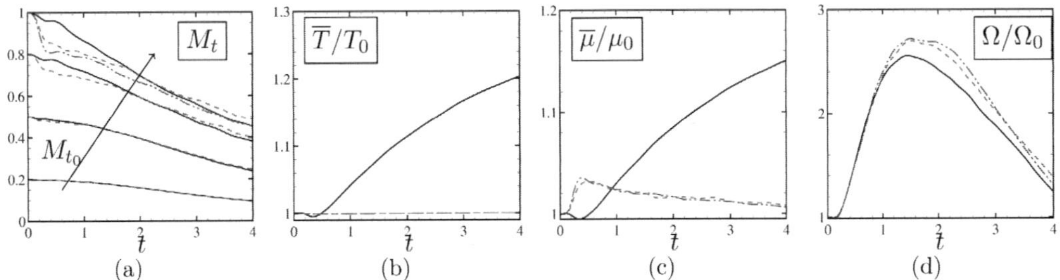

Figure 1: Decaying CHIT results. Time histories of the turbulent Mach number M_t at various M_{t_0} (panel a). Panels (b) to (d): normalized averaged temperature T/T_0, normalized viscosity $\overline{\mu}/\mu_0$ and normalized enstrophy Ω/Ω_0 at $M_{t_0} = 1$. ——: Air; - - - - -: PP11 IC1; -·-·-·-: PP11 IC2.

Finally, time integration is carried out by means of a low-storage six-step Runge-Kutta scheme optimized in the wave number space [11].

3. Decaying homogeneous isotropic turbulence

The CHIT decay problem is solved on a cubic domain of $[0, 2\pi]^3$. Periodic boundary conditions are applied in the three Cartesian directions. The initial turbulent kinetic energy spectrum is of Passot-Pouquet type, i.e., $E(k) = Ak^4 \exp\left[-2\left(k/k_0\right)^2\right]$, where k_0 is the peak wavenumber and A is a constant that depends on the initial amount of kinetic energy. The initial velocity field is purely solenoidal and the r.m.s. of the thermodynamic variables are set to zero. Compressible turbulence decay is governed by the turbulent Mach number $M_t = \sqrt{\overline{u_i^2}}/\overline{c}$ and the Reynolds number $Re_\lambda = u'\lambda\overline{\rho}/\overline{\mu}$, based on the Taylor microscale λ. We performed direct simulations of dense-gas CHIT at various initial turbulent Mach numbers, namely $M_{t_0} = 0.2$, 0.5, 0.8 and 1. The initial Reynolds number is set to $\mathrm{Re}_\lambda = 200$ and the initial peak wavenumber k_0 is equal to 2 for all simulations. The decay is simulated up to $t = 4$ (with t a non-dimensional time scale based on the initial large eddy turnover time). All simulations were carried out on a regular Cartesian grid of 512^3. In the most severe case (i.e. $M_{t_0} = 1$), the smallest resolved wavelength $k_{max}\eta$ (being k_{max} the largest wavenumber and $\eta = (\overline{\mu}^3/\overline{\rho}^2\overline{\epsilon})^{1/4}$ the Kolmogorov lengthscale [12]) is always greater than $\approx 6.5 - 7$, and the Kolmogorov lengthscale is discretized by at least 2 mesh points, which corresponds to a very good resolution. Dense gas simulations are reported for two choices of the initial thermodynamic state, corresponding to a point in the dense gas region ($\rho_0/\rho_c = 1.618$, $p_0/p_c = 1.02$, $T_0/T_c = 1.01$, $\Gamma_0 = 0.1252$, referred-to as IC1) and one inside the inversion zone ($\rho_0/\rho_c = 1.618$, $p_0/p_c = 0.98$, $T_0/T_c = 1.001$, $\Gamma_0 = -0.093$, IC2), for which the flow may exhibit BZT effects. Since entropy can only increase during the flow decay, no BZT effects can appear for the first choice of conditions. Results were systematically compared with those of a diatomic perfect gas (air). Compressibility effects are first investigated by comparing the time evolutions of relevant flow statistics for PFG and dense-gas simulations at various M_{t_0} (figure 1a). The PFG and DG solutions are practically superposed at $M_{t_0} = 0.2$ and progressively deviate as M_{t_0} increases, due to the different compressibility and transport properties of the two fluids. For the sake of brevity, in the following we restrict our analysis to the most representative case, i.e., $M_{t_0} = 1$. The average temperature (figure 1b) increases significantly in time for PFG, and viscosity (figure 1c) follows the same behavior. On the contrary, for DG cases the temperature remains approximately constant due to the much higher specific heat, and the time evolution of viscosity (which depends on density variations) exhibits a peak at the beginning of the decay and decays afterwards, i.e. behaves at the opposite of the PFG. The normalized turbulent kinetic energy K/K_0 (not reported for brevity) and enstrophy Ω/Ω_0

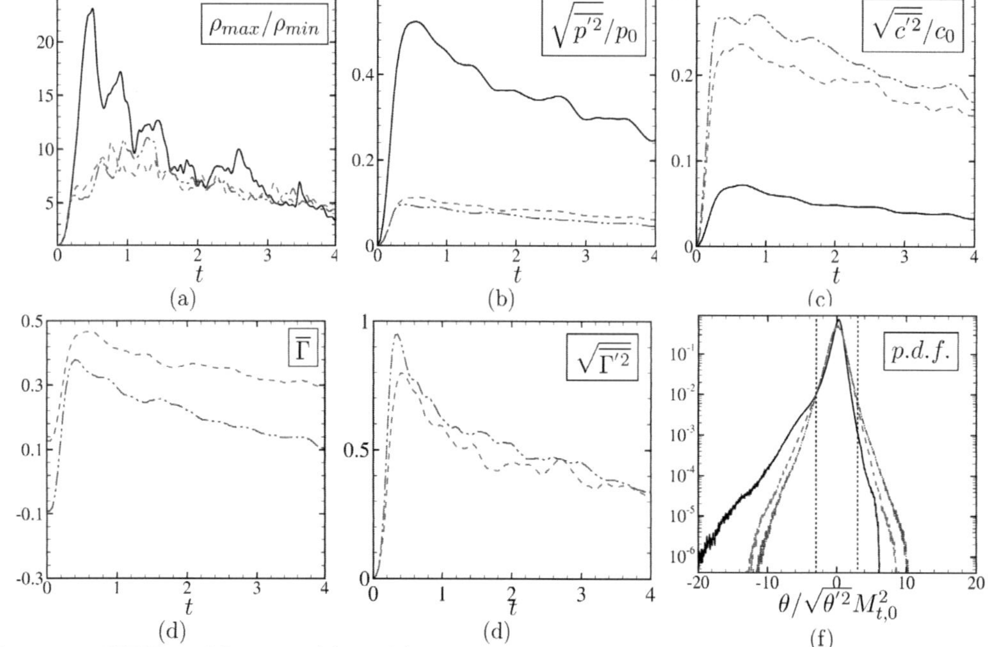

Figure 2: CHIT at $M_{t_0} = 1$. (a) to (c): time evolutions of the maximum to minimum density ratio ρ_{max}/ρ_{min}, the r.m.s. pressure $\sqrt{\overline{p'^2}}/p_0$ and the r.m.s. speed of sound $\sqrt{\overline{c'^2}}/c_0$; (d) to (f): time evolutions of the average $(\overline{\Gamma})$ and r.m.s. $(\sqrt{\overline{\Gamma'^2}})$ of the fundamental derivative , and p.d.f. of the normalized dilatation $\theta/\sqrt{\theta'^2}M_{t,0}^2$ at nondimensional time $t = 2$. Legend in figure 1.

(figure 1d) are similar for PFG and DG cases. An enstrophy peak is observed for $t \approx 1.4 - 1.5$, which is slightly higher for the dense gas, due to the lower average viscosity compared to PFG. Similar behaviours are observed for IC1 and IC2.

Time histories of the fluctuating thermodynamic quantities are presented in figure 2 (top line). The maximum to minimum instantaneous density ratio increases during the initial supersonic phase, driven by vortex-stretching mechanisms with the production of strong velocity gradients, reaching a maximum for a nondimensional time comprised between 1 and 2. In the dense gas, the maximum value is however about 2.5 times lower than in the perfect gas due to the formation of weaker eddy shocklets. Then the ratio decays and both fluids exhibit similar behaviours at later times. Similarly, pressure fluctuations (figure 2b) in the dense gas are about five times lower for PP11, independently on the choice of the initial thermodynamic state. Conversely, the r.m.s. values of the speed of sound are higher for the dense gas. For this fluid, $c = c(\rho, T)$ follows essentially the density variations (the temperature being almost constant throughout the decay). This mechanism is deeply different from the one observed for PFG, where the speed of sound increases due to friction heating. Time evolutions of the average and r.m.s fundamental derivative of gas dynamics are also reported for PP11 at various initial conditions. The average value $\overline{\Gamma}$ (figure 2d) follows the average density variations. For both IC1 and IC2, the initial development of compressible modes increases the average Γ, which however remains below 1 throughout the evolution (implying that the speed of sound tends to decrease in compressed regions) and then decays. The r.m.s. of Γ is as high as 1 at $t \approx 1$ meaning that, for high Mach number flows, the pointwise values of Γ are widely spread around the average and may be locally greater than 1 or, for IC2, lower than 0. Despite the different local behaviour of Γ, the macroscopic flow statistics are little affected and tend to behave similarly for both initial conditions.

Figure 3 shows the distribution of the flow thermodynamic states in the $p-v$ diagram at time $t = 2$, for IC1 at various M_{t_0} and for IC2 at $M_{t_0} = 1$. Viscosity and shock losses being small, all states tend to distribute along the initial isentrope. The thermodynamic region spanned by the $p - v$ states throughout the decay increases with M_{t_0}. For $M_{t_0} = 1$, the slight change in the initial thermodynamic conditions drastically changes the maximum pressure achieved. The specific volume, indeed, is bounded by a lower limit corresponding to the liquid/gas transition (respectively, an upper limit for density), as the flow undergoes strong compressions. On the other hand, strong expansions are favoured by the dense gas behaviour, since the speed of sound is increased and compressibility decreases. The dilatation regions are classified according to the local level of normalized dilatation $\theta/\sqrt{\theta'^2}M_{t_0}^2$ [13]. Regions with $0 < |\theta/\sqrt{\theta'^2}M_{t_0}^2| < 2$ are referred to as weak expansions/compressions, whereas $|\theta/\sqrt{\theta'^2}M_{t_0}^2| > 2$ denotes strong dilatation/compression motions. Additionally, it is commonly acknowledged that compression shocks may be formed for $\theta/\sqrt{\theta'^2}M_{t_0}^2 < -3$ [14]. The differences in the Clapeyron diagram are confirmed by the probability distribution functions (p.d.f.s) of the normalized dilatation shown in figure 2f. The dense gas exhibits a much more symmetric p.d.f. than PFG, which is characterized by a long left tail, according to similar results in the literature [15]. The greater balance between the left and right tails of the p.d.f. in the dense gas are due to the significant weakening or suppression of compression shocks in flow regions where Γ is close or less than zero; on the other hand, expansions are favoured, leading to a heavier right tail. This is more developed for IC2, due to the possibility of BZT effects leading to the appearance of expansion shocklets. At $t = 2$, approximately one half of the flow volume is characterized by thermodynamic states enclosed under the transition line. Among these states, approximately 36% consist of weak dilatation regions and ≈ 6% of strong expansion regions. Globally, strong compression regions decrease from 3% to 2% whereas strong expansions increase from 1.7% to 2.6%. More details about dense gas CHIT and its small-scale behaviour are given in [16].

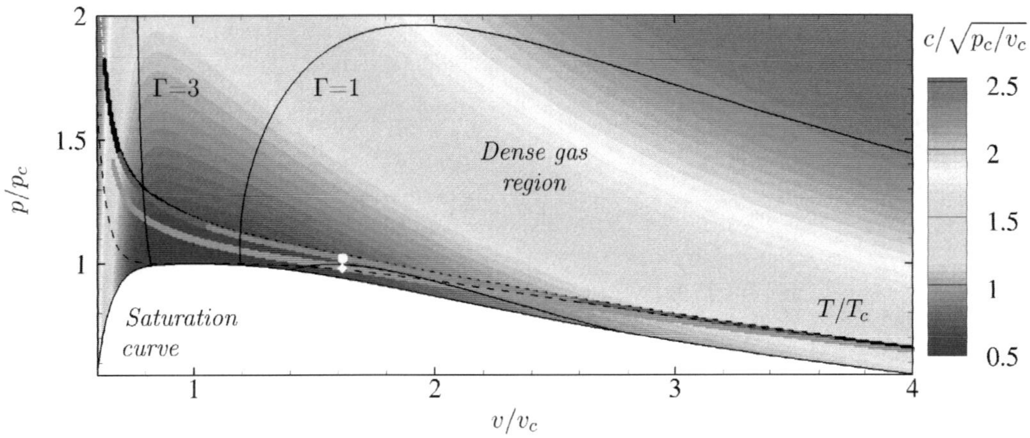

Figure 3: Distributions of the thermodynamic states in the Clapeyron diagram for PP11 at $t = 2$ for different M_{t_0} and thermodynamic initial conditions. Isocontours of the normalized speed of sound are plotted. White symbols denotes initial conditions: ○ PP11 IC1, ◇ PP11 IC2. For condition IC1: ● $M_{t_0} = 0.5$, ● $M_{t_0} = 0.8$, ● $M_{t_0} = 1$. For condition IC2: ● $M_{t_0} = 1$.

Table 2: Numerical parameters and DNS results for air and PP11 cases.

	air				PP11			
$M_{B,w}$	1.50		3.0		1.50		3.0	
$Re_{B,w}$	3000	7000	7000	12000	3000	7000	7000	12000
Name	AM1R3	AM1R7	AM3R7	AM3R12	PM1R3	PM1R7	PM3R7	PM3R12
Line	-·-·-·	········	-··-··-	——	-·-·-·	········	-··-··-	——
N_x	512	768	1024	1536	512	768	768	1280
N_y	256	384	512	768	256	384	384	768
N_z	256	512	768	1024	256	512	512	1024
Re_τ	218.7	466.5	627	1017	191.3	401.6	425.1	692.8
Re_{τ^*}	147.1	314.8	199.7	324.7	196.4	412.1	492.1	800.2
Δx^*_c	7.2	10.3	4.9	5.3	9.7	13.5	16.1	15.7
Δz^*_c	3.6	5.7	1.6	2.0	4.8	5.1	6.1	4.9
$\Delta y^*_{c,w}$	0.52	0.79	0.25	0.26	0.79	0.66	0.79	0.8
Δy^*_{CL}	2.2	5.5	1.8	2.0	2.7	5.0	6.0	4.3
$\overline{M}_{CL}$	1.5	1.47	2.18	2.16	1.62	1.60	2.61	2.58
$\overline{Re}_{CL}$	2740	6319	4035	6903	3659	8423	10256	17635
$\overline{T}_{CL}/\overline{T}_w$	1.39	1.39	2.59	2.58	1.00	1.00	1.01	1.01
$\overline{\rho}_w/\overline{\rho}_B$	1.36	1.37	2.49	2.51	1.05	1.05	1.25	1.25

4. Supersonic Turbulent channel flow

The TCF configuration is now considered. Periodic conditions are applied in the streamwise (x) and spanwise (z) directions. Isothermal no-slip wall conditions are applied on the lower and upper walls. In order to counteract viscous friction and maintain the target bulk mass flow, a spatially constant body force is applied in the streamwise direction [17]. In the following, the subscripts $(\cdot)_B$, $(\cdot)_w$ and $(\cdot)_{CL}$ denote time and space averaged values over the channel cross-section, at the wall and at the centerline, respectively; $\overline{(\cdot)}$ indicates Reynolds averaging and $(\cdot)'$ Reynolds fluctuations; similarly, $\widetilde{(\cdot)}$ and $(\cdot)''$ denote Favre averages and fluctuations. Flow conditions are defined by imposing the bulk Reynolds number $Re_{B_w} := \frac{\overline{\rho}_B \widetilde{u}_B h}{\overline{\mu}_w}$, and bulk Mach number $M_{B_w} := \frac{\widetilde{u}_B}{\overline{c}_w}$. Since isothermal conditions are applied at the walls, $\overline{T}_w = T_w$ =const. For the PFG model, fixing the reference wall temperature allows to fix the sound speed and transport properties accordingly. For dense gas cases, the latter quantities depend on both temperature and density, and thus their values change during the simulation because $\overline{\rho}_w$ cannot be fixed a priori. For this reason we used an iterative procedure on the wall temperature in order to match the prescribed Reynolds and Mach numbers. A parametric study is carried out at various bulk Reynolds numbers (Re_{B_w} = 3000, 7000 and 12000) and two bulk Mach numbers (M_{B_w} = 1.5 and 3.0). The computational domain has dimensions $L_x \times L_y \times L_z = 8\pi h \times 2h \times 2\pi h$ (being h the channel half-height); the selected computational grids ensure a spatial resolution comparable to other DNS of TCF [18, 19, 20, 21, 17], i.e., $\Delta x^* \in [5, 16]$, $\Delta y^*_w \in [0.5, 0.8]$, $\Delta y^*_{CL} \in [2, 6]$, $\Delta z^* \in [2, 6]$. The overall number of grid points varies between $33 \cdot 10^7 \div 1.2 \cdot 10^9$ for the various cases. The main numerical parameters and global DNS results are summarized in table 2. Starred quantities refer to the empirical semi-local scaling initially proposed by Huang *et al* [19] for compressible flows, which corrects the usual wall scaling with centerline quantities. Specifically, $y^* = \frac{\overline{\rho}(y) u^*_\tau y}{\overline{\mu}(y)}$ and $Re_{\tau^*} = Re_\tau \sqrt{\frac{\overline{\rho}_{CL}}{\overline{\rho}_w}} \frac{\overline{\mu}_w}{\overline{\mu}_{\mathrm{CL}}}$, with $u^*_\tau := \sqrt{\frac{\overline{\tau}_w}{\overline{\rho}(y)}}$ the semi-local friction velocity. This mixed scaling (recently discussed in [22]) has proven to give quite satisfactory results in collapsing first- and second-order moments [21] obtained from a wide range of M_{B_w}. We recall that in case of an incompressible flow, $Re_{\tau^*} = Re_\tau$.

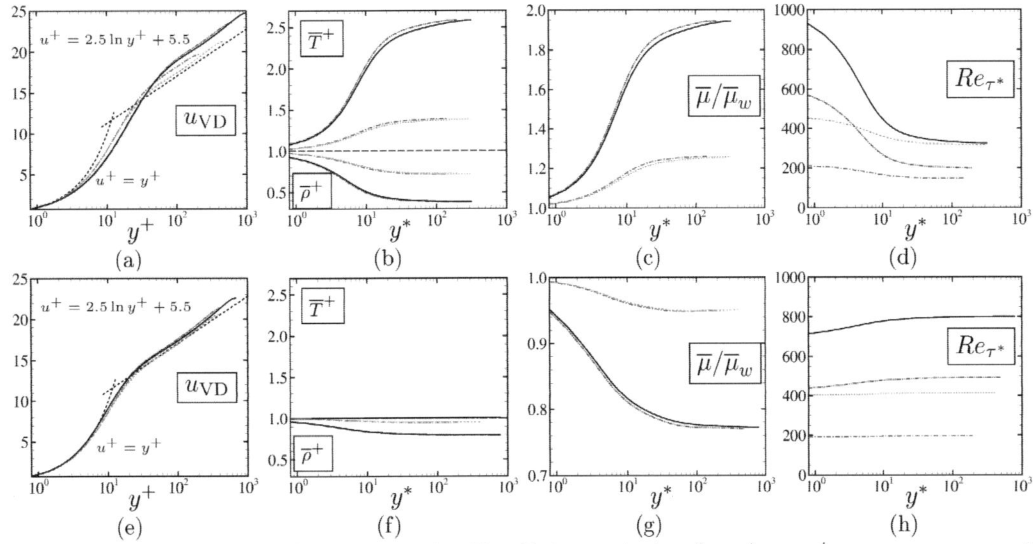

Figure 4: From left to right: streamwise Van-Driest velocity (u_{VD}) vs y^+, temperature and density normalized with the wall values ($\overline{T}^+$ and $\overline{\rho}^+$) vs y^*, normalized viscosity ($\overline{\mu}/\overline{\mu}_w$), and semi-local friction Reynolds number (Re_{τ^*}). Top: air; bottom: PP11. Legend as in table 2.

Figure 4 (panels a and e) shows the profiles of streamwise Van-Driest velocity in wall scaling. The Van-Driest velocity is introduced in the literature to improve scaling in the buffer layer and the outer region of compressible boundary layers by taking into account mean density variations. Despite considerable improvement is achieved over the classical incompressible scaling (with friction velocity u_τ), the profiles depart from the incompressible log law $(\ln y^+)/0.4 + 5.5$ when M_{B_w} increases. On the contrary, quite satisfactory results are obtained for PP11 at any Mach number. In some sense, the dense gas behaves similarly to a liquid with variable properties. For air, the average temperature $\widetilde{T}^+$ grows rapidly with M_{B_w}, due to the significant increase of friction heating, especially for the lowest Re_{B_w} (panel b). For PP11, due to the large specific heat of the fluid $\widetilde{T}^+$ is almost constant across the channel for any choice of the Mach and Reynolds numbers (panel f), and the centerline temperature differs less than 1% from $\widetilde{T}_w$. Decoupling of dynamic and thermal effects in the dense gas leads to smaller density variations across the channel. For air, the centerline density is up to 60% lower than $\overline{\rho}_w$ (at $M_{B_w} = 3$) whereas variations below 20% are observed for PP11. Panels c and d show profiles of the normalized viscosity $\overline{\mu}/\overline{\mu}_w$. This quantity follows temperature variations for air, so that $\overline{\mu}_{CL}/\overline{\mu}_w \approx 2$ for case AM3R3. For PP11, the viscosity exhibits a liquid-like behaviour at the considered thermodynamic conditions and takes lower values at the centerline. Unlike liquids, however, this is not due to a temperature rise, but to the reduced density. The preceding differences explain the opposite behaviour of the friction Reynolds number for the two fluids (panels d and h). For air, Re_{τ^*} decreases rapidly up to the buffer layer and more slowly up to centerline. For a given Re_{B_w}, higher Mach numbers enhance temperature gradients leading to the drop of Re_{τ^*} in the inner layer. For PP11, Re_{τ^*} increases monotonically up to the buffer layer; afterwards, its value is approximately constant. Dense gas effects become more evident at higher M_{B_w}, since the local thermodynamic states spread over a wider range. In general, $Re_{\tau^*} < Re_\tau$ for air (gas-like behaviour) and $Re_{\tau^*} > Re_\tau$ for PP11 (liquid-like behaviour). In the wall region, Re_{τ^*} is $\approx$ 20% lower in PP11 (leading to a smaller friction coefficient), whereas it is up to two times higher at centerline, that is then characterized by smaller and more erratic turbulent structures. In spite of the liquid-like behaviour of some properties in the dense gas, compressibility effects are not suppressed. On the contrary, the centerline Mach number is 20 to 30% greater in the dense gas than in air, where friction heating reduces the local Mach number with respect to the bulk one.

The distribution of the $p - v$ thermodynamic states for case PM3R12 is reported in figure 5 as a function of the distance to the wall in semi-local units. The diagram is coloured with the local Prandtl number. Moving from the wall to the centerline, a smaller thermodynamic region is spanned, while the average entropy tends to increase. We observe that, since the flow never crosses the transition line, no BZT effects are involved. Additionally, the maximum turbulent Mach number, approximately 0.5, is low enough to neglect the influence of compressibility on turbulence. On the other hand, the Prandtl number undergoes significant variations, especially in the near-wall region, which makes the use of a constant-Prandtl assumption quite inaccurate to predict dense gas flows characterized by a near-critical wall temperature.

Figure 6 displays the the r.m.s. values of density, the Reynolds stresses and shear stresses $\overline{\rho u_i'' u_j''}^+ = \tau_w^{-1} \overline{\rho u_i'' u_j''}$ in semi-local scaling and the ratio of the turbulent kinetic energy production to dissipation. Profiles of the r.m.s. density for the dense and the perfect gas (panels a and e) are deeply different. Previous reference results [17] have shown that for air cases, $O(\overline{T}^{-1}\sqrt{\overline{T'^2}}) = O(\overline{\rho}^{-1}\sqrt{\overline{\rho'^2}}) = O(\overline{p}^{-1}\sqrt{\overline{p'^2}})$, regardless of the y^* coordinate, the Mach and Reynolds number considered. The same behavior has been observed for both forced [23] and decaying highly-compressible homogeneous turbulence. The current results for air agree with reference trends; the r.m.s. density and temperature fluctuations are higher in the buffer layer, with a peak at $y^* \approx 10$, whereas pressure fluctuations (not shown) are approximately constant up to the logarithmic region. For PP11, relative density and pressure fluctuations are of the same order of air flows, whereas temperature fluctuations (not reported) are nearly two orders of magnitude lower. Nevertheless, $\overline{\rho'^2}$ decrease monotonically from wall to centerline as suggested from the smaller dispersion of the $p - v$ states), remaining approximately constant in the logarithmic layer. The observed behaviour is due to the fact that, as the average density decreases with the wall-normal coordinate, the fluid compressibility becomes lower (the average thermodynamic state moving away from the critical-point region), and density fluctuations decay. On average, the relative density fluctuations are smaller than those observed for air at $M_{B,w} = 1.5$, whereas an opposite result is obtained at $M_{B,w} = 3$. In other terms, density

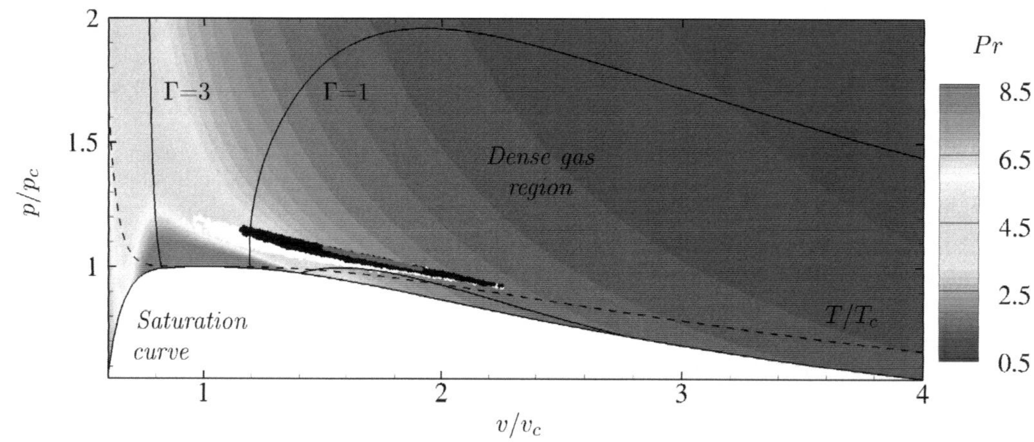

Figure 5: Distributions of the instantaneous thermodynamic states in the Clapeyron diagram for the case PM3R12 (fluid PP11) for different distances from the wall. Isocontours of the Prandtl number Pr are plotted. ○: $y^* \approx 10$, ●: $y^* \approx 50$, ●: $y^* \approx 800$ (centerline).

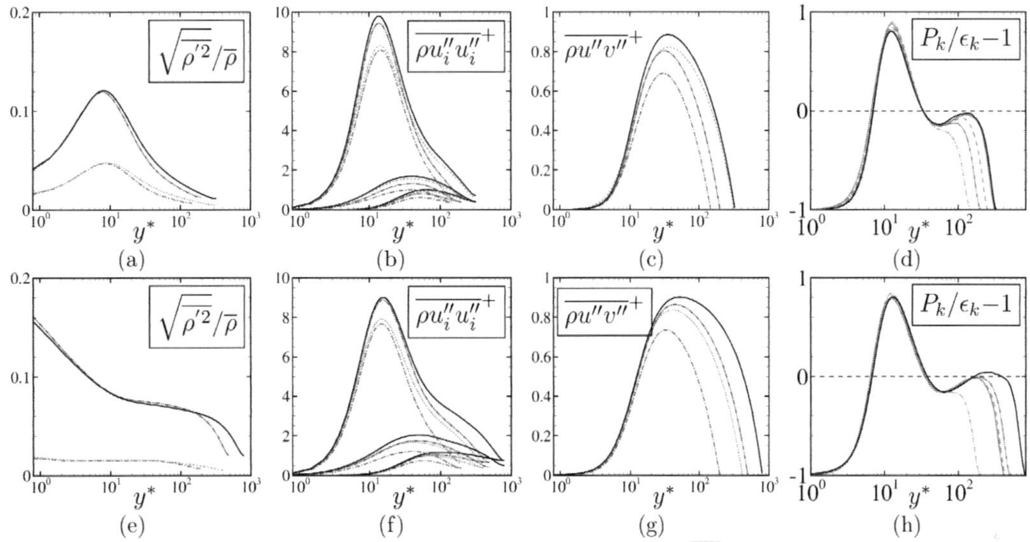

Figure 6: From left to right: y^*-profiles of the r.m.s. density ($\sqrt{\overline{\rho'^2}}/\overline{\rho}$), normal Reynolds stresses ($\overline{\rho u_i'' u_i''}^+$), Reynolds shear stresse ($\overline{\rho u'' v''}^+$), and production-to-dissipation ratio of turbulent kinetic energy ($P_k/\epsilon_k - 1$). Top line: air; bottom line, PP11. Legend as in table 2.

fluctuations grow more quickly with $M_{B,w}$ in PP11 than in air. Nevertheless, Morkovin's hypotesis remains quite well satisfied, even at the highest Mach number.

Despite the striking differences in the thermodynamic behaviour, the Reynolds stress profiles are similar for both fluids. Specifically, the dense gas profiles agree with those observed in other studies [22] for low-Mach TCF with temperature-dependent transport properties. The liquid-like behaviour of the viscosity leads to an increase of the spanwise, wall-normal and shear Reynolds stresses with respect to the corresponding incompressible evolution, whereas the streamwise one decreases, and this effect tends to be reinforced when increasing the Mach number (panels f and g). An opposite behaviour is observed for air (panels b and c). For instance, case AM1R7 ($Re_{\tau^*} \approx 315$) is characterized by a semi-local friction Reynolds number higher than case AM3R7 ($Re_{\tau^*} \approx 200$). Nevertheless, being the temperature-dependent viscosity much higher throughout the channel in the latter case, $\overline{\rho v'' v''}$, $\overline{\rho w'' w''}$ and $\overline{\rho u'' v''}$ are lower with respect to AM1R7. Finally, panels d and h show the profile of the ratio of production to dissipation term P_k/ϵ_k of the turbulent kinetic energy budget. The semi-local scaling, built with $\tau_w^2/\overline{\mu}(y)$, behaves better than both the bulk scaling [20] $\tau_w \overline{u}_B/h$ and the wall scaling $\overline{\rho}_w u_\tau^4/\overline{\nu}_w$. Previous works [24] showed that the main effect of compressibility is a reduction of both production and dissipation; this is confirmed for both air and PP11. All air and PP11 cases exhibit the usual production peak in the inner region, located approximately at $y^* \approx 12$. For the higher Reynolds PP11 case, a second production peak is observed in the outer region, like in high-Re incompressible flow, which is not the case for air at the same Re_{B_w}. This effect is due to the liquid-like behaviour of viscosity in PP11, leading to reduced dissipation when approaching the centerline.

5. Conclusions

Direct numerical simulations of the decay of compressible homogeneous isotropic turbulence (CHIT) and plane turbulent channel flow (TCF) of dense gases have been performed. Parametric studies with respect to the characteristic Mach and Reynolds numbers have been carried out for both configurations. The results were systematically compared to those obtained for air. In

CHIT simulations, important differences are observed for flows with a high initial turbulent Mach number. Different trend for the r.m.s. values of the thermodynamic quantities and transport properties greatly modify the turbulence evolution. In the dense gas, compressibility effects, and specifically eddy shocklets, are weaker. A tendency to symmetrization is observed for the p.d.f. of the normalized velocity divergence, due to the weakening of compression shocklets and the sharpening of rarefaction waves, leading possibly to the formation of rarefaction eddy shocklets for BZT fluids at suitable operating conditions. For TCF, the classical y^+ scaling based on the friction velocity fails to collapse thermodynamic profiles and Reynolds stresses when high M_{B_w} are considered and corrected scalings accounting for variations of the fluid properties have to be considered. The semi-local scaling, based on a mix of wall and local thermodynamic quantities, is shown to be the most suitable choice for collapsing Reynolds stresses and energy budgets. Nevertheless, the maximum turbulent Mach number is approximately 0.5 despite the high bulk Mach number, and turbulence structure is indeed little affected by compressibility effects. Additionally, the coupling between dynamic and thermal effects is very small for dense fluid characterized by high specific heats, contrary to air which undergoes significant friction heating (typical of highly compressible flows). For these reasons, for PP11 many of the quantities of interest follow an evolution similar to that observed in incompressible turbulence. For the reference thermodynamic conditions adopted, the transport properties have a liquid-like behavior and the actual Reynolds numbers are found to be much higher with respect to air. Of course, compressibility may become important if higher bulk Reynolds or Mach number are considered.

The numerical simulations reported in the paper have been performed for PP11, but preliminary simulations carried out for other perfluorocarbons (i.e. PP9 and PP10) have shown a similar behavior. Analyses are also underway for siloxanes (D5 and D6) and refrigerants (R245fa and R134a). Also note that simulations for a BZT van der Waals gas (not reported) exhibit a similar qualitative behavior. This suggests that, although the magnitude of the observed dense gas effects depends on the extent of the dense gas region for a given fluid and on the presence (or not) of an inversion zone, the qualitative behavior is expected to be similar for other dense gases. Results with other working fluids will be presented in a forthcoming paper.

References

[1] Anderson K W 1991 *AIAA J.* **29** 2179
[2] Kirillov N 2004 *Chem. Petr. Eng.* **40** 401
[3] Brown B and Argrow B 2000 *J. Prop. Power* **16** 1118
[4] Zamfirescu C and Dincer I 2009 . *Therm. Acta* **488** 66
[5] Thompson P A 1971 *Phys. Fluids (1958-1988)* **9** 1843
[6] Cramer M S and Kluwick A 1984 *J. Fluid Mech.* **142** 9
[7] Cramer M S 1991 *Nonlinear Waves in Real Fluids* 91
[8] Martin J J and Hou Y C 1955 *AIChE J.* **1** 142
[9] Chung T H, Ajlan M, Lee L L and Starling K E 1988 *Ind. Eng. Chem. Res.* **27** 671
[10] Bogey C, De Cacqueray N and Bailly C 2009 *J. Comp. Phys.* **228** 1447
[11] Bogey C and Bailly C 2004 *J. Comp. Phys.* **194** 194
[12] Donzis D A and Jagannathan, S 2016 *J. Fluid Mech.* **733** 221
[13] Sciacovelli L, Cinnella P, Content C and Grasso F 2016 *J. Fluid Mech.* **800** 140
[14] Samtaney R, Pullin D and Kosovic, B 2001 *Phys. of Fluids* **13** 1415
[15] Wang J, Shi Y, Wang L, Xiao Z, He X and Chen S 2012 *J. Fluid Mech.* **713** 588
[16] Sciacovelli L, Cinnella P and Grasso F 2016. Submitted to *J. Fluid Mech.*
[17] Gerolymos G A and Vallet I 2014 *J. Fluid Mech.* **757** 701
[18] Coleman G N, Kim J and Moser R D 1995 *J. Fluid Mech.* **305** 159
[19] Huang P G, Coleman G N and Bradshaw P 1995 *J. Fluid Mech.* **305** 185
[20] Lechner R, Sesterhenn J and Friedrich R 2001 *J. Turb.* **2** 001
[21] Foysi H, Sarkar S and Friedrich R 2004 *J. Fluid Mech.* **509** 207
[22] Patel A, Peeters J W R, Boersma B J and Pecnik R 2015 *Phys. Fluids* **27** 095101
[23] Donzis D A and Jagannathan S 2013 *J. Fluid Mech.* **733** 221
[24] Sarkar S 1995 *J. Fluid Mech.* **282** 163

IOP Conf. Series: Journal of Physics: Conf. Series **821** (2017) 012019 doi:10.1088/1742-6596/821/1/012019

Shock-induced energy transfers in dense gases

Nicolas Alferez
Department of Mechanical Engineering, Imperial College London, London, SW7 2AZ, UK
E-mail: n.alferez@ic.ac.uk

Emile Touber
Department of Mechanical Engineering, Imperial College London, London, SW7 2AZ, UK
E-mail: e.touber@ic.ac.uk

Abstract. Dense gases are characterised by molecules featuring large numbers of active degrees of freedom (quantified by the c_v/R ratio). The isentropes in such gases have the distinct property of following rather closely the isotherms (the two become identical in the limit of c_v/R going to infinity). Near the liquid-vapour critical point, this makes the isentropes very shallow and possibly concave (in the pressure-specific volume diagram). Whilst shallow isentropes are desirable when designing expanders (*i.e.* a large specific-volume increase may be achieved for virtually no pressure drop), could such extreme compressibility effects modify turbulence in a profound manner? This paper discusses two particularly interesting aspects: (i) shock-refraction properties (*i.e.* the way a shock can redistribute the energy of incoming perturbations), (ii) enstrophy production in homogeneous turbulence. A linear interaction analysis (LIA) is conducted on the shock configuration for which the incoming perturbation is decomposed into linear modes of the compressible Euler equations. The transmission coefficients relative to each eigen modes are solved analytically and results are compared against fully non-linear compressible direct numerical simulation reproducing the weak perturbation of an isolated two-dimensional compression shock wave. The linear analysis is found to be capable of predicting the shock-induced redistribution of the energy of the incoming perturbation between the different eigen modes. Non-ideal gas effects are observed both analytically and numerically with especially an unusual selective response for some particular choice of incoming Mach number. A two-dimensional isotropic turbulence configuration is then numerically investigated for the case of an inviscid compressible dense-gas flow close to the liquid-vapour critical point. Strong non-ideal-gas effects on enstrophy production are observed with the formation of eddy shocklets. In both cases non-convex isentropes close to the liquid-vapour critical point are extremely influential in letting both the shock and the turbulence redistribute any supply of turbulence kinetic energy in ways which are simply not observable in ideal gases. This will hopefully spark enthusiasm amongst turbulence modellers (and their end users?).

1. Introduction

Turbulence is the process by which any supplied amount of energy is redistributed across a broad range of length and velocity scales, constituting the turbulence kinetic energy (TKE). It is well accepted that the TKE cascades from large to small scales, down to scales for which friction forces dominate, ultimately dissipating the TKE into heat. For incompressible flows, this is the complete picture. If in addition fluid parcels can compress (*i.e.* change volume), the work done by the pressure forces converts some amount of TKE into turbulence internal energy (TIE). In the absence of viscosity, the conversion is isentropic and the TIE can be converted back to TKE. When the velocity fluctuations of the turbulence are of the order of the sound speed, localised shockwaves will appear (called eddy shocklets). Eddy shocklets immediately and irreversibly redistribute the energy across its various forms. Both vorticity and entropy are produced as the result of shocklets and their interactions. The entropy generated represents a leakage of TKE which is no longer recoverable – this is a direct loss. The vorticity production at small scales induced by the shocklets precipitates the cascade and speeds up the TKE dissipation (through friction forces at small scales) – this is an indirect loss.

The ability of eddy shocklets to significantly alter the cascade is conditioned to the actual thermo-physical properties of the fluid. For fluids comprised of molecules solely undergoing elastic collisions, *i.e.* ideal gases, pressure is linearly related to density and temperature. Thus, density fluctuations necessarily scale with that of pressure. Some fluids of practical interests (in the area of energy production, aerodynamics and propulsion) are made of molecules undergoing long-range interactions and therefore exhibit inelastic collisions. A particular occurrence is near the gas-liquid thermodynamic critical point (TCP), which is characterised by the presence of a saddle point on the critical isotherm (in the pressure/specific-volume phase diagram), a feature which is impossible to achieve in ideal gases. The presence of a saddle point on the isotherms is foot-printed on the isentropes (the two coincide for molecules with an infinite number of active degrees of freedom). Thus the isentropes can be made arbitrarily shallow near the TCP. Shallow isentropes translate in low sound speeds and mechanically promote the formation of eddy shocklets. Thus, for a given amount of energy, TKE transfers near the TCP will exhibit exacerbated compressibility effects relative to an ideal gas.

This paper aims at exploring some of the mechanisms by which shocks and shocklets affect the evolution of the turbulence kinetic energy when the fluid is governed by locally-concave isentropes. Only elementary interactions are considered: (i) the refraction properties of a compression shock following the impingement by an entropy (density) perturbation, (ii) the production of vorticity (and enstrophy) in two-dimensional turbulence.

2. Compression-shock refraction properties

2.1. Admissible compression shocks

Discontinuous solutions to the Euler equations are found by solving the well-known Rankine–Hugoniot (RH) algebraic system. The solutions are conveniently seen in the pressure/specific-volume diagram as the intersection points between the Rayleigh line ($\mathcal{R}$-line, along which mass is conserved and momentum balanced by the pressure forces) and the Hugoniot line ($\mathcal{H}$-line, along which changes in internal energy are balanced by the work done by the averaged pressure), therein joining all reachable thermodynamic states across the discontinuity:

$$\begin{aligned} p_2 &= p_1 - j^2(\vartheta_2 - \vartheta_1), \\ p_2 &= p_h(\vartheta_2, \vartheta_1, p_1). \end{aligned} \tag{1}$$

Notations $(.)_1$ and $(.)_2$ refer to pre- and post-shock quantities (in a reference frame attached to the shock), respectively, with p the pressure, ϑ the specific volume, and $j = \rho u$ the mass flow rate (u being the fluid velocity normal to the shock). In *classical* gas dynamics (*e.g.* when an ideal-gas equation of state is assumed) the $\mathcal{H}$-lines are convex and only two intersection points are at most solutions to equation (1). Admissible discontinuities are selected based on the classical Lax condition [1]. Dense gases modelled with a non-ideal equation of state can have non-convex $\mathcal{H}$-lines close to the thermodynamic critical point and equation (1) admits more

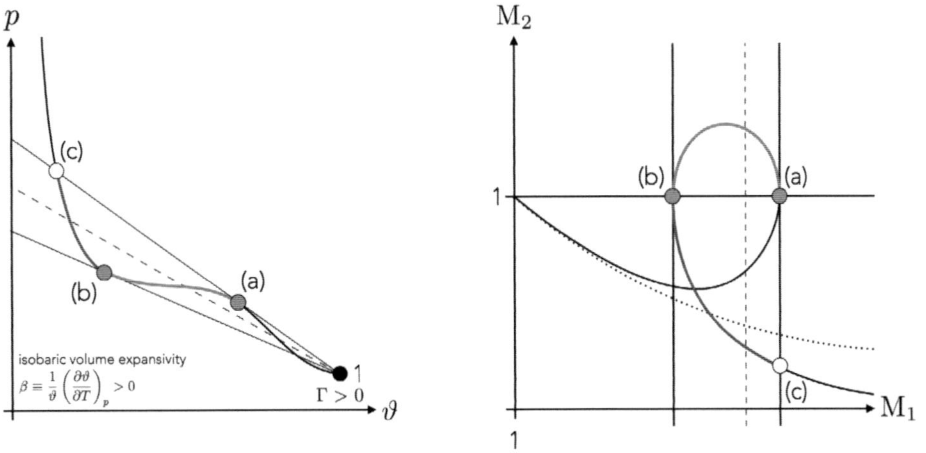

Fig. 1: Admissibility of compression shocks with locally-concave shock adiabats. The branch coloured in red is not Landau admissible. The branch coloured in blue is not Lax–Oleinik admissible. The branches in black are admissible and give rise to self-similar solutions. The equivalent ideal-gas post-shock Mach number evolution is shown in black-dotted line.

than two solutions, thus requiring additional selection criteria. This occurs when the $\mathcal{R}$-line intersects with the $\mathcal{H}$-line in regions of mix-convexity (see figure 1). The entropy-increase and Landau-stability (*i.e.* $M_2 < 1$) criteria are then supplemented by the Liu–Oleinik condition, which requires that no intermediate point on the $\mathcal{H}$-line region situated between the pre- and post-shock states be reachable through discontinuities with higher mass-flow rates [2]. Region **1-(c)** in figure 1 illustrates this so-called *mix-nonlinearity* problem for compression shocks in gases with positive isobaric volume expansivity. Landau's criterion is geometrically assessed for $\mathcal{H}$-lines steeper-down than $\mathcal{R}$-lines [3] (and is violated in the red-coloured **(a)-(b)** branch). A sonic post-shock state is reached for tangent Hugoniot and Rayleigh lines (points **(a)** and **(b)**). Branch **(b)-(c)** fails to fulfil the Liu–Oleinik constraint. The remainder of the $\mathcal{H}$-line (black-coloured branches) are admissible post-shock states. Thus, the admissible part of the $\mathcal{H}$-line may be discontinuous. Consequently, compression-shock properties may be discontinuous functions of the pre-shock Mach number M_1 as illustrated by the $M_2(M_1)$ curve in figure 1.

2.2. Linear perturbations in one dimension

Linear perturbations of a uniform solution to the Euler equations can be projected on three different modes: (1) an entropy mode (pure temperature and density disturbance fields); (2) two irrotational acoustic modes and; (3) a solenoidal vorticity mode [4]. In the linear regime all three modes propagate independently and may interact only at boundaries such as shockwaves. The ability of compression shocks to transfer energy from one mode to another is investigated close to the TCP.

2.2.1. Analytical results

An entropy mode is injected on the supersonic side of a compression shock close to the TCP. Perturbations normal to the shock front are considered. Thus, no vortical mode may contribute and our focus is on energy transfers between the entropy and acoustic modes at the shock. Reference thermodynamic values are taken at the TCP ($p_c^\star$, $\rho_c^\star$, $T_c^\star$), providing a velocity scale ($u^\star = \sqrt{p_c^\star/\rho_c^\star}$) and allowing the Euler equations to be written in non-dimensional form. The reference length scale ($\lambda^\star$) is prescribed by the perturbation. Dimensionless quantities are

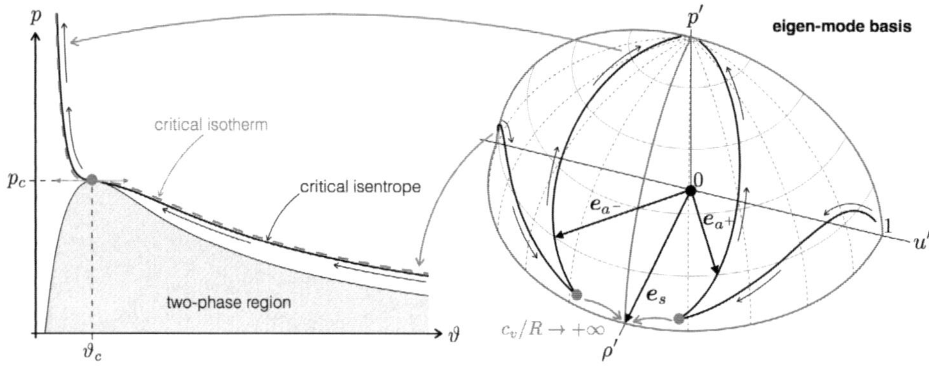

Fig. 2: Critical isotherm and isentrope for PP10 modelled with a van der Waals EOS (left). Kovasznay's eigen-mode basis evolution along the critical isentrope (right).

written without $(.)^\star$.

Small perturbations of a one-dimensional base flow of the Euler equations, noted with $\bar{(.)}$, are projected on Kovasznay's eigen-mode basis. The eigen values and corresponding unit eigen-vectors written in primitive variables (ρ, u, p) are:

$$\left\{\sigma_s = \bar{u}, \boldsymbol{e}_s = \begin{bmatrix}1\\0\\0\end{bmatrix}\right\}, \left\{\sigma_{a\pm} = \bar{u} \pm \bar{c}, \boldsymbol{e}_{a\pm} = \frac{1}{\sqrt{1+\bar{c}^2(\bar{c}^2+1/\bar{\rho}^2)}}\begin{bmatrix}1\\ \pm\bar{c}/\bar{\rho}\\ \bar{c}^2\end{bmatrix}\right\}, \tag{2}$$

respectively for the entropy mode, noted with $(.)_s$, and the upstream/downstream propagating acoustic modes, noted with $(.)_{a^+}/(.)_{a^-}$. Harmonic disturbances of pulsation ω are thus expressed as:

$$\boldsymbol{q}' = \begin{bmatrix}\rho'\\u'\\p'\end{bmatrix} = \xi_s \boldsymbol{e}_s \mathrm{e}^{\mathrm{i}(k_s x - \omega t)} + \xi_{a^+} \boldsymbol{e}_{a^+} \mathrm{e}^{\mathrm{i}\left(k_{a^+} x - \omega t\right)} + \xi_{a^-} \boldsymbol{e}_{a^-} \mathrm{e}^{\mathrm{i}\left(k_{a^-} x - \omega t\right)}, \tag{3}$$

with ξ_j and k_j the corresponding eigen-modes amplitudes and wave numbers defined as $\sigma_j = \omega/k_j$ with $j \in \{s, a^+, a^-\}$. In equation (3) the variable x is the space coordinate in the direction aligned with the mean flow.

The particular choice of equation of state (EOS) appears in equation (2) through the sound speed (noted c) which depends on the thermodynamic variables. Figure 2 (left) illustrates the convergence of the critical isentrope towards the critical isotherm in the (p, ϑ)-diagram for fluids composed of molecules featuring a large number of degrees of freedom [5]. Thus, a significant decrease of the sound speed is expected near the isotherm saddle point. This may come with a degeneracy of Kovasznay's eigen-mode basis in which case a strong bias towards density fluctuation is expected, see figure 2 (right). This is a unique feature of non-ideal gas dynamics.

Compression-shock refraction properties are investigated using the aforementioned linear framework. The incoming supersonic pure-entropy disturbance is projected on Kovasznay's eigen-mode basis, giving:

$$\boldsymbol{q}_1' = \begin{bmatrix} \rho_1' \\ u_1' \\ p_1' \end{bmatrix} = \xi_{s_1} \boldsymbol{e}_{s_1} \mathrm{e}^{\mathrm{i}(k_{s_1}\eta - \omega t)}, \tag{4}$$

where $\eta = x - x_s$ is the space coordinate centred around the mean shock position. Refracted entropy and acoustic disturbances are then produced and projected onto Kovasznay's eigen-mode basis:

$$\boldsymbol{q}_2' = \begin{bmatrix} \rho_2' \\ u_2' \\ p_2' \end{bmatrix} = \xi_{s_2} \boldsymbol{e}_{s_2} \mathrm{e}^{\mathrm{i}(k_{s_2}\eta - \omega t)} + \xi_{a_2^+} \boldsymbol{e}_{a_2^+} \mathrm{e}^{\mathrm{i}(k_{a_2^+}\eta - \omega t)}, \tag{5}$$

with similar notations as in equation (3). The refraction properties are discussed through the entropy-mode amplification ($\chi_s = \xi_{s_2}/\xi_{s_1}$) and acoustic-mode generation ($\chi_{a^+} = \xi_{a_2^+}/\xi_{s_1}$) coefficients. Analytical expressions are determined by solving a boundary value problem at the shock front (using the linearised RH conditions). These expression are obtained for a general equation of state:

$$\begin{cases} \chi_s = \left[\dfrac{\bar{\vartheta}_1}{\bar{\vartheta}_2}\right]^2 \dfrac{j^2 + M_2^2 \dfrac{\partial \bar{p}_2}{\partial \bar{\vartheta}_2} + \dfrac{\partial \bar{p}_2}{\partial \bar{\vartheta}_1}(M_2+1)^2}{j^2 - \dfrac{\partial \bar{p}_2}{\partial \bar{\vartheta}_2}(2M_2+1)} \\ \chi_{a^+} = -\left[\dfrac{\bar{\vartheta}_1}{\bar{\vartheta}_2}\right]^2 M_2^2 \dfrac{\dfrac{\partial \bar{p}_2}{\partial \bar{\vartheta}_2} + \dfrac{\partial \bar{p}_2}{\partial \bar{\vartheta}_1}}{j^2 - \dfrac{\partial \bar{p}_2}{\partial \bar{\vartheta}_2}(2M_2+1)}. \end{cases} \tag{6}$$

This result will be illustrated here using the van der Waals equation of state close to the TCP, and compared with fully nonlinear numerical simulations.

2.2.2. *Numerical method*

All numerical results presented in this paper are based on shock-resolved numerical simulations of the Euler equations. The equations are integrated in time explicitly using a 3rd-order TVD Runge–Kutta scheme. Spatial derivatives are evaluated with a 4th-order centred finite difference dispersion-relation-preserving scheme [6] using a 13-point stencil (a 4 points per wavelength cut-off is used for the optimisation). Numerical stability is ensured using a centred 13-point 8th-order explicit discrete filter optimised in spectral space following the methodology developed by Bogey *et al.* [7] (a 5 points per wavelength cut-off is used). This allows for accurate propagation of large-wavelength content (both from a dispersion and dissipation point of view) whilst minimising small-scale dissipation. Stretching of the grid is performed using a coordinate transform.

Shockwaves are captured using a localised artificial diffusivity method [8]. An additional viscous term is added to the Euler equations where a discontinuity is detected. This method captures shocks on four grid points with pre- and post-shock spurious oscillations reduced to about 1% of the shock strength [9]. Whilst being robust and easy to implement, this formulation does not explicitly set the shock thickness and a reduction of oscillations below a level of one percent requires tedious case-dependant adjustments of the model parameters. For simulations of isolated-shock refraction properties, an alternative and more flexible strategy has been developed to directly control shock-induced spurious oscillations down to a level of 0.01% of the shock strength. This was achieved by explicitly fixing the artificial viscosity magnitude (evaluated from a weak-shock approximation theory [5]) allowing a direct control of the shock resolution (set up to 30 points in the present work).

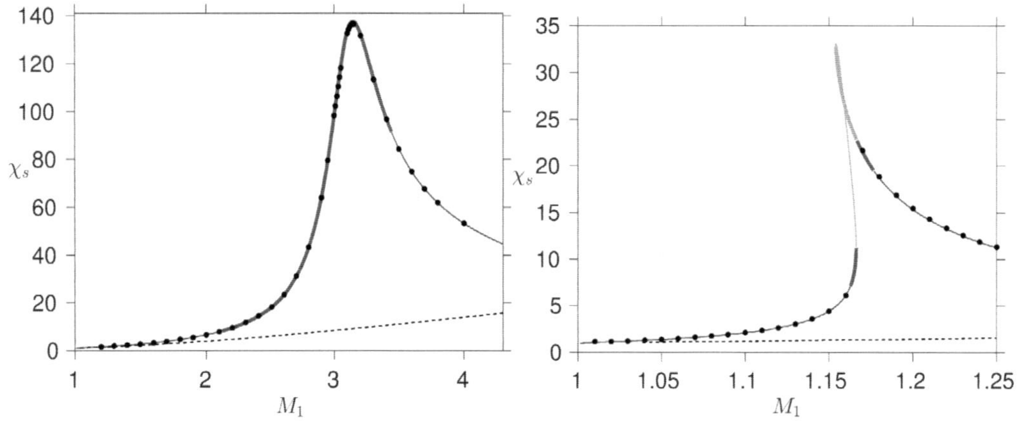

Fig. 3: Entropy-mode amplification factors for continuous (left, $p_1 = 0.13$, $T_1 = 1.00$) and discontinuous (right, $p_1 = 0.65$, $T_1 = 1.00$) shock adiabats for PP10. DNS data are shown with closed symbols (•). The thick lines represent shocks satisfying the D'yakov–Kontorovich criterion for spontaneous acoustic emission. The light blue portions identify non-admissible solutions. The dashed lines depict the equivalent ideal gas (same γ).

2.2.3. Results

Figure 3 illustrates the peculiar shock-refraction properties of PP10 modelled as a van der Waals gas ($\gamma = 1.013$) following the impingement of an entropy mode. Two pre-shock conditions are considered: (1) $p_1 = 0.13$, $T_1 = 1.00$ and (2) $p_1 = 0.65$, $T_1 = 1.00$. The $\mathcal{H}$-line is continuous in the former case, and discontinuous in the latter case. For both pre-shock conditions there exists mass flow rates for which the post-shock states are in the vicinity of the TCP where Kovasznay's acoustic eigen-vectors tend to align towards $[\rho', 0, 0]$. Pre-shock Mach numbers for which the so-called D'yakov-Kontorovich (DK) spontaneous acoustic emission [10, 11], specific to non-ideal gas dynamics, are also shown.

For both pre-shock conditions χ_s has a large local maximum with respect to M_1 (corresponding to conditions for which Kovasznay's acoustic eigen-vectors in the post-shock states are mostly biased towards the density component). Furthermore, χ_s can be several orders of magnitude higher than that of the equivalent ideal gas. This makes such shockwaves highly selective with respect to small changes in upstream conditions (*e.g.* the Mach number M_1), particularly when the $\mathcal{H}$-line is discontinuous. The analytical results are in very good agreement with numerical simulations of fully nonlinear one dimensional entropy-mode shock perturbation (for wich $\xi_{s_1} = 0.001\rho_1$).

These peculiar one dimensional shock-refraction properties match the M_1 range for which the two dimensional DK spontaneous emission is expected. The multidimensional shock-perturbation problem is then expected to be far from the classical ideal-gas case.

2.3. Refraction of a density pulse

Two-dimensional refraction properties are numerically investigated through the interaction of a density pulse with shocks of varying strengths. Special attention is paid to the shock-selectivity properties found in the one dimensional analysis.

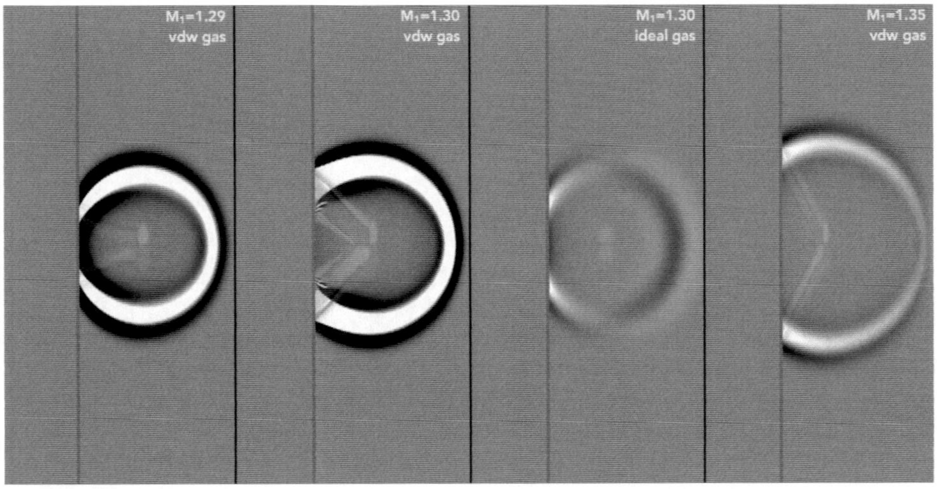

Fig. 4: Refraction of a low-density pulse (Gaussian perturbation in density) across van-der-Waals shocks following minute changes in the shock speed. The black and white contours represent dilatational rates and the coloured contours the vorticity. The same contour levels are shown throughout for relative comparison.

Impingements of isolated shockwaves by small amplitude ($0.01\rho_1$) Gaussian density pulses are computed. Inflow/outflow characteristic boundary conditions are set in the flow direction while periodic boundary conditions are used in the transverse direction. A fine grid with 3600×3024 points is used for all computations, thereby ensuring accurate resolution of the refracted content. Grid stretching at the shock location is applied so as to enforce a large separation of scales between the viscous-shock structure (shock capturing) and the initial pulse width (a ratio of 10^2 is used).

Results for base flows with pre-shock conditions at $p_1 = 0.55$, $T_1 = 1.00$ for $\gamma = 1.013$ and for which the $\mathcal{H}$-line is discontinuous at $M_1 = 1.29$ are shown in figure 4. Three different values of upstream Mach numbers (M_1) are considered around the discontinuity. The corresponding ideal-gas case (same γ) at $M_1 = 1.30$ is shown for comparison, and found to be in accordance with the linear interaction analysis (LIA) performed by Fabre *et al.* [12].

The linear interaction generates a cylindrical acoustic wave, a vorticity (shear) wave through baroclinic effects and a stretched refracted density pulse. The critical LIA angle for which the refracted shear wave reaches a maximum amplitude can be seen in all vorticity fields for the cases presented. A stronger acoustic intensity (shown by the dilatation field) is expected near the shock position due to the additional contribution of the non-propagative acoustic [12]. The generated acoustic wave for the $M_1 = 1.29$ case has an unusual circular shape due to the nearly sonic post-shock Mach number (M_2) close to the $\mathcal{H}$-line discontinuity. The vorticity production can be significantly larger in the non-ideal gas case, relative to the corresponding ideal-gas case (same pre-shock Mach number and thermodynamic conditions).

The strong-selectivity effect of the discontinuous $\mathcal{H}$-lines is made clear when comparing the refracted fields between the three van der Waals cases. Indeed, unlike for ideal gases, it is now possible to channel the incoming-perturbation energy preferably into the acoustic or the vorticity modes for virtually no change in the upstream Mach number. This new degree of

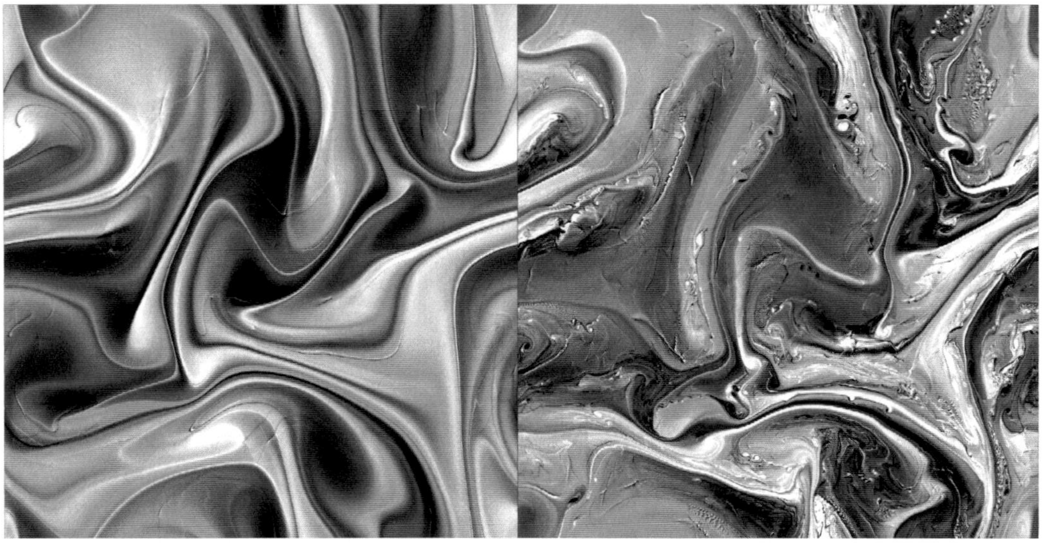

Fig. 5: Vorticity fields ($\omega_z \in [-15, 15]$) at $tU_0/L = 1$ (early stage of transition to "turbulence"): ideal gas (left) and van der Waals (right) for $\gamma = 1.001$ and $T_0 = 1.001$, $p_0 = 1.004$ for $k_0 = 2$.

freedom and associated selectivity effect should be considered in turbulence models and/or for practical application involving non-ideal gas compressible dynamics of dense vapours close to critical conditions.

3. Decaying isotropic turbulence

This section reports early results illustrating the importance of shocks in dense-gas turbulence in close vicinity to the critical point. It rests on the assumption that the slowing down of the sound speed in this region of the phase diagram allows for the formation of eddy shocklets, which are effective at redistributing the turbulence kinetic energy. The objective here is not to produce a realistic flow but rather to explore how some of the mechanisms associated with shock-refraction properties fit within a turbulent-like flow. In particular, we are interested in the production of enstrophy. For that purpose, simulations of decaying two-dimensional isotropic turbulence are performed in the absence of viscosity and conductivity apart from those associated with the explicit filter (thereby ensuring the stability of the simulation) and shock capturing.

The flow is initialised with a solenoidal velocity field with energy spectrum $E(k) = Ck \exp[-(k/k_0)^2/2]$, where C is a constant to obtain a maximum velocity of $U_0 = 1$. The pressure fluctuations are computed by solving the Poisson equation for pressure, and added to the chosen pressure p_0. The temperature is set uniform to T_0 and the density field computed so as to satisfy the thermal equation of state (van der Waals or ideal). The computational domain is a double periodic box of size $L \times L$ with $L = 2\pi$, discretised with 1920×1920 points. This choice of initial condition sets the dilatational velocity to zero and is of interest for studies at constant initial turbulence intensities but varying thermal equations of state.

Figure 5 gives the vorticity fields in the early stage of transition to "turbulence" for both the ideal-gas and van der Waals equations of state for a gas with $\gamma = 1.001$. The same solenoidal

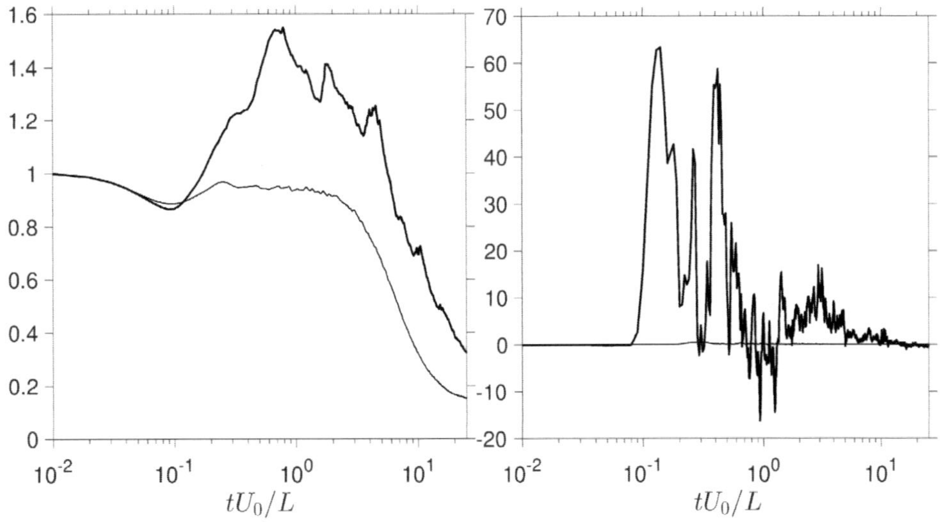

Fig. 6: Time evolutions of the space-averaged enstrophy $\langle \omega_z^2 \rangle / \langle \omega_z^2 \rangle \big|_{t=0}$ (left) and baroclinic-production term $\langle \boldsymbol{\omega} \cdot [\boldsymbol{\nabla} \rho \times \boldsymbol{\nabla} p]/\rho^2 \rangle$: ideal (thin line) *vs* van der Waals (thick line) gas.

field was used in both cases and explains why large-scale structures in both cases are similar early on in the simulations. This approach highlights the production of many more vortical structures in the van der Waals gas. In the ideal gas, small-scale vortical structures are found to emerge primarily as a consequence of shock-shock interactions. Indeed, shock-shock interactions produce slip lines which develop vortical structures through Kelvin–Helmholtz instabilities. In the van der Waals gas, the exacerbated compressibility of the gas near the critical point gives birth to more shocks (diverse in nature, *e.g.* expansion shocklets). When curved, these shocks produce significant amounts of vorticity through the baroclinic production term. When impinged by disturbances (*e.g.* vortex filaments), these shocks are also effective at producing vorticity as demonstrated with the shock/pulse interactions in the previous section.

Figure 6 gives the time-evolution of both the space-averaged enstrophy and baroclinic-production term. Given the incompressible initial solution, an overall expansion is expected at the early stage of the simulation. In the absence of the baroclinic-production term and the limit of no dissipation, the space-averaged enstrophy is then expected to decay in time. This is indeed observed for $tU_0/L < 0.1$. From $tU_0/L \approx 0.1$, eddy shocklets have formed in the flow. In the ideal-gas case, the shocks fail to deform and produce significant amounts of vorticity. In the van der Waals case however, the emergence of the shocks coincides with a sharp rise of the enstrophy-production term, two orders of magnitude larger than in the ideal gas. Consequently, the space-averaged enstrophy grows in time and exceeds its initial value.

Although limited in scope, this section confirms that dense gases near the critical point can experience an unusual production of vorticity. We will provide more details about the precise mechanisms responsible for the baroclinic-production term in future communications but hope that the present results demonstrate the need to derive new and suitable turbulence models, either for LES or RANS purposes, so as to predict TKE levels in component designs such as ORC expanders.

4. Summary

This paper explores some of the non-isentropic energy-transfer mechanisms associated with shockwaves in fluids modelled with a non-ideal equation of state. The focus is on dense gases formed by molecules featuring a large number of degrees of freedom when brought near their thermodynamic critical point.

An analysis of increasing complexity is proposed. First, the refraction properties of compression shockwaves are investigated for supersonic (in the shock reference frame) entropy perturbations normal to the shock front. Analytical expressions for the amplitudes of both the refracted entropy and the generated acoustic contents are derived for a general equation of state. A selective response is demonstrated for which significant changes in the refracted content can be achieved through minute changes of the incoming Mach number. This property is exacerbated for discontinuous Hugoniot lines featuring a non admissible part. The aforementioned one-dimensional selective response is then confirmed numerically for the two-dimensional linear interaction of a density pulse with an isolated shockwave near a discontinuous adiabat. The ability to redistribute a given supply of internal energy into either acoustic or vorticity modes in a way not possible in ideal gas is elucidated in multiple dimensions. Eventually, these energy-transfer mechanisms are investigated in the context of decaying two dimensional "turbulence" in a dense gas modelled with a van der Waals equation of state. The production of enstrophy through baroclinic effects associated with eddy shocklets is found to be significantly larger than for the equivalent ideal gas with same initial turbulence intensity.

This study is a first step towards the understanding of the complex energy transfers in turbulence in non-ideal gases. It sheds light on new non-isentropic energy pathways associated with shockwaves leading to a richer dynamic than in standard ideal-gas turbulence, eventually requiring a revamp of existing turbulence models.

Acknowledgements

This work was supported by the UK Engineering and Physical Sciences Research Council (EPSRC) grants EP/L021676/1 and EP/K503733/1.

References

[1] Lax P D 1957 *Communications on Pure and Applied Mathematics* **10** 537–566
[2] Liu T P 1975 *Journal of Differential Equations* **18** 218–234
[3] Cramer M S 1989 *Journal of Fluid Mechanics* **199** 281–296
[4] Kovásznay L 1953 *Journal of the Aeronautical Sciences* **20** 657–674
[5] Thompson P A and Lambrakis K C 1973 *Journal of Fluid Mechanics* **60** 187–208
[6] Tam C K W and Webb J C 1993 *Journal of Computational Physics* **107** 262–281
[7] Bogey C and Bailly C 2004 *Journal of Computational Physics* **194** 194–214
[8] Cook A W and Cabot W H 2004 *Journal of Computational Physics* **195** 594–601
[9] Kawai S and Lele S 2008 *Journal of Computational Physics* **227** 9498–9526
[10] D'yakov S P 1954 *Zh. Eksp. Teor. Fiz* **27** 288–295
[11] Kontorovich V M 1957 *Zh. Eksp. Teor. Fiz* **33** 1525–1526
[12] Fabre D, Jacquin L and Sesterhenn J 2001 *Physics of Fluids* **13** 2403–2422

Numerical Simulation of Jet Injection and Species Mixing under High-Pressure Conditions

Aswin Gnanaskandan[1] and Josette Bellan[1,2]
[1]California Institute of Technology, Pasadena, CA 91125
[2]Jet Propulsion Laboratory, California Institute of Technology, CA 91109
E-mail: Josette.Bellan@jpl.nasa.gov

Abstract. Direct Numerical Simulation (DNS) and Large Eddy Simulation (LES) are performed of a round fluid jet entering a high-pressure chamber. The chemical compositions and temperatures of the jet and that of the fluid in the chamber are initially prescribed. The governing equations consist of the conservation equations for mass, momentum, species and energy, and are complemented by a real-gas equation of state. The fluxes of species and heat are written in the framework of fluctuation-dissipation theory and include Soret and Dufour effects. For more than two species, the full mass diffusion and thermal diffusion matrices are computed using high-pressure mixing rules which utilize as building blocks the corresponding binary diffusion coefficients. The mixture viscosity and thermal conductivity are computed using standard mixing rules and corresponding states theory. To evaluate the physical model and numerical method, LES is employed first to simulate a supercritical N_2 jet injected into N_2. Time averaged results show reasonable agreement with the experimental data. Then, DNS is conducted to study the spatial evolution of a supercritical N_2 jet injected into CO_2. Time averaged results are used to compute the length of the potential core and the species diffusion characteristics. Spectral analysis is then applied on a time series data obtained at several axial locations and a dominant frequency is observed inside the potential core.

California Institute of Technology, Pasadena, CA 91125, USA
E-mail: `Josette.Bellan@jpl.nasa.gov`

1. Introduction

Numerous propulsion devices - diesel, rocket, gas turbine and HCCI engines - rely on mixing and combustion of chemical species under high-pressure conditions. What distinguishes high-pressure (high-p) fluid behavior from its low-pressure counterpart is the nature of the molecular interactions; under high-p the molecules are much tightly packed and may experience van der Waals interactions and caging effects. Macroscopically, this means that the perfect gas equation of state (EOS) is no longer appropriate to describe this fluid and should be replaced by a real-gas EOS. Moreover, under high-p conditions the fluxes of mass and heat must assume their full form given by the fluctuation-dissipation theory [1], that includes Soret and Dufour effects. For multi-species situations the mass diffusion and thermal diffusion matrices must be fully populated to account for interactions among species rather than interactions between species or interactions of one species with its global complement in the mixture. Clearly, simulating such flows requires complex models. Evaluating such models and simulations with experimental data is crucial in order to gain confidence that the model captures the essential aspects of the situation and that the numerical method is appropriate. Despite the definite interest in understanding multi-species situations, experimental data for only single and binary species exist so far. Selected representative experimental studies are those of Oschwald and Schik [2] and Meyer et al. [3] who obtained data for N_2-into-N_2 injection; Oschwald et al. [4] who performed experiments with N_2 injection into H_2; Chehroudi et al. [5] who injected N_2 into N_2, O_2, He or a 50%/50% mixture of CO and N_2; Segal and Polikhov [6], Roy and Segal [7] and Roy et al. [8] all of whom obtained data for a fluoroketone jet injected into N_2; and Falgout et al. [9, 10] who observed the injection of n-dodecane in air containing 21% O_2, and butanol, n-dodecane and n-hexadecane in N_2, respectively. No experiments with multi-species (i.e. more than two species) mixing exist, except those of Manin et al. [11] for n-heptane, n-dodecane and n-hexadecane injected into a hot fluid mixture resulting from the burning of C_2H_2 and H_2 in air, but the percent composition of this mixture is not documented although it can generally be assumed to be of H_2O, CO_2 and N_2 in unknown proportions. Most experimental observations are of visual type trying to unravel the species-specific conditions separating the single-phase and two-phase regimes. Quantitative data for validation are rare and restricted to few studies. We ask here whether the available data is sufficient to validate models, i.e. distinguish between the predictions of models that are different and indicate the model which best represents the physics. Failing to do would mean that experimental data of different type than one acquired so far must be obtained, and would open up the discussion regarding which data would be appropriate for model evaluation and whether the diagnostics to acquire this data in an incontrovertible way are available. Among the above-cited investigations, we are particularly interested in the data of Mayer et al. [3] for N_2-into-N_2 injection and of forthcoming data for N_2 injected into CO_2. This last case is partially relevant to engines using exhaust gas re-circulation and of paramount importance for understanding conditions encountered by a lander on Venus.

This paper is organized as follows: First, the previously derived governing equations in the most general form [12] are described, and we refer to other publications for the detailed computation of transport properties. Then, we conduct both LES to evaluate the model by comparison with experimental data and DNS to understand the details of flow evolution in a spatial configuration rather than the temporal configuration of the past [12]. Model evaluation must be necessarily conducted in a spatial rather than temporal realm. Since all our previous studies (e.g. [12]) were performed in the context of a temporal mixing layer, the configuration, numerical method and boundary conditions including inflow conditions are described in detail. The description and analysis of the results follows. A summary and conclusions are finally offered.

2. Description of the governing equations

The governing equations are the conservation equations - continuity, momentum, species and total energy - complemented by the Peng-Robinson (PR) real-gas equation of state (EOS). In the conservation equations, the fluxes of species and heat are written according to fluctuation-dissipation theory [1]. The species fluxes consist of the multicomponent species mass diffusion with a fully populated diffusion matrix and the Soret effect; the heat flux contains the Fourier term, Dufour effects and the enthalpy transported by the species fluxes. All transport properties - viscosity, mass diffusion coefficients, thermal diffusion coefficients and thermal conductivity - are computed using high-pressure-valid mixing rules [13], and thus transport properties are functions of all thermodynamic variables, (p, T, X_α) where T is the temperature and X_α is the mole fraction of species α. The PR EOS is well known to be inaccurate for species other than hydrocarbons and therefore the volume correction of [14] computed from a reference Gibbs free energy is used to enhance its accuracy to the same level as the Lee-Kesler EOS. Details on the entire formulation can be found in [12].

According to the 'alternate approach' suggested by Jordan [15], to obtain the present LES

DNS/ LES	Jet comp.	Chamber comp.	Re_{jet}	Re_{ch}	T_{jet} (K)	T_{ch} (K)	ρ_{jet}/ρ_{ch}
LES	N_2	N_2	1.6×10^5	0	140	298	3.43
DNS	N_2	CO_2	1.0×10^3	0	750	450	0.33

Table 1. Initial conditions and other characteristics of the simulations. The abbreviation "comp." denotes the chemical composition.

conservation equations, the governing differential equations are first transformed into curvilinear coordinates and then filtered with a lowpass filter rather than opposite (i.e. the filtering occurs before the transformation); the two approaches lead to the same mathematical equations. This filtering results in the formation of unclosed terms in the conservation equations that contain the subgrid activity; these are the subgrid-scale (SGS) terms. The LES equations have been presented in general form in [16], and contain the conventional subgrid terms as well as unconventional subgrid terms arising from the steep thermodynamic-variable gradients which are inherent in the flow ([12]). Since LES is only here used in the context of single fluid composition, the unconventional SGS terms identified for a mixture of species initially at different temperatures and the unconventional SGS terms arising from the species fluxes are neglected. Similarly, since the available data is only global, that is, the spatial details of the flow cannot be validated, we also neglect, in a first approximation, the SGS terms arising from the pressure gradient in the momentum equation and the heat flux in the energy equation that were shown to be important to recover in LES the spatial flow details [17, 18]. The constant-coefficient Smagorinsky model is used for the SGS conventional fluxes since it has been shown that if the interest is only in the mean flow, the nature of the SGS model is not as important as the thermodynamics [19].

3. Configuration, numerical method and boundary conditions

The configuration of interest is that of a three-dimensional round jet injected into a chamber at high-p conditions. The chamber dimensions for DNS in the (x_1, x_2, x_3) directions are $25D$, $10D$ and $10D$ respectively where D is the diameter of the jet, here set to be 0.005m. The domain extent for LES are $50D$, $25D$ and $25D$ in the (x_1, x_2, x_3) directions respectively. The domain is large enough to assume that the injected fluid is effectively a free jet. The boundary conditions in the x_2 and x_3 directions are of non-reflecting type for real-gases [20]. In the x_1 axial direction, the boundary conditions are inflow and outflow type. Acoustic reflections due to non–orthogonal wave impingement on the boundaries is prevented by applying sponge boundary conditions [21] and by artificially increasing viscosity near the outflow boundary; this region is not used when presenting statistical averaging results. The jet inflow velocity profile is given by

$$u_1 = \frac{U_j}{2}\left[1 - \tanh\left(\frac{r - r_j}{2\delta}\right)\right] \tag{1}$$

where U_j is the specified jet centerline velocity, $r = \sqrt{x_2^2 + x_3^2}$, r_j is the jet radius and δ is the initial shear layer thickness chosen to be $0.05r_j$. Similar hyperbolic tangent profiles are used at the inflow for the temperature and the species mass fraction. For the DNS, the domain is initialized using a method previously adopted by Nichols et al. [22] for simulating variable density jets at atmospheric pressure. In this method, the same profile as that of the inflow (i.e. $x_1 = 0$) for velocity, temperature and mass fraction is initially specified at all x_1 locations throughout the domain. Random perturbations of amplitude 5% of U_j, in the form of white noise are added to the initial velocity field in the shear layer. After this initial forcing, the flow

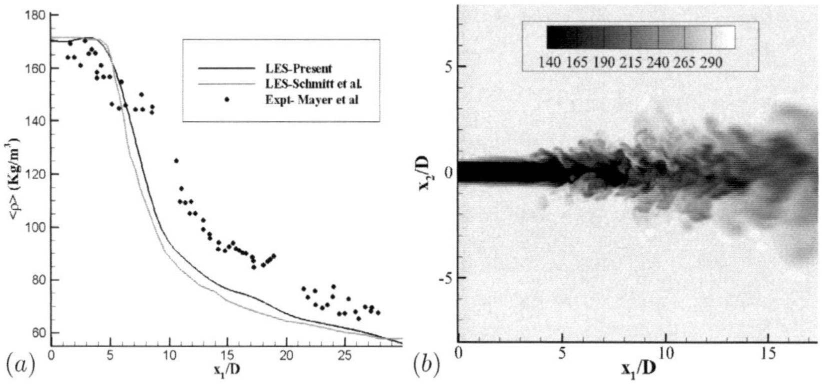

Figure 1. Results from LES of N_2 injection into N_2. (*a*) Time averaged density variation along jet center line. (*b*) Instantaneous temperature contours (in K) in the $x_3 = 0$ plane at $t = 0.0416$ s.

evolves with no further perturbations being added. For the LES, the inflow perturbations are provided using the method adopted by Freund [23].

The governing equations are temporally discretized using a fourth-order explicit Runge-Kutta time integration and spatially discretized utilizing a sixth-order compact scheme. In both DNS and LES, stability was achieved by filtering the conservative variables using a tenth-order filter [24]. The filtering was performed every 4 time steps in DNS and every time step in LES; the rationale for these filtering frequencies was explained in detail elsewhere [17].

4. Results

4.1. LES of N_2/N_2 jet

LES of N_2 supercritical jets injected into a chamber containing N_2 were performed in the past. Zong et al. [25] performed two-dimensional jet simulations and compared the radial variation of the density profile at two axial locations to the experimental data of Chehroudi et al. [26]; the agreement with the experimental data was reasonable despite the lack of portraying stretching/tilting effects which are typically considered important in vorticity production. Schmitt et al. [27] performed three-dimensional LES of supercritical jets using a Wall Adapted Local Eddy viscosity(WALE) model and the centerline density profile was compared with the experimental data of Mayer et al. [3] with good agreement. More recently Müller et al. [19] performed LES of Mayer's experiments using the Smagorinsky and Vreman models and found good agreement for the first order statistics between the models and a good agreement with the experiment for the centerline density profile. Considering that different models led to good agreement with data indicates that the data so far available is insufficient to distinguish between, or among, models - a model based on appropriate and relevant physics versus a model containing inappropriate simplifications - and therefore none of the favorable comparisons with available data can be considered a validation of that model. Thus, we propose that good agreement with existing data is only a necessary, but not sufficient condition, for model validation. In fact, we propose that one of the roles of simulations should be to highlight quantities which would constitute discriminating tests able to select the best model to describe high-pressure mixing and combustion.

With the goal of showing the necessary evaluation with the data, LES of a supercritical N_2 jet into a N_2-filled chamber is performed and compared to the experiments of Mayer et al. [3].

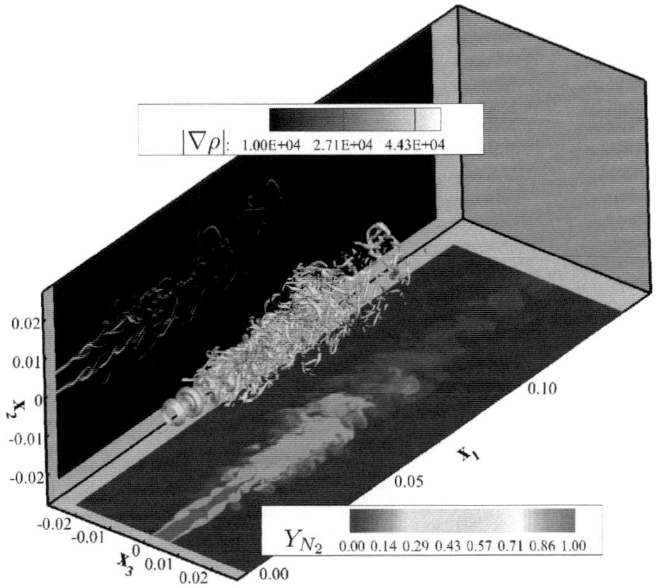

Figure 2. Instantaneous ($t = 0.02650$ s) results showing isocontours of Q-criterion (in s^{-2}), the second invariant of the velocity gradient tensor, contours of $|\nabla\rho|$ (in kg m^{-4}) in the $x_1 - x_2$ plane and contours of N_2 mass fraction in the $x_1 - x_3$ plane.

The jet exit conditions and chamber conditions are listed in Table 1. Figure 1(a) shows the x_1 variation of the time averaged density along the jet centerline obtained from experiments and simulation. The time averaged quantities (denoted by <> hereafter) is defined over a fixed time interval as

$$< \phi >= \frac{1}{t'} \int_{t_1}^{t_2} \phi(t)dt \quad (2)$$

where $t' = t_2 - t_1$ is the integration time. The jet centerline is defined as the line along which $x_2 = 0$ and $x_3 = 0$, and the values along this line are computed by interpolating from the nodes closest to the line at all x_1 locations. The simulation results agree reasonably with the experimental results. We also compare our results with those of Schmitt et al. [27] and a good agreement is obtained between the two simulations that use very different numerical schemes and turbulence models. This corroborates the observation of Müller et al. [19] that the first order statistics are generally insensitive to the numerical schemes and turbulence models and further reiterates the need for experimental data pertaining to higher order statistics. Figure 1(b) depicts the instantaneous ($t = 0.0416$ s) temperature contours in the $x_3 = 0$ plane. The presence of small scales downstream of the potential core is evident. Stretched entities emanating from the potential core can be observed that are characteristic of a single-phase mixing layer observed at the edge of supercritical jets.

4.2. DNS of N_2/CO_2 jet

DNS of a supercritical N_2 jet injected into a chamber containing CO_2 is performed at $Re_j = 1000$. To our knowledge, this is the first DNS of a jet having a specified composition that is injected into a chamber having a different composition. The jet exit and chamber conditions are listed in Table 1. Figure 2 shows the isocontours of the Q-criterion which is a commonly used quantity

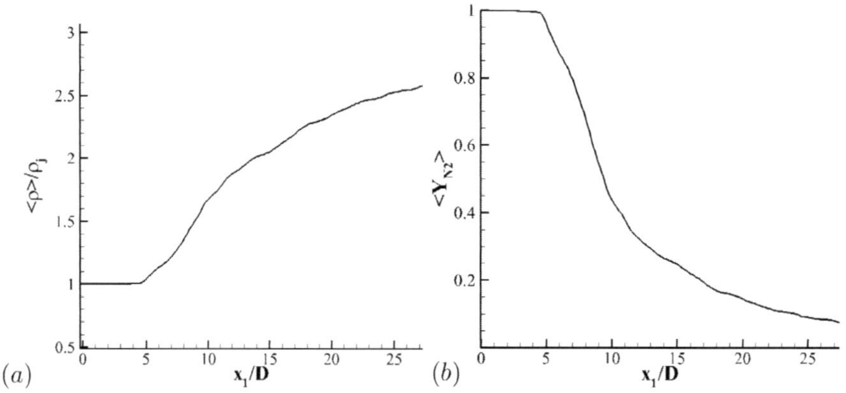

Figure 3. Results from DNS of N_2 injection into CO_2. (*a*) Time averaged density variation along the jet center line, (*b*) Time averaged N_2 mass fraction variation along the jet center line.

to visualize vortical structures,

$$Q = \frac{1}{2}\left[\frac{1}{2}\omega_i\omega_i - (S_{ij}S_{ij})\right], \tag{3}$$

where ω_i is the ith component of the vorticity vector and S_{ij} is the strain-rate tensor. The isocontours show the vortex ring near the inflow that further breaks up to produce a range of small scale vortices. Figure 2 shows the $|\nabla\rho|$ contours in the $x_1 - x_2$ plane at $x_3 = 0$. These high-value $|\nabla\rho|$ regions are a significant feature of high-p flows and representing them accurately is an important criterion measuring the success of subgrid scale models in the context of high-p flows. As a manifestation of mixing due to molecular diffusion enhanced by turbulence, shown on the $x_1 - x_3$ plane (at $x_2 = 0$) are the Y_{N_2} contours, $\langle Y_{N_2}\rangle$, where Y is the mass fraction. Since the injected N_2 fluid is lighter than the surrounding, the encountering of the heavier fluid leads to a rapid three-dimensional breakdown of the jet and therefore to enhanced mixing. The edges of mass fraction contours also show evidence of a single-phase mixing layer which is characteristic of supercritical jets, as opposed to spherical droplets formed due to surface tension at subcritical pressures. Quantifying this aspect requires further investigation.

To examine if a potential core exists, statistics are obtained by performing time averaging for approximately $100tU_j/D$. Figure 3(a) shows the variation of $\langle\rho\rangle/\rho_j$ along the jet center line. A time averaged potential core length $x_1/D = 5$ is obtained after which rapid mixing of the jet is observed. Figure 3(b) shows $\langle Y_{N_2}\rangle$ which also exhibits a potential core length that extends 5 diameters downstream of the injector; the decreased $\langle Y_{N_2}\rangle$ past $x_1/D = 5$ is due to mixing with the heavier CO_2 that is responsible for the increase of $\langle\rho\rangle/\rho_j$ past that axial location. To further obtain an understanding of the flow, we investigate time-averaged statistics along the radial direction at four different streamwise locations $x_1/D = 2$, 10, 15 and 20. Figure 4(a) illustrates the variation of the mean axial velocity, non-dimensionalized by the jet centerline velocity, along the radial direction at different streamwise locations. The $x_1/D = 2$ location is inside the potential core and hence the profile retains a shape similar to the inlet velocity profile. However, at further downstream locations, rapid mixing causes the profile to flatten out as the jet fluid mixes with the ambient fluid and the jet spreads out. Figure 4(b) illustrates the variation of $\langle Y_{N_2}\rangle$ along the radial direction. By comparing the mean velocity and mean mass fraction profiles it can be observed that the decay rate of the jet centerline value is larger for the species than that of axial velocity; however the spreading rate in the radial direction

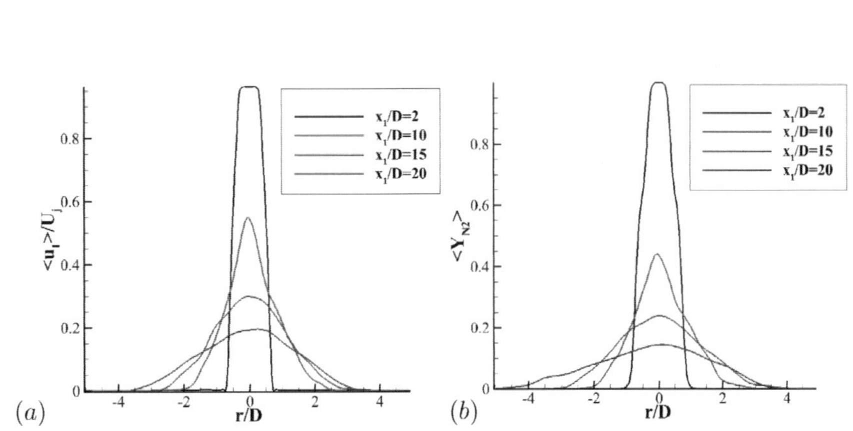

Figure 4. Results from DNS of N_2 injection into CO_2. (*a*) Variation of time averaged axial velocity (u_1) along the radial direction, (*b*) Variation of time averaged N_2 mass fraction along the radial direction.

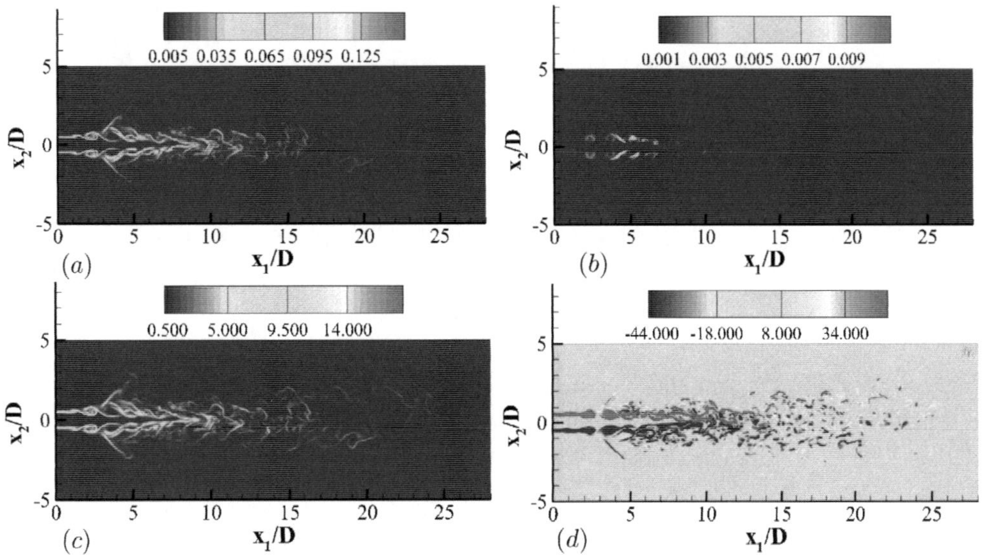

Figure 5. Results from DNS of N_2 injection into CO_2. Species mass flux for CO_2 (*a*) Contours of $|\rho Y_\alpha D_{T,\alpha} \frac{\nabla T}{T}|$ (kg/s m^2), (*b*) Contours of $|\rho Y_\alpha D_{p,\alpha} \frac{\nabla p}{p}|$ (kg/s m^2), (*c*) Contours of $|D'_{\alpha\beta} \frac{m_\alpha}{m_\beta} \nabla Y|$ (kg/s m^2) (*d*) Contours of ω_3 (s^{-1}). $t = 0.0265$ s.

remains approximately the same. Whether this axial evolution and radial spreading comparison are universal, and if so, the reasons for these effects should be investigated in future work.

At atmospheric p, it is generally considered that mixing occurs primarily due to molecular mass diffusion enhanced by turbulence. Under high-pressure conditions, the Soret effect may also play an important role in species diffusion. The species α mass flux is given by

$$J_{\alpha j} = -\rho \left[Y_\alpha \left(D_{T,\alpha}\right) \frac{1}{T} \frac{\partial T}{\partial x_j} + Y_\alpha \left(D_{p,\alpha}\right) \frac{1}{p} \frac{\partial p}{\partial x_j} + \sum_{\beta=1}^{N-1} \left(D'_{\alpha\beta} \frac{m_\alpha}{m_\beta} \right) \frac{\partial Y_\beta}{\partial x_j} \right] \tag{4}$$

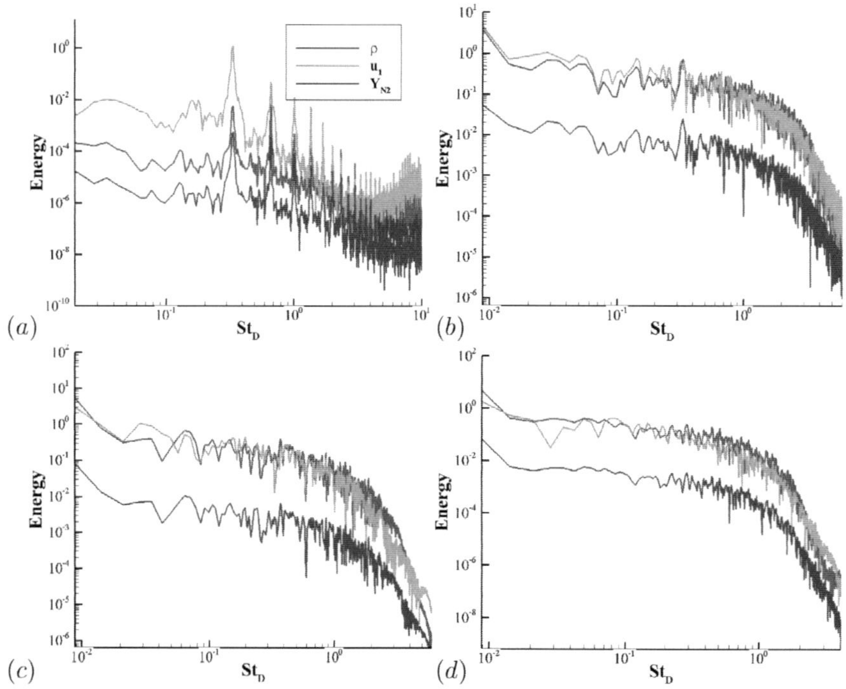

Figure 6. Spectra of ρ, u_1 and Y_{N_2} along the jet center line at $(a) : x_1/D = 4.0$, $(b) : x_1/D = 8.0$, $(c) : x_1/D = 12.0$ and $(d) : x_1/D = 16.0$. The units of Energy (energy spectral density) are JHz^{-1}).

where $D_{T,\alpha}$, $D_{p,\alpha}$ and $D'_{\alpha\beta}$ are diffusion coefficients and m is the molar mass (see [12] for details). The contribution of the Soret term (the first term in Eqn. 4) along with the other two terms are plotted in Figure 5. Since a statistical representation of these terms would involve computing budgets, a calculation which is beyond the scope of this investigation, we defer that calculation to future study and present here only the magnitude of the instantaneous values on $x_3 = 0$ plane. The largest contribution to J_α comes from the mass fraction gradient term (Figure 5(c)) followed by the Soret term (Figure 5(a)) the magnitude of which is approximately two orders of magnitude smaller than the largest term. The contribution of ∇p (Figure 5(b)) is negligible for the current case, as generally accepted. By comparing the spatial distribution of these terms with that of ω_3 contours in Figure 5(d), we can conjecture that these diffusion effects play an important role in the formation of small scales through mixing induced formation of large magnitude $|\nabla\rho|$ regions which act as a solid boundary at which vorticity forms. A detailed evaluation is necessary to further shed light on this aspect. To understand the turbulent characteristics of the flow, a spectral analysis is performed by applying a Fourier transform to ρ, Y_{N_2} and u_1 at four locations ($x_1/D = 4, 8, 12$ and 16) along the jet centerline. A dominant frequency at a Strouhal number ($St_D = fD/U_j$, where f is the frequency ($f = 1/t$)) of 0.33 is obtained. This is the frequency at which a continuous series of vortex rings are generated and is particularly visible as peaks in the spectra at $x_1/D = 4$ in Figure 6(a). As x_1/D increases, (e.g. $x_1/D = 8, 12$ and 16) in Figures 6(b), (c) and (d), the spectra becomes smoother, a fact which is manifestation of a flow with turbulent characteristics. The spectrum shows a monotonic fall in the energy at higher frequencies indicating that all relevant frequencies are captured in the simulation.

5. Summary and Conclusions

Direct Numerical and Large Eddy Simulations are used to study high-pressure jet injection and mixing. LES is first used to evaluate the physical model and numerical method for a cryogenic N_2 jet injected into a N_2-filled chamber. The time averaged density variation along the streamwise direction is compared to the experimental data and a reasonable agreement is obtained therefore fulfilling a necessary condition for suggesting the validity of the physical model and numerical method. The first DNS of a jet of a specified chemical composition injected into a chamber filled with a fluid of different composition is conducted: the DNS is of a supercritical N_2 jet injected into a CO_2-filled chamber. Instantaneous results show that the employed grid and model are able to capture the high density gradient magnitude regions that are characteristic of a high-p flow. The radial variation of mean axial velocity and mean N_2 mass fraction indicate that the centerline decay rate of species concentration is larger than the axial velocity decay rate; whether this fact is universal and its causes will be the subject of future research. The contribution of the Soret term is approximately two orders of magnitude smaller than the mass fraction gradient term; the spatial distribution of all diffusion terms indicate their important role in the formation of smaller turbulent scales. Statistical results show that the mean core length is about $5D$. Spectral analysis shows a dominant frequency of $St_D = 0.33$, the frequency at which vortex rings are produced. Spectra from locations downstream of the potential core show a broadband nature, characteristic of a turbulent flow.

Acknowledgments

This study was conducted at the California Institute of Technology under the sponsorship of the Department of Energy (DoE), Basic Energy Sciences with Drs. Wade Sisk and Mark Pederson as Program Managers and by the Army Research Office with Dr. Ralph Anthenien as Program Manager. Computational time was provided by the NASA Ames Supercomputing Center under the Transformational Tools and Technologies (T3) Project directed by Dr. Michael Rogers and by NERSC under DoE sponsorship.

References

[1] Keizer, J., 1987 *Statistical Thermodynamics of Nonequilibrium Processes* Springler-Verlag, New York.
[2] Oschwald, M. and Schik, A., 1999. Supercritical nitrogen free jet investigated by spontaneous Raman scattering. Experiments in Fluids, 27(6), pp.497-506.
[3] Mayer, W., Tellar, J., Branam, R., Schneider, G. and Hussong, J., 2003 Raman measurements of cryogenic injection at supercritical pressure, Heat and Mass Transfer, 39(8-9), pp. 709-719.
[4] Oschwald, M., Schik, A., Klar, M. and Mayer, W., 1999. Investigation of coaxial LN2/GH2-injection at supercritical pressure by spontaneous Raman scattering. In 35th AIAA/ASME/SAE/ASEE Joint Propulsion Conference and Exhibit.
[5] Chehroudi, B., Talley, D. and Coy, E., 2002. Visual characteristics and initial growth rates of round cryogenic jets at subcritical and supercritical pressures. Physics of Fluids (1994-present), 14(2), pp.850-861.
[6] Segal, C. and Polikhov, S.A., 2008. Subcritical to supercritical mixing. Physics of Fluids (1994-present), 20(5), p.052101.
[7] Roy, A. and Segal, C., 2010. Experimental study of fluid jet mixing at supercritical conditions. Journal of Propulsion and Power, 26(6), pp.1205-1211.
[8] Roy, A., Joly, C. and Segal, C., 2013. Disintegrating supercritical jets in a subcritical environment. Journal of Fluid Mechanics, 717, pp.193-202.
[9] Falgout, Z., Rahm, M., Wang, Z. and Linne, M., 2015. Evidence for supercritical mixing layers in the ECN Spray A. Proceedings of the Combustion Institute, 35(2), pp.1579-1586.
[10] Falgout, Z., Rahm, M., Sedarsky, D. and Linne, M., 2016. Gas/fuel jet interfaces under high pressures and temperatures. Fuel, 168, pp.14-21.
[11] Manin, J., Pickett, L. M. and Crua, C., 2015. Microscopic observation of miscible mixing in sprays at elevated temperatures and pressures. In ILASS meeting, Rayleigh, NC.
[12] Masi, E., Bellan, J., Harstad, K.G. and Okong'o, N.A., 2013. Multi-species turbulent mixing under supercritical-pressure conditions: modelling, direct numerical simulation and analysis revealing species spinodal decomposition. Journal of Fluid Mechanics, 721, pp.578-626.

[13] Harstad, K.G. and Bellan, J., 2004. Mixing rules for multicomponent mixture mass diffusion coefficients and thermal diffusion factors. The Journal of chemical physics, 120(12), pp.5664-5673.

[14] Harstad, K.G., Miller, R.S. and Bellan, J., 1997. Efficient high–pressure state equations. AIChE Journal, 43(6), pp.1605-1610.

[15] Jordan, S.A., 1999. A large-eddy simulation methodology in generalized curvilinear coordinates. Journal of Computational Physics, 148(2), pp.322-340.

[16] Borghesi, G. and Bellan, J., 2015. A priori and a posteriori investigations for developing large eddy simulations of multi-species turbulent mixing under high-pressure conditions, Phys. Fluids, 27, 035117 (35 pages).

[17] Taşkinoğlu, E.S. and Bellan, J., 2010. A posteriori study using a DNS database describing fluid disintegration and binary-species mixing under supercritical pressure: heptane and nitrogen. Journal of Fluid Mechanics, 645, pp.211-254.

[18] Taşkinoğlu, E.S. and Bellan, J., 2011. Subgrid scale models and Large Eddy Simulation of oxygen stream disintegration and mixing with a hydrogen or helium stream at supercritical pressure. Journal of Fluid Mechanics, 679, pp156-193.

[19] Muller, H., Niedermeier, C.A., Matheis, J., Pfitzner, M. and Hickel, S., 2016. Large-eddy simulation of nitrogen injection at trans-and supercritical conditions. Physics of Fluids (1994-present), 28(1), p.015102.

[20] Okong'o, N. and Bellan, J., 2002. Consistent boundary conditions for multicomponent real gas mixtures based on characteristic waves. Journal of Computational Physics, 176(2), pp.330-344.

[21] Bodony, D.J., 2006. Analysis of sponge zones for computational fluid mechanics. Journal of Computational Physics, 212(2), pp.681-702.

[22] Nichols, J.W., Schmid, P.J. and Riley, J.J., 2007. Self-sustained oscillations in variable-density round jets. Journal of Fluid Mechanics, 582, pp.341-376.

[23] Freund, J.B., 2001. Noise sources in a low-Reynolds-number turbulent jet at Mach 0.9. Journal of Fluid Mechanics, 438, pp.277-305.

[24] Kennedy, C.A. and Carpenter, M.H., 1994. Several new numerical methods for compressible shear-layer simulations. Applied Numerical Mathematics, 14(4), pp.397-433.

[25] Zong, N., Meng, S-Y. and Yang, V., 2004 A numerical study of cryogenic fluid injection and mixing under supercritical conditions, Phys. Fluids, 16(12), pp. 4248-4261.

[26] Chehroudi, B., Cohn, R., Talley, D. and Badakhshan, A., 2002 Cryogenic shear layers: Experiment and phenomenological modeling of the initial growth rate undr subcritical and supercritical conditions, Int. J. Heat Fluid Flow, 23, pp. 554-563.

[27] Schmitt, T., Selle, L. Ruiz, A. and Cuenot, B., 2010 Large-eddy simulation of sueprcritical-pressure round jets, AIAA Journal, 48(9), pp. 2133-2144.

[28] Hannoun, I.A., Fernando, H.J.S. and List, E. J., 1988 Turbulence structure near a sharp density interface, J. Fluid Mech. 189, pp. 189-209.

The effect of dense gas dynamics on loss in ORC transonic turbines

FJ Durá Galiana[1], APS Wheeler[2], J Ong[3] and CA de M Ventura[2]

[1]Aerodynamics and Flight Mechanics, Engineering and the Environment, University of Southampton, Southampton, SO17 1B, UK
[2]Whittle Laboratory, Department of Engineering, Cambridge University, Cambridge, UK
[3]GE Global Research, Munich, Germany

E-mail: fjdg1g08@soton.ac.uk

Abstract. This paper describes a number of recent investigations into the effect of dense gas dynamics on ORC transonic turbine performance. We describe a combination of experimental, analytical and computational studies which are used to determine how, in-particular, trailing-edge loss changes with choice of working fluid. A Ludwieg tube transient wind-tunnel is used to simulate a supersonic base flow which mimics an ORC turbine vane trailing-edge flow. Experimental measurements of wake profiles and trailing-edge base pressure with different working fluids are used to validate high-order CFD simulations. In order to capture the correct mixing in the base region, Large-Eddy Simulations (LES) are performed and verified against the experimental data by comparing the LES with different spatial and temporal resolutions. RANS and Detached-Eddy Simulation (DES) are also compared with experimental data. The effect of different modelling methods and working fluid on mixed-out loss is then determined. Current results point at LES predicting the closest agreement with experimental results, and dense gas effects are consistently predicted to increase loss.

1. Introduction

Organic Rankine Cycle (ORC) engines are widely used in power generation and heat recovery industrial processes. This is due to the good performance of ORC engines when producing power from low temperature heat sources. In order to increase the power density produced by the engine they commonly have a high pressure single stage turbine. Due to the configuration of the turbine, flows typically reach high Mach numbers at the trailing-edge of the vane resulting in large losses. Moreover the fluids used in ORC engines are typically Dense gases which present strong real gas and dense gas effects within the high pressure turbines.

The study of dense gas flow within the turbine is an emerging field of study. Only a few studies have looked at viscous effects in dense gas flows; these generally tend to show reduced viscous effects within the dense gas region [1, 2, 3, 4, 5]. However, it is unclear whether this is due to an increase of Reynolds number within the dense gas region. Generally, these studies do not consider trailing-edge loss which is a major input to turbine loss. In a single stage transonic turbine up to a third of the loss produced is originated downstream of the vane trailing-edge [6].

Recent investigations [7, 8] have shown that within the dense gas region the size of the base flow separation at the trailing-edge is significantly reduced as compared to gases outside this region. It was concluded that, while the viscous contribution to loss seemed to be significantly

The effect of dense gas dynamics on loss in ORC transonic turbines

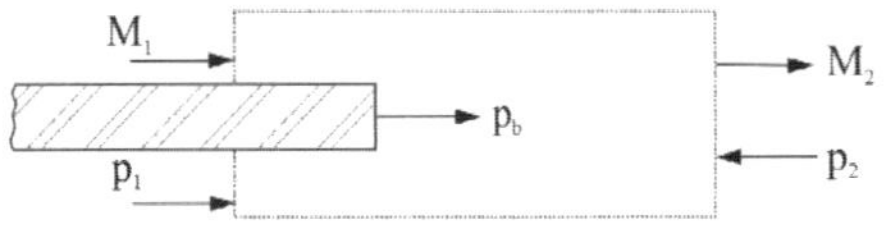

Figure 1. Control volume of a flat plate in a wind tunnel.

reduced within the dense gas region, the loss was dependent on the viscous-inviscid interaction, and this was largely dependent on the fundamental derivative ($\Gamma = 1 + \frac{\rho}{a}\left(\frac{\partial a}{\partial \rho}\right)_s$). This paper aims to determine how we can accurately model these flows and investigates the effect of trailing-edge shape on turbine vane performance. This investigation is divided into three parts. First, an analytical investigation of real gas effects on trailing-edge flows; secondly, an experimental validation of a computational simulation for dense gas trailing-edge flows; and, lastly, an investigation on the effects of trailing-edge shape on an ORC turbine vane.

2. 1D inviscid analysis of trailing edge flows

Denton & Xu [9] show that it is possible to predict the loss of a transonic trailing-edge in air using an inviscid Euler computation, as long as it predicts correctly the average suction surface pressure. In this section the authors extend this theory to Dense gas flows. In the Denton & Xu analysis of a transonic flow over a flat plate, they look at the control volume shown in Fig. 1. The Mach number upstream of the trailing edge is $M_1 = 1$, the inlet pressure is kept constant and the downstream pressure is varied to achieve a range of pressure ratios and downstream mixed-out states. Assuming an adiabatic choked flow, where the mass is conserved and viscous effects on the control volume boundaries are neglected it can be shown that for an ideal gas $P_{o2}/P_{o1} = (A_1F(M_1))/(A_2F(M_2))$, where $F(M)$ is the non-dimensional mass flow as a function of Mach number, and total temperature is conserved ($T_{o1} = T_{o2}$). However, these temperature and pressure relations are only valid for an ideal gas. For any real-gas where the equation of state is known, the same conservation laws can be used to determine the downstream stagnation pressure and temperature, if the downstream static pressure is specified. Using these results, the base pressure can be determined from the momentum balance as:

$$p_b t_{t.e.} = m(V_2 - V_1) + p_2A_2 - p_1A_1. \tag{1}$$

Using these same assumptions and conservation of mass, momentum and energy, and using real gas thermodynamic models, it is possible to determine the mixed-out loss ($\zeta = \frac{H_2 - H_{2S}}{H_{o1} - H_{2S}}$) for the case of a dense-gas flow over a trailing-edge. Hence a 1-dimensional study on a transonic trailing-edge flow over a flat plate using these relations is presented; ideal gas relations were used for air, and thermodynamic properties for dense gases were produced using the National Institute of Standards and Technology (NIST) real gas routines (REFPROP) [10]. All predictions are for single phase gas flows. The selected trailing-edge blockage was chosen to match that of the experiments carried out here ($A_2/A_1 = 0.237$). All dense gas flows have been expanded from initial reduced conditions of $p/p_c = 0.3245$ and $T/T_c = 0.952$. For these calculations the trailing-edge total pressure, $p_{o,in}$, and temperature and the Mach number, M_1, are set and the static properties of the flow are obtained from these assuming isentropic expansion from stagnation conditions. Following this, a range of downstream mixed-out static pressures, p_2, is set and flow properties are calculated for the whole range using the relations previously described.

Figure 2 shows the relationship between loss coefficient and base pressure as the downstream Mach number (M_2) is increased from a subsonic to supersonic value. Moving along these lines from the subsonic branch ($M_2 < 1$) to the supersonic branch ($M_2 > 1$), as the downstream Mach number is increased, the base pressure drops and the loss rises up to a peak maximum. However, the peak loss does not occur where the base pressure is lowest; the minimum base pressure occurs when $M_2 = 1$, while the peak loss occurs in the region of $M_2 = M_{2\text{peak}} = 0.78 - 0.81$ and

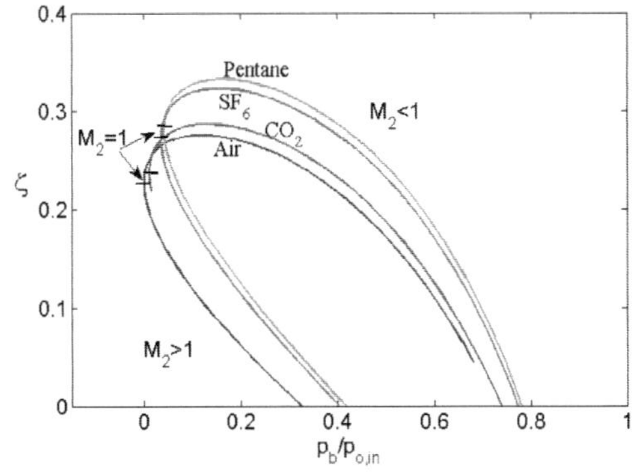

Figure 2. Loss coefficient variation with base pressure for air, CO_2, SF_6 and pentane with a trailing-edge Mach number of $M_1 = 1$.

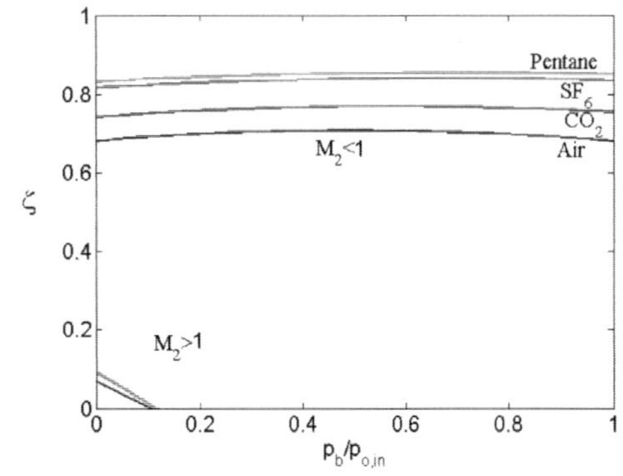

Figure 3. Loss coefficient variation with base pressure for air, CO_2, SF_6 and pentane with a trailing-edge Mach number of $M_1 = 2$.

$p_b = 0.12 - 0.16$. Thus, there is a small region between the maximum M_2 and $M_2 = 1$, where the loss reduces as base pressure drops. The results show that in general, dense gas effects tend to increase base pressure but at the same time raise trailing-edge loss for a given base pressure.

The same analysis was performed on flows for which the inlet Mach number is higher than unity. Figure 3 shows the variation in loss coefficient with base pressure for a case where the trailing-edge Mach number was $M_1 = 2$ for Air, CO_2, SF_6 and pentane. Here, there is a range of mixed-out conditions (M_2 and p_2) for which base pressure and loss coefficient achieve values which are not physically possible, and there is a discontinuity in the physical solutions on the loss-p_b curve. This discontinuity separates the subsonic and supersonic branches of the curve. For both cases ($M_1 = 1$ and $M_1 = 2$), there is a significant range of base pressure for which two solutions of the loss coefficient could exist. As a result of this, loss coefficient cannot be determined based solely on the base pressure. It is also shown here that dependence of loss on base pressure is greatly reduced for the subsonic branch as compared for the first case where $M_1 = 1$. Generally, dense gas effects increase loss, except for a small region in the supersonic branch where $M_2 > 2.1$.

These results show that for solutions along the subsonic branch, the maximum loss when $M_1 = 1$ is typically 60% lower than the $M_1 = 2$ case. However, a 20% reductions in fundamental derivative will typically raise maximum loss by 20% both when $M_1 = 2$ and $M_1 = 1$. Hence, the sensitivity of trailing-edge loss to dense gas effects is weakly dependent on the choice of vane-exit Mach number for the range of chosen fundamental derivatives.

In order to determine the dependence of trailing-edge loss on dense gas behaviour further, a set of experiments and CFD studies are presented next.

3. Experimental set-up

The experimental results have been obtained using a Ludwieg tube facility capable of producing high speed flows at a range of temperatures [7, 8]. The Ludwieg tube has three main parts: a 7.5m charge tube (⌀50mm), a 0.4m test section and a 150 litre reservoir tank. In a Ludwieg tube configuration the diaphragm is placed between the test section and the reservoir tank downstream; consequently, the charge tube and test section are pressurized together. The diaphragms have been calibrated to burst at desired pressures.

SF_6 is the chosen dense gas for the study, as it has similar thermodynamic properties and behaviour to pentane, which is a widely used gas in ORC systems [7, 8]. For the experiments

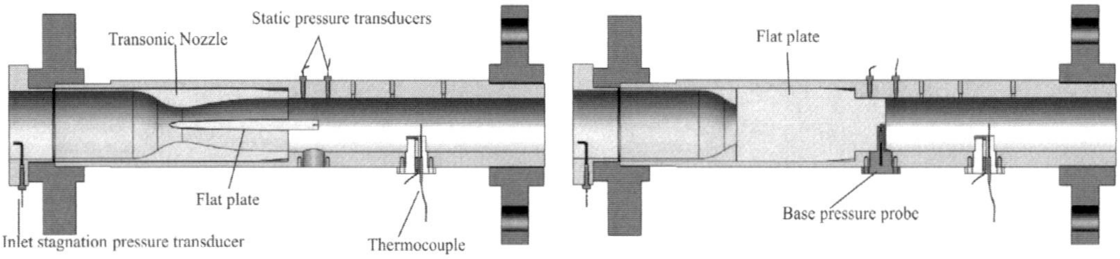

Figure 4. Test section and nozzle with embedded plate modified for base pressure measurements in free-stream static pressure (left) and base pressure (right) measurement configurations.

performed with SF_6 a "DILO B052R11 SF_6-gas" service cart was used to recover in full the used gas. All gases were tested at $M \approx 2.0$ and a Reynolds number based on the trailing-edge conditions and thickness of the flat-plate trailing-edge ($Re \approx 1 \times 10^6$) obtaining a range of fundamental derivative values; $\Gamma_{\text{Air}} = 1.2$, $\Gamma_{\text{CO2}} = 1.13$-$1.15$ and $\Gamma_{\text{SF6}} = 0.99$-$1.04$. Γ values represent the range values of Γ from the throat of the nozzle to the flat-plate trailing-edge at the specified conditions for each gas; these where determined using REFPROP for CO_2 and SF_6. Air is assumed to be an ideal gas.

The experiment designed for this study is a continuation of the work previously done on the study of trailing-edge wake flows [7, 8]. Similarly, a supersonic trailing-edge flow was produced by a flat plate embedded in a transonic nozzle which accelerates isentropically the flow up to an exit Mach number of 2 (see Fig. 4). The nozzle has been produced by an in-house code which accounts for the dense gas effects in supersonic flows and has been previously validated [7, 8]. The flat-plate was designed to include a base pressure measurement as shown in Fig. 4.

A "Kulite XCL-62" pressure transducer was used to capture the base pressure and "Kulite XTEL-140" transducers were used to measure wall static and inlet total pressures. All pressure transducers are fast-response transducers sampled at 100kHz using the National Instruments' "USB-6356 X-series DAQ" data acquisition system. All probes were calibrated and tested using the high precision modular pressure controller GE PACE5000. Due to the short run times (t_{run} = 30-100ms) temperature was difficult to measure accurately within the runs, so pre-run measurements were taken using a fast response needle type "OMEGA HYP0" probe.

4. Computational set-up

Three different computational studies were produced; the first study is a CFD simulation of the flat plate in the Ludwieg tube where the results where compared with the experimental results. The second computational study is an ORC vane flow. Lastly a trailing-edge shape study was conducted on the ORC vane geometry with the aim to determine how shape affects the base flow and loss. The software used for flow solving was FLUENT v14.5. Steady and unsteady simulations were performed using the density based solver. Simulations were performed using the third order MUSCL implicit algorithm and unsteady simulations were performed using a dual-time-stepping method. Air was computed assuming an ideal gas behaviour. The equations of state for Pentane, CO_2 and SF_6 were given by the NIST real gas data base within ANSYS.

RANS modelling was used for steady simulations using the vorticity based Spalart-Allmaras (SA) turbulence model; Other RANS models were also compared, but no significant differences in wake prediction were observed. LES were performed for the unsteady simulations. For this the Smagorinsky-Lilly model was employed, where $C_s = 0.1$ and $Pr_{wall} = 0.85$. A mesh and time-step dependency study is discussed later. DES simulations were also performed using the vorticity based SA RANS model; this is due to uncertainty of the impact of using wall functions in the described flows. Flat plate DES simulations were only performed using air.

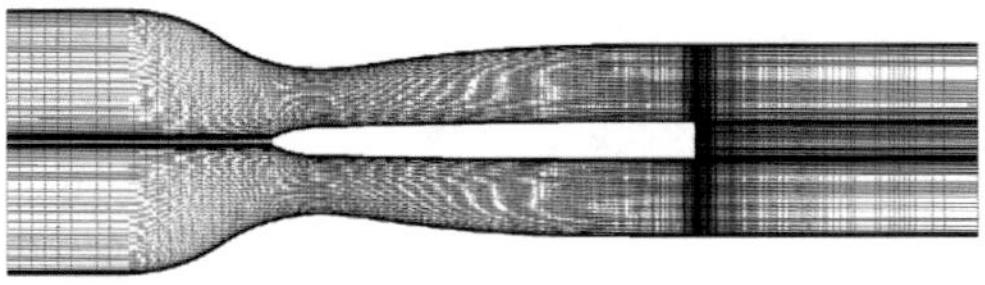

Figure 5. Nozzle with embedded plate mesh (symmetry plane).

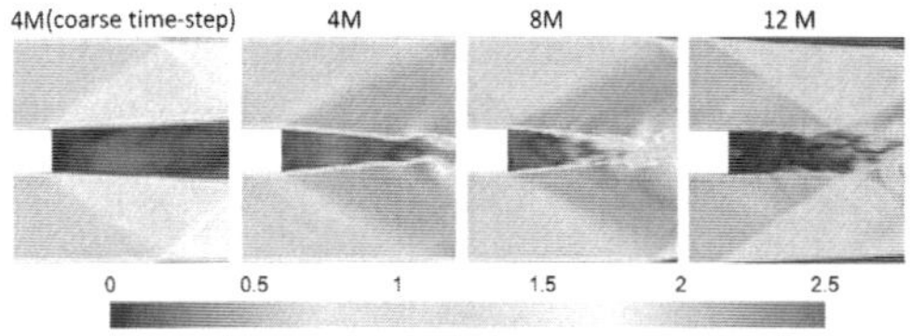

Figure 6. Instantaneous LES predicted change in flow in the base region with varying mesh size in an air flow.

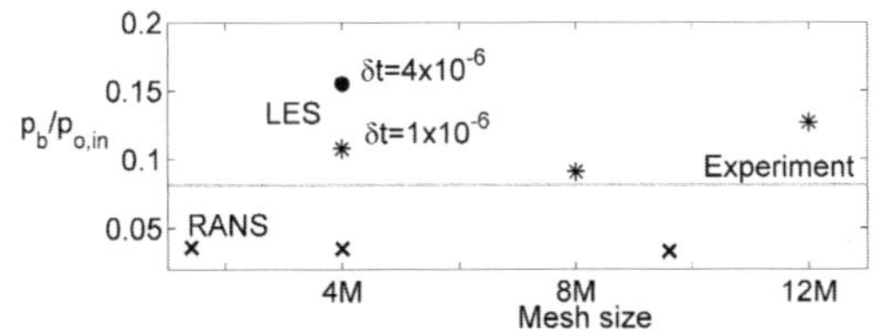

Figure 7. (above) Predicted base pressure for various RANS (×) and LES (•,∗) mesh and time-step sizes compared to measured base pressure for the air case.

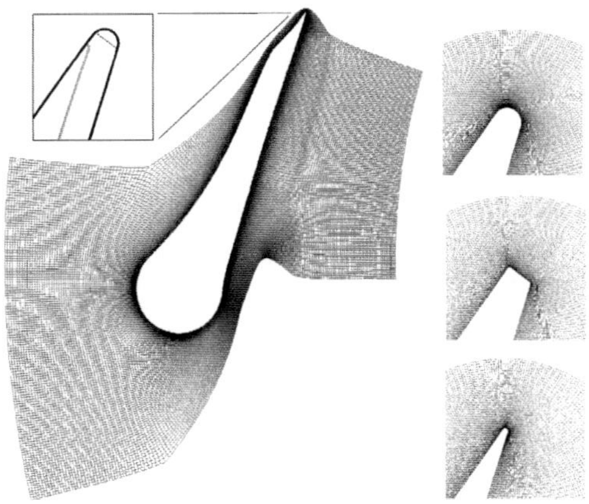

Figure 8. (right) ORC turbine vane baseline mesh and trailing-edge geometry variations.

The geometrical model for the flat plate study (see Fig. 5) is a representation of the geometry in the test section. Mesh sensitivity studies were carried out using RANS and LES with mesh sizes ranging from 0.7 to 12 million points. Over the plate surface, the near wall grid sizes were (on the region close to the trailing-edge for the air 8M mesh) $y^+ \approx 30$, $x^+ \approx 48$ and $z^+ \approx 88$. y^+ was selected to be consistent with the wall-function used within the solver. The use of wall functions was necessary due to the high Reynolds numbers and thus the prohibitively high computational cost of resolving the viscous sublayer. Furthermore, the focus of this work was the base region and not the surface flow.

Figure 6 shows how the trailing-edge flow changes with changes in time step size and mesh density. From left to right it can be seen that the size of the base region and wake width gets reduced as the time step size decreases and the mesh density increases. For the last case (12 million cells) it can be seen that the base region increases again to a rectangular shape. In this case, the boundary-layers on the outer walls of the domain appear to have increased significantly in thickness and the additional blockage is likely to be affecting the trailing-edge flow.

Figure 7 shows the experimentally measured and CFD predicted base pressure for air for different mesh and time-step sizes. It can be seen here that the accuracy of the base pressure is highly dependent on the selection of time-step size and mesh density for LES. The time-step size was initially defined to be 10 times smaller than the flow time-scale; this was estimated using the domain mass-averaged flow velocity and the domain length. The time-step size study determined that the optimum time-step size was $\delta t = 1\mu s$; this gave good agreement with the experiment results, while keeping computation costs to an acceptable level. An increase in mesh density improves the prediction initially, but it can cause a divergence from the experimental

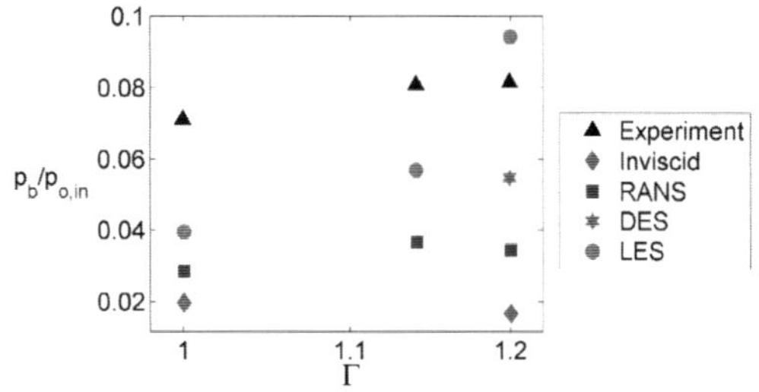

Figure 9. Base pressures for SF_6 ($\Gamma \approx 1$), CO_2 ($\Gamma \approx 1.14$) and air ($\Gamma \approx 1.2$).

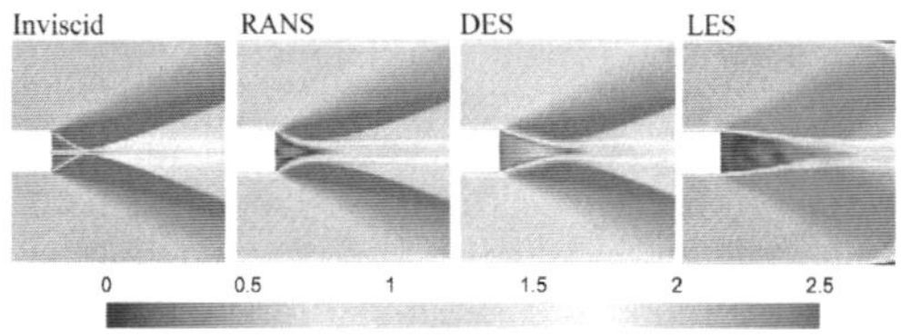

Figure 10. Predicted time-averaged Mach contours for air by inviscid, RANS, DES and LES CFD calculations.

results as seen for the 12 million cells mesh. This is likely to be due to the over-prediction of the outer wall boundary-layer blockage.

The final investigations involve the study of a typical turbine vane flow with pentane as the working fluid. The baseline geometry (Fig. 8) was designed in a previous ORC turbine study [6]. The vane was designed for a radial turbine operating with pentane with a mass flow rate of $2kg/s$, a temperature ratio of $T_{o,in}/T_{exit} = 1.15$, rotational speed of 26500RPM, power output of 155kW and a pressure ratio $p_{o,in}/p_{exit} = 11$ across the turbine. The inlet stagnation reduced conditions of the stage were $T_{o,in}/T_c = 0.96$ and $p_{o,in}/p_c = 0.59$, resulting in a range of fundamental derivative of $0.75 < \Gamma < 0.95$ across the vane.

For the current study the shape of the trailing-edge was modified to three different designs (Fig. 8) while the baseline model was kept unchanged. The three trailing edges used are: the original round trailing-edge, a square-cut trailing-edge, and a radial-cut trailing-edge. The latter is produced by a radial cut-away of the suction surface of the turbine vane. The square-cut trailing-edge is produced by cutting away the round trailing-edge producing a surface perpendicular to the pressure surface of the vane.

The meshing of the vane geometries (Fig. 8) was performed using similar mesh resolution for the flat-plate study as this is the mesh that has been validated to give acceptable results. Hence the wall y^+ along the surface of the vane has been set so that it is generally $y^+ \gtrsim 30$ and wall functions are used, and a similar mesh density has been produced at the trailing-edge, with 50 points along the trailing-edge height ($t_{\mathrm{t.e.}} = 0.5\%c$) and 40 points along the span (span$= 5\%c$). The vane was produced using a Matlab script for the geometry and ANSYS turbogrid v14.5 for the meshing. The square and radial-cut trailing edges included a small corner radius of $r_{\mathrm{corner}} = 0.04$mm. This is 14% of the round trailing edge thickness, $t_{\mathrm{t.e.}}$.

5. Comparison of experiments and CFD results

Figure 9 shows base pressure CFD predictions and experimental measurements. Both the CFD and experimental data show that the base pressure reduces with decreasing fundamental derivative, this effect is specially seen between the CO_2 and the SF_6. Comparing the different CFD models it can be seen that, while the Inviscid and RANS modelling predictions give close results between them and agree with the trend seen in the experiments, the LES modelling gives a closer agreement with experiments. It can also be observed that DES for air predicts a base pressure in between RANS and LES.

Figure 10 shows contours of Mach number in the base region of the flat plate for the air case. Here it can be seen how as viscous modelling is introduced from Inviscid to RANS the size of the base region increases, as well as the wake width. Similarly, when scale-resolving methods are included with DES and LES the base region increases again. By comparing these plots to the values of base pressure given by each simulation, it can be seen that an increase in the size

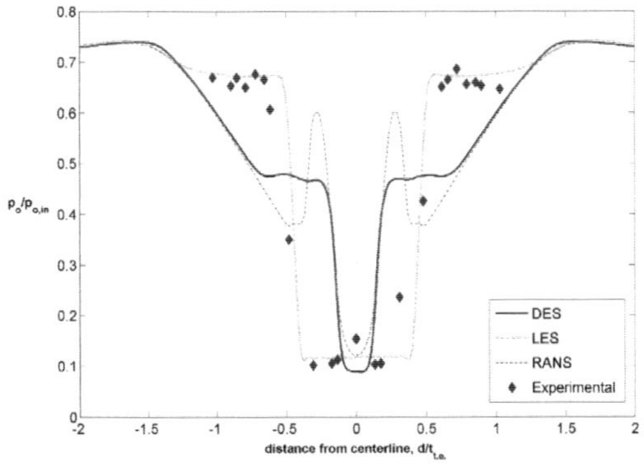

Figure 11. Measures and predicted shocked pressure profiles across the wake at a distance of $1.5t_{\mathrm{t.e.}}$ downstream of flat-plate trailing-edge.

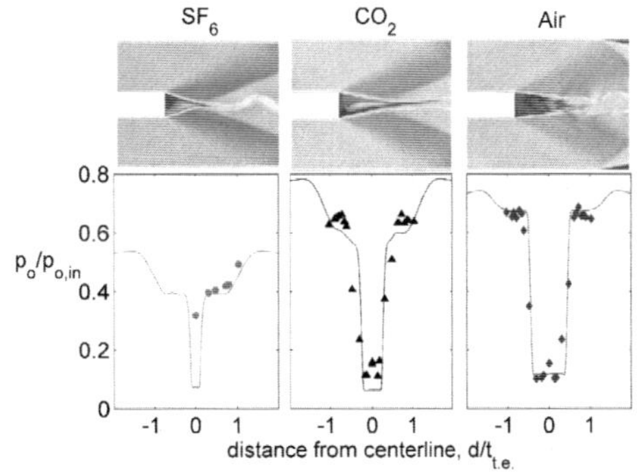

Figure 12. Measures and predicted shocked pressure profiles across the wake at a distance of $1.5t_{\mathrm{t.e.}}$ downstream of flat-plate trailing-edge and contours of instantaneous Mach number at the trailing-edge.

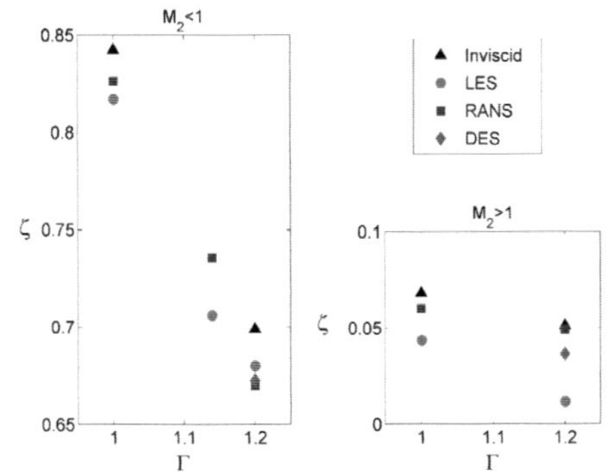

Figure 13. Predicted possible mixed-out loss coefficient for the three gases depending on the mixed out velocity M_2.

of the base region is accompanied by an increase in the base pressure.

Figure 11 shows a comparison of the measured experimental and CFD predicted wakes of air at a distance of $1.5t_{t.e.}$ downstream of the flat plate trailing-edge. The values of pressure in the profiles shown have been corrected to account for the total pressure drop across a normal shock and will be referred to as *shocked total pressures*. *Shocked total pressures* predicted by RANS, LES and DES modelling are shown. The figure shows that LES gives the closest agreement with the experimental measurements.

Figure 12 shows how a reduction in fundamental derivative produces a reduction in the size of the base region, which in-turn produces a reduction in base pressure as shown in Fig. 9. The smaller base region is produced as a result of an increasingly larger wake turning angle of the supersonic slipstream. This is directly related to the reduced Fundamental derivative in dense gases as shown in [7, 8]. Figure 12 also shows a comparison of the measured experimental and LES predicted wakes. As it can be seen here, the LES predictions in general matches the experimental measurements for all the studied flows. The results show a clear reduction in the wake size as the fundamental derivative decreases.

Figure 13 shows the possible predicted mixed-out loss coefficients for the three cases. These were calculated by using the momentum flux computed at the trailing-edge plane from the CFD calculations and the 1D analysis of a flat-plate flow discussed in the beginning of this paper. Following this mixed-out and base pressures were calculated for a range of downstream

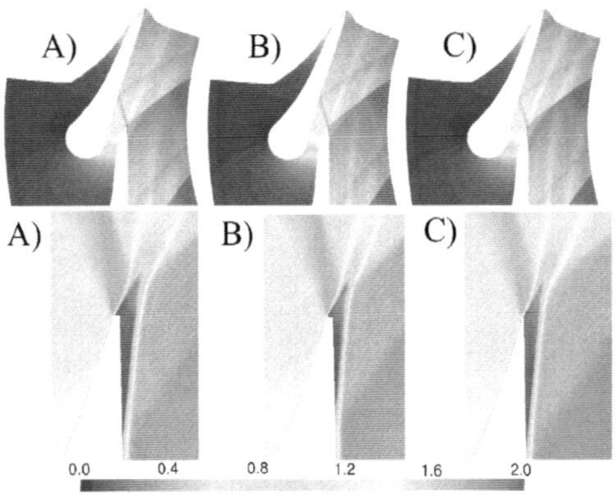

Figure 14. Predicted mean Mach numbers for pentane vane of various trailing-edge shapes: A) round, B) square-cut, C) radial-cut.

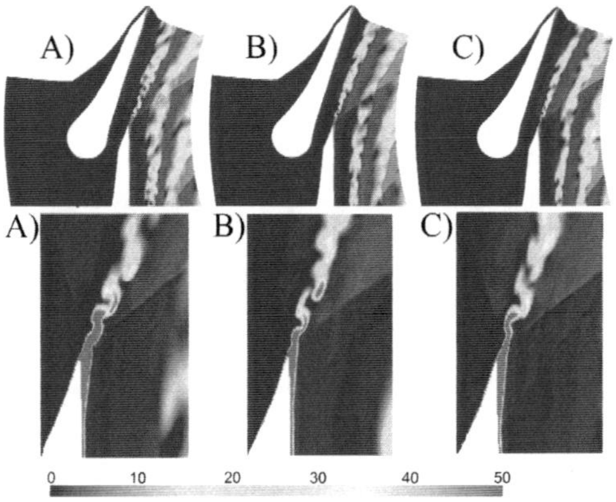

Figure 15. Predicted instantaneous entropy increase (J/kgK) for a pentane vane of various trailing-edge shapes: A) round, B) square-cut, C) radial-cut.

conditions. Finally, by using the CFD predicted base pressure, a set of possible mixed-out loss coefficients were calculated. As a result, two mixed-out loss coefficients are possible for the Air and SF_6, one of them belonging to the subsonic branch and the other one to the supersonic branch. On the other side, CO_2 only presented one possible solution of mixed-out loss coefficient with a subsonic mixed-out state.

The results show that the sensitivity of the predicted loss due to the viscous modelling is dependent on the final mixed-out Mach number; for subsonic solutions, the loss is much less dependent on viscous effects as compared to the supersonic solutions. In both cases, nonetheless, the trailing-edge loss is increased as the fundamental derivative is reduced.

6. The effect of trailing edge shape

The results obtained in the previous section are used here to provide an analysis of an ORC turbine vane. The emphasis of the study falls on the dependence of the flow features on the trailing-edge shape of a turbine vane.

Figure 14 shows a comparison of the time-averaged flowfields produced by the three different trailing-edge shapes. For all three cases the flow on the pressure side (P.S.) is expanded through the vane nozzle and around the trailing-edge ending up in a trailing-edge shock that impinges on the suction side (S.S.) of the adjacent vane. On the suction side of the vane there is a boundary-layer separation which occurs prior to the trailing-edge, and an oblique compression shock is generated. Further downstream, this merges with the compression shock formed at the confluence region of the base flow. The point of separation and strength of the subsequent oblique shock appears to be modified by the trailing-edge shape, and thus the size of the base region is dependent on the trailing-edge shape.

In the case of the square-cut trailing-edge, the point of separation occurs further downstream as compared to the other two cases (see Fig. 14). The base region produced by the square-cut geometry is also the smallest of the three cases. In the case of the radial-cut trailing-edge the point of separation occurs further upstream as compared to the other two cases, but the size of the base region is comparable to the separation produced by the round trailing-edge.

The major difference between the three cases is the over-expansion produced by the radial-cut trailing-edge. Since the radial cut modifies part of the suction surface as well as the trailing-

The effect of dense gas dynamics on loss in ORC transonic turbines

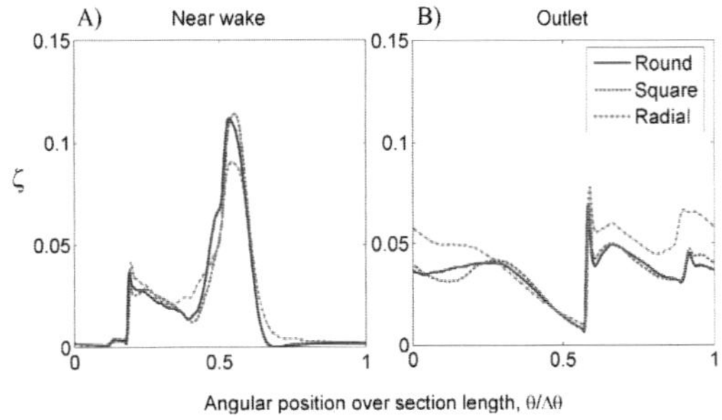

Figure 16. LES predicted time-averaged mass-weighted loss coefficient (ζ) for a pentane vane of various trailing-edge shapes at $0.25c_r$ (A) and $1c_r$ (B) radial distances downstream of trailing-edge. c_r is the radial length of the vane and $\zeta = \frac{T_{2s}\Delta s}{h_{o1}-h_{2s}}$.

edge, it produces an over-expansion ahead of the separation point. This expansion will lead to a greater shock strength as discussed later.

Figure 15 shows instantaneous contours of the entropy increase for the studied flows. Here it can be seen that an entropy increase is generated both by the base region and the trailing-edge shocks, with the base region producing the largest part of the entropy increase. The entropy rise within the wake is amplified by the trailing-edge shock, this can be observed as an enlargement of the wake width as it passes the trailing-edge shock.

As discussed before, the over-expansion of the radial-cut trailing edge produces a stronger trailing-edge shock. This can be observed clearly on Fig 15, as the entropy rise produced across the shock is larger as compared to the other two cases.

The mass-averaged loss coefficient has been calculated at specified planes of constant radial coordinates. These values have been then used to determine the loss increase throughout the domain. Loss coefficients have been calculated using the local entropy increase and the mass-weighted enthalpies and isentropic temperature, where local values have been mass-averaged at each plane.

Figure 16 shows two loss coefficient profiles along two selected radial planes. One was chosen to be in the near wake region and the second one on the outlet boundary of the domain (A and B respectively). The profile on the left of the figure shows the mass-averaged loss coefficient in the near wake. Here it can be seen how there is a sharp increase in loss coefficient across the trailing edge shock followed by a large increase in entropy across the wake. The profiles produced by the square-cut and round trailing-edge vanes are similar here, but the profile produced by the radial-cut trailing-edge presents a stronger shock and a shallower wake. On the right figure, the outlet profile can be seen. Here the wake has been diffused by the flow mixing and the interaction with the shock. Again, there is a sharp increase in loss produced by the trailing-edge shock downstream. The overall increased loss observed in the radial-cut trailing-edge case is predominately due to the increase in shock loss.

The mas-averaged loss coefficient values integrated over each plane show that both the round and square-cut trailing-edges produce loss coefficients of $\zeta_A = 0.02$ and $\zeta_B = 0.035$ at the near wake and outlet planes respectively. Comparatively, the radial-cut trailing-edge vane produces loss coefficients a 30% higher ($\zeta_A = 0.021$ and $\zeta_B = 0.044$), and this is due to the over-expansion and subsequent increased shock strength at the trailing-edge. For the other two cases, it appears that the trailing-edge shape does not affect the total loss, and this is consistent with the 1D analysis presented in this paper and in the previous work of Denton and Xu. That is to say that unless the trailing-edge shape is changing significantly the aft suction-surface loading, the trailing-edge shape will not have a significant effect on the total loss produced.

The effect of dense gas dynamics on loss in ORC transonic turbines

7. Conclusions

This paper presented an investigation of dense gas effects and trailing-edge geometry on loss. A 1D analysis of base pressure and mixed-out loss in supersonic wake flow was performed and this found that dense effects tend to increase loss.

Experiments of supersonic wake flows were performed in a Ludwieg tube using a range of gases at similar Mach and Reynolds numbers. The experimental results were used to validate a computational model for such flows. The validation of the CFD models showed that, while RANS simulations could capture changes in the flow produced by different fundamental derivatives, these failed at predicting the base region and near-wake profile. On the other hand, LES modelling showed to highly improve the agreement of the predictions with the experimental data. DES showed an improvement on the flow prediction over RANS, but still underperformed in comparison to LES results.

The results of the experiments and CFD validation showed that both the base regions and base pressure were reduced within the dense gas region. It was demonstrated that a correct modelling of the shear-layer unsteadiness in the base region is important to predicting the trailing-edge flow structure and base pressure.

Finally, a study of the effect shape of trailing-edge shape on vane wake flow and loss was performed using the validated method. This showed that, unless the trailing-edge shape modified the suction-side loading, there appeared to be little effect of trailing-edge shape on loss.

Acknowledgments

Project supported by the EPSRC (EP/J006394/1, EP/L027437/1) and GE Global Research.

Nomenclature

A	Area	s	Entropy
c, c_r	Chord, radial chord	T, $T_\circ$	Temperature, total temperature
h, $h_\circ$	Enthalpy, total enthalpy	T_c	Critical temperature
M	Mach number	$t_{t.e.}$	Trailing-edge thickness
m	Mass flow rate	t, δt	time, time-step size
p, $p_\circ$	Static pressure, total pressure	V	Flow velocity
p_b	Base pressure	Γ	Fundamental derivative
p_c	Critical pressure	ζ	Loss coefficient
r_{corner}	corner radius	θ	Angular coordinate

References

[1] Anderson W 1991 *AIAA Journal* **29** 2179–2180
[2] Anders J, Anderson W and Murthy A 1999 *Journal of Aircraft* **36** 957–964
[3] Kluwick A 2000 *Philosophical Transactions of the Royal Society of London Series A - Mathematical Physical and Engineering Sciences* **358** 3169–3192
[4] Kluwick A 2004 *Acta mechanica* **169** 123–143
[5] Cinnella P and Congedo P 2007 *Journal of Fluid Mechanics* **580** 179–217
[6] Wheeler A and Ong J 2014 *ASME Turbo Expo* GT2014-25475 (Dusseldrof, Germany)
[7] Durá Galiana F J, Wheeler A and Ong J *ASME Turbo Expo* GT2015-42920 (Montreal, Canada)
[8] Durá Galiana F J, Wheeler A and Ong J 2016 *Journal of Turbomachinery* **138**
[9] Denton J D and Xu L 1990 *Journal of Turbomachinery* **112** 277–285
[10] Lemmon E and Span R 2006 *Journal of Chemical Engineering Data* **51** 785–850

IOP Conf. Series: Journal of Physics: Conf. Series **821** (2017) 012022 doi:10.1088/1742-6596/821/1/012022

Preliminary characterization of an expanding flow of siloxane vapor MDM

A. Spinelli[1], F. Cozzi[1], G. Cammi[1], M. Zocca[2], P. Gaetani[1], , V. Dossena[1] and A. Guardone[1]

[1] Department of Energy, Politecnico di Milano
Via Lambruschini 4, 20156, Milano, Italy
[2] Department of Aerospace Science and Technology, Politecnico di Milano
Via La Masa 34, 20156, Milano, Italy

E-mail: `andrea.spinelli@polimi.it, alberto.guardone@polimi.it`

Abstract. The early experimental results on the characterization of expanding flows of siloxane vapor MDM ($C_8H_24O_2Si_3$, octamethyltrisiloxane) are presented. The measurements were performed on the Test Rig for Organic VApors (TROVA) at the CREA Laboratory of Politecnico di Milano. The TROVA test-rig was built in order to investigate the non-ideal compressible-fluid behavior of typical expanding flows occurring within organic Rankine cycles (ORC) turbine passages. The test rig implements a batch Rankine cycle where a planar converging-diverging nozzle replaces the turbine and represents a test section. Investigations related to both fields of non-ideal compressible-fluid dynamics fundamentals and turbomachinery are allowed.The nozzle can be operated with different working fluids and operating conditions aiming at measuring independently the pressure, the temperature and the velocity field and thus providing data to verify the thermo-fluid dynamic models adopted to predict the behavior of these flows. The limiting values of pressure and temperature are 50 bar and 400 °C respectively. The early measurements are performed along the nozzle axis, where an isentropic process is expected to occur. In particular, the results reported here refer to the nozzle operated in adapted conditions using the siloxane vapor MDM as working fluid in thermodynamic regions where mild to medium non-ideal compressible-fluid effects are present. Both total temperature and total pressure of the nozzle are measured upstream of the test section, while static pressure are measured along the nozzle axis. Schlieren visualizations are also carried out in order to complement the pressure measurement with information about the 2D density gradient field. The Laser Doppler Velocimetry technique is planned to be used in the future for velocity measurements. The measured flow field has also been interpreted by resorting to the quasi-one-dimensional theory and two dimensional CFD viscous calculation. In both cases state-of-the-art thermodynamic models were applied.

1. Introduction

Non-ideal compressible flows are present in a variety of industrial processes, among which energy plants based on the organic Rankine cycle (ORC) technology. Expanding flows which typically occur within ORC turbine passages are highly non-ideal and supersonic. The non-ideality is related to the proximity of the expansion processes to the critical point and to the vapor saturation curve of the working fluid, while the high Mach numbers result from the high molecular complexity and molecular mass exhibited by the organic working fluids employed in

ORC applications. Currently, no experimental data are available to characterize these non-ideal flows and to verify the accuracy of the numerical tools that are used in order to design and optimize ORC turbine blades. These are Computational Fluid Dynamic (CFD) codes implementing state-of-the-art thermodynamic models to predict the thermodynamic properties of the working fluids [1, 2, 3, 4, 5, 6]. The accuracy of these codes needs to be assessed, especially within close to saturation thermodynamic regions. Aiming at providing experimental data to improve the fundamental knowledge of non-ideal compressible flows and test cases to be compared with CFD calculations, especially for ORC applications, the *Test Rig for Organic VApors* (TROVA) has been built at the CREA Laboratory of Politecnico di Milano [7]. Expanding flows of different organic compounds in non-ideal conditions can be investigated within the TROVA by independent measurements of pressure, temperature and velocity. The facility implements a batch organic Rankine cycle (either sub-critical or super-critical), where the expansion process take place within a nozzle replacing the turbine. A straight axis, planar converging-diverging nozzle has been selected as the test section, being suitable to perform experiments of significance for both non-ideal gas-dynamics fundamentals and ORC turbine flows. The nozzle size (20 mm is the equivalent diameter at the geometrical throat) is large enough to preserve a large isentropic core of the expanding flow, therefore the temperature and pressure fields can be gathered without the use of calibrated probes, see [8]. Total pressure and temperature are measured in the settling chamber ahead of the nozzle inlet; static pressure taps at different sections along the nozzle axis are used to follow the flow evolution as it expands from rest to supersonic conditions.

The paper reports the early experimental observation of the expanding flow of siloxane MDM vapor at mild and medium non-ideal conditions. Section 2, presents the general set-up of the test-rig and an overview of the measurement techniques adopted. Within section 3, experimental results for siloxane vapor MDM ($C_8H_{24}O_2Si_3$, octamethyltrisiloxane) are reported in terms of pressure and temperature measurements and of Schlieren visualization. Experimental data have been also compared with two-dimensional viscous flow calculation. Final remarks and observations are reported in section 4.

2. TROVA: set-up and measurement techniques

The TROVA operates as a blow-down wind tunnel, namely, in a discontinuous way, to reduce the power requirements. The fluid under test is stored in a high pressure vessel (HPV) (see figure 1) and isochorically heated up to saturated, superheated, or supercritical conditions (point 4 in figure 1) at a pressure P_4 and temperature T_4 above the nozzle stagnation conditions (point 6). The heating system consist of electrical bands and wires assembled on the external surface of the HPV and of the high pressure pipeline (up to valve V4, see 1). A fast operating control valve (MCV) regulates the flow rate, thus the total pressure P_{T6} at the inlet of the nozzle during the entire test. The vapor is then expanded (to state 7) through the nozzle, where wall pressure measurements are performed. The vapor is discharged into a large area pipe (point 8) and then brought in a low pressure vessel (LPV, state 9) where the fluid is condensed (state 1). The loop is closed by the liquid compression to the HPV (point 2), performed by a membrane pump. Further details on the test rig design and arrangement are in [7].

Within the TROVA test section, the expansion flow of different organic vapors can be investigated at different operating conditions, providing that a proper nozzle profile is designed for each test [9]. The initial test scheduled concern the expansion flow of siloxane MDM, a fluid is of particular interest for industrial ORC plant exploiting relatively high temperature sources (e.g. biomass, solar radiation). The operating conditions of interest for current and future tests are summarized in table 1, as computed using the Span-Wagner thermodynamic model of MDM (see [10]) implemented in FluidProp [11]. Test labeled as MDM_{SC} and MDM_{SH} are the TROVA design cases, here reported for completeness; they allow to investigate, respectively,

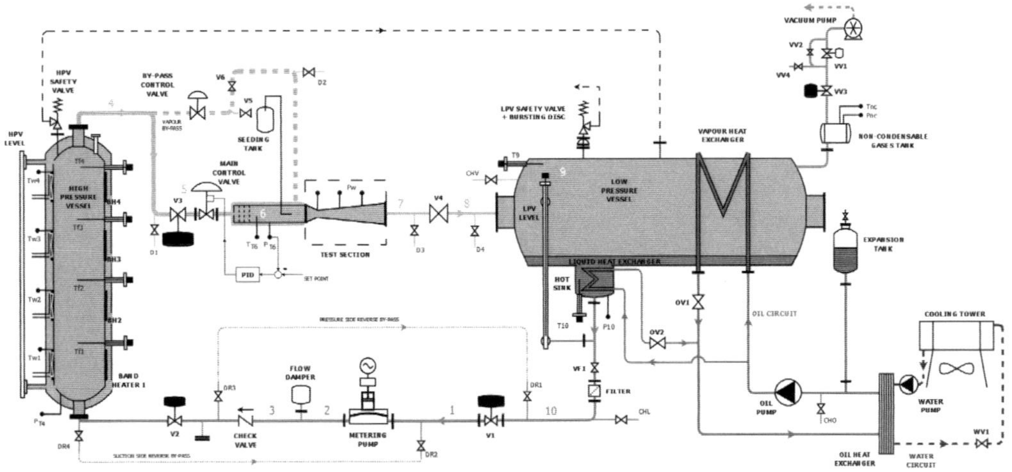

Figure 1: Scheme of the TROVA test-rig. The total conditions in the settling chamber are measured by sensors P_{T6} and T_{T6}, respectively. Static pressure taps in the test section are indicated by P_w.

Siloxane fluid MDM ($C_8H_{24}O_2Si_3$, octamethyltrisiloxane)					
TEST	P_{T6} [bar]	T_{T6} [°C]	Z_{T6}	β	M_7
MDM_0	3.15	246.0	0.88	10.08	2.12
MDM_1	4.57	245.7	0.82	9.80	2.10
MDM_2	4.40	240.4	0.82	9.81	2.10
MDM_{SH}	*25.0*	*310.3*	*0.31*	*25.0*	*2.25*
MDM_{SC}	*10.0*	*276.9*	*0.62*	*10.0*	*2.05*

Table 1: Operating conditions for the early test performed MDM_0, MDM_1, MDM_2 and for the TROVA design cases MDM_{SH}, MDM_{SC}. The expansion ratio β and exit Mach number M_7 refer to adapted conditions.

flows representative of supercritical and subcritical ORC turbine expansions. In both cases strong non-ideal effects are expected, as indicated by the value of the compressibility factor $Z = Pv/RT$, which is unity for ideal gas conditions. Three tests are reported in this paper, labeled as MDM_0, MDM_1 and MDM_2. The initial conditions of test MDM_0 correspond to mild non-ideal gas state, while the staring thermodynamics points of tests MDM_1 and MDM_2 are in medium non-ideal conditions. All test have been designed in order to reach moderate maximum values of temperatures (to avoid the risk of fluid decomposition) keeping at the same time appreciable non-ideal effects. The expansion processes described are reported in the temperature-specific entropy diagram ($T - s$) for MDM (see figure 2) referring to the initial time of each test.

2.1. Test section and instrumentation

The test section consist of a planar converging-diverging nozzle (see figure 3); the diverging portion of the nozzle profile was designed by applying the method of characteristics coupled with state-of-the-art thermodynamic models for the siloxane MDM ([9],[10]), while the converging

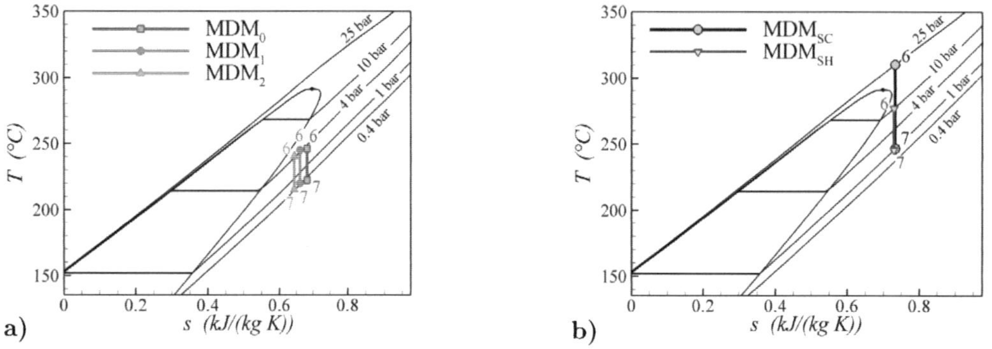

Figure 2: Overview of the expansion processes of interest for the MDM vapor. *a)* Actual expansion performed in the early tests. *b)* TROVA design expansion processes.

Figure 3: Overview of the test section (a) and details of the nozzle (b). The flow is from left to right. The settling chamber is on the left. The front window allows for optical access, the back closure is mirror polished (not installed to take the picture). Static pressure taps are located along the symmetry axis and are marked with red dots. The measuring point for total conditions (P_T6, T_T6) and for the downstream static pressure (P_v) are also indicated.

portion was realized was using a 5th order polynomial profile At the geometrical throat a recessed step of 0.1 mm depth (1.2% of the nozzle semi-height at the throat) was machined on the contoured profiles in order to fix the location of the minimum nozzle area, independently from possible boundary layer unsteadiness.

The frontal optical access is guaranteed by a planar quartz window, while 9 pressure taps along the nozzle axis are machined on the rear steel plate. Each tap is connected with the sensing element, a piezo-resistive pressure transducer, via a 25 mm long pneumatic line-cavity system exhibiting a 0.3 to 1.5 mm diameter. The volume of the capacities ahead of the transducer sensor is extremely small ($\sim 45\ mm^3$). A natural frequency of about 900 Hz has been estimated for the transmission lines by applying a viscous model developed for an arbitrary complex system of lines and cavities [12]; the non-ideal behavior of the flow has been here neglected.

In order to perform Schlieren visualization in a double passage configuration (see 2.2), the plate surface has been mirror polished. The stagnation conditions are measured in the settling chamber ahead of the test section. A wall pressure tap/line/transducer system is used also for the total pressure, due to the very low flow velocity in the chamber; the total temperature is measured by a K type thermocouple whose hot junction is located at the chamber axis.

All pressure sensors are piezo-resistive transducers (Kulite XTEH-7L series) operating at high temperature (up to 343 °C). The data acquisition (DAQ) system consist of analog modules for

Property	**Sensor**	**Type**	**CR P (bar)**	**CR T (C)**	**U_2 P (%FS)**	**U_2 T (C)**
T_T	Thermocouple	K (Ni/Cr - Ni/Al)	-	25-300	-	1.0
P_T, P_w	Piezo-resistive	Kulite XTEH	1-FS (3.5-40)	25-300	~ 0.07	-

Table 2: Type, calibration range (CR) and expanded uncertainty (U_2) of the instruments employed for pressure and temperature measurements. FS is the transducer full scale.

signal conditioning and of a high speed 16 bit ADC data acquisition board. The K thermocouple has been calibrated in the temperature range 25-300 °C . Due to the large sensitivity of the pressure transducers to temperature variation they were calibrated both in pressure and in temperature in the range of 1 bar to full scale for the pressure and of 25-300 °C for the temperature. The final accuracy of each sensor is summarized in table 2.

2.2. Schlieren set-up

The adopted Schlieren system is of double passage type, with the emitting and receiving optical components on the same optical bench. This solution entails the indubitable advantage of easy alignmentThe light emitted from a 100 W Hg arc-lamp is focused by a F/1.5 silica lens and then collimated by a Schlieren lens head of 150 mm diameter and 1000 mm. The parallel light rays traverse the test section and are then reflected back to the Schlieren head by the polished nozzle back wall (where also the pressure taps are located) and focused on the knife edge The knife orientation is vertical, thus allowing to visualize the density gradient along the horizontal nozzle axis. A semi-reflecting prism split the light beam originated by the light source and the reflected one. Finally, a real image of the test section is formed on the sensor of a high speed CMOS camera, focused by a lens of 75 mm focal lenght and 50 mm diameter) located behind the knife. The CMOS camera resolution and frame rate were respectively set to 1936 x 1216 pixels and to 20 fps (frames per second), while the exposure time was set to 1 ms. For further details on the Schlieren bench refer to [13].

3. Non-ideal nozzle flows of MDM vapor

Results from test MDM_0 (see table 1) are here reported in detail and then compared with the results of tests MDM_1 and MDM_2. Each test is triggered by the opening of the valve V3 (figure 1). The main control valve MCV has been set at a fixed position corresponding to 100% opening, in order to leave the total conditions P_{T6}, T_{T6} varying according to the HPV emptying and, therefore, exploring a wider range of inlet thermodynamic conditions and, possibly, of intensity of non-ideal behavior.

The measured test conditions in the settling chamber at time $t = 0$ s, are reported in table 1. Time $t = 0$ s has been selected at the conclusion of the initial transient. The measured value of the pressure P_{T6} and temperature T_{T6} are used to compute the value of the compressibility factor Z_{T6} and of the fundamental derivative of gasdynamics Γ_{T6} from the thermodynamic model [10]. These quantities are assumed to be the total ones, due to the very low flow velocity in the settling chamber (around 1 m/s). From time $t = 0$ s to time $t = 33$ s the nozzle operated in supersonic adapted conditions. This time interval is considered as the reference test duration for the data analysis. For the three cases, the explored regions in terms of stagnation conditions is represented in Figure 4 a) in the temperature-entropy plane.The entropy is computed from the thermodynamic model using the measured values of the total pressure and temperature.

All the investigated expansion processes are found to occur in non-ideal compressible-fluid region, as it is shown by the computed value of compressibility factor Z and of fundamental

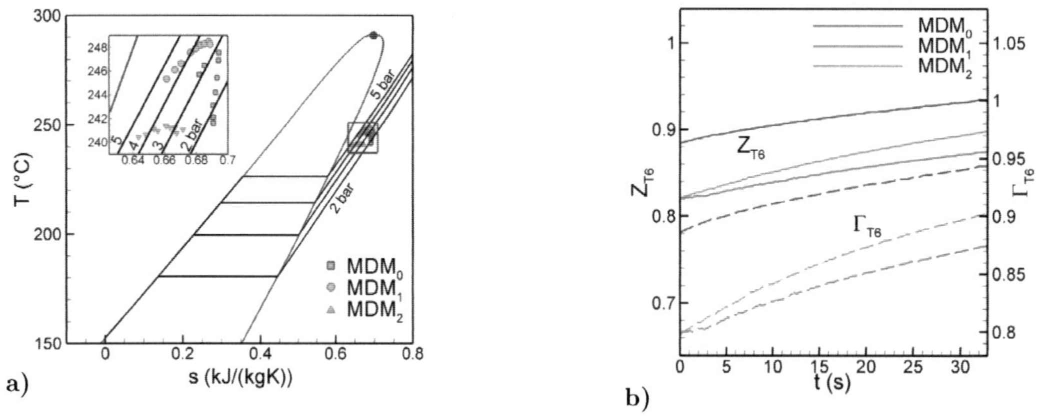

Figure 4: *a)* Thermodynamic states in the settling chamber during each test run. The entropy is computed from the thermodynamic model using the measured values of the total pressure and temperature. *b)* Computed values of the compressibility factor Z and of the fundamental derivative of gasdynamics in the settling chamber.

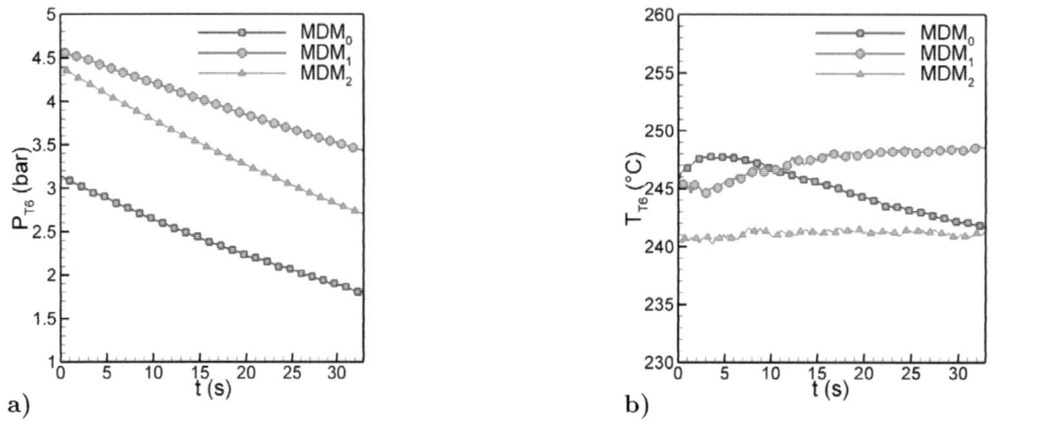

Figure 5: Measured total condition at the nozzle inlet of the three test as a function of time from $t = 0$ s and $t = 33$s. *a)* Total pressure P_{T6}. *b)* Total temperature T_{T6}.

derivative of gasdynamics Γ in the settling chamber (subscript $T6$); see figure 4 b). Mild non-ideal conditions characterize the MDM_0 expansion, while medium non-ideal states are explored during MDM_1 and MDM_2 tests.

Figure 5 shows the measured total pressure P_{T6} and temperature T_{T6} as a function of time from $t = 0$ s and $t = 33$s for the presented tests. The total pressure always decreases because the control valve is left fully open during the test run and the HPV empties. Correspondingly, the static pressure at the geometrical throat (not shown in the picture) is found to decrease with time. On the other hand the total temperature T_{T6} exhibits a non-monotonic trend, which is reasonably related to a non-uniform temperature distribution within the HPV and the downstream pipeline (up to valve V_3, see figure 1). In the following, a detailed discussion of

t [s]	P_{T6}/P_v	P_{T6} [bar]	T_{T6} [C]	Z_{T6}	γ_{T6}	Γ_{T6}	P_{th} [bar]	Z_{th}	γ_{th}	Γ_{th}	$Re_{x,th}$
0.0	6.0	3.146	245.7	0.88	1.033	0.89	2.027	0.93	1.027	0.93	1.22×10^7
9.5	5.6	2.666	246.9	0.90	1.030	0.91	1.710	0.94	1.025	0.95	1.02×10^7
20.5	5.5	2.203	244.2	0.92	1.027	0.93	1.415	0.95	1.024	0.96	8.48×10^6
30.4	5.5	1.885	242.1	0.93	1.026	0.94	1.192	0.96	1.023	0.97	7.29×10^6

Table 3: Measured stagnation conditions (subscript $T6$) and throat conditions (subscript th) for the MDM_0 test. P_{T6}/P_v represents the nozzle expansion ratio. The values of the compressibility factor Z, of the specific heat ratio γ, of the fundamental derivative of gasdynamics Γ, and of the Reynolds number $Re_{x,th}$ (based on the nozzle axial distance at the throat) are estimated from the thermodynamic and transport property models [10].

the results obtained for the MDM_0 is presented. Results of tests MDM_1 and MDM_2 are finally presented by comparison with those obtained for MDM_0.

Table 3 reports relevant flow quantities at four different time levels, namely, for $t = 0.0$ s, $t = 9.5$ s, $t = 20.5$ s and $t = 30.4$ s. The ratio of the total pressure P_{T6} to the downstream pressure P_v is such that the nozzle is choked for the entire test. At the geometrical throat, the model input are the measured static pressure and the stagnation entropy, which is assumed to be constant during the expansion (isentropic expansion along the nozzle axis) and it is therefore computed from P_{T6} and T_{T6}. The computed compressibility factors, at both the settling chamber and the geometrical throat, are increasing with time, as expected, since the HPV is emptying towards dilute conditions. At both locations, the calculatedf Z and Γ, confirm that the flow start in mild non-ideal conditions and then approaches more dilute states.

The pictures in figure 6 are the Schlieren visualizations at the four time levels listed in table 3. Notice that, due to a thermal dilation of the nozzle gaskets, the nozzle profile is not accurately visible in the images; therefore, a detailed picture of the visualized flow field at the geometrical throat at time $t = 0$ s is reported (Figure 6 e)). The actual nozzle profile (orange line) is overlapped to the dilated gasket profiles; the singular points (A, B and C) which define the shape of the recessed step, are also highlighted. The knife edge in the Schlieren bench is aligned in such a way that compressions appear as darker regions and expansions are associated to lighter regions. As observed already in preliminary experiments using air in dilute conditions [13], the presence of the recessed step (0.1 mm height) results in the occurrence of a complex wave system that is clearly visible in the divergent, supersonic portion of the nozzle. In particular, an expansion wave originates at the step edge (point A of Figure 6 e)) and it is immediately followed by a compression shock from the flow reattachment point (downstream point B). A second rarefaction fan is visible at the end of the recessed portion of the nozzle (point C), where a discontinuity in the surface geometry occurs.

In the first two frames in figure 6, the two expansions in the vicinity of the nozzle throat appear as dark regions, possibly due to the large offset of the light beam with respect to the receiving optics. A similar limitation of the adopted Schlieren technique was already observed for dilute air [13] close to the step edge. Indeed, the displacement of the light beam depends on the density gradient, which is larger in the vicinity of the step edge, where the density difference across the rarefaction wave occurs along a shorter flow path. The flow conditions in figures 6a) and 6b) are non-ideal and the fluid compressibility is almost one order of magnitude larger than in the ideal case of air and such are the density gradients. Therefore, the occurrence of the observed dark regions which extend further towards the axis of the nozzle is compatible with the fluid non-ideality. In figures 6c) and 6d), which correspond to fluid states closer to ideal conditions, the Schlieren image correctly represents the expansion regions as light grey

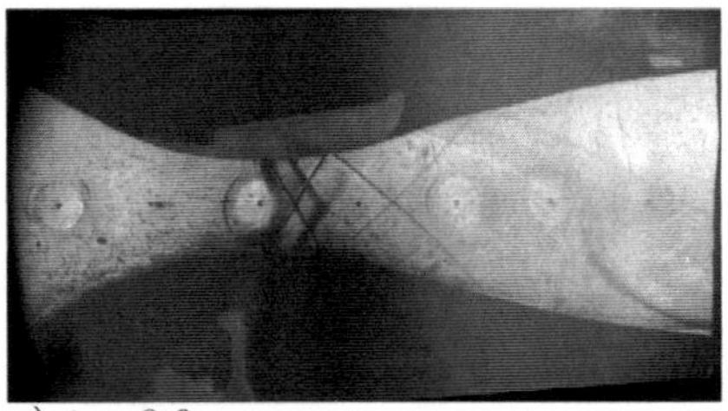

a) $t = 0.0$ s

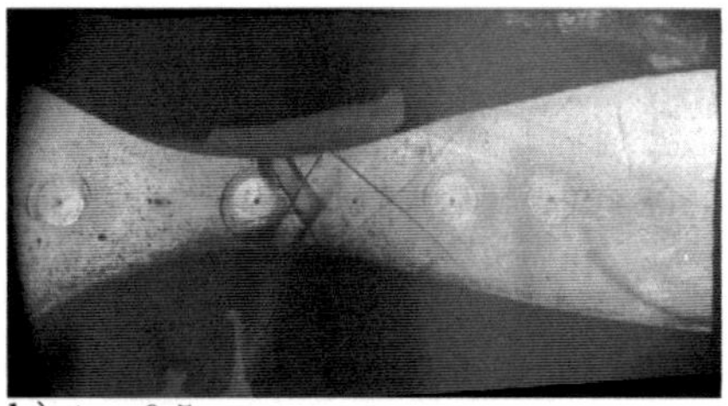

b) $t = 9.5$ s

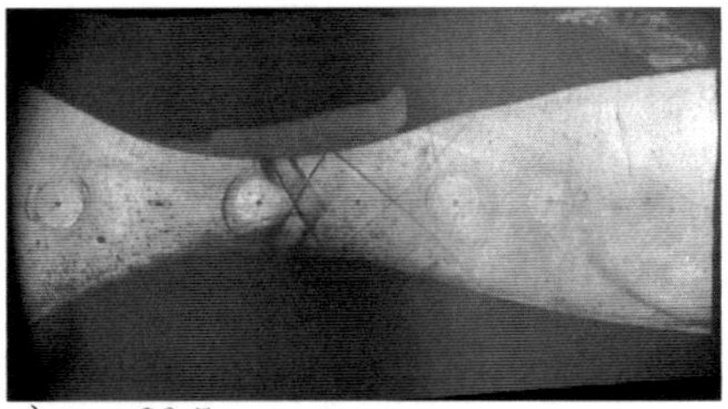

c) $t = 20.5$ s

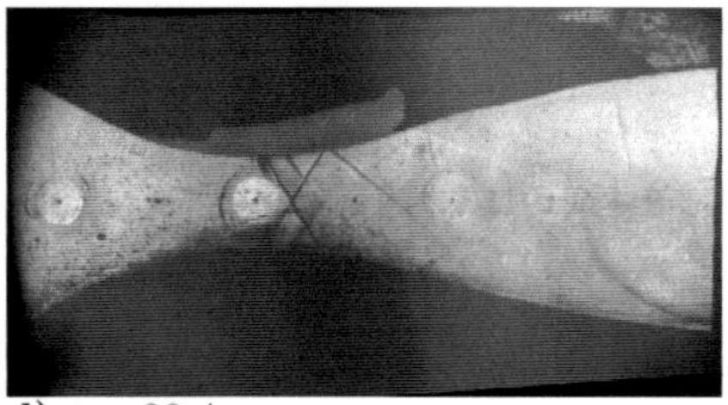

d) $t = 30.4$ s

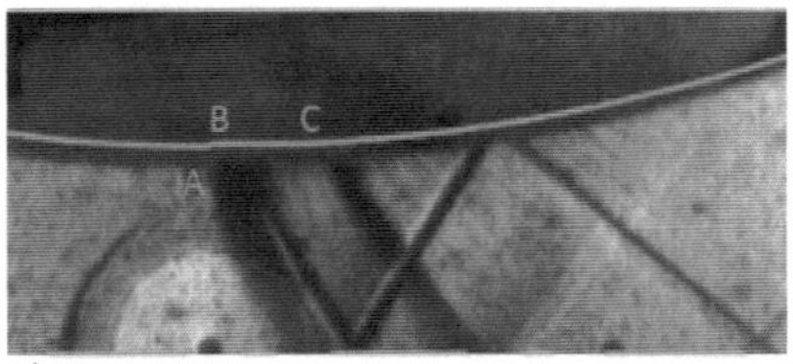

e) $t = 0.0$ s

Figure 6: Schlieren visualization of the flow field at time $t = 0.0$ s *(a)*, $t = 9.5$ s *(b)*, $t = 20.5$ s *(c)* and $t = 30.4$ s *(d)*. *(e)* Detailed view of the flow field at time $t = 0$ s. The actual nozzle profile is superimposed to the gasket profiles thermally dilated. The line segments A, B and C define the recessed step geometry.

regions and the size and shape of the dark regions closely resemble those observed in ideal gas experiments with air.

Figure 7 a) reports the comparison between the MDM run and a test with air in dilute conditions in terms of the ratio of the static pressure P_{th} at the geometrical nozzle throat to the total pressure P_{T6} from $t = 0$ s and $t = 33$s. For air in dilute conditions, the ratio is found to be constant in time, in accordance with well known results for adapted nozzles operated with ideal gases. On the contrary, for the non-ideal run with MDM, the ratio P_{th}/P_{T6} is predicted to decrease as more dilute conditions are approached. For the test MDM_0, though the nozzle is clearly working in adapted conditions in the considered time interval, this non-ideal effect is not clearly visible, since the small expected decrease (related to the mild non-ideal conditions explored) cannot be detected due to the transducer uncertainty. Therefore, in order to catch this effect, test MDM_1 and MDM_2 have been performed in higher non-ideal conditions and by using lower full scale transducers, with consequent smaller uncertainty.The trend exhibited by the P_{th}/P_{T6} ratio is now clearly decreasing as predicted by non-ideal thermodynamic models. Moreover, by comparing the P_{th}/P_{T6} trend of the three test, it can be seen that higher values are obtained in thermodynamic regions exhibiting higher non-ideal effect, namely smaller Z and Γ (see figure 4 b)), also in accordance with model predictions.

Finally the measured values of P_{th}/P_{T6} at $t = 0$ are compared with the one obtained with both inviscid and viscous CFD calculation, performed at the boundary conditions of each test at time

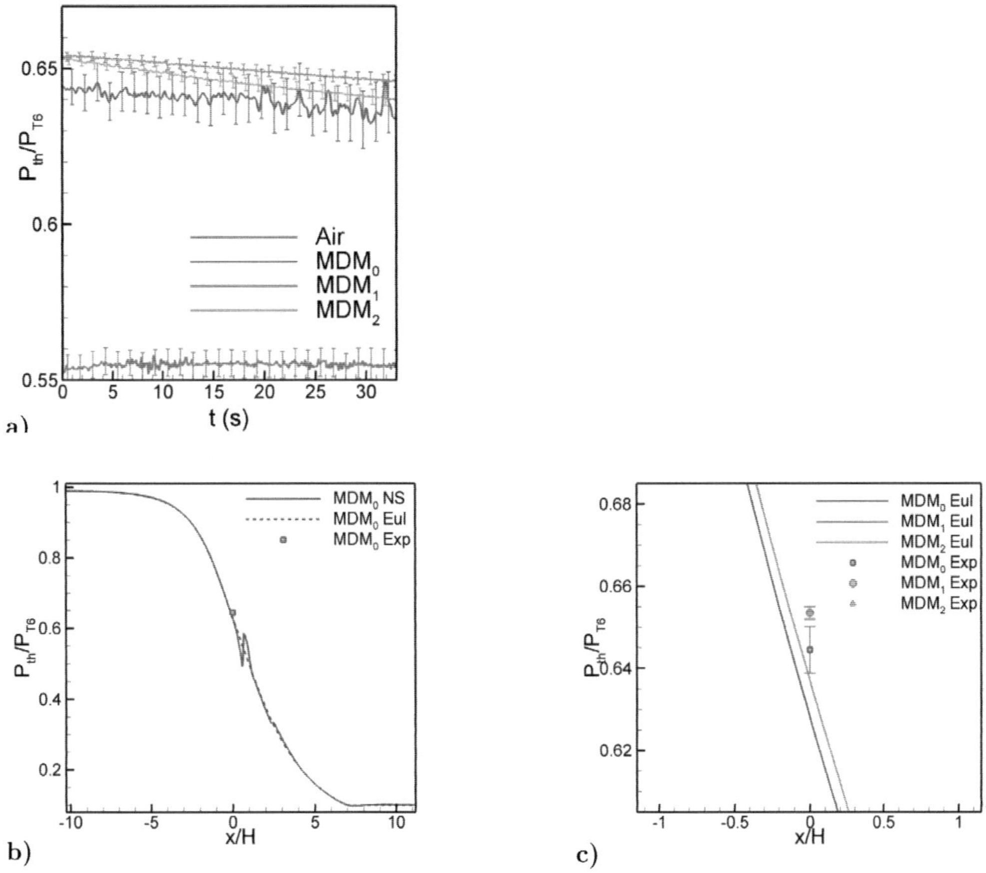

Figure 7: *a)* Static-to-total pressure ratio P_{th}/P_{T6} at the geometrical throat from $t = 0$ s and $t = 33$s for all MDM test run and for air in dilute conditions. Uncertainty bars are also reported. *b)* Comparison between experimental data (Exp) and CFD results at the nozzle axis line for the MDM_0 run at time $t = 0$. x represents the nozzle axial coordinate, H the nozzle semi-height at the throat. Inviscid (Eul) and viscous (NS) calculation have been performed, giving the same results at the throat. *c)* Detail of the comparison between experimental data and CFD inviscid calculation at the geometrical throat for the three MDM test run at time $t = 0$. MDM_1 and MDM_2 CFD plots are overlapped since the two expansion occur in the same non-ideal conditions

$t = 0$. In figure 7 b) for MDM_0 test, the experimental value (labeled Exp) of P_{th}/P_{T6} is compared with the one extracted along the nozzle axis from two two-dimensional CFD steady calculation; one inviscid (labeled Eul) and one viscous (labeled NS). In both cases, the thermodynamic model is PRSV (Peng Robinson Stryjek Vera), while the viscous calculation adopts the (k-ω SST) turbulence model implemented in the SU2 solver. The comparison shows that the inviscid calculation is able to catch the static-to-total pressure ratio up to $x/H = 0.6$, where the shock departing from the re-attachment point downstream of the step intersects the nozzle axis. For this reason in figure 7 c) the experimental values of P_{th}/P_{T6} (together with their respective uncertainty bands) obtained for the three tests are compared with the results of the inviscid calculation. The agreement between the experimental and the numerical results is very good, being the difference of about 2.8% of the measured values for test MDM_0 and of about 2.5% for tests MDM_1 and MDM_2. This differences are probably due to a slight inaccuracy in the modeling of the boundary layer.

4. Conclusions

Preliminary results from the Test-Rig for Organic Vapors (TROVA) at Politecnico di Milano were reported here for siloxane fluid MDM ($C_8H_{24}O_2Si_3$, octamethyltrisiloxane). A subsonic-to-supersonic expansion in a converging-diverging nozzle was observed during the experimental runs. A double-passage Schlieren technique was used to visualize the streamwise density gradient and to locate the salient wave structures in the supersonic flow region. Total pressure and temperature were measured in the settling chamber. Static pressure measurement were also carried out along the nozzle axis.

The observation of shock waves induced by the recessed step confirms the occurrence of a supersonic nozzle flow of MDM in the vapor phase. The static-to-total pressure ratio measured at the geometrical throat is in accordance with values predicted by resorting to the theory of non-ideal compressible-fluid flows. Further measurement campaign are foreseen in order to complete the MDM flow characterization, they include static pressure and direct velocity measurements along the entire nozzle axis. Such experimental results can represent a useful database to verify CFD codes for non-ideal compressible flow.

References

[1] Colonna, P., and Rebay, S., 2004. "Numerical simulation of dense gas flows on unstructured grids with an implicit high resolution upwind Euler solver". *Int. J. Numer. Meth. Fluids*, ***46***(7), pp. 735–765.
[2] Guardone, A., 2007. "Three-dimensional shock tube flows of dense gases". *J. Fluid Mech.*, ***583***, pp. 423–442.
[3] Cinnella, P., and Congedo, P. M., 2007. "Inviscid and viscous aerodynamics of dense gases". *J. Fluid Mech.*, ***580***, pp. 179–217.
[4] Colonna, P., Harinck, J., Rebay, S., and Guardone, A., 2008. "Real-gas effects in organic rankine cycle turbine nozzles". *J. Prop. Power*, ***24***, March-April, pp. 282–294.
[5] Hoffren, J., Talonpoika, T., Larjola, J., and Siikonen, T., 2002. "Numerical simulation of real-gas flow in a supersonic turbine nozzle ring". *J. Eng. Gas Turbine Power*, ***124***, pp. 395–403.
[6] Pini, M., Spinelli, A., Persico, G., and Rebay, S., 2015. "Consistent look-up table interpolation method for real-gas flow simulations". *Computers and Fluids*, ***107***, pp. 178–188.
[7] Spinelli, A., Pini, M., Dossena, V., Gaetani, P., and Casella, F., 2013. "Design, simulation, and construction of a test rig for organic vapours". *ASME Journal of Engineering for Gas Turbines and Power*, ***135***, p. 042303.
[8] Spinelli, A., Dossena, V., Gaetani, P., Osnaghi, C., and Colombo, D., 2010. "Design of a test rig for organic vapours". In Proceedings of ASME Turbo Expo, Glasgow, UK.
[9] Guardone, A., Spinelli, A., and Dossena, V., 2013. "Influence of molecular complexity on nozzle design for an organic vapor wind tunnel". *ASME Journal of Engineering for Gas Turbines and Power*, ***135***, p. 042307.
[10] Colonna, P., Nannan, R., Guardone, A., and Lemmon, E. W., 2006. "Multi-parameter equations of state for selected siloxanes". *Fluid Phase Equilib.*, ***244***, pp. 193–211.
[11] Colonna, P., and van der Stelt, T. P., 2004. FluidProp: a program for the estimation of thermo physical properties of fluids. software.
[12] Antonini, C., Persico, G., and Rowe, A. L., October 13, 2008. "Prediction of the dynamic respinde of complex transmission line system for unsteady pressure measurament". *Measurament Science and Technology*, ***19***(12).
[13] Cozzi, F., Spinelli, A., Carmine, M., Cheli, R., Zocca, M., and Guardone, A., 2015. "Evidence of complex flow structures in a converging-diverging nozzle caused by a recessed step at the nozzle throat". In 45th AIAA Fluid Dynamics Conference, Dallas, TX, USA, 22-26 June.

Acknowledgment

The research is funded by the European Research Council under ERC Consolidator Grant 2013, project NSHOCK 617603. The initial layout of the plant was funded by Turboden S.r.l..

Design and test of a 10kW ORC supersonic turbine generator

J R Seume[1], M Peters[1], H Kunte[1]

[1]Institute of Turbomachinery and Fluid Dynamics, Leibniz Universitaet Hannover, Appelstrasse 9, 30167 Hannover

peters@tfd.uni-hannover.de

Abstract: Manufactures are searching for possibilities to increase the efficiency of combustion engines by using the remaining energy of the exhaust gas. One possibility to recover some of this thermal energy is an organic Rankine cycle (ORC). For such an ORC running with ethanol, the aerothermodynamic design and test of a supersonic axial, single stage impulse turbine generator unit is described. The blade design as well as the regulation by variable partial admission is shown. Additionally the mechanical design of the directly coupled turbine generator unit including the aerodynamic sealing and the test facility is presented. Finally the results of CFD-based computations are compared to the experimental measurements. The comparison shows a remarkably good agreement between the numerical computations and the test data.

1. Introduction

Automotive manufacturers are permanently searching for possibilities to increase the efficiency of internal combustion engines to reduce the fuel consumption and the emission of greenhouse gases. The investigation of energetic losses in a combustion engine showed that about one-third of the chemical energy is lost as thermal energy through the exhaust gas [1]. A possibility to recover some of this thermal energy is an organic Rankine cycle (ORC).

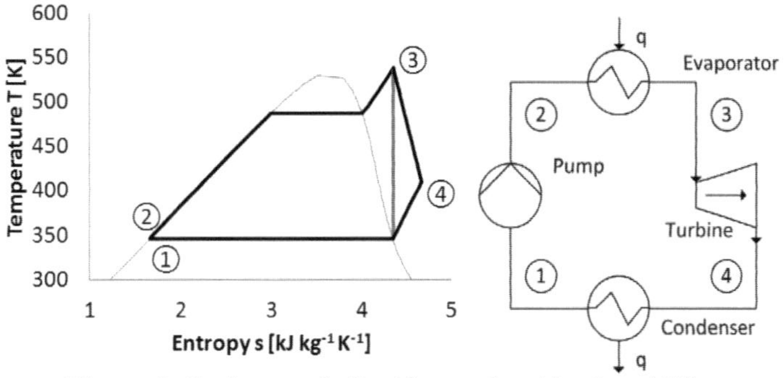

Figure 1: Basic organic Rankine cycle with ethanol [2]

The four working processes of an ORC are shown in figure 1. First a pump raises the pressure of the liquid working fluid ①→②. Afterwards the fluid is heated, evaporated und superheated in an evaporator ②→③. The superheated steam is expanded in a turbine ③→④, generating the desired power. At last the fluid enters a condenser where it is liquefied and returned to the pump ④→①.

2. Operating conditions of the ORC

The intention of this investigation is the development of a turbine generator unit for a waste heat recovery system of a 12.8 l diesel engine with a maximum power output of 375 kW. The design point is a medium engine load point with an exhaust gas mass flow of 0.249 kg/s at a temperature of 615 K. Table 1 shows the design boundary conditions for ethanol heated by this exhaust gas flow. Ethanol was chosen as working fluid for this ORC cycle because a prior investigation of Kunte and Seume [3] showed that this is the most promising fluid for this boundary conditions.

Table 1. Design boundary conditions for ethanol

Parameter	Units	Target
Fluid		Ethanol
$\dot{m}_{S,in}$	kg/s	0.052
$T_{S,in}$	K	530
$p_{t,S,in}$	bar	40
$p_{R,out}$	bar	0.81

In contrast to the prior investigation a mass fraction of 5 % water is added to the ethanol in order to improve the corrosion resistance of the titanium rotor. The water changes the fluid properties and therefore the boundary conditions of the ORC have to be adjusted. The new boundary conditions of the ORC for a mixture of ethanol and water are summarized in table 2.

Table 2. Design boundary conditions for a mixture of ethanol and water

Parameter	Units	Target
Fluid	Ethanol (95%mass), Water (5%mass)	
$\dot{m}_{S,in}$	kg/s	0.045
$T_{S,in}$	K	539
$p_{t,S,in}$	bar	40
$p_{R,out}$	bar	0.81

The temperature of the fluid at the turbine inlet must be raised from 530 K to 539 K to avoid condensation and droplet impact erosion which would damage the turbine rotor. Due to the increased process temperature, the higher thermal capacity, and the higher enthalpy for the vaporization of the water the mass flow is reduced to 0.045 kg/s. The maximum inlet and outlet pressure are mainly depended on the turbine design and remain unchanged.

3. Design of the turbine generator unit

The boundary conditions of the ORC are quite ambitious for turbines due to the combination of a low mass flow with a very high pressure ratio. The high pressure ratio results in high flow velocities and dissipative compression shocks if expanded in a single stage turbine. Whereas the low mass flow requires small wheel diameters and low blade heights, which lead to increased blade losses such as wall friction and tip gap losses. In addition the small turbine diameter combined with the high flow velocity results in a high rotational speed which increases the complexity of the bearing and the coupling of the machine. Together with the demand for a compact design and low production cost with regard to the automotive application, the right choice of the turbine type is crucial to the overall performance of the system. In a prior study Sourell et al. [4] investigated different turbine types for these boundary conditions. They concluded that an axial impulse turbine is best suited for this application. The advantages of an axial impulse turbine compared to other turbine designs, e.g. a radial inward turbine, are comparable low speeds which simplify the coupling and bearing of the generator, simple manufacturing due to the 2D blade design and the capability to expand high pressure ratios in one stage which allows a compact design. Furthermore an axial impulse turbine can be used with partial admission.

Especially the partial admission is advantageous in this application. Partial admission means that just a part of the circumference is provided with flow and the rest of the circumference merely ventilates. On the one hand the partial admission causes additional losses, which decreases the turbine efficiency. On the other hand partial admission allows increased blade heights which decreases the relative gap height and hence the tip gap losses. As a result the additional partial admission losses are usually overcompensated by the reduced tip gap losses, especially for low mass flows. However the main advantage is that the partial admission can be used to regulate the turbine mass flow rate. The presented turbine design features a variable partial admission, which can be changed between 20%, 40%, 60%, and 80% of the turbine circumference. As a result, the turbine operating range and therewith the range of high efficiency is significantly extended. Without a regulation it would be impossible to reach high efficiencies for all desired operating points of the truck engine. The regulation strategy of the turbine is shown on left side of figure 2. The degree of partial admissions depends on the number of opened stator passages. For low mass flows of less than about 0.03 kg/s only two stator passages are opened. This accounts for 20% partial admission. Every time the stator inlet pressure exceeds 40 bar two additional stator passages are opened until all eight passages respectively 80% of the circumference are opened.

Eventually it was decided that a single stage axial impulse turbine with partial admission is best choice for the described boundary conditions. In an impulse turbine the conversion from pressure energy to kinetic energy only takes place in the stator. Hence, there is no pressure gradient in the rotor and the flow is only redirected. This means that the stator has to expand a pressure ratio of up to 49 with the described boundary conditions which leads to a supersonic flow at the outlet of the stator. To minimize the expansion losses the stator passages are formed like Laval nozzles. The working principle of a Laval nozzle can be described using the right side of figure 2. The flow enters the Laval nozzle at subsonic velocity ① and is accelerated until it reaches the speed of sound in the throat ②. Afterwards the flow is accelerated to supersonic speed in the divergent part of the nozzle ③. The Laval nozzle is chocked for all examined operating points. Hence the mass flow through the nozzle is only dependent on the inlet temperature, the inlet pressure, and the cross-section of the nozzle throat. Components or fluid conditions downstream of the nozzle, including the rotational speed of the turbine do not influence the mass flow. The shape of the Laval nozzle stator are designed with the help of an in house code which uses the typical equations for Laval nozzles to calculate the cross sections of the trapezoid shaped passages. The rotor blades are designed with a commercial design tool. The cross sections of the rotor passages are constant from inlet to outlet to avoid any acceleration. The radii of the blade leading and trailing edges are thickened to 0.2 mm to improve the mechanical integrity against abrasion. The blade profiles shapes of the stator and the rotor are shown in figure 2.

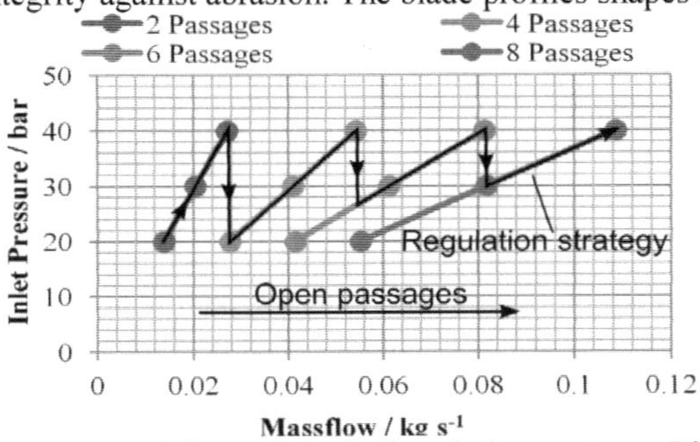

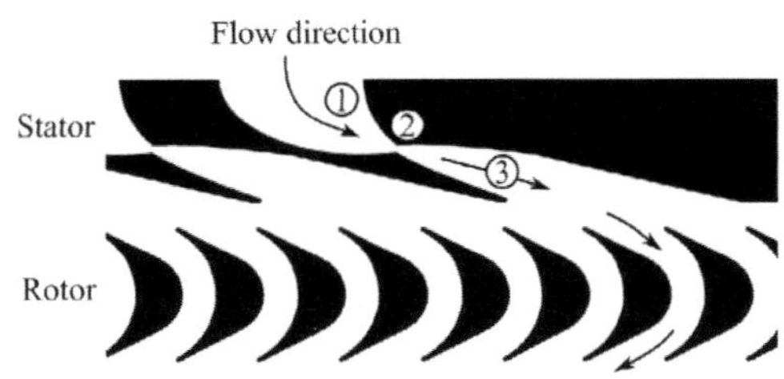

Figure 2: Left: Regulation strategy; Right: Blade profile of the stator and rotor

The geometrical parameters of the single stage impulse turbine, designed with the process described above, are listed in table 3. The turbine is designed for a partial admission of 40% at a rotational speed of 100,000 min^{-1}. At this operating point the predicted turbine power output is 8.0 kW. With the maximum degree of partial admission of 80% a turbine power of 18.3 kW is achieved at the maximum rotational speed of 110,000 min^{-1}. The rotor has 33 blades whereas the stator has 8 passages. The blade height is only 3.4 mm with a shroud diameter of about 63 mm.

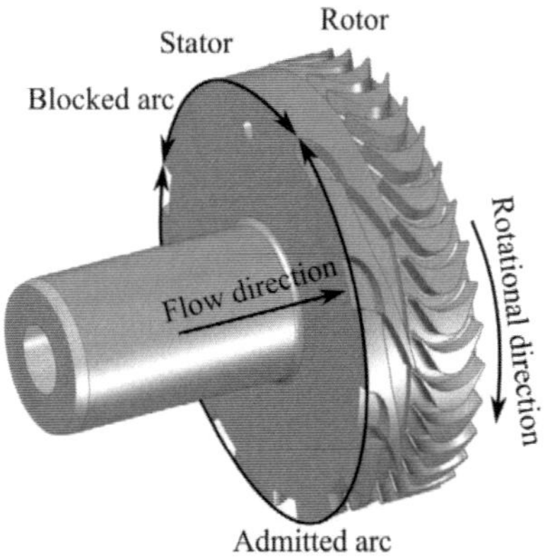

Table 3. Parameters of the investigated turbine

Parameter	Unit	Value
D_{Shroud}	m	0.0631
h	m	0.00343
δ	m	0.00013
$n_{Operation}$	min^{-1}	60,000...110,000
n_{Design}	min^{-1}	100,000
N_S		8
ε_{Design}		40%
ε		20%, 40%, 60%, 80%
N_R		33

Figure 3: Model of stator and rotor

The turbine consists of the a stator with Laval nozzles and a turbine wheel with impulse blades, as shown in figure 3. The stator is made of steel and the rotor blisk is made of titanium. The lighter weight of the titanium is beneficial for the rotor dynamics compared to a steel blisk. To avoid corrosion a small part palladium is added to the titanium. Both turbine parts are made via milling.

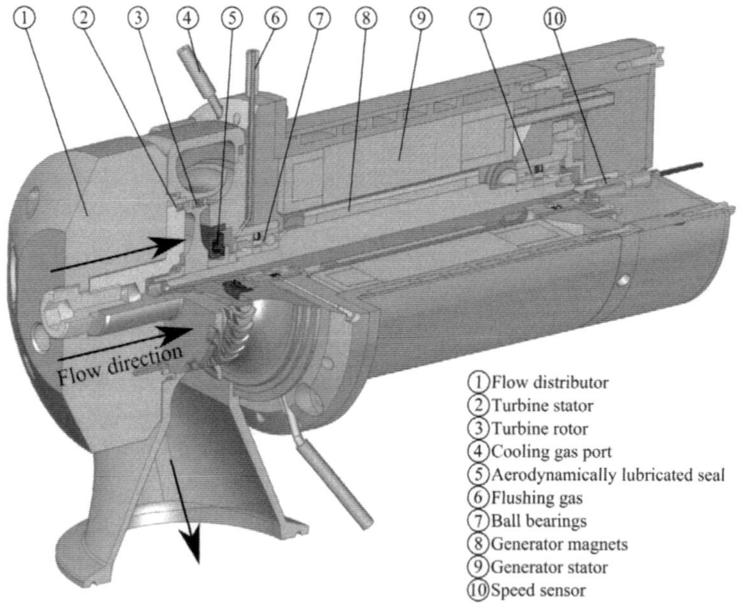

Figure 4: Cut section view of the mechanical design

A cut section view of the mechanical design of the turbine generator unit is shown in figure 4. The ethanol steam enters the flow distributor ① through three individual pipes which are necessary for the different degrees of partial admission. Inside the distributor are channels which guide the flow to the individual stator nozzles ②. Each nozzle is sealed by an O-Ring to avoid bypass flow. The flow is accelerated inside the stator and enters the rotor ③ with supersonic speed. The rotor converts the kinetic energy of the flow into rotational energy of the shaft. Downstream of the rotor the flow is redirected radially inside a diffuser. Finally the flow is collected in a plenum which guides the flow to the outlet.

The turbine rotor is mounted in an overhung position on a shaft equipped with permanent magnets ⑧. Through the magnetic field the mechanical power is transferred to the generator stator ⑨. The shaft is supported by two pairs of high precision ball bearings ⑦ with lifetime lubrication. In order to avoid a contamination of the ethanol cycle with air the shaft is equipped with an aerodynamically lubricated face seal ⑤. This face seal transports small amounts of ethanol from the cycle into a collecting space. The collecting space is permanently flushed with nitrogen through a port ⑥.

4. Testing of the ORC turbine generator

The designed and manufactured ORC turbine generator unit is tested on a multipurpose test bench of the University of Hannover. The basis of test bench, which is shown in figure 5, is a turbocharger test bench with a dual burner for the examination of passenger vehicle and commercial vehicle turbochargers. For ORC measurements an ORC-module is connected to the large burning chamber which provides the hot exhaust gas for the cycle. The design of the ORC-module is focused on versatility and robustness rather than size and weight. Therefore only approved components that fulfill all governmental regulations are used instead of small and lightweight parts for automotive applications. However, the modular design of the test bench allows an easy exchange of components meaning that not only ORC turbines but all parts of an ORC cycle can be tested under different conditions. For this purpose the test bench can be operated with several working fluids and not only the turbine but also each component of the cycle is highly instrumented.

Figure 5: ORC test bench of the University of Hannover

Nevertheless, the major task of the ORC test bench is the performance evaluation of expansion machines for ORCs. A simplified scheme of the test bench including the and the main measurement locations is shown in figure 6.

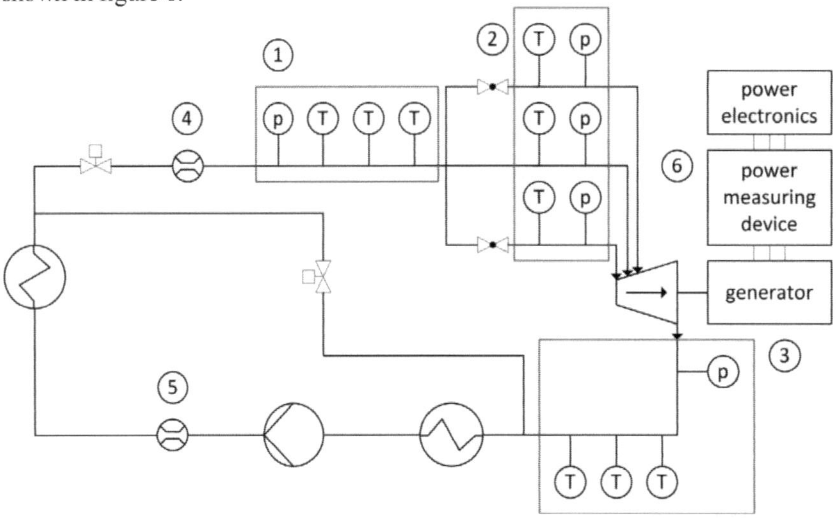

Figure 6: Simplified scheme of the test bench and the main measurement positions [2]

Instrumented tubes at the inlet ① and outlet ③ of the turbine generator unit are used to measure the temperature and pressure of the fluid. Additional temperature and pressure probes are implemented inside the turbine directly in front of the stator ② to reduce the influence of thermal losses, which are impairing the measurement accuracy of the efficiency. The mass flow is determined in the vapor phase right in front of the turbine ④ with a flow meter and in the liquid phase behind the pump with a mass flow meter ⑤. The test bench is equipped with a bypass for the start procedure as well as for security reasons. The bypass is located between the evaporator and the turbine and splits the flow into two parts. One part flows through the turbine and the other part flows directly to the condenser. The flow meter in front of the turbine measures only the flow which passes through the turbine whereas the mass flow meter in the liquid phase measures the complete mass flow from the pumps. If the bypass is completely closed both meters measure the same turbine flow.

The electric power of the generator is measured with a high accuracy power analyzer for high frequencies ⑥. The analyzer is located between the generator and the power electronics which allows the determination of the turbine isentropic efficiency, if the mechanical efficiency of the expansion machine is known.

All shown experiments are conducted with a constant static outlet pressure of 0.81 bar. The inlet temperature is set in dependency of the variable inlet pressure to prevent condensation. This is achieved by setting the inlet temperature 10 K above the temperature needed to avoid condensation at the turbine outlet, assuming an ideal isentropic expansion inside the turbine. The rotational speed can be set freely, however only the results of the rotational speed with the highest efficiency are shown in this paper.

5. Numerical simulations

For the numerical performance prediction of the ORC turbine a combination of CFD calculations and empirical methods is used. For the first step the aerodynamic performance is calculated by means of CFD simulations. Secondly empirical calculations are used to take losses due to the partial admission into account.

The steady-state CFD calculations for the performance evaluation of the turbine are performed with Ansys CFX (ver. 14.5) using the SST turbulence model. The fluid is a mixture of 95% ethanol and 5% water which is modeled by the Aungier Redlich Kwong real gas model of Ansys. The computational domain consists of one stator and three rotor passages followed by an axial-radial diffuser. To match the pitch of all components the rotor has been geometrical scaled so that all parts cover 36° of the circumference. The stator and rotor and the rotor diffuser domain are connected with an axisymmetric frozen rotor interface which allows the interaction of the shocks between rotor and stator. The alternative mixing plane interface was not used because a test showed non-realistic reflections of the shocks at the interface. Mass flow rate and total temperature are used as inlet boundary conditions and static temperature is used as outlet boundary condition. All values for the boundary conditions including the rotational speed of the turbine are derived from the experimental results (see section 4).

The grid generation is performed with Ansys ICEM CFD for all parts. All numerical grids are structured grids with a y-plus value of about 30 so that the near wall velocity profile is computed using wall functions. A grid-independence study is performed with grids of 0.19 million, 2.6 million and 9.13 million nodes. The study shows that the efficiency deviation of the medium grid compared to the fine grid is 0.2%, which is assumed to be sufficient for this investigation.

As a result of the rotational periodicity condition of the numerical model no partial admission losses occur in the CFD calculations. To take the partial admission losses into account, the calculated efficiency values of the CFD are reduced using the empirical approach of Aungier [5]. The losses of the empirical model are composed of two parts. The first part is the windage loss that occurs in the non admitted arc of the rotor by reentry of the outlet flow into the inactive passages. The second part is the end of sector loss that occurs if a rotor blade passage enters or leaves the admitted part of the stator.

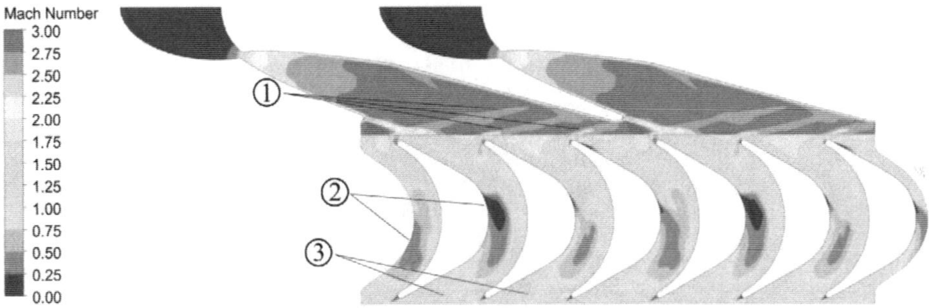

Figure 7: Mach Number at 50% span for an inlet pressure of 40 bar

Figure 7 shows the relative Mach number at 50% span of the stator and rotor at a rotational speed of 100,000 min^{-1}, which is the optimal speed for this operating point. The flow in the divergent part of the nozzle is highly supersonic (Mach number > 3). Close to the outlet of the nozzle several shocks ① occur which reduce the outlet velocity of the nozzle. The flow is even more decelerated by additional shocks at the leading edge of the rotor. This leads to a transonic flow with subsonic regions in the rotor ②. At the outlet of the rotor ③ the flow is slightly accelerated by a free expansion due to the remaining pressure gradient relative to the outlet pressure of 0.81 bar.

6. Comparison of experimental and numerical results

A comparison of the experimental and numerical results, achieved with the setup described above, is shown in the figures 8, 9, and 10 (for boundary conditions and experimental setup see section 4). All figures show the results for a degree of partial admission of 20% and 40%, which means that two respectively four passages of the turbine stator are opened. The results for two opened passages are shown in blue and the results for four passages in brown. The lighter colors represent the results of the CFD and the darker colors of the experiments.

In figure 8 the mass flow through the turbine is shown in dependency of the inlet pressure. The deviation between the numerical and experimental results is about 0.3% to 0.5% for two passages and 1.3% to 2.0% for four passages. The good agreement between both results is partly due to the behavior of the supersonic Laval nozzle used in the stator. For a given nozzle geometry the mass flow through the stator and with it also through the whole turbine is only depended on the inlet pressure and temperature. Whereas other values like the rotational speed do not influence the mass flow characteristic. For this reason the rotational speed of the turbine could be chosen to meet the highest power output and efficiency for each shown operating point without influencing the mass flow.

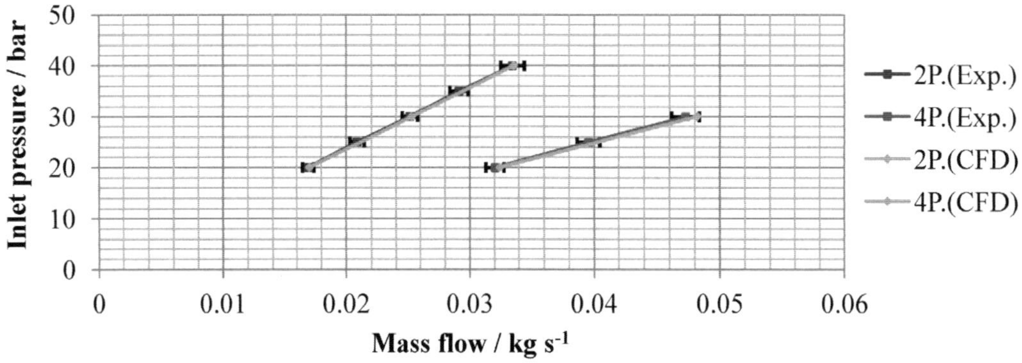

Figure 8: Mass flow of the turbine in dependency of the inlet pressure

Figure 9 shows the aerodynamic turbine power output. The turbine reaches a maximum power output of about 8 kW with four opened passages at an inlet pressure of 30 bar. With two passages the maximum power output is about 5 kW at an inlet pressure of 40 bar. The deviation from the turbine power between the numerical and experimental results is between 4.7% and -4.0% for two opened passages and between 3.8% and 2.9% for four opened passages. Based on these results it is estimated that the turbine reaches a maximum power output of up to 19 kW with all eight passages opened.

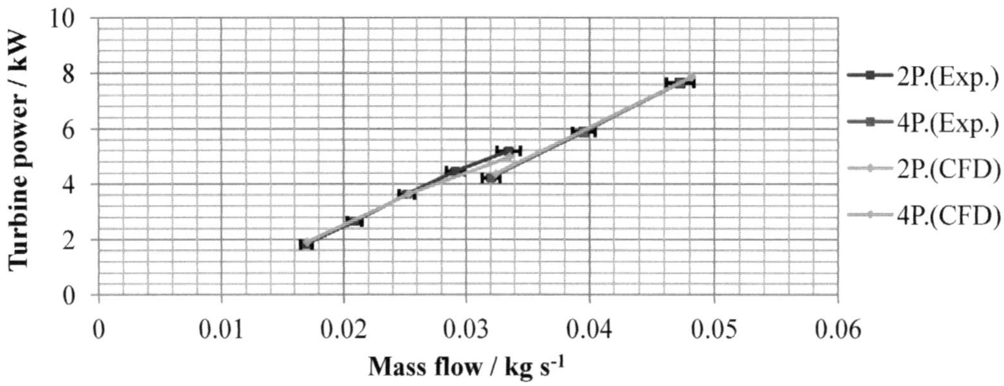

Figure 9: Turbine power output

The aerodynamic total-to-static efficiency of the turbine is shown in figure 10. The experimental total-to-static efficiency increases from 43.1% at the lowest mass flow and two opened passages to 57.0% at the highest mass flow and four opened passages. The deviation between the numerically and experimentally evaluated efficiencies is significantly larger for two passages with 2.1% to -3.3% than the deviation of four passages with 1.2% to 0.2%. Notably is the different trend of the numerical and experimental data for two opened passages. The numerical calculations predicted a decreasing efficiency of the turbine above an inlet pressure of 30 bar with two opened passages. Whereas the experimental data shows an increasing efficiency up to the maximum inlet pressure of 40 bar. No explanation for this discrepancy is found so far.

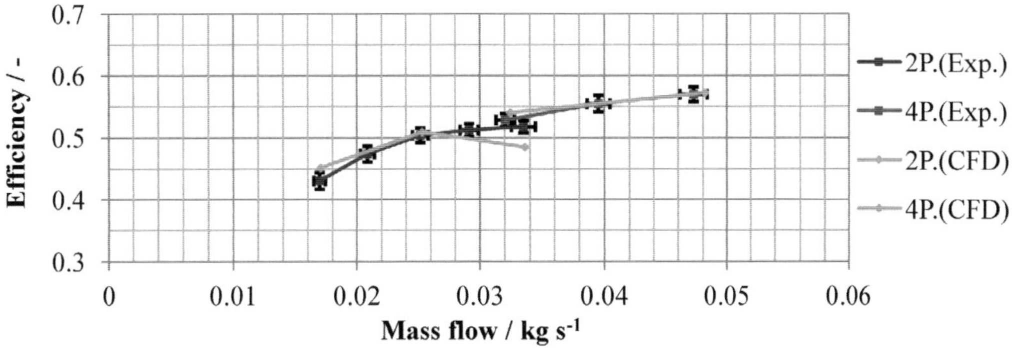

Figure 10: Efficiency of the turbine

7. Conclusions

An ORC supersonic turbine generator unit for an automotive applications has been developed at the TFD. The turbine is designed for challenging boundary conditions of a pressure ratio of up to 49 and a low mass flow of 0.045 kg/s. The turbine is designed as a single stage impulse turbine to achieve a compact design and to allow partial admission which is used to achieve feasible blade heights. The stator consist of Laval nozzles which accelerate the flow to supersonic speed. The rotor consist of an impulse turbine with zero degree of reaction. To cover a wide operating range the turbine is regulated using different degrees of partial admission between 20% and 80%.

A prototype was designed and manufactured. This prototype consists of two main components, the supersonic impulse turbine with variable partial admission and the high speed generator. The high speed generator and the turbine are mounted on one shaft. This leads to a compact and lightweight machine suitable for automotive applications.

The prototype was experimentally examined at the ORC test bench of the University of Hannover. The test bench is designed for a variety of ORC fluids and components. The objective of the experimental investigation was the validation of the numerical calculations and the functional verification of the ambitious design. The numerical calculations are a combination of CFD calculations and empirical methods to take into account the partial admission losses. The experimental results showed the proper functionality of the variable partial admission and of the high speed generator in a operating range between 2...8 kW.

The comparison between the numerical calculations and the experimental data showed a very good agreement. The deviation of the calculated mass flow is in average less than 1%. The measured turbine power output is up to 8 kW with a degree of partial admission of 40% and an inlet pressure of 30 bar. It is estimated that the turbine reaches a power output of up to 19 kW with its maximum degree of partial admission of 80%. Based on experimental measurements, the turbine reaches an aerodynamic total-to-static efficiency between 43.1% and 57.0%. Whereas the maximum deviation between the numerical and experimental efficiency is only 3.3%.

8. Acknowledgements

This investigation was funded by industry fees in the context of the Sonderforschungsprogramm CO2 of the Forschungsvereinigung Verbrennungskraftmaschinen e.V. (FVV) and the Forschungsvereinigung Antriebstechnik e.V. (FVA) which is gratefully acknowledged. Also the authors would like to thank the "Reginal Computer Center of Lower Saxony" (RRZN) for supporting this study

Substantial support was provided by Antriebstechnik und Entwicklungs GmbH, Leutkirch im Allgaeu/Germany, who contributed the rotor and stator of the generator, by Schaeffler, Herzogenaurauch/Germany, and its British subsidiary Barden Corp, Plymouth/UK, who contributed the design of the bearings and the bearing hardware, by Dr.-Ing. Martin Busch of Schaeffler for his support with the analysis and design suggestions regarding the rotor dynamics, and by Sieb & Meyer, Lueneburg/Germany, who furnished the power electronics. The authors also gratefully acknowledge the guidance of the industrial partners of the FVV Expansion Machine committee under the guidance of Dr.-Ing. Thomas Streule of Daimler and the valuable advice of Prof. Dr.-Ing. Bernd Ponick's team at Leibniz Universitaet Hannover Dept. of Electrical Engineering which helped to overcome challenges of the electrical generator.

The ORC Test Facility was partly financed by the European Community under an EFRE grand and co-financed by the Institute of Power Plant Technology at the Leibniz University Hannover, headed by Prof. Dr.-Ing. R. Scharf. The ORC test facility is located at the EFZN, an energy research center of Lower Saxony at Goslar/Germany.

The authors also highly recognize the support of the colleagues from the TFD during the experiments. A special thanks goes to Mr. Henning Raetz and Mr. Ole Willers. Without their help on the test bench and their guidance during the assembly and commissioning of the ORC turbine generator unit this work would not be possible.

References

[1] Bourhis G and Leduc P 2010 Energy and Exergy Balances for Modern Diesel and Gasoline Engines *Oil and Gas Science and Technology Vol. 65*

[2] Kunte H, Seume J R 2016 Design of a turbine-generator-unit for commercial vehicle ORC applications *12th International Conference on Turbochargers and Turbocharging* May 2016 London

[3] Kunte H and Seume J R 2013 Partial Admission Impulse Turbine fo Automotive ORC Application *11th International Conference on Engines & Vehicles* Capri/Neapal Italy

[4] Sourell R, Kunte H, Hütker J, Nikolov A, Eilts P, Seume J, Bruemmer A 2011 Expansionsmaschine (Vorstudie) Gegenüberstellung des Anwendungspotentials von Hubkolben-, Rotationsverdrängermaschinen und Strömungsmaschinen für Abwärmenutzung in einem Rankine-Kreislauf am Abgasstrang eines PKWs und eines NKWs *FVV Vorhaben Nr. 1060 R558(2012)* Frankfurt, Germany

[5] Aungier R H 2006 Turbine Aerodynamics: Axial-flow and radial-inflow turbine design and analysis *ASME Press New York* ISBN 0791802418

[6] Kunte H, Seume J R, Pautov D, Cherkasova M, Zabelin N, Olennikov S and Rassokhin V 2015 Supersonic Impulse Turbine for Automative Waste Heat Recovery: A Comparison of three Designs of Turbine Blade Profiles *11th European Turbomachinery Conference* 23-27 March 2015, Madrid Spain

[7] Kunte H and Seume J R 2015 Experimental Setup of a small supersonic Turbine for an Automotive ORC Application running with Ethanol *3rd International Seminar on ORC Power Systems* City of Brussels Belgium

IOP Conf. Series: Journal of Physics: Conf. Series **821** (2017) 012024 doi:10.1088/1742-6596/821/1/012024

Design and testing of high temperature micro-ORC test stand using Siloxane as working fluid

Teemu Turunen-Saaresti, Antti Uusitalo, and Juha Honkatukia

School of Energy Systems, Lappeenranta University of Technology, Finland

teemu.turunen-saaresti@lut.fi

Abstract. Organic Rankine Cycle is a mature technology for many applications e.g. biomass power plants, waste heat recovery and geothermal power for larger power capacity. Recently more attention is paid on an ORC utilizing high temperature heat with relatively low power. One of the attractive applications of such ORCs would be utilization of waste heat of exhaust gas of combustion engines in stationary and mobile applications. In this paper, a design procedure of the ORC process is described and discussed. The analysis of the major components of the process, namely the evaporator, recuperator, and turbogenerator is done. Also preliminary experimental results of an ORC process utilizing high temperature exhaust gas heat and using siloxane MDM as a working fluid are presented and discussed. The turbine type utilized in the turbogenerator is a radial inflow turbine and the turbogenerator consists of the turbine, the electric motor and the feed pump. Based on the results, it was identified that the studied system is capable to generate electricity from the waste heat of exhaust gases and it is shown that high molecular weight and high critical temperature fluids as the working fluids can be utilized in high-temperature small-scale ORC applications. 5.1 kW of electric power was generated by the turbogenerator.

1. Introduction

There is an inherent demand for using sustainable and renewable energy systems for an energy production in the future. One of the promising alternatives to fulfil this demand are decentralized energy systems in which the power output is usually low [1]. These decentralized power systems could be used to utilize solar energy, biomass, and waste heat e.g. from combustion engines in mobile and stationary applications. Organic Rankine Cycle (ORC) is a promising technology to be used in these applications because it is a versatile and proven system for larger power.

The high temperature and relative small amount of heat requires the use of a turbine as an expander and an organic working fluid which tolerates high temperature without decomposition. Characteristic for the working fluids to be used in these circumstances is a high molecular complexity which leads to a low speed of sound and therefore, supersonic flow and non-ideal compressible flow phenomena in the turbine. There is a clear need for the research facilities where non-ideal compressible flow phenomena can be studied. In addition, the performance prediction of the turbines is usually relying on measured data of turbine test sections or existing turbines [2,3].

However, the measured data is not available for the high pressure ratio supersonic turbines and there are no test setups available where this kind of turbines and complete ORC utilizing highly complicated working fluid(s) could be tested. On the other hand, a blowdown wind tunnel for real gases is presented in [4] and the first preliminary results of an expansion of siloxane MDM are shown in [5]. In addition, a hybrid test setup to conduct experimental campaigns in a nozzle and an expander is introduced in [6] and the test setup for an organic working fluid wind tunnel is presented in [7]. Both of these test setups are under construction. The test facility using ethanol as the working fluid and a partial admission axial turbine as an expander is described in [8]. To the best knowledge of the authors, the system is tested successfully.

The design procedure and the first test results of the micro-ORC power plant located in Lappeenranta University of Technology (LUT) are reported in the present article. The working fluid selection and process design are discussed in the next two sections. This is followed by a discussion of the first measured results of the ORC using the high pressure ratio supersonic turbine. The main focus of this article is to study the performance of a high temperature and high pressure ratio ORC suitable for small-scale applications. Especially, the use of a high-speed turbogenerator for producing electric power is discussed.

2. Working fluid selection

The selection of the working fluid is probably the most important step in designing ORC systems since it highly affects the component dimensioning, operational temperature and pressure, and cycle performance [9]. In general, the fluids having a high critical temperature can reach high thermodynamic efficiencies in cycles in which high turbine inlet temperatures can be adopted [10-12]. The relation of the molecular weight and the critical temperature as well as the relation of the critical temperature and the cycle efficiency for different kinds of working fluids are shown in Figure 1. Another advantage of using fluids with high molecular complexity is the relatively low enthalpy change when adopting high pressure ratios and high volumetric flow rates on the low pressure side of the process which enables the design of expanders with single or few stages. In addition, the optimal expander rotational speeds with this type of fluids are lower when compared to fluids with simple molecules which is beneficial especially in small-scale systems in which the expander rotational speeds tend to be impractically high [12]. The disadvantages of using fluids having a high critical temperature are the low speed of sound, which leads to highly supersonic flows when using high expansion ratios, and low condensing pressures, which introduces additional requirements for the system sealing during the operation. In addition, the low condensing pressure and high volumetric flow rate on the low pressure side leads to the use of relatively large sized heat exchangers, namely the recuperator and the condenser, when compared to fluids with lower critical temperature and simpler molecule.

The thermal and chemical stability of the fluid can be identified as important fluid selection criteria, particularly in the case of high-temperature ORCs. Erhart et al. [13] carried out an experimental study on siloxane decomposition in 7 large-scale high-temperature ORC plants. Their results indicated that decomposition of the fluid occurred over a long time period, and that the degradation of the fluid was caused by the use of petroleum-based lubricants in the system alongside the high working fluid temperatures in the process. Thus, the use of the working fluid lubricated bearings was evaluated to be beneficial in this type of systems.

After screening a large number of different fluid candidates, siloxane MDM was evaluated as the most suitable fluid for the high temperature experimental ORC system presented in this study. The selection of the working fluid was a compromise taking into account the cycle efficiency, process design, and turbine design. MDM was selected since it allows to design a turbine with a significantly lower rotational speed and larger turbine wheel when compared to the hydrocarbons and fluorocarbons and siloxane MM. In addition, the blade height at the rotor inlet is higher with MDM when compared to the hydrocarbons. The disadvantages of using MDM as a working fluid are the relatively low condensing pressure, high expansion ratio over the turbine which leads to a large variation on the rotor blade height between the rotor inlet and outlet and the low speed of sound. Also the high volumetric flow rate at the turbine outlet, which leads to the use of large sized heat exchangers and process piping on the low pressure side of the process and a moderate cycle efficiency when compared to more complex siloxanes such as D4 or hydrocarbons with a high critical temperature can be accounted as disadvantages. MDM is classified as a linear siloxane having the molecular formula of $C_8H_{24}O_2Si_3$, the critical pressure of 14.2 bar and the critical temperature of 290.9 °C.

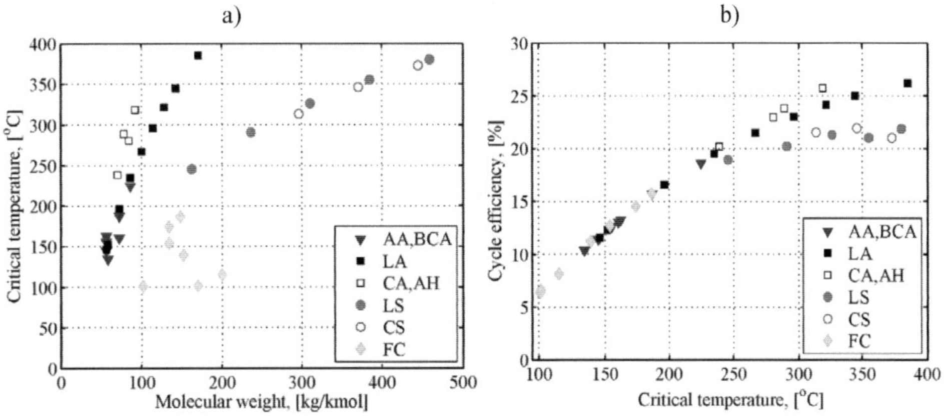

Figure 1. Relation between the molecular weight and critical temperature a) and relation between the critical temperature and cycle efficiency b). AA (acyclic alkane), LA (linear alkane), BCA (branched-chain alkane), CA (cyclic alkane), and AH (aromatic hydrocarbon) refer to different kinds of hydrocarbons, LS and CS refer to linear and cyclic siloxanes, and FC refers to fluorocarbons.

3. Process design

The aim of the process design was to achieve detailed information of the micro-ORC power plant using heat of the exhaust gases of a diesel generator as a heat source. Based on this process design, a power plant was built in Laboratory of Fluid Dynamics at LUT. The design objectives of the power plant were 10kW electric power output, hermetic construction, and the best possible conversion efficiency (>15%).

3.1. Cycle design

The experimental setup is designed for studying the performance of a small-scale and high-temperature ORC. In this ORC test stand (see Figure 2), the high-speed turbogenerator contains the turbine, generator and main feed pump that are coupled on the common shaft as schematically presented in Figure 3a, and the liquid working fluid is used to lubricate the bearings of the high-speed turbogenerator. The process at the design point is shown on T,s –diagram in Figure 3b. As the rotational speed of the turbogenerator shaft is relatively high, in the order of 31 000 rpm, a frequency converter is used between the generator and the grid. The exhaust gas used as a heat source in the ORC test stand is obtained from the diesel generator. In the evaporator, the liquid MDM is preheated, vaporized and superheated using the heat transferred from the hot exhaust gas. The superheated vapor is directed to the turbine that runs the generator and the main feed pump coupled on the common shaft.

At the turbine outlet, vapor MDM is highly superheated, and hence a recuperator is used to desuperheat the vapor before being directed to the condenser in where the vapor is further desuperheated and finally condensed to liquid. In the recuperator, heat is transferred from the vapor MDM exiting the turbine to the liquid MDM entering the evaporator. The pre-feed pump is used to prevent cavitation in the main feed pump as well as to supply liquid to the bearings of the turbogenerator.

The process values used in the design of the ORC experimental setup are presented in Table 1. The electric power output of the ORC of 10 kW was used as a target value in the design of the generator as well as in the selection of the frequency converter. In the turbine design, the mass flow rate of vapor MDM of 0.2 kg/s, the turbine inlet temperature of 265 °C and the turbine inlet pressure of 7.9 bar (abs.) were used. The turbine outlet pressure of 0.07 bar (abs.) used in the design depends on the condenser pressure as well as on the pressure loss between the turbine outlet and the condenser. Under the conditions presented in Table 1, MDM has a low condensing pressure, 0.03 bar (abs.), which sets demands for the sealing of the process.

Figure 2. Test facility at the laboratory of fluid dynamics, LUT.

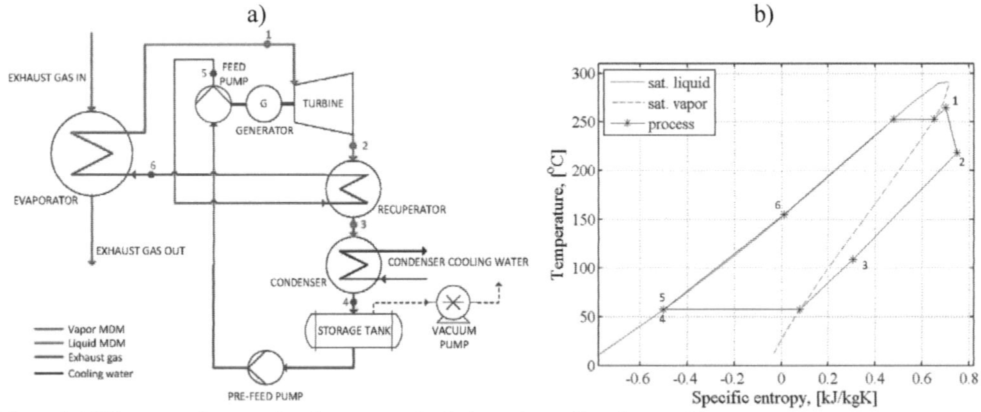

Figure 3. ORC process diagram a) and process at the design point on T,s –diagram b).

Table 1. Design values of the micro-ORC experimental facility. The pressures are absolute pressures.

Working Fluid	MDM	**Recuperator heat rate**	40 kW
Working fluid mass flow rate	0.2 kg/s	**Recuperator effectiveness**	0.68
Electric power output	10 kW	**Condensing temperature**	57 °C
Evaporator heat rate	67 kW	**Condensing pressure**	0.03 bar
Turbine inlet pressure	7.9 bar	**Turbine outlet pressure**	0.07 bar
Turbine inlet temperature	265 °C	**Conversion efficiency (heat-to-el.)**	15 %

3.2. Heat exchangers

The heat exchangers of the ORC experimental setup were designed to meet the desired process conditions of the turbine design. The temperature diagrams of the system heat exchangers are shown in Figure 4 and the main design values are presented in Table 2.

The evaporator is a plate-and-shell (PS) type once-through heat exchanger. The working fluid is preheated, evaporated and superheated directly by the exhaust gas. The exhaust gas temperature level is about 400 °C and the temperature of the superheated MDM is above 250 °C. The maximum working fluid temperature is limited to about 300 °C in order to avoid thermal decomposition of the fluid. One of the objectives for the evaporator design was the pressure drop on the exhaust gas side. The pressure drop cannot be large because it effects on the operation of the diesel engine. Therefore, the maximum accepted pressure drop on the exhaust gas side was set to 2 kPa.

The recuperator preheats the liquid working fluid by using the high-temperature and low-pressure MDM vapor exiting the turbine. The recuperator is a plate-and-shell type and the recuperation effectiveness of 0.68 was used in the design as presented in Table 1.

The condenser is a plate-and-shell type and uses cold water as a coolant for removing the heat from the system.

Table 2. Design values of the heat exchangers.

Heat exchanger	Type	A_{tot} [m^2]	U_{avg} [W/m^{2}K]	m [kg]	ΔT_{LM} [K]
Evaporator	PS	30.4	31	515	72
Recuperator	PS	16.0	43	218	59
Condenser	PS	7.0	201	168	40

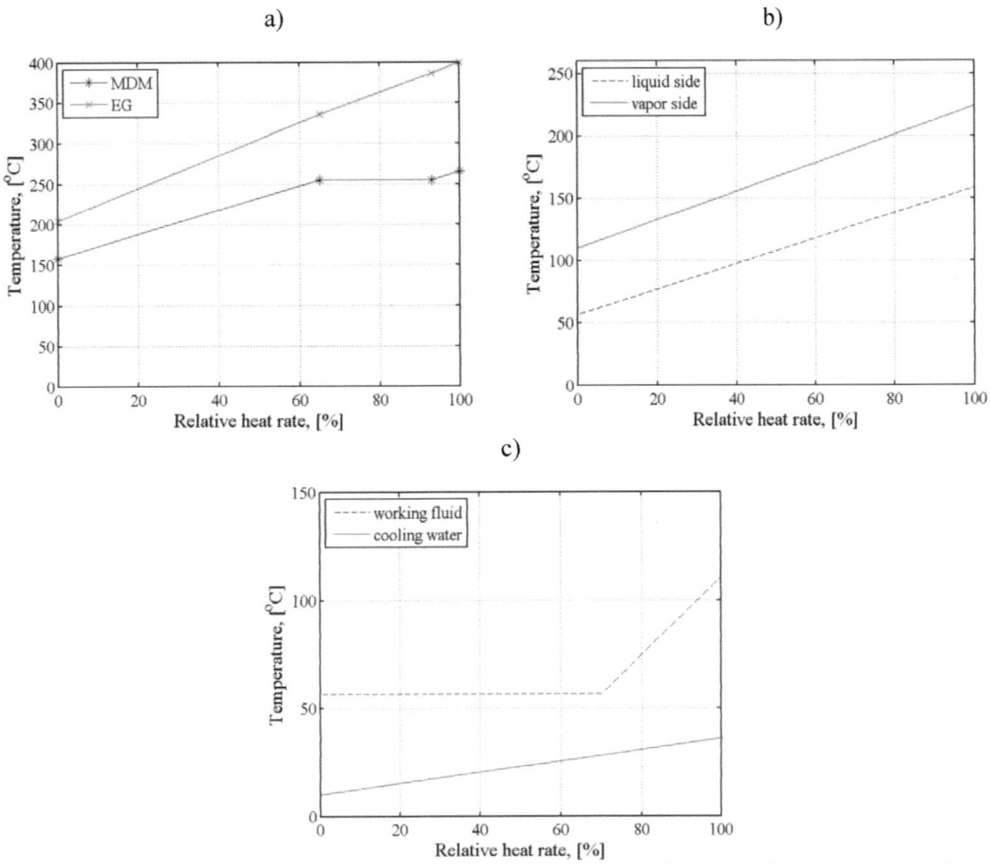

Figure 4. Temperature diagram of a) the evaporator, b) recuperator, and c) condenser at the design operating condition.

3.3. Turbogenerator

The most challenging task of the design process was the design of the turbogenerator. One of the design objectives was to have hermetic design and that can be achieved only if there are no shaft seals that are exposed to the environment. The turbogenerator consists of the turbine, generator and pump which are placed on the same shaft and using this kind of design shaft seals exposed to the environment are avoided. The designed turbogenerator is shown in Figure 5. The turbine is located on the left, the generator at the middle and the main feed pump on the right. In addition, the turbogenerator is equipped with the bearings lubricated with the working fluid in order to avoid the use of oil or other additional fluid in the system which could cause decomposition of the working fluid and/or deduction of the cycle efficiency. The bearings are located at the both sides of the generator.

The turbogenerator design consists of fluid dynamic design of the turbine and pump, electrical design of the generator, dynamic design of the bearings, mechanical design of the turbogenerator including dynamic and stress calculations, and design of cooling of the generator. All these design procedures are coupled to each other in order to achieve a functional and best possible outcome. Many iterations of the design procedure were needed to meet the goal. In addition, the process calculation of the ORC is coupled to the turbogenerator design because many details such as the turbine and generator efficiency, bearing losses and flows, mechanical losses, and generator cooling affect the process calculation.

Figure 5. Turbogenerator. The turbine is located on the left, generator in the middle and pump on the right.

3.3.1. Turbine

The designed turbine is a radial inflow turbine having a high pressure ratio over the turbine stator and rotor. The degree of reaction of less than 0.5 was adopted and thus, the fluid expands more in the stator vanes than in the rotor. The flow is supersonic at the stator outlet reaching the Mach number of about 2.1. The relative velocity at the rotor inlet was designed to be subsonic at the rotor inlet. The turbine design values are shown in Table 3 and the stator and the rotor of the turbine are shown in Figure 6. The turbine design methods are presented more in detail in [12].

Numerical modelling was carried out for five different supersonic stator geometries and three turbine rotors. The CFD calculations were carried out by using both k-ε and k-ω SST turbulence models and the gas properties of MDM were implemented to the solver by using polynomial fittings for the superheated gas region. The numerical methods are described more in detail in [12]. The final stator and rotor geometries were selected based on the results of predicted flow angles and velocities, efficiencies, and pressure ratios obtained with the different designs. In addition, the mechanical design, vibrations and forces acting on the turbogenerator shaft were taken into account when selecting the final geometries for the setup.

Table 3. Turbine design values.

Evaluated turbine efficiency	76 %
Specific speed	0.49
Specific diameter	3.45
Rotational speed	31000 rpm
Rotor diameter	144 mm
Stator blade height	2 mm
Mach number at the rotor inlet	2.18
Mach number at the rotor outlet	0.7

Figure 6. ORC turbine stator and rotor.

4. Measurements in the test facility

For evaluating the performance of the cycle, the temperatures and pressures at the turbine inlet and outlet (points 1 and 2 in Figure 3a) as well as the mass flow rate of liquid working fluid in the outlet pipe of the turbogenerator feed pump were measured (point 5 in Figure 3a). The mass flow rate of liquid working fluid was measured by oval gear flow meter the accuracy of which is ± 0.5 % of reading. The temperatures were measured by Pt100 temperature sensors and the pressures by pressure sensors including transmitters that have been connected to the data acquisition system. The electric power output of the ORC system was obtained from the frequency converter. The performance of the heat exchangers could be evaluated by temperature and pressure measurements in the process (points 1, 2, 4, 5 and 6 in Figure 3a) and by using the measured working fluid mass flow rate as well as by the temperature and pressure measurements of the exhaust gas and the temperature measurements of the cooling water to and from the condenser. In addition, the flow rate of the condenser cooling water was also measured by an oval gear flow meter.

5. Experiences and results of the first test runs

The test runs with the ORC experimental setup were performed in two phases. First, the test runs were made without the turbogenerator and second, with the turbogenerator installed in the ORC experimental setup.

In the first test runs, the turbogenerator was removed from the process and the vapor coming from the evaporator was expanded through an expansion valve instead of the turbine. Thus, a separate electric-driven main feed pump in series with the pre-feed pump was used. The main idea was first to clean the process by circulating the working fluid through the filter and to adjust the process parameters. Also, the process components were tested and suitable test run procedures were developed to ensure a safe operation of the process before the use of the turbogenerator. In addition, the performance and operation of the heat exchangers at different heat loads were investigated and these results are reported in [15].

In the second phase, the turbogenerator was installed in the process. The test run with the turbogenerator showed that electric power can be extracted with this ORC system, and the maximum electric power output of about 5.1 kW was measured. The turbogenerator was first rotated using the generator as an electric motor in order to achieve sufficient mass flow through the process and sufficient pressure at the evaporator and turbine inlet. The minimum rotational speed (12400 rpm) was used at the beginning and the evaporation of the working fluid was started at the corresponding rotational speed. The turbine was quickly accelerated to this rotational speed in order to avoid a wear of the bearings. At this point, the flow was bypassing the turbine and it was entering the turbine outlet pipe before being directed to the recuperator. As soon as there was working fluid vapor available the bypass valve was gradually closed and the turbine valve was gradually opened. There can be seen the peak of used power at this point (at 3000 s in Figure 7). This might be associated to the evaporation of the working fluid. When the bypass valve is open the flow was entering at the outlet of the turbine. It might generate forces to the turbine and therefore, there is a need for more power. However, the peak is quickly passed by when the turbine valve was opened (right after 3000 s in Figure 7).

At the minimum rotational speed the pressure at the turbine inlet and the mass flow were rather low and no power was yet produced. The consumed power was decreased quickly when the turbine valve was opened and bypass valve was closed. Gradually, the rotational speed of the turbogenerator was increased and correspondingly mass flow and pressure was increased which increased the power produced by the turbine.

At the rotational speed 14500 rpm, the turbine was generating more power than the pump and mechanical losses were consuming and the generator started to supply power to the grid (after 3500 s in Figure 7). The generated power increased as the rotational speed increased until maximum produced power of 5.1 kW was reached. At this operation point, the turbine was still operating at the off-design point in which the rotational speed and pressure ratio are lower than at the design point. Therefore, the electric power output reached in this operation point was lower than the design value of 10 kW used in the ORC experimental setup design. This can be partly explained by the elevated condenser pressure (see Figure 8) when compared to the design value of 0.03 bar (Table 1).

The rotational speed was further increased towards the design rotational speed (31 000 rpm) but the power generated by the turbogenerator decreased (Figure 7). However, at the same time condenser pressure (Figure 8) and mechanical losses increased which partly explains the lower rate of power.

In the test run, the condenser pressure was observed to be higher than the saturated pressure of MDM corresponding to the measured condenser temperature. This was estimated to be caused by the presence of small amounts of air in the process. There is also a possibility that the working fluid used in the test runs was not pure MDM but contained small amounts of other substances possibly having an influence on the condensing pressure.

In the future, the sealing of the process should be improved to prevent possible air leakages into the process and chemical analysis of the working fluid used in the ORC experimental setup should also be made. After improving the sealing of the process, further test runs will be carried out and a detailed analysis on the turbogenerator losses including the turbine, generator, main feed pump as well as the bearings will be done in order to estimate the chances to increase the electric power output of the ORC system.

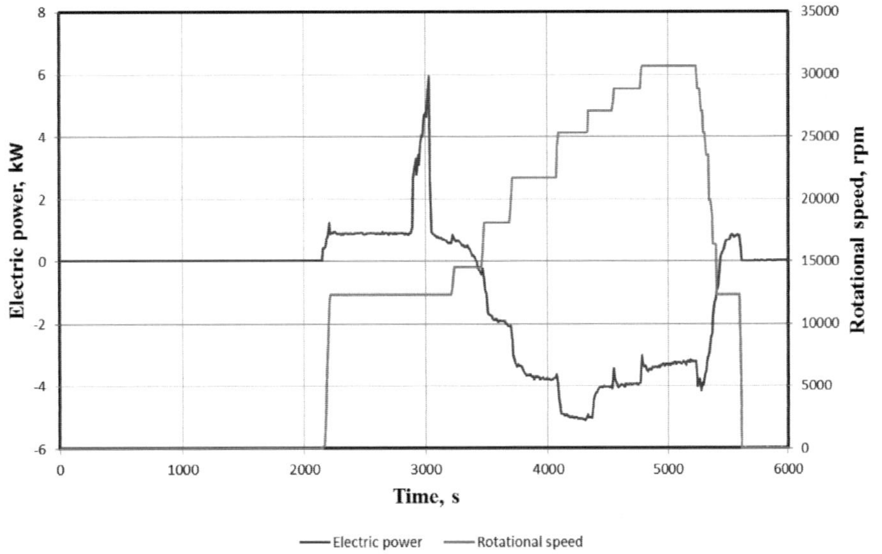

Figure 7. Measured rotational speed and electric power. Positive power means power consumed and negative power means generated power.

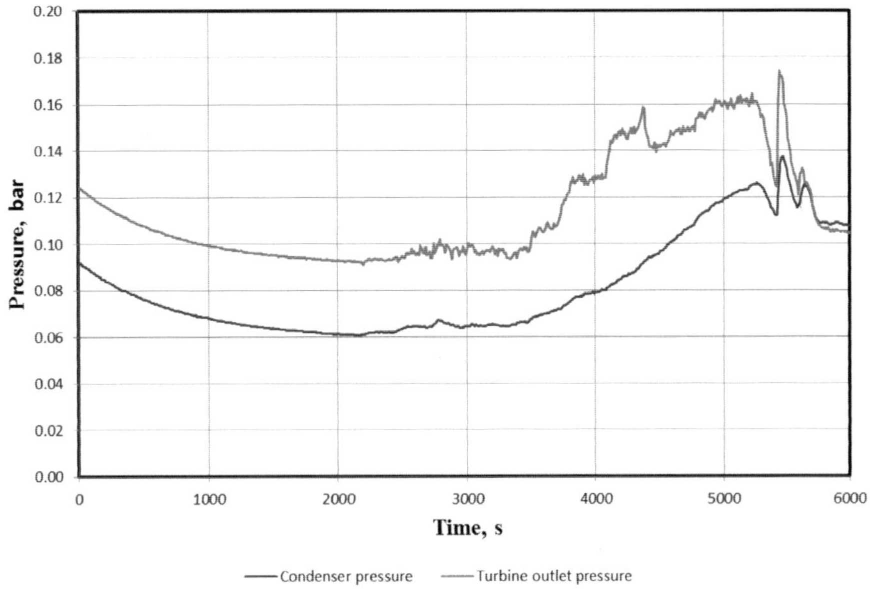

Figure 8. Measured pressure at the turbine outlet and in the condenser

6. Conclusions and discussion

Preliminary results from the micro-ORC test facility located in the Laboratory of Fluid Dynamics at Lappeenranta University of Technology are presented. The test facility is utilizing the exhaust gas of the diesel generator. The working fluid used in the test facility is siloxane MDM and the turbogenerator consisting of the radial inflow turbine, generator and pump is designed, manufactured and tested. The process was identified to be able to produce electricity from the diesel generator exhaust gas and the turbogenerator was confirmed to be mechanically functional.

The design operation point of the ORC was not reached because the condenser pressure low enough was not achieved. This was estimated to be due to small leakages of air into the process and the presence of non-condensable gases in the condenser. The maximum electrical power 5.1 kW was achieved.

The leakages of air into the process will be minimized in the future and the design point of the process will be measured. In addition, more detailed measurements in the turbine will be made in order to study flow phenomena occurring in the stator and rotor. The detailed study of losses in the turbogenerator will also be done.

References

[1] World Alliance for Decentralized Energy, 2006. World Survey of Decentralized Energy 2006.

[2] Rohlik, H., 1972, Radial Inflow Turbines, Chapter 10 in Glassman A. J., Turbine Design and Application: Volume 1-3, NASA, 388 pages.

[3] Balje, O. E., 1981, Turbomachines: A Guide to Design, Selection and Theory. John Wiley and Sons, New York.

[4] Spinelli, A., Pini, M., Dossena, V., Gaetani, P., and Casella, F., 2013. Design, simulation, and construction of a test rig for organic vapours. Journal of Engineering for Gas Turbines and Power, 135(4), p. 10 , doi:10.1115/1.4023114.

[5] Spinelli, A., Zocca, M., Cozzi, F., Guardone, A., Dossena, V., Gaetani, P., 2016. Experimental Investigation of a Non-Ideal Expansion Flow of Siloxane Vapor MDM. ASME Turbo Expo, June 13-17, Seoul, South Korea. GT2016-57357.

[6] Head, A., De Servi, C., Casati, E., Pini, M. and Colonna, P., 2016. Desing of the ORCHID: a Novel Hybrid Facility for Studying Non-Ideal Compressible Fluid Dynamics and Testing ORC Expanders. ASME Turbo Expo, June 13-17, Seoul, South Korea. GT2016-56103.

[7] Passmann, M., Reinker, F., Hasselmann, K., aus der Wiesche, S. and Joos, F., 2016. Development and Design of a Two-Stage Contraction Zone and Test Section of an Organic Rankine Cycle Wind Tunnel. ASME Turbo Expo, June 13-17, Seoul, South Korea. GT2016-56580.

[8] Kunte, H. S., and Seume, J. R., 2015. Experimental setup of a small supersonic turbine for an automotive ORC application running with ethanol. ASME ORC2015 - 3rd International Seminar on ORC Power Systems, October 12-14, Brussel, Belgium.

[9] Colonna, P., Casati, E., Trapp, C., Mathijssen, T., Larjola, J., Turunen-Saaresti, T., and Uusitalo, A., 2015. Organic Rankine cycle power systems: from the concept to current technology, applications and an outlook to the future. Journal of Engineering for Gas Turbines and Power, 137(10), p. 19, doi:10.1115/1.4029884.

[10] Fernández, F. J., Prieto, M. M., and Suárez, I., 2011. Thermodynamic analysis of high-temperature regenerative organic Rankine cycles using siloxanes as working fluids. Energy, 36(8), pp.5239-5249.

[11] Uusitalo, A., Turunen-Saaresti, T., Honkatukia, J., Colonna, P., and Larjola, J., 2013. Siloxanes as working fluids for mini-ORC systems based on high-speed turbogenerator technology. Journal of Engineering for Gas Turbines and Power, 135(4), p.9, doi:10.1115/1.4023115.

[12] Uusitalo, A., 2014. Working fluid selection and design of small-scale waste heat recovery systems based on organic Rankine cycles. Doctoral thesis, Lappeenranta University of Technology, Finland.

[13] Erhart, T., Gölz, J., Eicker, U., and van den Broek, M., 2015 Fluid stability in large scale ORCs using siloxanes: long-term experiences and fluid recycling. Proceedings of 3rd International Seminar on ORC Power Systems, October 12-14, Brussels, Belgium.

[14] Uusitalo, A., Turunen-Saaresti, T., Guardone, A., and Grönman, A., 2014. Design and Flow Analysis of a Supersonic Small Scale ORC Turbine Stator With High Molecular Complexity Working Fluid. ASME Turbo Expo, June 16-20, Düsseldorf, Germany. GT2014-26204.

[15] Uusitalo, A., Honkatukia, J. and Turunen-Saaresti, T., 2016, Experimental and Numerical Study of Small Scale Waste Heat Recovery. Submitted for publication.

IOP Conf. Series: Journal of Physics: Conf. Series **821** (2017) 012025 doi:10.1088/1742-6596/821/1/012025

Non-equilibrium condensation of supercritical carbon dioxide in a converging-diverging nozzle

Alireza Ameli[1], Ali Afzalifar[1] and Teemu Turunen-Saaresti[1]

[1]Laboratory of Fluid Dynamic, Lappeenranta University of Technology, Finland.

E-mail: Teemu.Turunen-Saaresti@lut.fi

Abstract. Carbon dioxide (CO_2) is a promising alternative as a working fluid for future energy conversion and refrigeration cycles. CO_2 has low global warming potential compared to refrigerants and supercritical CO_2 Brayton cycle ought to have better efficiency than today's counter parts. However, there are several issues concerning behaviour of supercritical CO_2 in aforementioned applications. One of these issues arises due to non-equilibrium condensation of CO_2 for some operating conditions in supercritical compressors. This paper investigates the non-equilibrium condensation of carbon dioxide in the course of an expansion from supercritical stagnation conditions in a converging-diverging nozzle. An external look-up table was implemented, using an in-house FORTRAN code, to calculate the fluid properties in supercritical, metastable and saturated regions. This look-up table is coupled with the flow solver and the non-equilibrium condensation model is introduced to the solver using user defined expressions. Numerical results are compared with the experimental measurements. In agreement with the experiment, the distribution of Mach number in the nozzle shows that the flow becomes supersonic in upstream region near the throat where speed of sound is minimum also the equilibrium reestablishment occurs at the outlet boundary condition.

NOMENCLATURE

H, h	total and static enthalpy
J	nucleation rate
k	Boltzmann constant
Kn	Knudsen number
L	Latent heat
m	molecular mass
N	droplet number per unit of volume
p	pressure
r	droplet radius
S_1, S_2	source terms
T	temperature
ΔT	supercooling degree
t	time
u	velocity
w	specific lost work
X	Cartesian coordinate
Z	compressibility factor

GREEK SYMBOLS

α	volume fraction
ζ	Markov loss coefficient
λ	vapor thermal conductivity
θ	divergence angle
ρ	density
σ	surface tension
τ	viscous stress tensor

SUBSCRIPTS

c	critical
r	reduced property
s	saturation

1. Introduction

Supercritical Brayton cycles have several advantages over conventional ones. For Supercritical Brayton cycles, depending on the working fluid, a higher cycle efficiency can be achieved with the relatively low temperature at the turbine inlet.

A previous study by Angelino and Invernizzi [1] indicated that the efficiency of the supercritical compressor and therefore, supercritical Brayton cycle becomes higher when the compressor inlet operation condition approaches the critical point. Previous studies showed that for low flow coefficient compressors operating in the vicinity of the critical point, due to the flow acceleration near the suction side of the main blade leading edge, the fluid state crosses the saturation curve and therefore, there is the possibility of spontaneous condensation [2-3]. Accordingly, an investigation of the non-equilibrium condensation of the supercritical carbon dioxide (SCO_2) is essential. Near the critical point, fluid has non-linear behavior and the behavior of the fluid is very sensitive to temperature and pressure. Previous studies discussed the difficulties of the compressor design and numerical calculations near the critical point [2, 4-5].

Numerical and thermophysical investigations of the formation of the dispersed phase in the rapid expansion of SCO_2 were studied in literature [6,7] and it was concluded that the most important factor which has a significant effect on the pressure and flow fields inside a supercritical nozzle is the inlet pressure. Moreover, the pressure at the inlet can alter the location at which the pressure goes below the critical pressure and consequently it changes location changes of onset of dispersed phase formation.

Furthermore, the impact of the supercritical condensation inside a centrifugal compressor was studied [8] by a non-dimensional criterion, which is related to the nucleation time and expansion rate. It was reported that condensation might not occur when the inlet boundary condition is far from the critical point. However, near the critical point, the effects related to phase transition are expected to become more significant.

The non-equilibrium condensation for high-speed flows in converging-diverging nozzles has been investigated for different fluids in many works such as [9-13]. The latent heat release due to condensation leads to a pressure drop or rise in the converging or diverging part of a nozzle, respectively. To isolate the unnecessary difficulties of modelling the condensation in supercritical fluids and focus on the essential elements of the model, the non-equilibrium condensation in a converging-diverging nozzle is selected for investigation to keep the flow complexity to a minimum. It is hoped that the current work provides a validation basis for modeling condensation in centrifugal compressors in future.

This work applies the numerical model developed by Gerber [14] to predict the flow field in the course of spontaneous condensation. The focus is on the prediction of the location of condensation and its effects on the flow field.

2. Experimental, theoretical and numerical procedures

2.1. Experimental setup

In the experimental measurements (which numerical results in this paper are compared with) performed by Nakagawa et al. [11] different converging-diverging nozzles with the same divergence angles and different boundary conditions were used. The shape and dimensional parameters of the nozzle are shown in Figure 1. The nozzle was rectangular due to ease construction and the heat transfer through the walls was reported to be too small to affect the experiment. The nozzle with the closest inlet conditions to the critical point is selected for the numerical investigation in this study, which has a divergence angle θ of 0.48°.

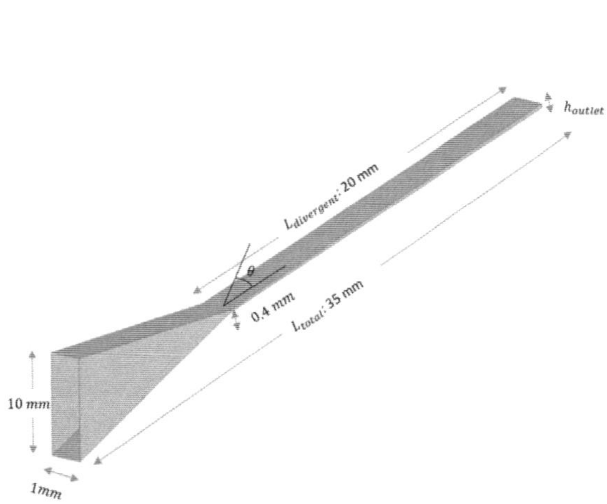

Figure 1. Dimensional parameters of the nozzle.

2.2. Non-equilibrium model

The foundations for almost all non-equilibrium phase change models are laid upon the prediction of the new phase evolution in forms of two consecutive stages described by the nucleation and droplet growth theories. It is believed that initially the condensation is favored by nucleation providing the first droplets of the liquid phase. Then, the phase transition is governed by the growth of supercritical droplets, which ideally quenches the nucleation and reestablishes the equilibrium [15]. The supercritical term is referred to the droplets larger than a critical radius, r^*, which is calculated according to the theory developed by Gibbs and Thomson for the equilibrium between a vapor with supersaturation ratio S and a liquid droplet. Here by use of Clausius-Clapeyron equation [16], critical radius r^* is related to the supercooling degree, ΔT, and the specific enthalpy of evaporation, L, as below,

$$r^* = \frac{2\sigma}{\rho_l R ln(S)} \approx \frac{2\sigma T_s}{\rho_l L \Delta T}, \Delta T = T_s(\rho_v) - T_v \quad (1$$

where T_s is the saturation temperature at vapor pressure, p_v, ρ_l is the liquid density and σ is surface tension. The calculation of the surface tension of the liquid carbon dioxide is based on a semi-empirical formulae proposed by Brock and Bird [17] using critical pressure, temperature and volume to express the liquid's surface tension as follows

$$\sigma = \left(-0.951 + \frac{0.432}{Z_c}\right)(1 - T_r)^{11/9} (P_c^2 T_c)^{1/3} \quad (2$$

where T_r and Z_c are reduced temperature and critical compressibility factor, respectively. These parameters are given as follow

$$T_r = T/T_c \quad (3$$

$$Z_c = \frac{1}{R}\left(\frac{P_c V_c}{T_c}\right) \quad (4$$

where subscript $c$refers to the critical state of the variables. Several expressions of surface tension equation are available in the literature [18-20]. A comparison between these surface tension models and experimental data was conducted by Jianxin and Yigang [21] and it was concluded that close to the critical point and in the range of interest in this article, Brock and Bird model has the highest accuracy. Therefore, this model was implemented to the flow solver by means of a user defined CFX Expression Language (CEL).

Despite decades of research after formulating the classical nucleation theory by Becker and Döring [22] and Zeldovich [23], the classical nucleation theory is still the most popular approach to model the nucleation process. Also, here the formation rate of critical droplets per unit volume J is computed by the Becker and Döring expression for the classical theory of the homogenous steady nucleation as follows

$$J = \frac{\rho_v^2}{\rho_l}\left[\frac{2\sigma}{\pi m^3}\right]^{1/2} \exp\left(-\frac{4\pi r^{*2}\sigma}{3kT_v}\right). \quad (5$$

A growing droplet absorbs monomers of the vapor phase through molecular collisions. Simultaneously, the droplet needs to give away the latent heat released by the condensation of monomers into the droplet. Due to large amount of the latent heat, of the droplet growth is dictated by the rate of heat transfer between the droplet and the vapor [24]. As the droplet temperature is unknown, one needs to iteratively solve the heat and mass transfer equations to obtain the growth rate. Although, the iterative method can be discarded applying the following estimation for a droplet temperature T_l sized r [25].

$$T_l = T_s - \Delta T\,\frac{r^*}{r} \quad (6$$

The details of the non-equilibrium model were provided by Gerber, in [26]. Here only a general description about the model in accordance with choices and assumptions made in this work are given. For the continuous and dispersed phases, the discretized transport equations are solved in an Eulerian reference frame. The model can also benefit from the so-called Source Specific capability to divide the liquid phase into separated groups based on their locations of nucleation/introduction. The mass and energy transfer between the dispersed phases and the continuous one are handled through source terms. Similar to the continuous phase, the dispersed phases are also evolving in the form of volume fractions. Thus, the conservation equations for the continuous and the dispersed phases are expressed using their volume fractions respectively denoted by α_v and α_l, along with the constraint $\alpha_v + \sum_{l=1}^{n_l}\alpha_l = 1$

$$\frac{\partial(\alpha\rho)_v}{\partial t} + \frac{\partial(\alpha\rho u_j)_v}{\partial x_j} = -\sum_{l=1}^{n_l} S_{1,l} - \sum_{l=1}^{n_l} S_{2,l} \quad (7$$

$$\frac{\partial(\alpha\rho)_l}{\partial t} + \frac{\partial(\alpha\rho u_j)_l}{\partial x_j} = S_{1,l} + S_{2,l}. \quad (8$$

Above equations define the mass conservation for each phase and they are connected by means of source terms $S_{1,l}$ and $S_{2,l}$. The mass transfer from the continuous phase to the dispersed one, induced by nucleation is computed by $S_{1,l}$, and $S_{2,l}$ accounts for the growth/decay of a dispersed phase as follows

$$S_{1,l} = m^*\alpha_v J \quad (9$$

$$S_{2,l} = \left(\frac{3\rho\alpha}{r}\frac{dr}{dt}\right)_l, r_l = \left(\frac{3\alpha_l}{4\pi N_l}\right)^{\frac{1}{3}} \quad (10$$

where m^* is the mass of a critical droplet, α is the volume fraction. Moreover, the droplet number N_l of each dispersed phase must be conserved,

$$\frac{\partial(\alpha N)_l}{\partial t}+\frac{\partial(\alpha N u_j)_l}{\partial x_j}=S_{1,l} \quad (11$$

In the same manner as with the mass conservation equations, by neglecting the relative acceleration between phases and discarding energy equations for dispersed phase, the conservation of momentum and energy for the continuous phase yield as below

$$\frac{\partial(\alpha\rho u_i)_v}{\partial t}+\frac{\partial(\alpha\rho u_j u_i)_v}{\partial x_j}=-\alpha_v\frac{\partial p}{\partial x_j}+\frac{\partial(\alpha\tau_{ij})_v}{\partial x_j}-\sum_{l=1}^{n_l}u_i S_{2,l} \quad (12$$

$$\frac{\partial(\alpha\rho H)_v}{\partial t}+\frac{\partial(\alpha\rho u_j H)_v}{\partial x_j}=-\alpha_v\frac{\partial p}{\partial t}+\frac{\partial(\alpha\lambda T)_v}{\partial x_j}+\frac{\partial(\alpha u_j\tau_{ij})_v}{\partial x_j}-\sum_{l=1}^{n_l}LS_{2,l}. \quad (13$$

2.3. Numerical procedure
Commercial Navier-Stokes flow solver ANSYS CFX 17.0 [27] was used for the steady states simulations in this article. Furthermore, RANS equations were closed through the two-equation $k-\omega$ SST turbulence model of Menter [28]. Convergence criteria of the simulation were based on the reduction of Root Mean Square (RMS) momentum, mass and energy residuals below 10^{-4}, reduction and stability of the imbalance (difference between inlet and outlet) of mass, volume, energy and momentum below 10^{-3}%, and constant volume fractions for both phases and droplet numbers at the outlet. Turbulence Intensity of 5% was prescribed at the inlet and reference pressure was set to zero. Total energy model, including the viscous work term was defined and all walls were assumed as non-slip. Total pressure and temperature were defined as the inlet boundary conditions and the static pressure was set as the outlet boundary condition. According to the different experimental measurements at the different boundary conditions, the closest inlet boundary condition to the critical point was selected to investigate the difficulty of the simulation and have a similar inlet boundary condition as one could have in a supercritical centrifugal compressor. Pressure and temperature at the inlet are 9 MPa and 40°C, respectively. Static pressure 3.89 MPa was set as the outlet boundary condition.

The structured 3D mesh was generated using a sufficiently fine grid for one fourth of the whole geometry.. Grids were denser near the walls to ensure the values of y^+ close to unity. The quality of the structured grid was tested by Jacobian and skewness factors and most of the cells had skewness below 0.11. Grid dependency test was carried out and it was seen that a number of grids around 300000 cells is sufficient to ensure the grid-independent pressure trend and drag coefficient on the walls.

Fluid near its critical point has non-linear behavior and properties change rapidly. Figure 2 shows the specific heat in constant pressure at different pressure and temperature near the critical point. Sharp variation of the fluid properties causes high instabilities in the simulation. In order to calculate with high accuracy, an external look-up table of properties is coupled with the flow solver. Real Gas Properties (RGP) table includes nine main properties of fluid such as entropy, enthalpy, speed of sound, specific heat in constant pressure and volume and their derivatives. Properties of both vapor and liquid components were defined in a single RGP table.

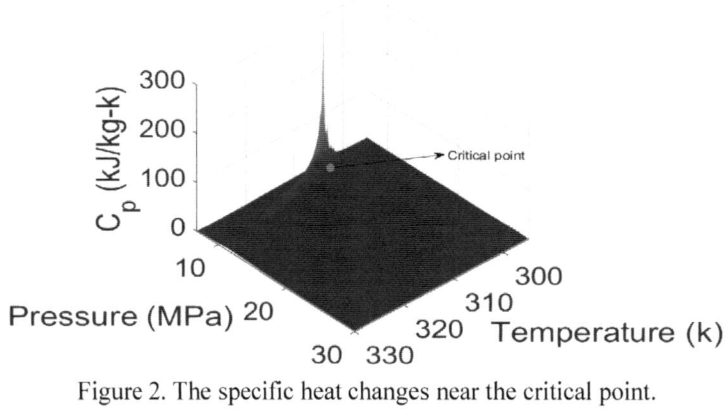

Figure 2. The specific heat changes near the critical point.

Among different real gas Equation of States (EOS) equations, Span and Wagner (SW) [29] has been proposed especially for CO_2. This EOS was previously tested for simulations near the critical point and shown to be more accurate compared to other EOSs [2, 30]. SW EOS is formed in terms of Helmholtz energy and covers thermodynamic properties of CO_2 from the triple point up to 1100 K and 800 MPa for temperature and pressure, respectively. Full details on SW EOS and its derivatives can be found in the Ref. [31]. The fluid properties for the external look-up tables were calculated using the multipurpose NIST REFPROP 9 [31] database. These data were stored in the separate tables includes, liquid, vapor, metastable and saturated regions. Range of the RGP table is from 0.7 MPa and 230 K to 50 MPa and 730 K for pressure and temperature, respectively. Range of the table was chosen wider than the operating condition to prevent extrapolating or clipping methods by the solver, which decreases the accuracy of the numerical simulation. RGP resolution dependency test was done previously by authors [3] and it was concluded that intervals of 0.26 K and 0.05 MPa for temperature and pressure, respectively, are sufficient to have a stable and accurate simulation.

The vapor properties in the metastable region, i.e. between the saturation and the upper spinodal curves in the Figure 3, are given from a bilinear cubic extrapolation of the saturation properties to the spinodal curve. The extrapolation procedure is done by an in-house FORTRAN code.

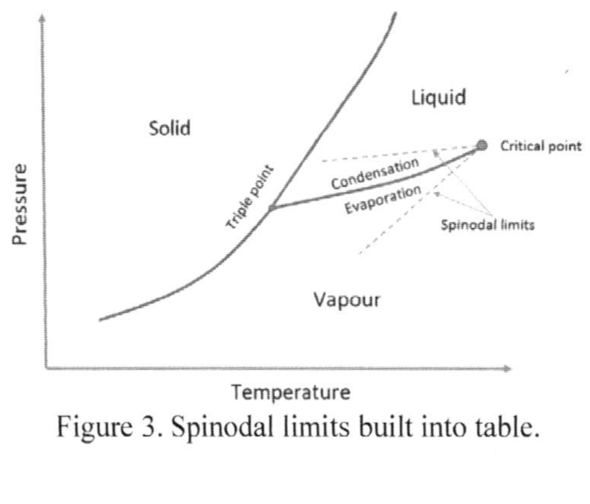

Figure 3. Spinodal limits built into table.

2.4. Numerical difficulties

In the non-equilibrium condensation model in Ansys CFX, critical radius is calculated according to the following equation [32],

$$r^* = \frac{2\sigma}{\rho_l \Delta G_v} \quad (14$$

where, ΔG_v is Gibbs free energy change of the vapor. Therefore, by comparing the equation (14) with equation (1) it can be deduced that ΔG_v in Ansys CFX must be calculated as

$$\Delta G_v = \frac{L\Delta T}{T_s}. \quad (15$$

The Gibbs free energy change of the vapor near the critical point is very small which leads the equation to have an infinite value and causes error in the solver. In order to prevent this numerical instability, the solver clips the critical radius to a certain amount, as a default 1 mm. This extremely large default clipping value decreases the accuracy of the simulation and results in non-realistic droplet sizes. The problem is that it is not possible to change or modify the formula in the Ansys CFX by the user. To sidestep this problem, the critical radius was calculated according to the equation (1) in the post processing stage. It was observed that the critical radius, see Figure 4, changes very moderately also in the nucleation zone. Therefore, the radius of one nano meter has been estimated to be realistic for the critical droplet size in this study and in the calculations the critical radius is clipped, by a CEL, to one nano meter.

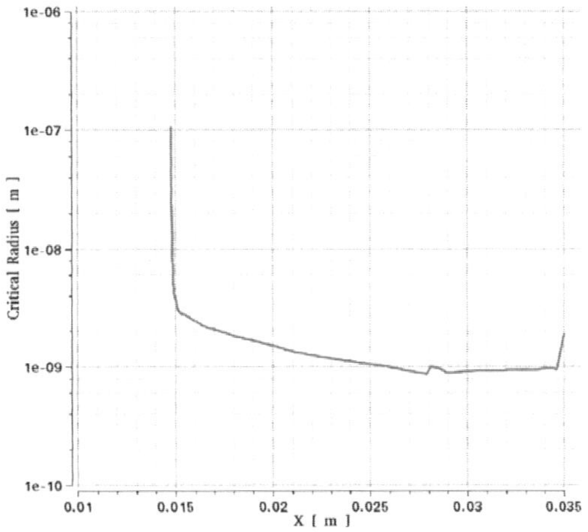

Figure 4. Critical radius distribution over the nozzle centerline.

3. Results and discussion

In order to prevent shock waves inside the nozzle, shorter length (8.38 mm) of the diverging part of the nozzle was assumed in experimental pressure measurements [11]. Thermocouple channels were implemented into the drilled holes along the nozzle wall to measure the temperature profile. Through the nozzle, saturated pressures where obtained according to the measured temperature. In figure 5, experimental saturation pressure were connected using dash lines.

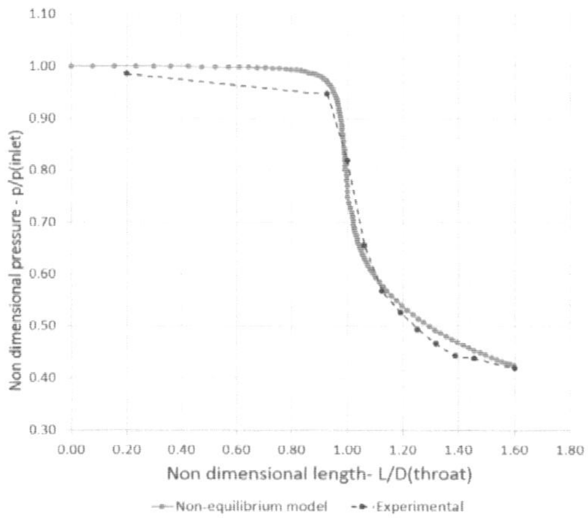

Figure 5. Pressure profiles along the nozzle.

Saturation pressure in numerical calculation is in a good agreement with the experimental data in most of the locations, although near the outlet of the nozzle due to the droplet condensation, slight deviation has been observed. This is mostly because of lack of accuracy of the classical nucleation theory in supercritical region, experimental uncertainties and numerical errors.

Figure 6 shows liquid mass fraction, supercooling, surface tension and Mach number inside the nozzle. Maximum mass fraction of the liquid carbon dioxide is observed near the outlet of the nozzle. As mentioned in the experimental procedure, near the outlet, shock has been occurred which is in agreement with the numerical results. Furthermore, by investigating the Mach number inside the converging-diverging nozzle, it can be noticed that the onset of phase transition occurs near the outlet, but the flow becomes supersonic in the upstream region near the throat where the speed of sound is minimum in that region.

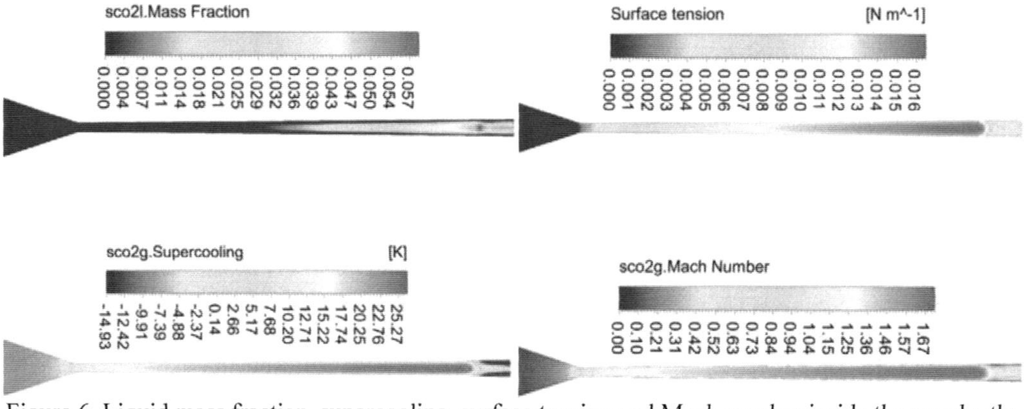

Figure 6. Liquid mass fraction, supercooling, surface tension and Mach number inside the nozzle, the converging part of the nozzle is cut partly to concentrate more on the diverging part.

4. Conclusion

In this paper, non-equilibrium condensation of supercritical CO_2 has been modelled. An external look-up table of properties was coupled with the solver to accurately and efficiently calculate thermodynamic properties near the critical point. Bilinear cubic extrapolation of the saturation properties to the spinodal curves was done by an in-house FORTRAN code to calculate the fluid properties at the metastable region. In order to prevent clipping near the critical point, an external function was defined to estimate more accurate critical radius for CO_2. Shock was seen near the outlet of the nozzle in a same location as mentioned in the experimental data. Around 5.7% mass fraction of the liquid was observed near the outlet and the flow becomes supersonic in upstream region near the throat where the speed of sound is minimum. Further investigations will be conducted by the authors to implement the same approach in turbomachines especially in the centrifugal compressors to predict the effect of the condensation on the blade loading and compressor performance. Also, more studies will be done in order to increase the accuracy of non-equilibrium condensation near the critical point by modifying the nucleation and droplet growth theories.

Acknowledgement
Authors gratefully acknowledge the financial contribution of the Graduate School of Lappeenranta University of Technology.

References

[1] G. Angelino and C. Invernizzi, "Real gas Brayton cycles for organic working fluids," *Proceedings of the Institution of Mechanical Engineers, Part A: Journal of Power and Energy*, vol. **215**, no. 1, pp. 27-38, February 1. 2001.

[2] N. D. Baltadjiev, An Investigation of Real Gas Effects in Supercritical CO2 Compressors, Master Thesis, Massachusetts Institute of Technology, 2012.

[3] A. Alireza, T.-S. Teemu and B. Jari, "Numerical Investigation of the Flow Behavior inside a Supercritical CO2 Centrifugal Compressor," in *ASME Turbo Expo*, Seoul, 2016.

[4] Y. Wang, G. Guenette, P. Hejzlar and M. J. Driscoll, "Compressor Design for the Supercritical CO2 Brayton Cycle," in *2nd International Energy Conversion Engineering Conference, no.* **5722** *in AIAA, pp. 1599-1611*, 2004.

[5] G. K. Seong, L. Jekyoung, A. Yoonhan, l. L. Jeong, A. Yacine and K. Bockseong, "CFD Investigation of a Centrifugal Compressor Derived from Pump Technology for Supercritical Carbon Dioxide as a Working Fluid," *The Journal of Supercritical Fluids,* vol. **86**, pp. 160-171, 2014.

[6] J. Liu, G. Amberg and M. Do-Quang, "Numerical Simulation of Particle Formation in the Rapid Expansion of Supercritical Solution Process," *The Journal of Supercritical Fluids,* vol. **95**, pp. 572-587, 2014.

[7] S. Yamamoto and T. Furusawa, "Thrmophysical Flow Simulations of Rapid Expansion of Supercritical Solutions (RESS)," *The Journal of Supercritical Fluids*, vol. **97**, pp. 192-201, 2015.

[8] C. Lettieri, D. Yang and Z. Spakovszky, "An Investigation of Condensation Effects in Supercritical Carbon Dioxide Compressors," in *The 4th International Symposium - Supercritical CO2 Power Cycles*, Pittsburgh, Pennsylvania, 2014.

[9] A. Guha, " Thermal Choking Due to Nonequilibrium Condensation," *Journal of Fluids Engineering,* vol. **116**, no. 3, pp. 599-604, 1994.

[10] N. Masafumi, S. B. Menandro and K. Akinori, "Supersonic Two-phase Flow of CO2 through Converging-diverging Nozzles for the Ejector Refrigeration Cycle," *International Journal of Refrigeration,* vol. **32**, pp. 1195-1202, 2009.

[11] M. Nakagawa, M. S. Berena and A. Harada, "Shock Waves in Supersonic Two-Phase Flow of CO2 in Converging-Diverging Nozzles," in *International Refrigeration and Air Conditioning Conference*, Purdue, 2008.

[12] G. Gyarmathy, "Nucleation of Steam in High-Pressure Nozzle Experiments," *Proceedings of the Institution of Mechanical Engineers, Part A: Journal of Power and Energy,* vol. **219**, pp. 511-521, 2005.

[13] M. Yazdani, A. A. Alahyari and T. D. Radcliff, "Numerical Modeling and Validation of Supersonic Two-Phase Flow of CO2 in Converging-Diverging Nozzles," *Journal of Fluids Engineering,* vol. **136**, p. 014503, 2014.

[14] A. G. Gerber, "Inhomogeneous Multifluid Model for Prediction of Nonequilibrium Phase Transition and Droplet Dynamics," *Journal of Fluids Engineering,* vol. 130, no. 3, p. 031402, 2008.

[15] K. Arthur, "Nucleation in Very Rapid Vapor Expansions," *The Journal of Chemical Physics,* vol. **19**, no. 9, pp. 1097-1100, 1951.

[16] J. B. Joung, "Two-Dimensional, Nonequilibrium, Wet-Steam Calculations for Nozzles and Turbine Cascades," *Journal of Turbomachinery,* vol. **114**, no. 3, pp. 569-579, 1992.

[17] J. R. Brock and R. B. Bird, "Surface tension and the principle of corresponding states," *AIChE Journal,* vol. **1**, no. 2, pp. 174-177, 1955.

[18] K. S. Pitzer, Thermodynamics, vol. **43**, New York: McGraw-Hill, 1995, p. 285 pages.

[19] L. Riedel, "Kompressibilität, Oberflächenspannung und Wärmeleitfähigkeit im flüssigen Zustand. Untersuchungen über eine Erweiterung des Theorems der übereinstimmenden Zustände. Teil IV.," *Chemie Ingenieur Technik,* vol. **27**, no. 4, pp. 209-213, 1955.

[20] D. I. Hakim, D. Steinberg and L. I. Stiel, "Generalized Relationship for the Surface Tension of Polar Fluids," *Ind. Eng. Chem. Fundamen.,* vol. **10**, no. 1, pp. 174-175, 1971.

[21] J. Peng and Y. Lu, "Estimation of the surface tension of liquid carbon dioxide," *Physics and Chemistry of Liquids: An International Journal ,* vol. **47**, no. 3, pp. 267-273, 2009.

[22] R. Becker and W. Döring, "Kinetische Behandlung der Keimbildung in übersättigten Dämpfen," *Annalen der physik,* vol. **416**, no. 8, pp. 719-752, 1935.

[23] J. Zeldovich, "Theory of the Formation of a New Phase," *Zh Eksp Theor Fiz,* vol. **12**, pp. 525-538, 1942.

[24] J. B. Young, "Spontaneous Condensation of Steam in Supersonic Nozzles," *Physicochemical Hydrodynamics,* vol. **3**, pp. 57-82, 1982.

[25] G. Gyarmathy, Grundlagen einer Theorie der Nassdampfturbine, ETH Zürich, Juris-Verlag Zürich. English Translation: USAF-FTD-TT-63-785, 1962.

[26] A. G. Gerber, " Inhomogeneous Multifluid Model for Prediction of Nonequilibrium Phase Transition and Droplet Dynamics," *Journal of Fluids Engineering,* vol. **130**, no. 3, p. 031402, 2008.

[27] ANSYS® Academic Research, ANSYS CFX User Guide, Release 17.0, ANSYS, Inc, 2015.

[28] F. R. Menter, "Two-Equation Eddy-Viscosity Turbulence Model for Engineering Applications," *AIAA Journal,* vol. **32**, no. 8, pp. 1598-1605, 1994.

[29] R. Span and W. Wagner, "A New Equation of States for Carbon Dioxide Covering the Region from the Triple-Point Temperature to 1100 K at Pressure up to 800 MPa," *J. Phys. Chem,* vol. **25**, no. 6, pp. 1509-1596, 1996.

[30] N. D. Baltadjiev, C. Lettieri and S. Z. Spakovszky, "An Investigation of Real Gas Effects in Supercritical CO2 Centrifugal Compressors," *Journal of Turbomachinery,* vol. **137**, no. 9, p. 091003 (13 pages), 2015.

[31] E. Lemmon, M. Huber and M. McLinden, "NIST Standard Reference Database 23: Reference Fluid Thermodynamic and Transport Properties-REFPROP, Version 9.0," National Institute of Standards and Technology, Standard Reference Data Program, Gaithersburg, 2012.

[32] ANSYS, "Private communication," 27.05.2016.

Design of an efficient space constrained diffuser for supercritical CO_2 turbines

Joshua A. Keep [1], Adam J. Head [2] and Ingo H. Jahn [1]

The University of Queensland, Brisbane, Australia [1], Delft University of Technology, Delft, The Netherlands [2]

E-mail: `j.keep@uq.edu.au`, `a.j.head@tudelft.nl`, `I.jahn@uq.edu.au`

Abstract. Radial inflow turbines are an arguably relevant architecture for energy extraction from ORC and supercritical CO_2 power cycles. At small scale, design constraints can prescribe high exit velocities for such turbines, which lead to high kinetic energy in the turbine exhaust stream. The inclusion of a suitable diffuser in a radial turbine system allows some exhaust kinetic energy to be recovered as static pressure, thereby ensuring efficient operation of the overall turbine system. In supercritical CO_2 Brayton cycles, the high turbine inlet pressure can lead to a sealing challenge if the rotor is supported from the rotor rear side, due to the seal operating at rotor inlet pressure. An alternative to this is a cantilevered layout with the rotor exit facing the bearing system. While such a layout is attractive for the sealing system, it limits the axial space claim of any diffuser. Previous studies into conical diffuser geometries for supercritical CO_2 have shown that in order to achieve optimal static pressure recovery, longer geometries of a shallower cone angle are necessitated when compared to air. A diffuser with a combined annular-radial arrangement is investigated as a means to package the aforementioned geometric characteristics into a limited space claim for a 100kW radial inflow turbine. Simulation results show that a diffuser of this design can attain static pressure rise coefficients greater than 0.88. This confirms that annular-radial diffusers are a viable design solution for supercritical CO_2 radial inflow turbines, thus enabling an alternative cantilevered rotor layout.

1. Introduction

The inclusion of a suitable diffuser in a radial turbine system is essential for the efficient operation of the turbine system in order to recover exhaust kinetic energy as static pressure. Several studies into supercritical CO_2 power cycles show that cycle efficiency is maximised for turbine inlet pressures between 20 and 30 MPa and turbine exhaust pressures immediately above the critical pressure for CO_2 in the vicinity of 8 Mpa [1, 2]. For these pressure ratios, it is feasible to employ a singe stage radial inflow turbine as the work extraction method. While this may be feasible from an aerodynamic standpoint, this can lead to a sealing challenge if the rotor is supported from the rear in a cantilevered arrangement as shown in Fig. 1a. If the rotor layout is inverted such that the rotor exit is facing the bearing system as in Fig. 1b, the conditions for any sealing system are improved with a lower static pressure and temperature.

While such a layout is favourable for a shaft sealing system, it axially limits the extent of any diffuser due to shaft bending and constrains its entrance to an annular geometry. In effect, the diffuser must adopt either a cropped annular geometry, or annular-radial geometry.

Within the power conversion and turbomachinery group at the University of Queensland a prototype design of a 100kW radial inflow supercritical CO_2 turbine has been completed

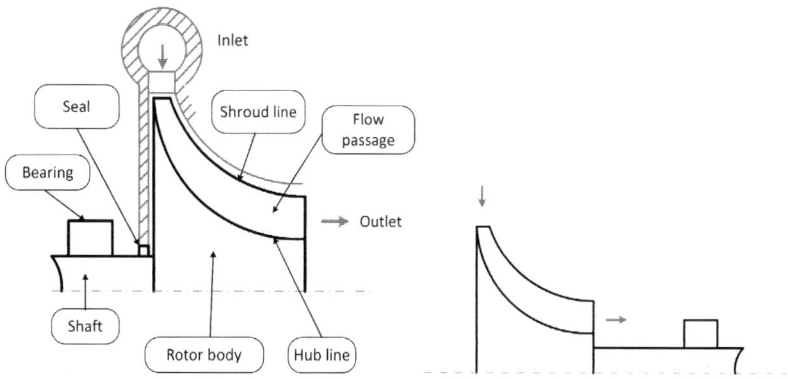

(a) Conventional arangement. (b) The proposed arangement.

Figure 1: Cantilevered rotor layouts.

Table 1: Geometric constraints for 100kW radial inflow turbine diffuser.

Dimension	Maximum (mm)	Constraint
Axial length	30	Shaft bending
Inner diffuser diameter	15.2	Shaft bending + Rotor hub diameter
Radial extent	150	Pressure casing design

including a shaft layout as described in Fig. 1b. The design of the diffuser to be incorporated in this turbine design will be the focus of this paper. The diffuser geometric constraints are listed in Table 1.

This paper presents a theoretical design of a swirl robust combined annular-radial diffuser. A parameter study on the influence of diffuser geometric features on aerodynamic performance for fixed flow conditions is detailed.

2. Theory and background

The aim of a diffuser is to convert kinetic energy into a pressure rise. Diffuser performance is typically described by the pressure rise coefficient, C_p defined in Eqn. 1.

$$C_{\mathrm{p}} = \frac{P_2 - P_1}{P_{1,\ \mathrm{total}} - P_1} \tag{1}$$

Where subscripts 1 and 2 refer to inlet and outlet respectively. For a diffuser of an arbitrary annular geometry as described in Fig. 2, the ideal pressure rise coefficient is defined according to Eqn. 2 [3].

$$C_{\mathrm{p,i}} = 1 - (\frac{r_1}{r_2})^2 \frac{\tan^2 \alpha_1 + (\frac{h_1^2}{h_2^2})}{\tan^2 \alpha_1 + 1} \tag{2}$$

Where α_1 is the swirl angle. For a diffuser without swirl, this reduces to Eqn 3.

$$C_{\mathrm{p,i}} = 1 - \frac{1}{Ar^2} \tag{3}$$

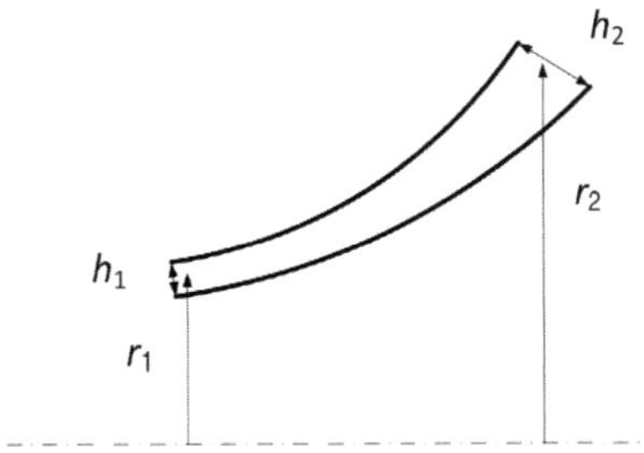

Figure 2: A diffuser with arbitrary annular geometry.

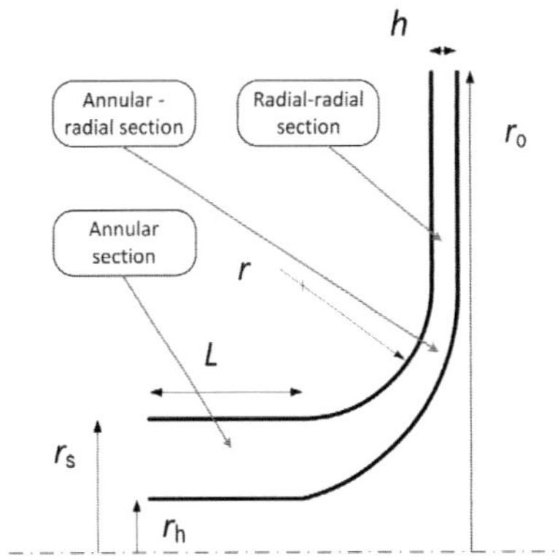

Figure 3: Sectional view of proposed geometry for analysis. N.b. hub centre body curve approximated as Bezier curve.

Table 2: Aerodynamic parameters of influence.

First order	Second order
Inlet Mach number (M_1)	Turbulence length scale and intensity
Inlet Reynolds Number (Re_1)	Inlet mass flow distribution
Inlet total temperature and pressure ($T_{1,\text{total}}, P_{1,\text{total}}$)	
Inlet aerodynamic blockage	
inlet swirl angle (α_1)	

It is recognised by several authors that aerodynamic parameters have the most significant impact on diffuser performance and hence are classified as first order parameters [3, 4, 5]. These parameters are listed in Table 2 along with second order parameters.

Whilst these flow parameters are significant to the performance of any diffuser, the impact of their variation on performance is not explicitly considered in the current study, instead they are considered fixed and to be representative of nominal turbine outlet conditions.

Monje et.al. [4] completed a CFD study of conical diffuser geometries with both CO_2 and air as the working fluids. The authors replicated some of the geometries and earlier experimental studies of Dolan and Runstadler [6] with air as the working fluid. This study indicated that the frictional effects for supercritical $\mathrm{CO_2}$ were less pronounced than for air, leading to optimal CO_2 geometries possessing a narrower cone angle and longer passage length. Considering this, a combined annular-radial diffuser geometry is investigated (see Fig. 3) as a means to package a large mean length in a short axial space with annular inflow from the turbine. A key study into annular diffusers was that of Sovran and Klomp in the 1960's [5] which identified annular diffusers as industrially significant for turbomachinery applications and considered performance impact of non-curved geometry variations.

Diffuser inlet swirl is anticipated in operation of the turbine due to required off-design operation and uncertainty in the rotor outlet velocity direction at the on-design condition and also during off-design operation. Hence any diffuser considered for this application should be swirl robust. From observing Eqn. 2, the influence of swirl on pressure recovery in an annular diffuser can be reduced by limiting change in mean radius (r) or passage width (h) over the diffuser. Considering this, there are two simple cases that are swirl robust. Firstly, purely axial flow (r=constant), which dictates no change in passage area, and hence limited diffusion, as diffusion is primarily area governed [3]. Secondly, purely radial flow (h=constant), where passage area increases with mean radius, leading to diffusion.

In considering the transition from an annular to a radial geometry, limited pressure recovery is possible due to the risk of flow separation [3]. A key study into a combined axial and radial diffuser by Moller [7] presented a design method for a bend with no change in passage area along the flow path.

Previous studies on axial-radial and radial-radial diffusers gave strong consideration to the impact of wall contouring and wall spacing on static pressure rise with respect to swirl. An experimental study into wall taper in a radial-radial diffuser by Yingkang and Sjolander [8] showed that parallel walled diffusers produced the most consistent pressure recovery coefficient with varying swirl angle. In considering the passage width of the parallel radial section of an axial-radial diffuser, Moller [7] tested geometries with non-dimensional width h/d between 0.1 and 0.2 with optimal determined analytically as 0.143. In a separate experimental study presented by Japikse [3], De Kranski and Sarpal determined the optimal h/d as 0.15.

3. Nominal design

The nominal diffuser geometry was created based on the geometric characteristics reviewed in the previous section for swirl robustness and pressure rise, and size constrained by factors listed in Table 1.

Considering the impact of swirl on ideal performance, the annular inlet section of diffuser under consideration is of fixed mean radius with no passage area change. In the transition from annular to radial a constant passage area is maintained. For the radial-radial section, walls remain parallel with spacing scaled from axial studies to an annular geometry. Graphically, this geometry is presented in Fig. 3.

The geometric relations for determining spacing h in Fig. 3 presented by Moller [7] are based on the diameter of a cylindrical inlet. For an annular inlet, these are scaled based on equivalent diameter (d_{eq}) defined in Eqn. 4 as opposed to the conventional definition of hydraulic diameter (d_h) for an annular passage.

$$d_{\mathrm{eq}} = 2\sqrt{r_{\mathrm{s}}^2 - r_{\mathrm{h}}^2} \tag{4}$$

The passage width of the radial-radial section of the diffuser can be scaled to the equivalent inlet diameter according to Eqn. 5

$$\frac{h}{d_{\mathrm{eq}}} = a_1 \tag{5}$$

For the transition from annular to radial, the shroud curve can be prescribed as a circular section with radius, r. For bend inlet and outlet passage areas to be equivalent, r is mathematically described by Eqn. 6.

$$r = \frac{d_{\mathrm{eq}}}{8\,a} - r_{\mathrm{s}} \tag{6}$$

To define the hub curve, a similar method to Moller [7] was employed, where passage areas normal to the flow were fixed and passage width was numerically evaluated in order to prescribe coordinates of the hub centre body curve. For practical purposes it is sufficient to approximate this curve by a Bezier curve defined using 4 control points.

In considering the length, L, of the annular inlet, it should be noted that the primary influence is that upon boundary layer growth and its impact on aerodynamic blockage [4]. For the nominal case, a larger value of L within the geometric constraints was chosen.

To define the maximum radial extent of the diffuser, relations presented in Moller were adapted to an annular geometry. The ideal radial extent of an axial-radial diffuser is defined according to Eqn. 7.

$$r_o = 2.471\, r_{\text{in}}\, Re_{\text{d}}^{\frac{1}{5}} \tag{7}$$

For the nominal design under consideration, diffuser inlet flow conditions are approximated as the rotor outlet conditions from a mean-line design. Values are listed in Table 3. Considering inlet conditions, this would lead to a prohibitively large outer radius. This correlates with the work of Monje [4] where it was noted that supercritical CO_2 diffuser geometries posses the optimal pressure recovery for area ratios in the region of 20. For analysis presented in this paper, outer radius is constrained to 75 mm, see Table 1, leading to a cropped diffuser. A larger area ratio could be mapped to the geometry through increasing h, however this would dictate a large increase in passage area ratio on the bend, likely resulting in separation and consequently poor performance. Geometric parameters for the nominal diffuser design are presented in Table 4.

Table 3: Nominal diffuser inlet conditions.

Parameter	Value
$\dot{m}$	1.0 kg/s
$V_{1,n}$	50 m/s
ρ_1	65 kg/m^3
α_1	0
μ_1	3.2×10^{-5} Pa.s
Re_d	9.6×10^{6}
T	750 K

Table 4: Geometric parameters of nominal diffuser.

Name	Symbol	Nominal value (mm)
Shroud radius	r_{s}	12.3
Hub radius	r_{h}	7.6
Inlet axial length	L	15.0
Transition radius	r	4.6
Outlet radius	r_{o}	75.0
Radial passage scaling factor	a_1	0.143
Radial passage width	h	2.8

4. Modelling

The diffuser was modelled as a quarter circumferential section of a 3D geometry. The ANSYS CFX 16.1 solver was used for the high-fidelity calculations [9]. The three-dimensional Reynolds Averaged Navier Stokes (RANS) equations were closed with the Shear Stress Transport k-ω turbulence model.

The required thermodynamic and transport properties for the vapour phase are incorporated into the CFD solver through lookup tables. These properties were retrieved through an in-house software interface [10]. The generated Real Gas Property (RGP) tables allow for accurate property estimations which are bilinearly interpolated by the solver. The thermophysical library

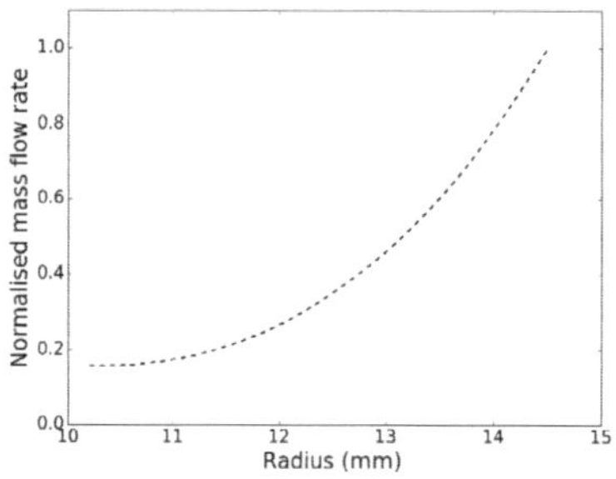

Figure 4: Non-dimensional radial distribution of mass flow at diffuser inlet.

used was RefProp, which implements NIST's formulation of the Span and Wagner Equation of State (EoS) model [11]. The tables were systematically refined until a sufficient accuracy was obtained in the solution.

Boundary conditions for simulations are specified in Table 5. The turbulence intensity was set as 10% based on work by Monje [4]. In prior turbine simulations it was noted that turbine exit mass flow distribution possessed a strong skewness towards the shroud line. This was incorporated in the diffuser inlet boundary conditions in order to more physically represent the operating environment. The two dimensional distribution of inlet mass flow rate boundary condition is prescribed according to Fig. 4.

Table 5: Boundary conditions.

Surface	Type	Value
Inlet	Mass flow	Radial variation described Fig. 4, total 0.25 kg/s
Outlet	Static pressure	8.9 MPa
Shroud and hub	No-slip wall	N/A
Edges	Rotational periodicity	N/A

4.1. Mesh

A structured 3 dimensional grid of approximately 1.0×10^6 cells was generated using Pointwise. A suitable first layer height was determined as 1.0×10^{-6} mm based on similarity of wall shear stress profiles to coarser and finer first layer heights, and maintaining y+ values in a range between 2 and 10.

In order to minimise skewness and area ratio, the nominal mesh was specified 140 nodes in the stream-wise direction,150 between the flow bounding surfaces, and 50 in the circumferential direction giving a total of 1.02×10^6 cells. Based on the nominal mesh a grid convergence study was conducted using a refinement ratio of 1.2. For this study both a fine and coarse mesh relative to the nominal were generated. Static entropy and total enthalpy were monitored as mass flow rate averages at inlet and exit of the geometry. Values for these are shown in Table 6. The nominal mesh sizing for this study was used for all geometry variants with the ability to coarsen if needed.

Table 6: Grid study of static entropy and total enthalpy, refinement ratio 1.2.

Mesh	s_1 (J/kg.K)	s_2 (J/kg.K)	$h_{1,\,\mathrm{total}}$ (J/kg)	$h_{2,\,\mathrm{total}}$ (J/kg)
Coarse	2797.63	2797.72	957635	957635
Nominal	2797.64	2797.72	957635	957635
Fine	2797.64	2797.72	957635	957635

5. Results and discussion

5.1. Performance of the nominal diffuser

Pressure rise coefficient, C_p was calculated according to Eqn. 1 as 0.89. In order to visualise sources of pressure losses and possible separation within the diffuser, pressure contours were plotted for a slice of the geometry in Fig. 5a. Further to this, pressure along shroud, hub and meridional (mean of shroud and hub) lines were also plotted in Fig. 5b. Meridional total pressure was also plotted. The meridional pressure lines give insight into the progression of inviscid flow within the diffuser.

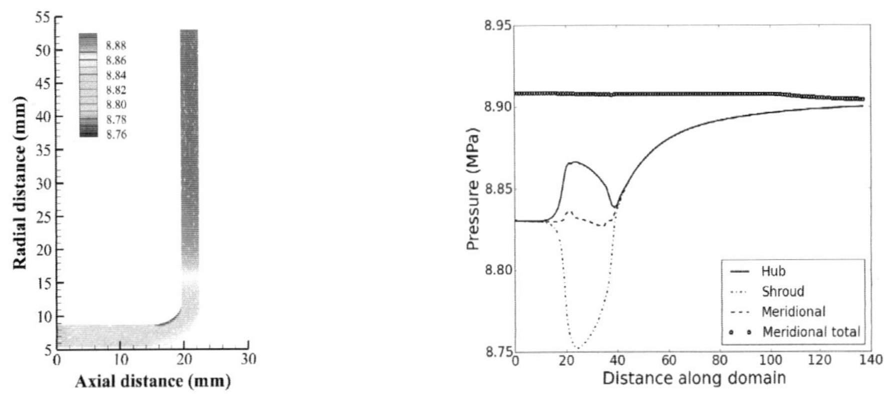

(a) Pressure contour (MPa).

(b) Pressure along hub, shroud and meridional lines.

Figure 5: Pressure distribution for nominal geometry.

From Fig. 5 it can be seen that there are high and low pressure regions on hub and shroud surfaces respectively, however there is no separation or re-circulation. There is no significant rise in meridional static pressure prior to the radial-radial section which matches design intent.

5.2. Effects of diffuser geometry

In order to understand the impact of geometry changes on performance and flow, a parameter study was conducted on the geometry whilst retaining boundary conditions and mesh density. Details of the geometries are summarised in Table 7.

Results of C_p for each geometry are presented in the last column of Table 7. Pressure contours are plotted for a slice of the geometry in Figures 6a - 8a. Static pressure for shroud, hub and meridional lines as well as meridional total pressure are plotted in Figures 6b - 8b.

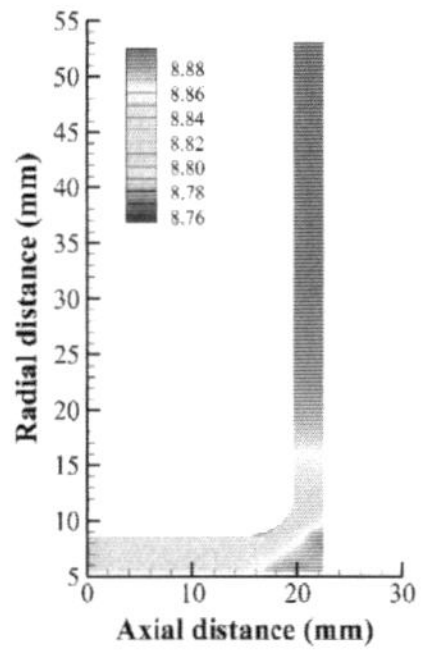

(a) Pressure contour (MPa)

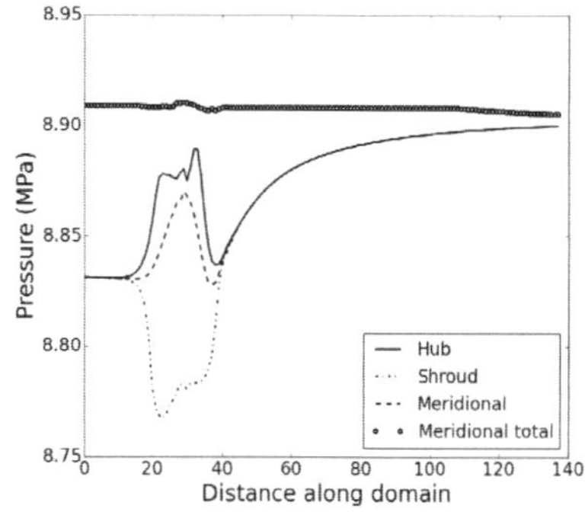

(b) Pressure along hub, shroud and meridional lines.

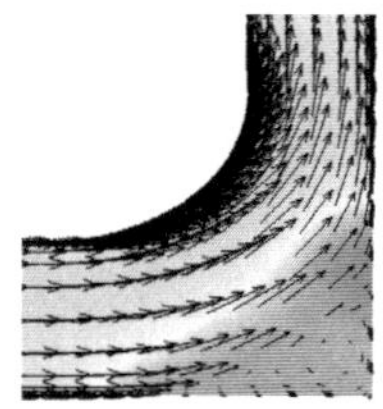

(c) Vector plot of re-circulation region.

Figure 6: Pressure distribution for geometry variation 1.

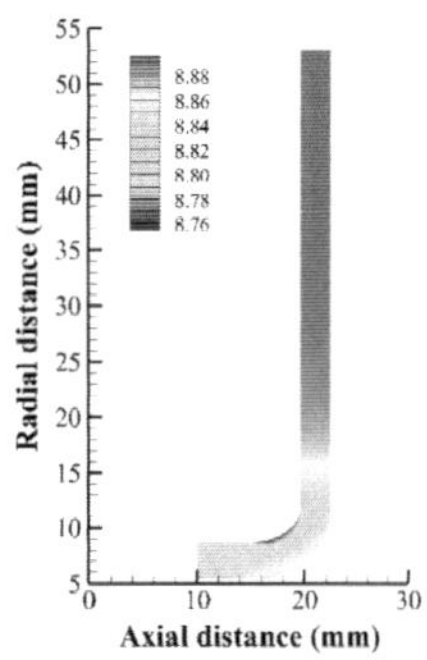

(a) Pressure contour (MPa).

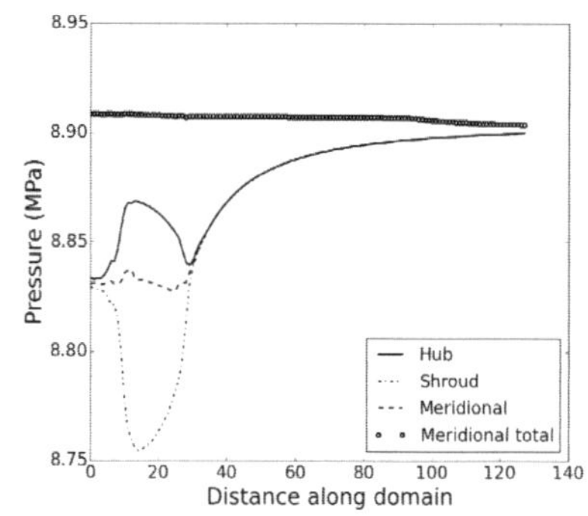

(b) Pressure along hub, shroud and meridional lines.

Figure 7: Pressure distribution for geometry variation 2.

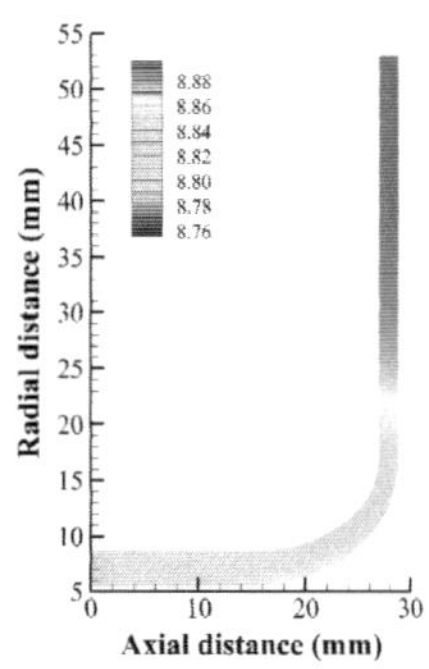

(a) Pressure contour (MPa).

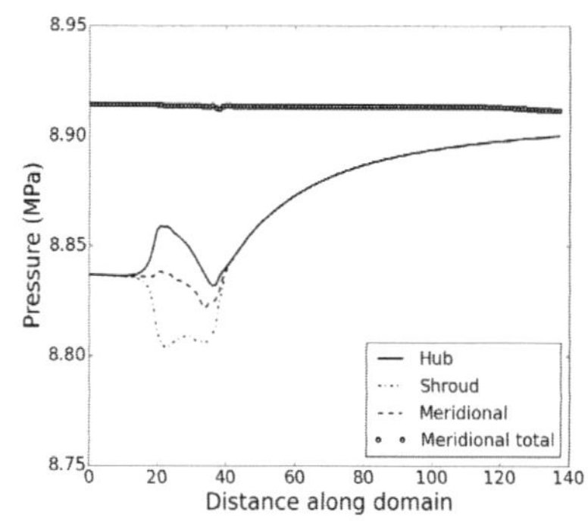

(b) Pressure along hub, shroud and meridional lines.

Figure 8: Pressure distribution for geometry variation 3.

Table 7: Details of geometries for parameter study.

I.D.	description	a_1	r_o	L	hub centre body	C_p
Nominal	nominal	0.143	75mm	15mm	Y	0.89
1	no hub centre body	0.143	75mm	15mm	N	0.88
2	short inlet	0.143	75mm	5mm	Y	0.89
3	narrow radial-radial	0.1	75mm	15mm	Y	0.82

5.3. Discussion

From observing Figures 5a-8a, all geometries show the majority of the static pressure rise in the radial-radial section, confirming design intent. This trend is also illustrated in meridional total pressure plots where decreases are present in all geometries towards the end of the radial-radial section. Flow separation is only observed in geometry variation 1, where a re-circulation region is present in place of the hub centre body. This re-circulation region is illustrated in Figure 6c. The nominal diffuser displays the highest pressure recovery, however removal of the hub centre body and shortening the inlet have only a small impact on the bulk pressure recovery.

Geometry variation 1 showed good attachment to the shroud surface, with a captive pocket in place of the hub centre body due to the discontinuous nature of the hub line. Overall pressure rise was not significantly influenced by this compared to the nominal case, however a spike in meridional static pressure is observed. This is due to the meridional line passing through the high pressure ciculation region. Compared with the nominal case geometry variation 2 shows a smaller pressure drop along the shroud line. Bulk pressure recovery however is lower, but is not affected significantly. Geometry variation 3 presents a similar flow distribution on the bend to the nominal case and geometry variation 2, however with a small difference between shroud and hub pressures, however overall pressure recovery was adversely effected in comparison.

From comparing the 4 geometries, it appears that the largest parameter of influence on pressure rise is a_1 (passage height to equivalent diameter ratio). A value close to that presented in the literature (0.15) being in the optimal region. Lower values, as illustrated by geometry 3, showed lower pressure recovery. Higher values of a_1 constrain r to be unsuitably small, and geometries were non-physcial. As a geometric alternative, lower values of a_1 may be suitable if the shroud surface of the bend in the diffuser is not constrained to a circular profile.

The impact of the diffuser inlet length appears to be minimal on overall pressure rise performance, however it remains uncertain if the inlet section needs to be sufficiently large to accommodate non-uniform mass flow rate from turbine exhaust. Further simulation work should be undertaken to verify this with shorter inlet lengths than simulated in geometry variation 2.

Overall, high values of pressure rise coefficient were observed when compared to radial-radial diffusers operating on air as presented by Japikse [3] in the range 0.7-0.8. A probable cause for this is the lower viscosity of CO_2, and hence lower viscous losses.

6. Conclusions and recommendations

Overall the design method adapted from Moller [7] appears to be suitable for the preliminary design of a annular-radial diffuser. Design intent is confirmed with pressure recovery focused in the radial-radial section of each geometry. Comparing values of pressure rise coefficient, the nominal design produced favourable results compared to geometry variations. Additionally, a similar trend was observed regarding the omission of the hub centre body to that presented in Moller [7].

Diffuser geometries with passage height to equivalent diameter ratios in the region of 0.15 showed the highest pressure recovery. Further nominal flow studies should be conducted with passage height to equivalent diameter ratios in this region in order to determine the optimum. The presence or absence of a hub centre body showed a minimal impact on pressure recovery, and it's impact on manufacturing and costs should be considered for future designs. Inlet length did not appear to be significant to pressure rise, and should be minimised in order to reduce shaft bending.

The key result from this work is that a short and simple diffuser design (nominal and geometry variations 1 and 2) can attain static pressure rise coefficients greater than 0.88. This confirms that annula-radial diffusers are a viable design solution for supercritical CO_2 radial inflow turbines, thus enabling an alternative cantilevered rotor layout.

In selection of a finalised design, robustness to off-design operating conditions should be quantified. Off-design diffuser inlet conditions should be considered with particular attention to swirl, and the impact on flow separation on the bend for such a design. To enhance the reliability of simulations, the hub line should be modelled as a no-slip wall rotating boundary in future work. Future work should focus on optimising the present nominal design with respect to off-design performance and manufacturing constraints.

References

[1] Dyreby, J., Klein, S., Nellis, G., and Reindl, D., 2014. "Design considerations for supercritical carbon dioxide brayton cycles with recompression". *Journal of Engineering for Gas Turbines and Power*, ***136***(10), Jul, p. 101701.

[2] Dostal, V., 2004. "A supercritical carbon dioxide cycle for next generation nuclear reactors". PhD thesis, Massachusetts Institute of Technology, http://hdl.handle.net/1721.1/17746.

[3] Japikse, D., and Baines, N. C., 1998. *Turbomachinery Diffuser Design Technology*. Concepts Eti.

[4] Monje, B., Sanchez, D., Chacartegui, R., Sanchez, T., Savill, M., and Pilidis, P., 2013. "Aerodynamic analysis of conical diffusers operating with air and supercritical carbon dioxide". *International Journal of Heat and Fluid Flow*, ***44***, Sep, pp. 542 – 553.

[5] Sovran, G., and Klomp, E., 1965. "Experimentally determined optimum geometries for rectilinear diffusers with rectangular, conical or annular cross-section". In Fluid mechanics of internal flow - Proceedings of the symposium of the fluid mechanics of internal flow, G. Sovran, ed.

[6] Dolan, F. X., and Runstadler, P. W., 1973. Pressure recovery performance of conical diffusers at high subsonic mach numbers. Technical Report NASA-CR-2299, NASA, Jul.

[7] Moller, P. S., 1966. "A radial diffuser using incompressible flow between narrowly spaced disks". *Journal of Basic Engineering*, ***88***(1), Mar., pp. 155–162.

[8] Zhu, Y., and Sjolander, S. A., 1987. "Effect of geometry on the performance of radial vaneless diffusers". *Journal of Turbomachinery*, ***109***(4), Oct, pp. 550–556.

[9] *ANSYS Academic Research, 2015, ANSYS Release 16.1, Theory Manual, Inc., Canonsburg, P.A.*

[10] Colonna, P., and Van der Stelt, T., 2016. Fluidprop: a program for the estimation of thermo physical properties of fluids.

[11] Lemmon, E. W., Huber, M. L., and McLinden, M. O., 2010. "Nist standard reference database 23: Reference fluid thermodynamic and transport properties - REFPROP". Version 9.0. National Institute of Standards and Technology, Standard Reference Data Program, Gaithersburg.

Preliminary design of a supercritical CO_2 wind tunnel

B Re[1], A Rurale[2], A. Spinelli[2] and A. Guardone[1]

[1] Department of Aerospace Science and Technology, Politecnico di Milano
Via La Masa 34, 20156, Milano, Italy
[2] Department of Energy, Politecnico di Milano, Via Lambruschini 4, 20156, Milano, Italy

E-mail: alberto.guardone@polimi.it

Abstract. The preliminary design of a test-rig for non-ideal compressible-fluid flows of carbon dioxide is presented. The test-rig is conceived to investigate supersonic flows that are relevant to the study of non-ideal compressible-fluid flows in the close proximity of the critical point and of the liquid-vapor saturation curve, to the investigation of drop nucleation in compressors operating with supercritical carbon dioxide and and to the study of flow conditions similar to those encountered in turbines for Organic Rankine Cycle applications. Three different configurations are presented and examined: a batch-operating test-rig, a closed-loop Brayton cycle and a closed-loop Rankine cycle. The latter is preferred for its versatility and for economic reasons. A preliminary design of the main components is reported, including the heat exchangers, the chiller, the pumps and the test section.

1. Introduction

Supercritical carbon dioxide (sCO_2) is currently being considered as working fluid in several industrial and power generation applications thanks to its relatively low critical pressure and its low critical temperature. The seminal work of Angelino and Feher points out the advantages of using sCO_2 for power generation [1, 2]. Indeed, sCO_2 Brayton power cycles can be used to exploit solar, geothermal and waste heat thermal sources. Advantages over e.g. steam include compactness of turbomachinery and high thermal efficiency at low temperature. Moreover, fewer corrosion issues may be encountered except than at high temperatures and pressures [3]. For nuclear power generation, sCO2 is preferred over steam for safety reasons in sodium cooled fast reactions [4]. Rapid expansion of supercritical solutions (RESS) of CO_2 is used in the chemical, pharmaceutical and food industry for particle generation or extraction of chemicals, see [5, 6, 7] and reference therein. Other applications include sterilization and cleaning [8].

The fluid dynamics of sCO_2 flows departs significantly from the well-known gas dynamics of dilute gases, such as air in standard conditions. For example, a relatively low speed of sound is observed at very high, liquid-like densities. Moreover, non-equilibrium shock structure is possibly observed due to the relaxation of the internal vibrational modes [9, 10]. In the supercritical and close-to-critical region, the heat transfer as well exhibits peculiar behavior [11, 12].

The dynamics of compressible fluids in the close proximity of the vapor-liquid saturation curve and critical point or within the supercritical region are referred to in the following as Non-Ideal Compressible-Fluid Dynamics (NICFD).

Despite its widespread usage, a comprehensive understanding of the fundamental properties of CO_2 flows in supercritical conditions is not available, yet. Preliminary theoretical and numerical studies using accurate equations of state [13] and non-ideal flow solvers [14, 15], are yet to

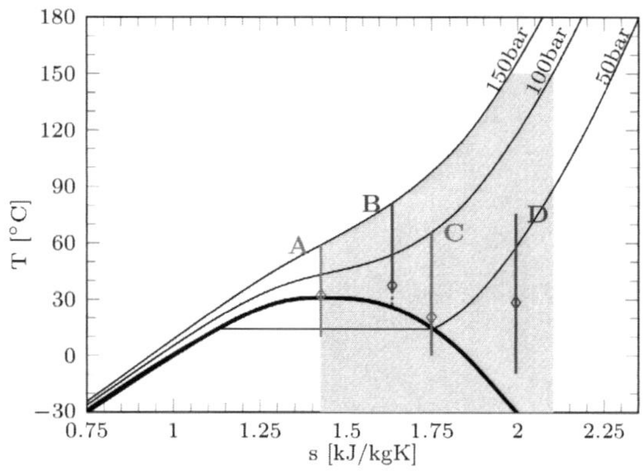

Figure 1: Thermodynamic plane temperature-entropy (T-s) with the four exemplary supersonic expansions to be realized in the SCO_2PRI facility. The region of interest (gray area), the saturation curve and three isobars are also reported. The symbol $\Diamond$ indicates the throat section, at $M = 1$.

be complemented with experimental data. Indeed, only recently experimental activities are being carried out to investigate the fundamentals of sCO_2 flows in supercritical conditions. For instance, Lettieri and collaborators at MIT assessed the condensation effects in sCO_2 compressors and defined a criterion to predict fluid condensation [16]. At KAIST, Lee's research team is developing an experimental facility to accurately take into account non-ideal gas effects during sCO_2 compressor design and performance analysis [17]. Finally, the university of Seville and Altran have designed a pressurized sCO_2 wind tunnel to improve the design of blade cascade of turbomachinery [18]. Diverse technology demonstrators of Brayton power cycles using sCO_2 are currently in operation in the USA [19, 20, 21], though the technology is not sufficiently developed to commercial exploitation [22]. For instance, heat transfer in sCO_2 cycles still raises some questions and requires further investigations to efficiently design heat exchangers [23, 24].

The design of a novel sCO_2 wind tunnel, named SCO_2PRI (Supercritical CO_2 for the PRocess Industry), is currently under-way at the CREA (Compressible-fluid dynamics for Renewable Energy Application) Laboratory of Politecnico di Milano. Fundamental studies of supersonic sCO_2 flows will be carried out in the close proximity of the critical point and the liquid-vapor saturation curve, where sCO_2 compressors for power production are designed to operate. Moreover, the test-rig will be used as a calibration tunnel for non-ideal flows pressure probes and optical measurement techniques, including Schlieren and Laser Doppler Velocimetry (LDV).

The present work outlines the preliminary design of the plant and the technical specifications of the relevant components. To attain supersonic speeds, the test section consists in a convergent-divergent nozzle followed by a rectangular-section chamber for flow visualization. Three possible configurations are initially taken into account to drive the fluid through the nozzle: an open-loop batch-operating system, an "inverse" Joule-Brayton cycle and a Rankine cycle. A preliminary analysis indicates that the last configuration is the most suitable for the operating conditions of interest. Then, the main components of the plant, namely the pump, the heater, the chiller and the heat exchangers are designed and a preliminary cost analysis is also carried out.

The present paper is structured as follows. In Section 2, the region of interest for the experimental observation of sCO_2 flows is determined, including conditions suitable for studying critical-point flows and condensing flows in compressors. The design constraints are also reported. The diverse test-rig configurations are presented and discussed in Section 3. The preliminary design of the main components of the chosen configuration, namely, closed-loop Rankine cycle, is reported in Section 4. In Section 5, final comments are given.

2. Region of interest and design constraints

The preliminary design of the SCO_2PRI test-rig starts with the definition of the thermodynamic region to be investigated, according to the relevant research and industrial applications of sCO_2

reported in the previous section.

Figure 1 depicts this region in the thermodynamic plane T-s, where T is the temperature and s is the specific entropy. Since the research interest is limited to NICFD flows, a first parameter taken into account to define the region of interest is the compressibility factor $Z = Pv/RT$, where v is the specific volume, P is the pressure and R is the gas constant. This quantity allows to estimate the deviation of the actual thermodynamic behavior from the ideal gas model, which predicts $Z \equiv 1$ [25]. Therefore, the region of interest is limited towards the vapor state by a maximum value of Z around $0.8 \div 0.85$. On the opposite side, it extends to the critical point. Moreover, the maximum pressure and temperature of $P_{\mathrm{max}} = 150$ bar and $T_{\mathrm{max}} = 150°\mathrm{C}$ are imposed as design constraints, to limit the cost of the test-rig. The lowest temperature is arbitrarily fixed at $-30°\mathrm{C}$ to include also a portion of the two-phase region, below the Vapor-Liquid Equilibrium (VLE) curve.

Within the region of interest, four exemplary isentropic expansions are defined and a convergent-divergent nozzle is adopted to reach supersonic speeds. The four expansions, labeled A, B, C and D, are depicted in the plane T-s in Figure 1. Initial computations of thermodynamic states are performed by means of the quasi-one-dimensional theory, which describes a steady, one-dimensional and isentropic flow in a duct neglecting viscous and thermal effects [26]. Thanks to this theory, it is possible to predict the inviscid core of the flow.

Expansion A starts at the maximum pressure $P = 150$ bar, it crosses the critical point and eventually ends inside the two-phase region. This expansion is probably the most challenging from the point of view of the control. At the exit of the nozzle, the pressure $P = 45$ bar is imposed, corresponding to a Mach number $M = 1.72$. The measurement of thermodynamic properties along the expansion process, and possibly in the close proximity of the critical point, would give a significant improvement in the fundamental knowledge of sCO_2 behavior in the NICFD regime. The second expansion (B) occurs completely in the supercritical region, starting from the maximum pressure of 150 bar down to 75 bar, namely to an isobar just above the critical one. The initial temperature is 81.1°C, so that a slightly supersonic expansion can be realized (exit Mach $M = 1.05$) and the compressibility factor at the nozzle inlet and outlet is around 0.5. Measurements of the fluid states through expansion B are relevant for investigating sCO_2 transport properties, for validating of CFD models and for assessing the isentropic expansion coefficient in non-ideal conditions. Finally, conditions that typically occurs in turbo-machinery are reproduced in expansions C and D. Condition C replicates the expansions that may occur in sCO_2 compressors near the leading edge of the impeller, which were investigated also by Lettieri et al. [16]. In this case, the design exit Mach number is $M = 1.49$. The fourth expansion (D) is representative, in terms of reduced conditions, of flows in Organic Rankine Cycle (ORC) turbine. The experimental investigation of this kind of flows is typically challenging because of the high temperatures, which might be close to the thermal stability limit of the fluid. However, the behavior of the organic fluid under consideration can be reproduced by an other fluid with a similar compressibility factor at the critical point, provided that their behaviors can be described by the same equation of state. Indeed, according to the principle of corresponding states, all fluids behave alike at the same thermodynamic conditions made dimensionless with respect to the critical-point values [25]. Thus, a flow of sCO_2 can reproduce, for instance, a siloxane flow in an ORC turbine in terms of reduced conditions at a lower temperature, which allows to measure more easily quantities of interest. In expansion D, the initial and final pressures are 62.1 and 20 bar, while the exit Mach number is $M = 1.42$.

To realize the outlined goal expansions, a plant suitable for feeding the converging-diverging nozzle has to be designed. The plant can possibly operate in a continuous fashion or as a batch facility, provided that the test time is sufficient to perform quasi-steady measurements. In this regard, some observations can be drawn according to the peculiar aim and scope of the test-rig. First, the SCO_2PRI is a research facility that is expected to operate discontinuously

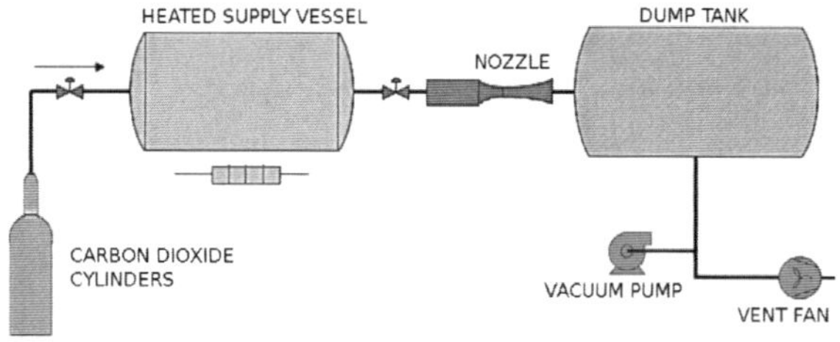

Figure 2: Sketch of the batch operating test-rig. A heated vessel, filled by CO_2 cylinders, discharges the fluid at supersonic speeds through a convergent-divergent nozzle into a low pressure tank.

Table 1: Test durations and mass discharged in batch-operating plant for expansion B. Results are reported for two throats (th) and three vessels (ves).

$\mathbf{A}_{\mathrm{th}}$ [mm^2]	$\mathbf{P}_{\mathrm{ves}}$ [bar]	$\mathbf{V}_{\mathrm{ves}}$ [m^3]	**t** [s]	$\mathbf{\Delta M}$ [kg]
200	200	5	30	266
200	250	3	32	284
200	250	5	54	473
20	200	2	100	86
20	250	1	100	86

and for few hours per year, therefore the initial and maintenance costs are more relevant than operating costs. Moreover, it should be installed in current available spaces within the CREA laboratory and this imposes limits on the size and on the power requirement of the test-rig, which roughly can not exceed 400 kW. Finally, the overall costs for designing and constructing the SCO_2PRI test-rig must be compatible with funding opportunities that are available for fundamental research. According to these peculiarities, three different plant configurations are investigated in the next section.

3. Assessment of the test-rig configurations

Three different test-rig configurations are now considered as possible solutions to realize the exemplary expansions A, B, C and D introduced in the previous section. These are an open-loop, batch operating test-rig and and two closed-loop, continuous operating plants. In the following subsections a brief description and the initial sizing of the main components are reported for each configuration. As a preliminary step, a nozzle throat area of 200 mm^2 was chosen. However, as explained in the following, the flow rate associated to such a relatively large throat area is not compatible with the design constraints and a more realistic throat area of 20 mm^2 is finally chosen.

3.1. Batch operation

The first configuration considered for the SCO_2PRI test-rig consists in the simple open-loop facility sketched in Figure 2. An heated pressurized vessel is filled with CO_2 cylinders up to a pressure greater than the one at the beginning of the expansions. Then, through a control valve, the selected expansion is realized in the nozzle, which discharges into a dump tank equipped with a vacuum pump to set the desired back pressure. Carbon dioxide is eventually released into the atmosphere. This set-up is very similar to the one currently available at MIT [16].

The main advantage of this solution is the simplicity and, thus, the low initial and maintenance costs. On the other hand, the batch operation imposes a limit to the experiment duration, which strictly depends on the volume and initial pressure of the vessel. A target test duration of 100 s is pursued. Table 1 reports the time and the mass released during the expansion B using different vessels and initial pressure, computed through the quasi-1D theory and assuming the process adiabatic. Unfortunately, with the initial nozzle with $A_{\mathrm{th}} = 200$ mm^2, it is possible to perform only short tests, even by using a quite large vessel of 5 m^3. According to these results and to the similar ones obtained for the other expansions but not reported here for brevity the flow rate is reduced by one-tenth so that $A_{\mathrm{th}} = 20$ mm^2. With this geometry, the target test duration can be achieved. A further drawback, inherently related to batch operation,

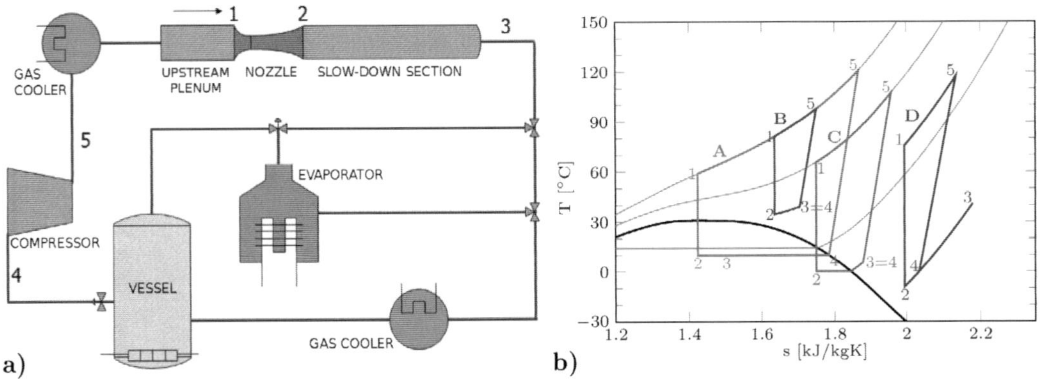

Figure 3: Compressor-based configuration: *a)* sketch of the plant, *b)* exemplary thermodynamic cycles for expansions (1-2) of interest. Points from 1 to 5 identify different steps: 1-2 nozzle expansion, 2-3 slowing down, 3-4 evaporation/cooling, 4-5 compression, 5-1 cooling.

	P_1 [bar]	T_1 [°C]	P_2 [bar]	T_2 [°C]	T_3 [°C]	T_4 [°C]	T_5 [°C]	W_c [kW]
A	150.0	58.9	45.0	10.0	10.0	10.0	120.8	121.8
B	150.0	81.1	75.0	34.5	39.0	39.0	97.7	64.0
C	100.0	65.6	35.0	0.1	5.6	5.6	107.5	52.2
D	62.1	75.5	34.0	-9.2	40.8	0.0	117.4	33.4

Table 2: Thermodynamic cycles for the four expansions in the "inverse" Joule-Brayton configuration. $\mathbf{W_c}$ is the estimated power required by the compressor for $A_{\text{th}} = 20$ mm^2.

concerns the difficulties in reaching steady state: this requires an accurate and fast control of the throttling valve, because of the emptying of the pressurized vessel and of the filling of the dump tank. The inaccurate control of the thermodynamic state in the steady flow could jeopardize the possibility of using the test-rig in the close proximity of the liquid-vapor critical point and its usage as a pressure probe calibration facility. In conclusion, the analysis of the strengths and limitations of this configuration suggests to discard batch-operation option and to investigate a different solution.

3.2. "Inverse" Joule-Brayton cycle

The first closed-loop configuration can be described as an "inverse" Joule-Brayton cycle, in which a compressor is used to restore the pressure after the nozzle expansion. Since the compressor represents the main (and the most expensive) component of the test-rig, the compressor is sized first. For simplicity, a single machine that permits to realize all expansions is preferred. For this reason, the limit temperatures $T_{\text{min}} = 0$°C and $T_{\text{max}} = 120$°C are imposed for the suction and discharge sections, respectively, and the end states of expansions A, C and D are chosen according to them. Moreover, for expansion A, a pressure $P_2 = 45$ bar is imposed to avoid excessively large compression and cooling power, while for expansion D, a gas cooler is added to reduce the inlet (and so the outlet) temperature of the compressor. Figure 3 displays the sketch of the plant along with the thermodynamic cycles in the T-s plane, which are detailed in Table 2. The table also reports the estimates of the electrical power consumption, computed assuming an isentropic efficiency $\eta = 0.6$ for the compressor and neglecting all other losses, for the reduced throat area. For the initial area $A_{\text{th}} = 200$ mm^2, the power required by the compressor considerably exceeds the available one.

The preliminary computations allow to define the operating range of the required compressor,

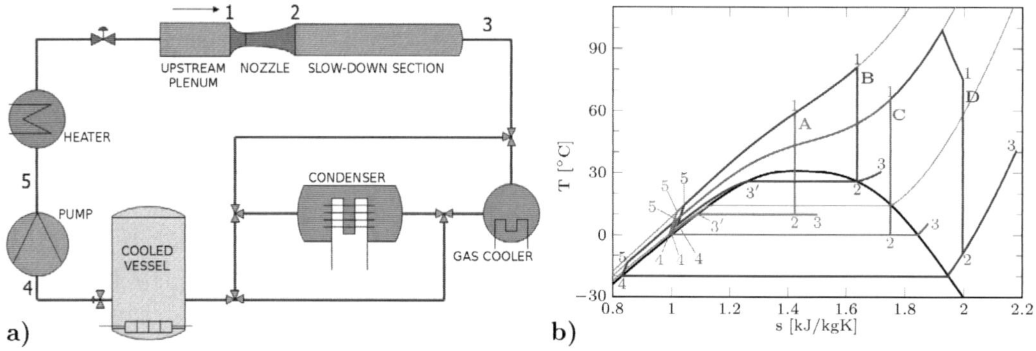

Figure 4: Pump-based configuration: *a)* sketch of the plant, *b)* exemplary thermodynamic cycles. Points from 1 to 5 identify different steps: 1-2 nozzle expansion, 2-3 slowing down, 3-4 de-superheating/condensation/cooling, 4-5 compression, 5-1 heating. For tests A and B, 3′-4 indicates the additional cooling to avoid cavitation issues.

	$\mathbf{P_4}$ [bar]	$\mathbf{T_4}$ [°C]	$\mathbf{T_5}$ [°C]	$\mathbf{W_p}$ [kW]	$\dot{\mathbf{Q}}_\mathbf{h}$ [kW]	$\dot{\mathbf{Q}}_\mathbf{c}$ [kW]	$\mathbf{W_c}$ [kW]	$\dot{\mathbf{m}}$ [kg/s]
A	45.0	0.0	11.1	29	200	229	65	1.59
B	65.7	5.0	14.3	17	221	239	68	1.16
C	35.0	0.2	7.4	8	159	167	88	0.7
D	20.0	-20	-13.6	5	120	125	94	0.36

Table 3: Thermodynamic cycles in the Rankine configuration, with $A_{th} = 20$ mm^2. $\mathbf{W_p}$ and $\mathbf{W_c}$ are the electrical powers required by the pump and the chiller unit, $\dot{\mathbf{Q}}_\mathbf{h}$ and $\dot{\mathbf{Q}}_\mathbf{c}$ are the thermal powers exchanged in the heating and cooling sections.

which is found to be a volumetric one, and to proceed with the selection of the machinery among the commercial available solutions. The machinery design is commissioned to a specialized company, which proposes an ad hoc reciprocating compressor, which a particular massive structure, since the high densities reached by the CO_2 in the specific operating conditions generate considerable axial stresses. The resulting compressor size is much larger than the one usually required by similar pressure ratios, flow rates and powers in a standard thermodynamic region, i.e. far from the critical point. Therefore, this extremely large (and noisy) machinery cannot be installed at the laboratory. Furthermore, the purchase of this component only would exhaust almost all the initial available fund. For these reasons, compressor-based test-rig is not deemed to be adequate.

3.3. Rankine cycle

A pump-based plant is now considered for the SCO_2PRI test-rig. This configuration results in a transcritical Rankine cycle with phase transition, as sketched in Figure 4. After the nozzle expansion and the subsequent deceleration, the flow is condensed and liquid CO_2 is pumped to the inlet nozzle pressure, then it is heated until the required temperature. The imposed nozzle exit states are the same used for the previous configurations (i.e. points 1, 2 and 3 in Table 2), except that expansion B has been extended until the VLE curve to make condensation possible ($T_2 = 25.9$°C). To realize expansion D, the pressure at the exit of the pump must be higher than the critical one, and a throttling process is performed before entering the nozzle to reach the correct pressure. All thermodynamic cycles are shown in Figure 4; relevant data are reported in Table 3.

First, the pump is characterized. As expected, the power consumptions, reported in Table 3 and computed assuming an isentropic efficiency $\eta = 0.6$, are much smaller than the ones computed for the compressor. Indeed, the specific enthalpy drop in the liquid region is smaller than the one in the gas region for the same pressure ratio. From a preliminary analysis, a volumetric pump is needed. With this regards, possible problems may occur due to the relatively high compressibility of sCO_2 with respect to standard liquids, especially in test A and B. For this reason, in these two cycles, an additional cooling is performed at the end of the condensation to reach a lower pump inlet temperature, and therefore reducing the liquid compressibility.

Differently from the "inverse" Brayton cycle, the heating and cooling processes are relevant both from the technical and the economic point of view, therefore a preliminary assessment of the test-rig configuration has to include them. Table 3 reports also the heating and cooling powers (included de-superheating and condensation). The results are reported only for the nozzle with $A_{\mathrm{th}} = 20\ \mathrm{mm}^2$ because the available power is not sufficient if the larger area of $A_{\mathrm{th}} = 200\ \mathrm{mm}^2$ is considered.

Despite the quite large thermal power involved, relatively standard components can be used to perform the required processes and no particular technical problems are expected in this regard. From a first estimate on the basis of the commercial available components similar to the ones required by this plant, the space required to house all the parts of the test rig is congruous with the laboratory capacity. Therefore, a more accurate investigation and a preliminary design of the components of the pump-based configuration are performed. As final remark, a further advantage provided by a phase-transition configuration concerns the possibility of carrying out research using different fluid phases, such as liquid or two-phase flows of CO_2.

4. Preliminary design of the SCO_2PRI test-rig

The previous analysis of the candidate plant configurations led to the selection of the pump-based closed-loop Rankine cycle. This section presents a preliminary design of the main components and of the test section of the SCO_2PRI. The components are selected among commercially available (possibly customized) solutions with the primary aim of realizing all four expansions by means of the same balance of plant. In this respect, it should be clarified that the different expansions will be investigated in separate experimental campaigns, therefore the possibility of connecting the different cycles is not required. For what concerns the test section, four different nozzles are designed and a modular solution is adopted so that they can be installed within the same arrangement.

4.1. Analysis of test-rig components

The first component that is analyzed is the pumping section. Since the flow rates in the four tests are rather different, a parallel configuration that includes two pumping groups working at constant speed is adopted and one device can be excluded through a by-pass valve. Each pumping group is composed by a plunger pump that has a maximum discharge pressure of 150 bar and a maximum inlet pressure of 65 bar, a 18.5 kW electrical motor, a pulsation damper and other safety accessories.

The heating section is now detailed. The temperature of the sCO_2 flow is increased by means of an heater exchanger that is fed on the hot-side by a thermal oil flow from a mono-bloc gas heater, already equipped with control and safety equipments. Plate & Shell Heat Exchangers (PSHE) are exploited to obtain benefits in terms of compactness and heat transfer efficiency at the operating temperatures and pressures. More specifically, a fully welded pack of circular plates of stainless steel with a thickness of 1.5 mm is contained in a carbon steel shell of approximately 0.13 m^3. Thanks to the peculiar design realized by a specialized company, this kind of heat exchangers can safely operate up to 170 bar. The issue of dealing with different flow rates and different inlet/outlet temperatures is solved by dimensioning all components on the most demanding case, e.g. test B. Since the described device cannot guarantee fine control of the

sCO_2 temperature at the nozzle inlet, which is of paramount important for expansion A, a 2 kW electrical heater is added before the nozzle for control purposes.

For the cooling section, the use of a chiller is mandatory due to the low temperatures. A direct cooling process with a chiller unit specially designed for the SCO_2PRI test-rig is used to perform de-superheating, condensation and sub-cooling processes, depending on the performed test. This technical solution requires the design of an additional cycle for the refrigerant fluid which, according to the required tasks and operating conditions, is ammonia. The design choice consists in the standard, widely-used single-stage vapor-compressor cycle, which is composed by a compressor, a condensing section, an expansion valve and an evaporator. The two main steps of the design process concern the definition of the condensation and evaporation temperatures, while the design of the refrigerant cycle is commissioned to a specialized supplier. The condensation temperature is set at 38°C and a water-cooled condenser is used to exploit a cooling tower already available at the laboratory. For the evaporation temperature, different values are established according to the test: for test A, B and C, a temperature difference of $10 - 15$°C is imposed between the two fluids; while in test D this difference is reduced to 2°C. The heat required by CO_2 de-superheating and condensation is absorbed by ammonia through a PSHE with a design thermal power of 400 kW, while a smaller dry expansion evaporator (design power of 50 kW) accounts for the additional cooling required in test A and B.

Finally, after a preliminary sizing of the SCO_2PRI plant, including also piping, the mass of CO_2 circulating within the plant is estimated. This quantity varies between 57 and 78 kg, depending on the considered test. Therefore, an accumulation tank of 50 L is included in the plant for storage purposes. Accounting also for this vessel, the area required to place all components is estimated to be less than 15 m^2. Moreover, the thermal oil heather and the chiller unit, which result to be the largest parts, can be safely placed in an external area adjacent to the laboratory.

4.2. Design of the test section

The test section, which includes the nozzle along with the downstream slow-down section, is now detailed. In particular, the geometry of the nozzles delivering the desired expansion ratios at the diverse operating conditions is designed. Since the exemplary supersonic expansions are defined considering chocked flows, four different convergent-divergent nozzles are required.

Before the detailed design of the geometry, the flow inside the nozzle is investigated more thoroughly. Figure 5 shows the variation of the Mach number with the density during all expansions, computed according to the quasi-1D theory for isentropic flows. The NIST databse REFPROP [27] is exploited to perform the thermodynamic evaluations by using the reference equation of state of Span and Wagner [13]. As expected, a non-monotone variation occurs during the test A. This behavior is typical of fluids characterized by a value of the fundamental derivative of gas-dynamics Γ between zero and one and it is related to close-to-critical-point effects. Expansions B, C and D lies in the thermodynamic region where $\Gamma > 1$ and the Mach number is monotonically increasing to supersonic values, similarly to what observed in the ideal flows of dilute gases. However, non-ideal compressible-fluid effects are relevant—including the non-ideal dependence of the sound speed on the density in isothermal transformations—and strongly influence the dynamics of the fluid in expansions B, C and D as well.

For the nozzle geometry design, a rectangular cross-section is adopted to produce a two-dimensional flow and to perform flow visualization more easily. With this regards, one wall of the test section consists in a planar glass windows. Since the throat of the nozzle is chocked, the thermodynamic state there is completely defined. Thus, the geometry of the nozzle can be computed in two independent steps. First, the methods of characteristics [28] is used to compute the length and the shape of the divergent section, in order to obtain an uniform flow at the exit with velocity parallel to the nozzle axis and the design pressure ratio. The state-of-the-art multi-

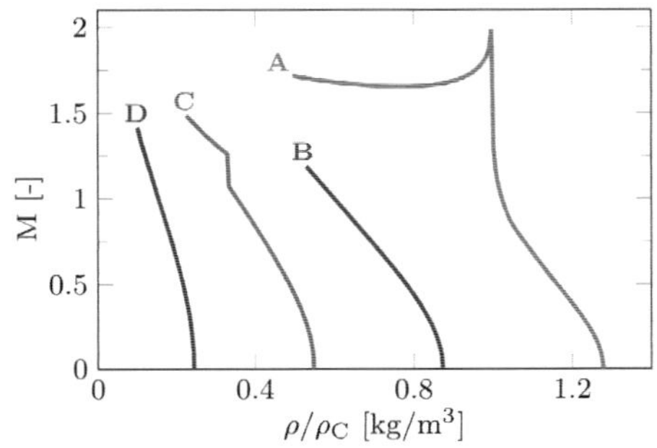

Figure 5: Variation of Mach number with density in the four exemplary expansions depicted in Figure 4. Density is scaled with respect to the critical value. As expected, a non-monotone variation occurs during Expansion A, where Γ is below 1.

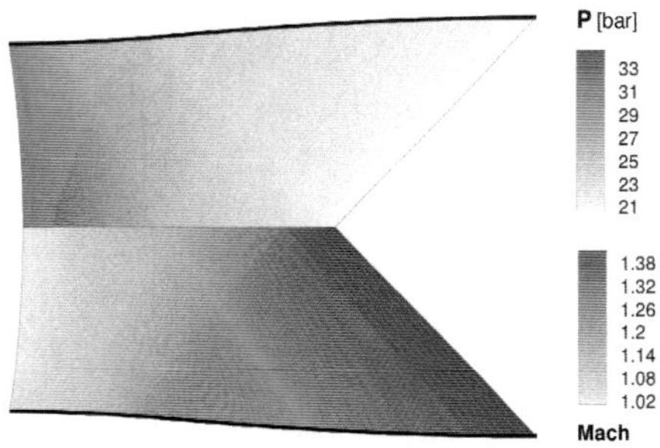

Figure 6: Divergent part of the nozzle for expansion D. Pressure and Mach contour plot computed by means of the method of characteristics, with the multi-parameter equation of state [13], are shown.

parameter thermodynamic model for CO_2 proposed in [13] is used. Additional details about the design procedure including code validation can be found in [29, 30]. Figure 6 displays the results obtained for test D, both in terms of pressure and Mach distribution. Then, the convergent part can be designed. Since this part of the geometry has a limited influence on the flow downstream the chocked throat, a polynomial profile is designed to realize the gradual area variation from the inlet section to the throat. To enhance flow regularity, the first and second derivatives of the divergent profile are matched at the throat. As a final remark, it should be pointed out that the geometry here presented refers to the inviscid core of the nozzle flow, thus additional CFD simulations will be performed to accurately predict the boundary layer in a further step of the plant design process.

5. Discussion and conclusions

Despite the wide use of sCO_2 in energy and chemical industry, the knowledge about carbon dioxide behavior in the supercritical region and near the critical point is still deficient, both in terms of fundamentals and of experimental data required for validation. In order to contribute to fill this lack, a new test-rig, named SCO_2PRI, is under-design at CREA Laboratory of Politecnico di Milano.

Within the non-ideal compressible fluid dynamics region, four different supersonic expansions through a convergent-divergent nozzle have been defined and different plant configurations have been taken into consideration to realize them. Due to the flow rates and the necessity of performing steady measurements, a closed-loop continuous-operating plant has been preferred. Moreover, the initial desired throat area had to be scaled to comply with the power availability of the laboratory. The "inverse" Joule-Brayton solution revealed to be inadequate due to the size and cost of the volumetric compressor, which has to operate in a region where the fluid density has high values, resulting in high stresses on the piston. Thus, the pump-based Rankine cycle configuration has been analyzed and it resulted to be feasible according to the economic and logistic constraints. Moreover, the choice of the Rankine cycle configuration results in a more flexible plant layout, which allows also for investigating the fluid flow at different conditions, including liquid and two-phase.

Similarly to the standard Rankine cycle, the main components of the SCO_2PRI test-rig are a pumping group, a heating section, the test section (expander) and the cooling section. The

IOP Conf. Series: Journal of Physics: Conf. Series **821** (2017) 012027 doi:10.1088/1742-6596/821/1/012027

availability of these components has been verified from suppliers and customized solutions have been designed when they were not commercially available. The pumping section is composed by two volumetric pumps which can operate independently to deal with the different flow rates that characterize the four expansions. On the contrary, heat exchangers in the heating and cooling sections have been sized on the most demanding case. An auxiliary cycle for the refrigerant (i.e. ammonia) was designed which includes a chiller. The chiller unit resulted to be the most expensive component in the test-rig, because of the low temperatures to be reached in evaporators. However, since low temperatures (less than 0°C) are required only in test conditions D, the feasibility of a less expensive test-rig in which only the chiller is sized on the three less demanding test cases was assessed. Such a pilot plant includes all the original components, except that the chiller unit that has been replaced by a smaller commercial available one.

In conclusion, the feasibility of the SCO_2PRI test-rig has been assessed for diverse test conditions and against different experimental techniques, including flow visualization and pneumatic measurements of non-ideal compressible fluid dynamics. The measurements of thermodynamic quantities along an expansion through the critical point, where non-monotone variation of the Mach number with density is expected, would supply important information that could fill the gap in sCO_2 fundamental knowledge. Future research activities will concern the design of all the other components of the test-rig, as well as dynamic simulations of its operation, including filling and starting. Moreover, the control and measurements systems will be detailed and designed.

References

[1] Angelino G 1969 *ASME Journal of Engineering for Power* **10** 272
[2] Feher E 1968 *Energy conversion* **8** 85–90
[3] Mahaffey J, Kalra A, Anderson M and Sridharan K 2014 *Supercritical CO2 Power Cycles Symp.*
[4] Ahn Y, Bae S J, Kim M, Cho S K, Baik S, Lee J I and Cha J E 2015 *Nucl. Eng. Technol.* **47** 647 – 661
[5] Reverchon E and Pallado P 1996 *J. of Supercritical Fluids* **9** 216–221
[6] Subramaniam B, Rajewski R A and Snavely K 1997 *J. Pharm. Sci.* **86** 885–890
[7] Sun X Y, Wang T J, Wang Z W and Jin Y 2002 *J. of Supercritical Fluids* **24** 231–237 ISSN 08968446
[8] Ito T, Otani Y and Inomata H 2004 *Sep. Purif. Technol.* **40** 41–46
[9] Zienkiewicz H K and Johannesen N H 2006 *J. Fluid Mech.* **17** 499
[10] Griffith W C and Kenny A 2006 *J. Fluid Mech.* **3** 286 ISSN 0022-1120
[11] Michels A, Sengers J V and van der Gulik P S 1962 *Physica* **28** 1216–1237
[12] Kim J K and Aihara T 1992 *International journal of heat and mass transfer* **35** 2515–2526
[13] Span R and Wagner W 1996 *J. Phys. Chem. Ref. Data* **25** 1509–1596
[14] Baltadjiev N, Lettieri C and Spakovszky Z 2014 *ASME Turbo Expo* **137** ISSN 0889-504X
[15] Rinaldi E, Pecnik R and Colonna P 2015 *J. Eng. Gas Turbines Power* **137** 1–7
[16] Lettieri C, Yang D and Spakovszky Z 2014 *Supercritical CO2 Power Cycles Symp.*
[17] Lee J, Cho S, Cha J and Lee J 2016 *Supercritical CO2 Power Cycles Symp.*
[18] Sánchez D, Monje B, Chacartegui R, Barragán J, Pajuelo E, Gómez J and Sánchez T 2013 *ASME Turbo Expo* (San Antonio, USA)
[19] Conboy T, Wright S, Pasch J, Fleming D, Rochau G and Fuller R 2012 *ASME Turbo Expo* pp 941–952
[20] Clementoni E, Cox T and Sprague C 2014 *J. Eng. Gas Turbines Power* **136** 1–6
[21] Moore J, Brun K, Evans N and Kalra C 2015 *ASME Turbo Expo*
[22] Rochau G E, Drennen T, Pasch J, Fleming D, Carlson M, Dawson L, Kruizenga A and Brooks R 2016 *Supercritical CO2 Power Cycles Symp.*
[23] Liao S M and Zhao T S 2002 *Journal of Heat Transfer* **124** 413–420
[24] Shiferaw D, Montero Carrero J and Le Pierres R 2016 *Supercritical CO2 Power Cycles Symp.*
[25] Callen H B 1985 *Thermodynamics and an introduction to thermostatistics* Second ed (Wiley)
[26] Thompson P A 1972 *Compressible-Fluid Dynamics* (McGraw-Hill Inc.)
[27] Lemmon E W, Huber M L and McLinden M O NIST Reference Fluid Thermodynamic and Transport Properties – REFPROP National Institute of Standards and Technology, Boulder, Colorado
[28] Zucrow M H and Hoffman J D 1976 *Gas dynamics* vol 1 (Wiley, Jhon & Son)
[29] Guardone A 2010 *7th HEFAT Conference* (Turkey)
[30] Guardone A, Spinelli A and Dossena V 2013 *J. Eng. Gas Turb. Power* **135** 042307–1